NATURAL HISTORY
UNIVERSAL LIBRARY

西方博物学大系

主编：江晓原

THE BIRDS OF NEW GUINEA AND THE ADJACENT PAPUAN ISLANDS

新几内亚鸟类志

[英] 约翰·古尔德 著

华东师范大学出版社

图书在版编目(CIP)数据

新几内亚鸟类志 = The birds of New Guinea and the adjacent Papuan islands：英文 /（英）约翰·古尔德著. — 上海：华东师范大学出版社，2018
（寰宇文献）
ISBN 978-7-5675-7720-6

Ⅰ.①新… Ⅱ.①约… Ⅲ.①鸟类-动物志-亚洲-英文 Ⅳ.①Q959.708

中国版本图书馆CIP数据核字(2018)第102596号

新几内亚鸟类志
The birds of New Guinea and the adjacent Papuan islands
（英）约翰·古尔德著

特约策划	黄曙辉　徐　辰
责任编辑	庞　坚
特约编辑	许　倩
装帧设计	刘怡霖

出版发行　华东师范大学出版社
社　　址　上海市中山北路3663号　邮编 200062
网　　址　www.ecnupress.com.cn
电　　话　021-60821666　行政传真　021-62572105
客服电话　021-62865537
门市（邮购）电话　021-62869887
地　　址　上海市中山北路3663号华东师范大学校内先锋路口
网　　店　http://hdsdcbs.tmall.com/

印刷者　虎彩印艺股份有限公司
开　本　16开
印　张　45
版　次　2018年6月第1版
印　次　2018年6月第1次
书　号　ISBN 978-7-5675-7720-6
定　价　1198.00元（精装全一册）

出版人　王　焰

（如发现本版图书有印订质量问题，请寄回本社客服中心调换或电话021-62865537联系）

总　目

《西方博物学大系》总序　（江晓原）	1
出版说明	1
the Birds of New Guinea Volume I	1
the Birds of New Guinea Volume II	141
the Birds of New Guinea Volume III	263
the Birds of New Guinea Volume IV	415
the Birds of New Guinea Volume V	543

《西方博物学大系》总序

江晓原

《西方博物学大系》收录博物学著作超过一百种，时间跨度为 15 世纪至 1919 年，作者分布于 16 个国家，写作语种有英语、法语、拉丁语、德语、弗莱芒语等，涉及对象包括植物、昆虫、软体动物、两栖动物、爬行动物、哺乳动物、鸟类和人类等，西方博物学史上的经典著作大备于此编。

中西方"博物"传统及观念之异同

今天中文里的"博物学"一词，学者们认为对应的英语词汇是 Natural History，考其本义，在中国传统文化中并无现成对应词汇。在中国传统文化中原有"博物"一词，与"自然史"当然并不精确相同，甚至还有着相当大的区别，但是在"搜集自然界的物品"这种最原始的意义上，两者确实也大有相通之处，故以"博物学"对译 Natural History 一词，大体仍属可取，而且已被广泛接受。

已故科学史前辈刘祖慰教授尝言：古代中国人处理知识，如开中药铺，有数十上百小抽屉，将百药分门别类放入其中，即心安矣。刘教授言此，其辞若有憾焉——认为中国人不致力于寻求世界"所以然之理"，故不如西方之分析传统优越。然而古代中国人这种处理知识的风格，正与西方的博物学相通。

与此相对，西方的分析传统致力于探求各种现象和物体之间的相互关系，试图以此解释宇宙运行的原因。自古希腊开始，西方哲人即孜孜不倦建构各种几何模型，欲用以说明宇宙如何运行，其中最典型的代表，即为托勒密（Ptolemy）的宇宙体系。

比较两者，差别即在于：古代中国人主要关心外部世界"如何"运行，而以希腊为源头的西方知识传统（西方并非没有别的知识传统，只是未能光大而已）更关心世界"为何"如此运行。在线性发展无限进步的科学主义观念体系中，我们习惯于认为"为何"是在解决了"如何"之后的更高境界，故西方的分析传统比中国的传统更高明。

然而考之古代实际情形，如此简单的优劣结论未必能够成立。例如以天文学言之，古代东西方世界天文学的终极问题是共同的：给定任意地点和时刻，计算出太阳、月亮和五大行星（七政）的位置。古代中国人虽不致力于建立几何模型去解释七政"为何"如此运行，但他们用抽象的周期叠加（古代巴比伦也使用类似方法），同样能在足够高的精度上计算并预报任意给定地点和时刻的七政位置。而通过持续观察天象变化以统计、收集各种天象周期，同样可视之为富有博物学色彩的活动。

还有一点需要注意：虽然我们已经接受了用"博物学"来对译 Natural History，但中国的博物传统，确实和西方的博物学有一个重大差别——即中国的博物传统是可以容纳怪力乱神的，而西方的博物学基本上没有怪力乱神的位置。

古代中国人的博物传统不限于"多识于鸟兽草木之名"。体现此种传统的典型著作，首推晋代张华《博物志》一书。书名"博物"，其义尽显。此书从内容到分类，无不充分体现它作为中国博物传统的代表资格。

《博物志》中内容，大致可分为五类：一、山川地理知识；二、奇禽异兽描述；三、古代神话材料；四、历史人物传说；五、神仙方伎故事。这五大类，完全符合中国文化中的博物传统，深合中国古代博物传统之旨。第一类，其中涉及宇宙学说，甚至还有"地动"思想，故为科学史家所重视。第二类，其中甚至出现了中国古代长期流传的"守宫砂"传说的早期文献：相传守宫砂点在处女胳膊上，永不褪色，只有性交之后才会自动消失。第三类，古代神话传说，其中甚至包括可猜想为现代"连体人"的记载。第四类，各种著名历史人物，比如三位著名刺客的传说，此三名刺客及所刺对象，历史上皆实有其人。第五类，包括各种古代方术传说，比如中国古代房中养生学说，房中术史上的传说人物之一"青牛道士封君达"等等。前两类与西方的博物学较为接近，但每一类都会带怪力乱神色彩。

"所有的科学不是物理学就是集邮"

在许多人心目中，画画花草图案，做做昆虫标本，拍拍植物照片，这类博物学活动，和精密的数理科学，比如天文学、物理学等等，那是无法同日而语的。博物学显得那么的初级、简单，甚至幼稚。这种观念，实际上是将"数理程度"作为唯一的标尺，用来衡量一切知识。但凡能够使用数学工具来描述的，或能够进行物理实验的，那就是"硬"科学。使用的数学工具越高深越复杂，似乎就越"硬"；物理实验设备越庞大，花费的金钱越多，似乎就越"高端"、越"先进"……

这样的观念，当然带着浓厚的"物理学沙文主义"色彩，在很多情况下是不正确的。而实际上，即使我们暂且同意上述"物理学沙文主义"的观念，博物学的"科学地位"也仍然可以保住。作为一个学天体物理专业出身，因而经常徜徉在"物理学沙文主义"幻影之下的人，我很乐意指出这样一个事实：现代天文学家们的研究工作中，仍然有绘制星图，编制星表，以及为此进行的巡天观测等等活动，这些活动和博物学家"寻花问柳"，绘制植物或昆虫图谱，本质上是完全一致的。

这里我们不妨重温物理学家卢瑟福（Ernest Rutherford）的金句："所有的科学不是物理学就是集邮（All science is either physics or stamp collecting）。"卢瑟福的这个金句堪称"物理学沙文主义"的极致，连天文学也没被他放在眼里。不过，按照中国传统的"博物"理念，集邮毫无疑问应该是博物学的一部分——尽管古代并没有邮票。卢瑟福的金句也可以从另一个角度来解读：既然在卢瑟福眼里天文学和博物学都只是"集邮"，那岂不就可以将博物学和天文学相提并论了？

如果我们摆脱了科学主义的语境，则西方模式的优越性将进一步被消解。例如，按照霍金（Stephen Hawking）在《大设计》（*The Grand Design*）中的意见，他所认同的是一种"依赖模型的实在论（model-dependent realism）"，即"不存在与图像或理论无关的实在性概念（There is no picture- or theory-independent concept of reality）"。在这样的认识中，我们以前所坚信的外部世界的客观性，已经不复存在。既然几何模型只不过是对外部世界图像的人为建构，则古代中国人干脆放弃这种建构直奔应用（毕竟在实际应用中我们只需要知道七政"如何"运行），又有何不可？

传说中的"神农尝百草"故事，也可以在类似意义下得到新的解读："尝百草"当然是富有博物学色彩的活动，神农通过这一活动，得知哪些草能够治病，哪些不能，然而在这个传说中，神农显然没有致力于解释"为何"某些草能够治病而另一些则不能，更不会去建立"模型"以说明之。

"帝国科学"的原罪

今日学者有倡言"博物学复兴"者，用意可有多种，诸如缓解压力、亲近自然、保护环境、绿色生活、可持续发展、科学主义解毒剂等等，皆属美善。编印《西方博物学大系》也是意欲为"博物学复兴"添一助力。

然而，对于这些博物学著作，有一点似乎从未见学者指出过，而鄙意以为，当我们披阅把玩欣赏这些著作时，意识到这一点是必须的。

这百余种著作的时间跨度为15世纪至1919年，注意这个时间跨度，正是西方列强"帝国科学"大行其道的时代。遥想当年，帝国的科学家们乘上帝国的军舰——达尔文在皇家海军"小猎犬号"上就是这样的场景之一，前往那些已经成为帝国的殖民地或还未成为殖民地的"未开化"的遥远地方，通常都是踌躇满志、充满优越感的。

作为一个典型的例子，英国学者法拉在（Patricia Fara）《性、植物学与帝国：林奈与班克斯》（*Sex, Botany and Empire, The Story of Carl Linnaeus and Joseph Banks*）一书中讲述了英国植物学家班克斯（Joseph Banks）的故事。1768年8月15日，班克斯告别未婚妻，登上了澳大利亚军舰"奋进号"。此次"奋进号"的远航是受英国海军部和皇家学会资助，目的是前往南太平洋的塔希提岛（Tahiti，法属海外自治领，另一个常见的译名是"大溪地"）观测一次比较罕见的金星凌日。舰长库克（James Cook）是西方殖民史上最著名的舰长之一，多次远航探险，开拓海外殖民地。他还被认为是澳大利亚和夏威夷群岛的"发现"者，如今以他命名的群岛、海峡、山峰等不胜枚举。

当"奋进号"停靠塔希提岛时，班克斯一下就被当地美丽的土著女性迷昏了，他在她们的温柔乡里纵情狂欢，连库克舰长都看不下去了，"道德愤怒情绪偷偷溜进了他的日志当中，他发现自己根本不可能不去批评所见到的滥交行为"，而班克斯纵欲到了"连嫖妓都毫无激情"的地步——这是别人讽刺班克斯的说法，因为对于那时常年航行于茫茫大海上的男性来说，上岸嫖妓通常是一项能够唤起"激情"的活动。

而在"帝国科学"的宏大叙事中，科学家的私德是无关紧要的，人们关注的是科学家做出的科学发现。所以，尽管一面是班克斯在塔希提岛纵欲滥交，一面是他留在故乡的未婚妻正泪眼婆娑地"为远去的心上人绣织背心"，这样典型的"渣男"行径要是放在今天，非被互联网上的口水淹死不可，但是"班克斯很快从他们的分离之苦中走了出来，在外近三年，他活得倒十分滋润"。

法拉不无讽刺地指出了"帝国科学"的实质："班克斯接管了当地的女性和植物，而库克则保护了大英帝国在太平洋上的殖民地。"甚至对班克斯的植物学本身也调侃了一番："即使是植物学方面的科学术语也充满了性指涉。……这个体系主要依靠花朵之中雌雄生殖器官的数量来进行分类。"据说"要保护年轻妇女不受植物学教育的浸染，他们严令禁止各种各样的植物采集探险活动。"这简直就是将植物学看成一种"涉黄"的淫秽色情活动了。

在意识形态强烈影响着我们学术话语的时代，上面的故事通常是这样被描述的：库克舰长的"奋进号"军舰对殖民地和尚未成为殖民地的那些地方的所谓"访问"，其实是殖民者耀武扬威的侵略，搭载着达尔文的"小猎犬号"军舰也是同样行径；班克斯和当地女性的纵欲狂欢，当然是殖民者对土著妇女令人发指的蹂躏；即使是他采集当地植物标本的"科学考察"，也可以视为殖民者"窃取当地经济情报"的罪恶行为。

后来改革开放，上面那种意识形态话语被抛弃了，但似乎又走向了另一个极端，完全忘记或有意回避殖民者和帝国主义这个层面，只歌颂这些军舰上的科学家的伟大发现和成就，例如达尔文随着"小猎犬号"的航行，早已成为一曲祥和优美的科学颂歌。

其实达尔文也未能免俗，他在远航中也乐意与土著女性打打交道，当然他没有像班克斯那样滥情纵欲。在达尔文为"小猎犬号"远航写的《环球游记》中，我们读到："回程途中我们遇到一群黑人姑娘在聚会，……我们笑着看了很久，还给了她们一些钱，这着实令她们欣喜一番，拿着钱尖声大笑起来，很远还能听到那愉悦的笑声。"

有趣的是，在班克斯在塔希提岛纵欲六十多年后，达尔文随着"小猎犬号"也来到了塔希提岛，岛上的土著女性同样引起了达尔文的注意，在《环球游记》中他写道："我对这里妇女的外貌感到有些失望，然而她们却很爱美，把一朵白花或者红花戴在脑后的髮髻上……"接着他以居高临下的笔调描述了当地女性的几种发饰。

用今天的眼光来看，这些在别的民族土地上采集植物动物标本、测量地质水文数据等等的"科学考察"行为，有没有合法性问题？有没有侵犯主权的问题？这些行为得到当地人的同意了吗？当地人知道这些行为的性质和意义吗？他们有知情权吗？……这些问题，在今天的国际交往中，确实都是存在的。

也许有人会为这些帝国科学家辩解说：那时当地土著尚在未开化或半开化状态中，他们哪有"国家主权"的意识啊？他们也没有制止帝国科学家的考察活动啊？但是，这样的辩解是无法成立的。

姑不论当地土著当时究竟有没有试图制止帝国科学家的"科学考察"行为，现在早已不得而知，只要殖民者没有记录下来，我们通常就无法知道。况且殖民者有军舰有枪炮，土著就是想制止也无能为力。正如法拉所描述的："在几个塔希提人被杀之后，一套行之有效的易货贸易体制建立了起来。"

即使土著因为无知而没有制止帝国科学家的"科学考察"行为，这事也很像一个成年人闯进别人的家，难道因为那家只有不懂事的小孩子，闯入者就可以随便打探那家的隐私、拿走那家的东西、甚至将那家的房屋土地据为己有吗？事实上，很多情况下殖民者就是这样干的。所以，所谓的"帝国科学"，其实是有着原罪的。

如果沿用上述比喻，现在的局面是，家家户户都不会只有不懂事的孩子了，所以任何外来者要想进行"科学探索"，他也得和这家主人达成共识，得到这家主人的允许才能够进行。即使这种共识的达成依赖于利益的交换，至少也不能单方面强加于人。

博物学在今日中国

  博物学在今日中国之复兴，北京大学刘华杰教授提倡之功殊不可没。自刘教授大力提倡之后，各界人士纷纷跟进，仿佛昔日蔡锷在云南起兵反袁之"滇黔首义，薄海同钦，一檄遥传，景从恐后"光景，这当然是和博物学本身特点密切相关的。

  无论在西方还是在中国，无论在过去还是在当下，为何博物学在它繁荣时尚的阶段，就会应者云集？深究起来，恐怕和博物学本身的特点有关。博物学没有复杂的理论结构，它的专业训练也相对容易，至少没有天文学、物理学那样的数理"门槛"，所以和一些数理学科相比，博物学可以有更多的自学成才者。这次编印的《西方博物学大系》，卷帙浩繁，蔚为大观，同样说明了这一点。

  最后，还有一点明显的差别必须在此处强调指出：用刘华杰教授喜欢的术语来说，《西方博物学大系》所收入的百余种著作，绝大部分属于"一阶"性质的工作，即直接对博物学作出了贡献的著作。事实上，这也是它们被收入《西方博物学大系》的主要理由之一。而在中国国内目前已经相当热的博物学时尚潮流中，绝大部分已经出版的书籍，不是属于"二阶"性质（比如介绍西方的博物学成就），就是文学性的吟风咏月野草闲花。

  要寻找中国当代学者在博物学方面的"一阶"著作，如果有之，以笔者之孤陋寡闻，唯有刘华杰教授的《檀岛花事——夏威夷植物日记》三卷，可以当之。这是刘教授在夏威夷群岛实地考察当地植物的成果，不仅属于直接对博物学作出贡献之作，而且至少在形式上将昔日"帝国科学"的逻辑反其道而用之，岂不快哉！

2018 年 6 月 5 日
于上海交通大学
科学史与科学文化研究院

《新几内亚鸟类志》 出版说明

约翰·古尔德
（1804-1881）

英国著名鸟类学家兼鸟类画家约翰·古尔德（John Gould）生于多塞特郡，其父是温莎城堡御花园的园丁主管。由于家庭关系，古尔德幼年常往来于御花园，十四岁时正式成为这里的园丁学徒。这份工作培养了他观察自然界生物的本领和剥制标本的手艺。1827年，古尔德即被位于伦敦的动物学会博物馆聘为首席标本剥制师。

新的职务给了古尔德接触国内顶尖博物学家的大好机会，送到动物学会的鸟类标本，也大都由他第一时间查验过目。1830年，他根据一批送到学会的喜马拉雅山鸟类标本，出版了著作《喜马拉雅百年鸟类志》。之后，查尔斯·达尔文完成第二次小猎犬号科考航行之后，也将收集到的鸟类样本交给古尔德研究，其成果刊发于达尔文主编的《小猎犬号科考动物志》（1839）。

自1838年起，古尔德对澳大利亚及南太平洋诸岛的鸟类进行了实地研究，也是第一个系统、完整记录该地区鸟类生态活动的科学家。为进行这项研究，古尔德和收藏家约翰·吉尔伯特一道先前往塔斯马尼亚岛，征求该地总督约翰·富兰克林爵士的许可。获准之后，二人在岛上进行了四个多月的调查、收集和研究。翌年2月，古尔德搭船前往悉尼进行了两个月的研究。同年5月，又到阿德莱德和探险家查尔斯·斯图尔特一起远行，1840年5月才回到英国。之后，他刊行了《澳大利亚鸟类志》，并投注大量精力绘制自己钟爱的蜂鸟及英国鸟类。

古尔德享寿77岁，一生共留下近三千幅插画，其中最被人称道的是鸟类画。这部《新几内亚鸟类志》是他晚年出版的著作，自1875年起，直到其身后的1888年才出齐全部五卷，历时十三年，几乎涉及新几内亚及附近巴布亚群岛的所有鸟类，特别是天堂鸟的生态。

在古尔德之前，法国探险家和一度支配这里的荷兰动物学家都在新几内亚鸟类研究领域收获颇丰，然而古尔德在本书中的描写，实地研究写生及借助英国本地、澳大利亚等地博物学家和博物馆提供的大量标本绘制的鸟类画，虽是后起之秀，却无疑是个中佼佼者。

THE
BIRDS OF NEW GUINEA

AND THE

ADJACENT PAPUAN ISLANDS,

INCLUDING MANY

NEW SPECIES RECENTLY DISCOVERED

IN

AUSTRALIA.

BY

JOHN GOULD, F.R.S.

COMPLETED AFTER THE AUTHOR'S DEATH

BY

R. BOWDLER SHARPE, F.L.S. &c.,

ZOOLOGICAL DEPARTMENT, BRITISH MUSEUM.

VOLUME I.

LONDON:

HENRY SOTHERAN & CO., 36 PICCADILLY.

1875–1888.

[*All rights reserved.*]

PRINTED BY TAYLOR AND FRANCIS,
RED LION COURT, FLEET STREET.

PREFACE.

In fulfilment of the promise made by the Publishers to the Subscribers, and in accordance with the intention of the late Mr. Gould, the present work has been brought to a close with the issue of the Twenty-fifth Part. For the last thirteen of these parts, which have appeared since the author's death, I have been solely responsible, as will be seen by the initials appended to the foot of each article; and I greatly regret that the pledges given to the Subscribers have necessitated the curtailment of the work at a period when the development of European enterprise in New Guinea and the neighbouring islands seems to point to the discovery of many new and highly interesting birds. The recent finding of two beautiful new Birds of Paradise in North-eastern New Guinea must be the forerunner of many other discoveries, and it is to be hoped that the Subscribers will feel sufficient interest in the subject to warrant the continuation of the present work, so that the Avifauna of Papuasia may be completely illustrated. Should such a wish be expressed, I shall be only too happy to continue the work.

R. BOWDLER SHARPE.

December, 1888.

INTRODUCTION.

The halo of romance which for nearly a century has centred round New Guinea and its animals does not get dimmed as time speeds on: indeed it shines more brightly than ever, and all naturalists who have travelled in the Moluccas have either paid a visit to that wonderful island or are looking forward to the time when they will be able to land upon its shores. Mr. Gould was perfectly right when he alluded to the Birds of Paradise as being one of the most remarkable families of birds inhabiting the Papuan Subregion, and their presence alone is sufficient to warrant our recognition of a zoological area distinct from that of the neighbouring Australian region. That there is a considerable affinity between the avifauna of Northern Australia and New Guinea is proved by the presence of a Cassowary in Queensland and of a *Malurus* in New Guinea; and this relationship is further accentuated by the presence in both subregions of such genera as *Ptilorhis, Microglossum, Tanysiptera*, and a large number of species of Meliphagidæ. Of the Lipotypes, or forms of birds conspicuous by their absence, in the Papuan and Australian Subregions, perhaps the non-appearance of any species of Woodpecker is the most remarkable.

No other country has produced in recent times more novelties in the way of birds than New Guinea, and every fresh expedition to the mountains of the interior appears to add to the number. Travellers of many nationalities have contributed to the disclosure of the hidden ornithological treasures which have been described during the last thirty years, as will be seen by the following brief *résumé* of the history of the subject.

In 1858, Dr. Sclater published, in the 'Proceedings' of the Linnean Society of London, a paper on the zoology of New Guinea. After visiting the museums of Paris and Leyden, and personally examining the specimens which they contained from this part of the world, he drew up a list of animals and birds known as inhabitants of the island, and he computed that of the former 10 species had been obtained, and of birds 170 species. The present extent of the described mammalian fauna I have no means of judging; but in 1865, Dr. Finsch, in his 'Neu-Guinea,' estimated the number of Mammalia at 15, and the birds at 252. Of course, as Count Salvadori's work deals with the ornithology of the Moluccas as well, no exact comparison of the avifauna of New Guinea, as known in 1858 and as at present known in 1888, is presented; but, as will be seen below, the various expeditions which have taken place have added enormously to the number of species, and we cannot suppose that the limit of our knowledge has yet been reached.

The area embraced by Count Salvadori's 'Ornitologia della Papuasia' is much the same as that admitted by Mr. Gould when he commenced the present book, as the few species from Australia which he included were only figured as a further Supplement to his 'Birds of Australia.' It may be remarked, however, that of the 1030 species enumerated by Count Salvadori, only 300 have been figured in the present volumes, and at least 700 remain to be described before a complete account of the ornithology of Papuasia will have been attained.

From Dr. Sclater's memoir before mentioned we take the following details of early exploration in New Guinea. Although Birds of Paradise were often sent, in a mutilated state, in early times, and some of them were figured in ancient books, very few other birds from this quarter of the globe were known to Linnæus, and our first real acquaintance with Papuan Ornithology is due to the French traveller Sonnerat, who went there in 1771, and procured some plants and animals, principally on the island of Jobi, in Northwestern New Guinea. The results of his discoveries were published in his 'Voyage à la Nouvelle Guinée,' in 1776. The French exploring-ship the 'Uranie' procured a few birds in 1824, and the 'Coquille' spent twelve days at a harbour in North-eastern New Guinea, which they named 'Havre Dorey'; about fifty species of birds were obtained, most of which were new to science and were described by Lesson. The voyage of the 'Astrolabe' also contains the description and figures of some Papuan birds, written by Messrs. Quoy and Gaimard, the naturalists attached to this expedition, as they had been to that of the 'Uranie.' In 1854 the volume on the 'Zoology' of the 'Voyage au Pôle Sud' contained some more descriptions of New-Guinea birds, and then French enterprise in Papuasia seems to have stood still till quite recently, when a young naturalist, M. Léon Laglaize, sent from that region some very interesting and important novelties in the way of birds.

The Dutch, however, to whom belongs the western half of the island at the present time, were very active in New Guinea during the second half of the century, and some admirable collections were made by Solomon Müller and Maklot. These formed the foundation of the splendid series of Papuan birds in the Leyden Museum, which is still one of its chief glories. Many of the species were figured by Temminck in the 'Planches Coloriées,' and also by Müller and Schlegel in the well-known 'Verhandelingen.'

While Dr. Sclater was writing his memoir, and placing the zoology of New Guinea as it were on a proper basis of knowledge, our great countryman, A. R. Wallace, was at work on that wonderful exploration of the Malay Archipelago which will render his name famous for all time. Most of Mr. Wallace's discoveries were epitomized and catalogued in the late George Robert Gray's 'List of the Birds of New Guinea' (1859); but many more species were described by the traveller himself, and a most complete list of the Mammals and Birds of Papuasia and the Moluccas was given by Dr. Otto Finsch in his 'Neu-Guinea und seine Bewohner' (1865).

Fired by the success of the English traveller Wallace in their East-Indian possessions, the Dutch nation provided a sum of money for several years to compass the zoological exploration of Papuasia, and, under the guidance of the late Professor Schlegel, several energetic collectors were despatched, Bernstein, Von Rosenberg, Hoedt, and others, who obtained extensive results, so that once more the scientific prestige of the Dutch nation was paramount in Papuasia, and many islands, hitherto unvisited by Europeans, yielded important novelties. Then came the celebrated expedition of Dr. A. B. Meyer, whereby a further number of new species were discovered, to be followed by the extensive explorations of Signor D'Albertis and Dr. Beccari. D'Albertis discovered *Drepanornis* in the Arfak Mountains in 1872, and visiting afterwards the southern and south-eastern portions of New Guinea, he met with *Harpyopsis* and other remarkable new genera of birds, many of which have been figured in the present work. Dr. Beccari sent some most interesting and wonderful birds from the Arfak Mountains and many of the islands of the Papuan Subregion, where also the hunters employed by an enterprising Dutch naturalist, Mr. Bruijn, have obtained many rare and new species. To this gentleman and to Mr. Riedel science is indebted for many important contributions to zoological science.

During the time which elapsed since Mr. Wallace's successful labours in the Malay Archipelago, Englishmen had been busy in exploring many of the outlying groups of islands to the eastward of New Guinea; and Captain Richards, Mr. Brazier, Mr. Cockerell, and other naturalists discovered many new species of birds in the Solomon Islands, and quite recently an energetic explorer, Mr. C. M. Woodford, has brought home a most interesting collection from this Archipelago. The Rev. George Brown and Mr. L. C. Layard have also done much to explore the natural history of New Britain and the adjacent islands, which were also visited by Dr. Finsch, to whom we owe much enduring work in the same locality and in the Caroline Islands. The collectors of the celebrated Hamburg firm, Messrs. Godeffroy, have also explored Ponapé and other islands to the west of New Guinea, and the names of Kubary, Graeffe, and Kleinschmidt will always remain famous amongst those of the explorers of these little-known and inaccessible localities. For the exploration of the Admiralty group of islands we are indebted to the 'Challenger' expedition. Nor have Englishmen been idle in their newly-acquired province of South-eastern New Guinea. Mr. Goldie, Mr. Octavius Stone, and the well-known missionaries the Rev. Mr. Lawes and the Rev. Mr. Macfarlane, have done wonders in procuring collections from the neighbourhood of Port Moresby, from whence also an interesting collection was sent by a young American explorer, Dr. James. The

INTRODUCTION.

Australians have also done much to increase our knowledge of the zoology of South-eastern New Guinea, and the collections of the 'Chevert' expedition and other explorers, Mr. Masters, Mr. Morton, Mr. Pettard, and Mr. Broadbent, have been described by Dr. E. P. Ramsay at Sydney, or by ourselves here in London. The Astrolabe Mountains have been visited by Mr. Goldie, Mr. Hunstein, and Mr. H. O. Forbes, and have yielded some surprising and beautiful novelties. Many of the species discovered originally in the Arfak Mountains have now been found in the Astrolabe Range, which, however, appears to possess a certain individual fauna, though we know so little of the mountain-ranges of the interior of New Guinea that it would be impossible to affirm that any species is peculiar to any portion of the mountain system and does not extend throughout its entire area.

Before concluding this sketch of zoological work in New Guinea and the Moluccas, we must allude to the excellent results obtained by Mr. H. O. Forbes and his heroic wife in the Tenimber Islands. They were the first Europeans to collect in the dreaded Timor Laut group, and though compelled to work, through the hostility of the surrounding natives, in a circumscribed space, the number of new species obtained reflected the greatest credit on the energy of these brave travellers. Mr. Riedel's hunters have also discovered a few new species on the Tenimber Islands.

In the pages of the present work frequent reference is made to the 'Ornitologia della Papuasia e delle Molucche' of Count Salvadori. The present writer knows how difficult, in these days of many books, is the task of the man who sets himself to write a monograph of any group of birds, and to write a complete account of the avifauna of any country is even more tedious. Although the collections stored in the Museo Civico at Genoa are most complete, the enthusiasm of the distinguished Director of that Museum, Marquis Doria, having drawn thereto the collections of the Italian travellers, as energetic and full of purpose as he is himself, yet the treasures in the other museums of Europe must be collated with the material accumulated by Italy, if a complete account of the ornithology of New Guinea has to be compiled. Travelling, therefore, from country to country, comparing the collections in his charge with those made by English, French, Dutch, and German travellers, Count Salvadori may well be congratulated on the result which his thoughtful earnestness obtained, and in the great work on Papuan Ornithology of which he is the author he has raised up for himself an imperishable reputation. The best tribute which the present writer can pay to his work exists in a reference to the number of times which he has been obliged to quote or to copy Count Salvadori's writings, because, on the subject of Papuan Ornithology, he left us little or nothing to add to the information given by him in the 'Ornitologia della Papuasia.'

R. BOWDLER SHARPE.

CONTENTS.

VOLUME I.

PREFACE.
INTRODUCTION.

Plate			Part	Date.
1.	Astur melanochlamys	Black-mantled Goshawk	XXII.	1886.
2.	Erythrotriorchis doriæ	Marquis Doria's Goshawk	XXI.	1886.
3.	Harpyopsis novæ guineæ	New-Guinea Harpy-Eagle	XXV.	1888.
4.	Baza gurneyi	Gurney's Cuckoo-Falcon	XXV.	1888.
5.	Ninox odiosa	New Britain Hawk-Owl	XVII.	1884.
6.	,, forbesi	Forbes's Hawk-Owl	XV.	1883.
7.	,, dimorpha	Salvadori's Hawk-Owl	XXII.	1886.
8.	Epimachus ellioti	Elliot's Promerops	XI.	1880.
9.	,, speciosus	Great Promerops	VII.	1878.
10.	Drepanornis cervinicauda	Bennett's Bird of Paradise	XVIII.	1884.
11.	,, albertisi	D'Albertis's Bird of Paradise	I.	1875.
12.	,, bruijnii	Bruijn's Bird of Paradise	XXI.	1885.
13.	Craspedophora magnifica	New-Guinea Rifle-bird	IX.	1879.
14.	Seleucides nigricans }	Twelve-wired Bird of Paradise	XII.	1881.
15.	,, ,, }			
16.	Paradigalla carunculata	Wattled Bird of Paradise	VII.	1878.
17.	Astrapia nigra	Gorget Paradise-bird	VIII.	1878.
18.	Lophorhina superba	Superb Bird of Paradise	VI.	1878.
19.	,, minor	Lesser Superb Bird of Paradise	XXIV.	1888.
20.	Diphyllodes respublica	Bare-headed Bird of Paradise	III.	1876.
21.	,, gulielmi III.	Waigiou Bird of Paradise	II.	1876.
22.	,, speciosa	Magnificent Bird of Paradise	II.	1876.
23.	,, chrysoptera	Golden-winged Bird of Paradise	II.	1876.
24.	Cicinnurus regius	King Bird of Paradise	III.	1876.
25.	Parotia sexpennis	Six-plumed Bird of Paradise	I.	1875.
26.	,, lawesi	Lawes's Bird of Paradise	XXIII.	1887.
27.	Paradisea decora	Grey-chested Bird of Paradise	XX.	1885.
28.	,, papuana }	Papuan Bird of Paradise	X.	1879.
29.	,, ,, }			
30.	,, apoda	Greater Bird of Paradise	IX.	1879.
31.	,, sanguinea	Red Bird of Paradise	IV.	1877.
32.	,, raggiana	Marquis de Raggi's Bird of Paradise	IV.	1877.
33.	Manucodia comrii	Curl-crested Manucode	V.	1877.
34.	,, chalybea	Green Manucode	V.	1877.
35.	Phonygama purpureoviolacea	Purple-and-Violet Manucode	XXIII.	1887.
36.	Lycocorax obiensis	Obi Paradise-Crow	XXIV.	1888.
37.	Ailurœdus stonii	Stone's Cat-bird	XII.	1881.
38.	,, maculosus	Queensland Cat-bird	I.	1875.
39.	,, melanotis	Black-cheeked Cat-bird	I.	1875.
40.	,, arfakianus	Arfak Cat-bird	I.	1875.
41.	,, buccoides	Barbet-like Cat-bird	I.	1875.
42.	,, melanocephalus	Black-naped Cat-bird	XXIV.	1888.
43.	Scenopœus dentirostris	Toothed-billed Bower-bird	X.	1879.
44.	Chlamydodera orientalis	Queensland Bower-bird	XI.	1880.
45.	,, occipitalis	Large-frilled Bower-bird	X.	1879.
46.	Amblyornis inornata	Gardener Bower-bird	IX.	1879.
47.	,, subalaris	Orange-crested Bower-bird	XXII.	1886.
48.	Xanthomelus aureus	Golden Bird of Paradise	VI.	1878.
49.	Oriolus decipiens	Deceptive Oriole	XVI.	1884.
50.	Chætorhynchus papuensis	Arfak Drongo	XI.	1880.
51.	Dicranostreptus megarhynchus	New-Ireland Drongo	VIII.	1878.
52.	Rectes leucorhynchus	White-billed Wood-Shrike	XX.	1885.
53.	,, cerviniventris	Fawn-breasted Wood-Shrike	XX.	1885.
54.	,, uropygialis	Rufous-and-Black Wood-Shrike	XII.	1881.
55.	,, jobiensis	Jobi-Island Wood-Shrike	XII.	1881.
56.	,, aruensis	Aru-Island Wood-Shrike	XXV.	1888.

ASTUR MELANOCHLAMYS.

ASTUR MELANOCHLAMYS.

Black-mantled Goshawk.

Urospizias melanochlamys, Salvad. Ann. Mus. Civic. Genov. vii. p. 905 (1875).—Id. op. cit. xii. p. 38 (1878).—Id. Orn. Papuasia e delle Molucche, i. p. 63 (1880).
Astur melanochlamys, Sharpe, Mitth. k. zool. Mus. Dresd. Heft iii. p. 355 (1878).

This fine species is at present only known from the Arfak Mountains in North-western New Guinea, whence the first specimens were sent to Europe by Mr. Bruijn and Dr. Beccari. We have also seen a specimen in Dr. Guillemard's collection from the same locality.

This Goshawk is a very distinct species, and belongs to the Austro-Malayan section of the genus *Astur*, which embraces the species which have a rufous collar round the hind neck. It differs from all its allies, however, by having the ear-coverts slaty black like the head and back, and by the deep vinous chestnut of the under surface.

The following description was taken by us from the type specimen belonging to the Genoa Museum, a sight of which was granted to us by Count Salvadori during his last visit to England:—

Adult female. General colour above deep black; head black like the back, from which it is separated by a broad well-defined collar of vinous chestnut; quills and tail black like the upper surface, the inner webs rather browner, barred with black; entire sides of face black like the crown; throat also black, but mottled with white bases to the feathers, many of the latter being white barred with black; remainder of under surface of body deep vinous chestnut, with slightly developed whitish bars on the flanks and lower abdomen; under wing-coverts and axillaries vinous chestnut, with remains of lighter cross bars, the greater series and the inner webs of the quills greyish white with a vinous tinge and barred with blackish. Total length 15·2 inches, culmen 1·1, wing 10·0, tail 7·8, tarsus 2·65.

The Plate represents an adult bird of about the size of life, with a smaller figure in the background. Both are drawn from a specimen procured by Dr. Guillemard in the Arfak Mountains, and kindly lent to us by that gentleman.

[R. B. S.]

ERYTHROTRIORCHIS DORIÆ.

ERYTHROTRIORCHIS DORIÆ.

Marquis Doria's Goshawk.

Megatriorchis doriæ, Salvad. & D'Albert. Ann. Mus. Civic. Genov. vii. p. 805 (1875).—Gurney, Ibis, 1877, p. 436.
—Id. Ibis, 1878, p. 87.—Salvad. Orn. Papuasia e delle Molucche, i. p. 41 (1880).—Gurney, List of Diurnal Birds of Prey, p. 45 (1884).
Erythrotriorchis doriæ, Sharpe, Journ. Linn. Soc., Zool. xvii. p. 406 (1884).

This is one of the finest discoveries made by the celebrated Italian traveller D'Albertis during his voyage to South-eastern New Guinea, and it is one of the peculiar forms which inhabit both that great island and Northern Australia.

The original specimen came from Hall Bay, in South-eastern New Guinea, and this is the bird which forms the principal figure in the Plate, and which we consider to be immature. The second specimen figured in the Plate is apparently a fully adult bird, and was obtained by Mr. Goldie on the Astrolabe Mountains in South-eastern New Guinea. We believe it to be an adult male, and the type to be an immature female, notwithstanding some discrepancies in the proportions of the quills and tail-feathers exhibited by the two specimens. If we are correct in thus assigning the relation of these two individuals, then it is evident that the genus *Megatriorchis* is the same as the Australian genus *Erythrotriorchis*, of which *E. radiatus* is the representative species on the latter continent. A larger number of specimens is, however, requisite to settle the point satisfactorily.

The following are exact descriptions of the two specimens referred to:—

Adult. General colour above glossy black, with nearly obsolete rufous margins to the feathers of the upper surface; wing-coverts black, very plainly and broadly edged with chestnut; bastard-wing and primary-coverts black, with indistinct bars of ashy brown, nearly obsolete on the bastard-wing; quills black, barred across with brown, these bars less distinct on the inner webs; tail-feathers black, barred with ashy grey, about twelve bars being discernible on the central feathers; the outer tail-feathers with about the same number of brown bars, as well as one at the tip; crown of head uniform black, the nape crested; feathers below the eye and ear-coverts black, the feathers edged with rufous buff, imparting a streaked appearance; cheeks and throat rufescent buff, streaked with black down the centre of the feathers; remainder of under surface white, broadly streaked with black, and marked slightly with chestnut, especially on the sides of the body; the black bars rather broken up and in the form of large spots on the flanks, abdomen, and under tail-coverts; thighs broadly barred with black and white, with a chestnut spot in the centre of the black bars; under wing-coverts black in the centre, rufous or rufous white on the edges; the lower series blackish, barred with ashy or greyish white like the lower surface of the quills. Total length 20 inches, culmen 1·05, wing 12·8, tail 10·0, tarsus 3·1.

Immature. General colour above brown, regularly barred across with lighter brown, the edges of the feathers rather more rufous or fawn-coloured; head brown, streaked with rufous, the feathers edged with this colour and having creamy white bars; broad eyebrow of white feathers streaked with black; a small crest of pointed plumes; nape-plumes white, with a rufous tinge, and mesially streaked with black; hind neck like the back, but more mottled with black spade-shaped terminal spots to the feathers; wing-coverts ashy brown, tipped with fawn-colour, and broadly barred across with darker brown; quills ashy brown, broadly barred across with blackish brown for their entire length, the interspaces on the secondaries lighter, these quills being much paler tipped; upper tail-coverts and tail-feathers ashy brown, tipped with greyish, slightly tinged with rufous on the former, the tail-feathers crossed with twelve bars of blackish brown; sides of face and ear-coverts white, the feathers mesially streaked with dark brown, the ear-coverts tipped with blackish brown, forming a distinct patch; under surface of body creamy white, streaked with dark brown, very narrowly on the throat, broader on the abdomen and lower breast, the thighs and under tail-coverts with nearly obsolete mesial streaks; fore neck and breast broadly streaked with light rufous, browner on the former, the breast-feathers with a shaft-streak of dark brown; under wing-coverts and axillaries white, with a dark brown shaft-streak; quills ashy below, white near the base, barred with dark brown, the bars about nine in number: "bill black, the cere and eyelids ash-colour; feet whitish grey; iris chestnut-brown" (*D'Albertis*). Total length 7·5 inches, culmen 1·7, wing 13·7, tail 12·8, tarsus 3·4.

The principal figure in our Plate is drawn from the type specimen, *Megatriorchis doriæ*, which was kindly lent to me by the Marquis Doria. The hinder figure is taken from the specimen in the British Museum, which we have identified as the adult male of the species.

[R. B. S.]

HARPYOPSIS NOVÆ-GUINEÆ, Salvad.

HARPYOPSIS NOVÆ GUINEÆ, Salvad.

New-Guinea Harpy-Eagle.

Harpyopsis novæ guineæ, Salvad. Ann. Mus. Civic. Genov. vii. p. 682 (1875).—Id. & D'Albert. tom. cit. p. 805 (1875).—Salvad. op. cit. ix. p. 10 (1875), x. pp. 115, 117 (1877), xii. p. 36 (1878).—Gurney, Ibis, 1877, p. 435; 1878, p. 87.—Sharpe, Mitth. k. zool. Mus. Dresden, i. p. 355, pl. xxix. (1878).—D'Albert. & Salvad. Ann. Mus. Civic. Genov. xiv. p. 28 (1879).—Sharpe, Journ. Linn. Soc., Zool. xiv. p. 627 (1879). —Salvad. Orn. Papuasia e delle Molucche, i. p. 40 (1880), iii. App. p. 507 (1882).—D'Albert. New Guinea, i. p. 278, pl. x. (1880).—Sharpe, Journ. Linn. Soc., Zool. xvi. p. 424 (1882).—Gurney, List of Diurnal Birds of Prey, p. 46 (1884).—Finsch u. Meyer in Madarász, Zeitschr. ges. Orn. iii. p. 2 (1886).

This magnificent bird of prey is one of the most important and at the same time one of the most interesting of all the discoveries made by Signor D'Albertis during his travels in New Guinea. It is a veritable Harpy, like the Harpy-Eagle of South America (*Thrasaetus harpyia*), which it so closely resembles in outward form that we had great difficulty in finding characters for its generic separation from the South-American bird. It has the crest differently formed to that of the true Harpy, and the wing is shorter in proportion to the length of the tail; but there can be no doubt that the two forms are intimately allied. How such a close similarity has been obtained between two Eagles inhabiting such widely different localities as South America and New Guinea, the avifaunæ of which have, generally speaking, very little in common, is a problem of geographical distribution which our present knowledge has no means of explaining.

D'Albertis first met with the *Harpyopsis* at Andei, in North-western New Guinea, and here it was also procured by Dr. Meyer, whose specimen was beautifully figured by Mr. Keulemans in the 'Mittheilungen' of the Dresden Museum (*l. c.*). During his travels in South-eastern New Guinea D'Albertis again met with the species in Hall Bay and on the Fly River, and a figure, copied from the before-mentioned plate of Mr. Keulemans, is given in his work on New Guinea. Mr. Broadbent found the Harpy at Fairfax Harbour, Port Moresby, and it was also procured by Mr. Hunstein in a small island off East Cape.

Mr. H. O. Forbes has sent a beautiful example from the Sogeri district in the Astrolabe Mountains, and Mr. Goldie has met with it in the same region. He says that the native name is "Duna," and he also forwarded to England two white eggs supposed to be those of the Harpy, but they appeared to us to be the eggs of some large Hornbill rather than those of a bird of prey.

Mr. Broadbent describes it as a "scrub bird," but nothing definite of its habits has yet been published. Signor D'Albertis shot a specimen on a tree stump, in the act of devouring a kangaroo (*Macropus papuanus*).

The following is a description of the specimen procured by Dr. Meyer at Andei:—

Adult. General colour above brown, with a slight bronzy gloss in certain lights, most of the feathers with a dull whitish edging, more distinct on the feathers of the head and neck, which are edged with hoary whitish, and form rather a full crest; lesser wing-coverts conspicuously margined with white, all the feathers dark brown before the tip, the greater series with one or two dark bars; quills brown, regularly barred with darker brown, the terminal bar broader than the others; tail brown, mottled with whity brown towards the base, and crossed with seven bars of darker brown, all the feathers narrowly edged with whitish at the tip; lores and region of the eye bare, scantily haired; sides of face and sides of neck brown like the crest, with paler and more fulvescent margins; chin whitish; lower throat brown like the sides of the neck, and with the same pale edges; rest of under surface white, ashy on the chest, some of the flank-feathers slightly washed with brown; under wing-coverts white; lower surface of quills greyish white, with broad bars of dark brown, breaking up into mottlings towards the base of the feathers: "bill blackish horn-colour; feet very pale yellow; iris chestnut-brown." Total length 30·5 inches, culmen 2·5, wing 16·5, tail 14·6, tarsus 5·0.

In a male collected by Mr. Broadbent the eyes were recorded as dark brown, and the measurements were as follows:—Total length 31 inches, culmen 2·7, wing 18·6, tail 15·5, tarsus 5·2.

Mr. Forbes's specimen, a figure of which is given in the accompanying Plate, of about two thirds of the natural size, measures as follows:—Total length 30 inches, culmen 2·2, wing 16·0, tail 15·5, tarsus 5·5.

BAZA GURNEYI, Ramsay.

BAZA GURNEYI, *Ramsay*.

Gurney's Cuckoo-Falcon.

Baza reinwardti (nec Müll. & Schl.), Tristram, Ibis, 1882, pp. 133, 141.
Baza gurneyi, Ramsay, Journ. Linn. Soc. xvi. p. 130 (1883).—Salvad. Orn. di Papuasia e delle Molucche, iii. App., p. 506 (1882).—Grant, Proc. Zool. Soc. 1888, p. 188.

For a considerable time the Cuckoo-Falcon of the Solomon Islands was supposed to be specifically the same as *B. reinwardti*, a species of somewhat extended distribution in Papuasia, as it occurs over the greater part of New Guinea and Salwati, as well as in the islands of Mysol, Misori, Ceram, Amboina, and the Ké and Aru groups. The history of the separation of the Solomon-Island bird is not very clear, but appears to be somewhat as follows:—In 1880 Mr. E. P. Ramsay described Mr. Cockerell's collections from the Solomon group, and recorded an example of a *Baza* from "Cape Pitt," which he said agreed exactly with Port-Moresby specimens. Later, in 1882, he described the species from the Solomon Archipelago as *Baza gurneyi*, and he observes:—"When I first notified *B. reinwardti* from the Solomon Islands, I was under the impression I had a veritable Solomon-Island bird before me. It now turns out that such was not the case: hence the mistake." Mr. Ramsay omits to tell us where the supposed "Cape Pitt" specimen *really* came from after all, and as he gives the localities for his *Baza gurneyi* as "Ugi" (*Rev. G. Brown*) and "Cape Pitt" (*Cockerell*), it will be seen that he has left the subject in a state of considerable uncertainty. What seems certain is that *B. gurneyi* is confined to the Solomons, one of Lieut. Richards's skins from Russell Island being now in the British Museum (*cf.* Tristram, *l. c.*), and the same institution has recently received two specimens collected by Mr. C. M. Woodford in Guadalcanar; the two last-named localities can be depended upon, as well as that of the island of Ugi.

When writing his account of *Baza reinwardti* for his work on Papuan ornithology, Count Salvadori remarks on a specimen collected by the Rev. G. Brown, but to which no locality was attached. This specimen is in the Tweeddale collection, and agrees thoroughly with a New-Ireland specimen also obtained by Mr. Brown and now in the British Museum, so that there can be little doubt as to the habitat of New Ireland being correct for the first specimen, as suggested by Count Salvadori. Both these individuals likewise agree with the specimen described and figured in the present work, which is a male from New Britain, collected by the late Dr. Kleinschmidt. The truth is that *B. gurneyi* is only an insular form of Papuan *B. reinwardti*, distinguished by its light under surface, broader terminal black band to the tail, and almost pure white under wing-coverts. The birds from New Britain and New Ireland are, again, a paler edition of *B. gurneyi*, with the same light under surface and white under wing-coverts, but still further distinguished by the light grey bars of the under surface. They are quite as worthy of a name as *B. gurneyi*; and as the Germans have called these islands by the title of the Bismarck Archipelago, we will attach the name of the great Chancellor to the Cuckoo-Falcon of that locality, as we fully believe it to be distinct. In the event of our surmise proving correct, the Plate of the present work must be referred to *Baza bismarckii*.

The figure in the Plate is of the natural size and is drawn from the specimen mentioned above.

[R. B. S.]

NINOX ODIOSA, Sclater.

NINOX ODIOSA, *Sclater*.

New-Britain Hawk-Owl.

Ninox tæniata (?), Ramsay, Proc. Linn. Soc. N. S. W. i. p. 369 (1876).
Ninox odiosa, Sclater, P. Z. S. 1877, p. 108.—Salvad. Ann. Mus. Civic. Genov. xii. p. 41 (1878).—Ramsay, Proc. Linn. Soc. N. S. W. iii. p. 249 (1879).—Salvad. Orn. Papuasia, etc. i. p. 86 (1880).—Id. Atti R. Accad. Torin. xvi. p. 620 (1881).—Gurney, Ibis, 1882, p. 131.—Salvad. Orn. Papuasia, etc. iii. p. 511 (1882).

The original specimen of this Owl was sent from New Britain by the Rev. George Brown, and since the arrival of the first example, several more have been procured by Lieut. Richards, Mr. Kleinschmidt, Dr. Finsch, and others; but it appears to have been found only in New Britain, and not in any of the adjacent groups of islands.

As suggested by Dr. Sclater in his description of the species, there can be no doubt that in many respects the present bird is allied to *N. punctulata* of Celebes; but there are so many points of difference that they cannot be confounded together. In the Celebean bird the back is spotted, as well as the head, and there are none of the white marks on the scapulars and wing-coverts which are conspicuous in *N. odiosa*. Underneath, the finely striated breast and abdomen distinguish the latter species from the thickly mottled and barred under surface of *N. punctulata*.

The following is the description of the adult male:—

General colour above pale chocolate-brown, nearly uniform; the scapulars with concealed bars of white and longitudinal ovate markings of the same on the outer web; the rump and upper tail-coverts with a few minute spots or bars of pale ruddy brown; wing-coverts nearly uniform, with a few spots and small bars of white, larger and more conspicuous on the greater series, which have the same ovate markings as the scapulars; bastard wing pale chocolate-brown, the primary-coverts rather more dusky brown, both series being perfectly uniform; quills dusky brown, externally pale chocolate, with a few spots of white on the edge of the outer and inner webs; head and nape dull umber-brown, profusely spotted with rather narrow bars of brownish white; nasal plumes brown with blackish centres; above the eye a broad streak of white; feathers below the eye also whitish; the ear-coverts chocolate-brown, barred with blackish and having whitish shaft-lines; throat white, the feathers on the sides of the throat having blackish tips and forming a disk; remainder of under surface white, the feathers centred with narrow brown streaks, somewhat widening towards the ends; sides of breast chocolate-brown, mottled with spots and bars of white; thighs dull white, slightly streaked on the upper parts with chocolate; under wing-coverts white with chocolate-brown tips; axillaries and under wing-coverts white, mottled with chocolate-brown on the edge of the wing, the lower series dusky brown barred internally with yellowish white, thus resembling the lower surface of the quills; "bill ash-colour; feet drab; iris yellow" (*Richards*). Total length 10 inches, culmen 0·8, wing 6·4, tail 4·2, tarsus 1·3.

The specimen is one which Mr. Ramsay lent to us; it is a male procured by Captain Richards in New Britain on the 30th of July, 1879, and is represented in the Plate of the full size. The description is taken from the same specimen.

[R. B. S.]

NINOX FORBESI, *Sclater*.

NINOX FORBESI, *Sclater.*

Forbes's Hawk-Owl.

Ninox forbesi, Sclater, Proc. Zool. Soc. 1883, p. 52, pl. xi.

THE present species, which has been named in honour of Mr. H. O. Forbes, who discovered it in the Tenimber Islands, belongs to a little group of Hawk-Owls which have the head uniform. *N. squamipila* of Ceram has many points of resemblance to the subject of our present article, but is altogether of a darker rufous colour, has the toes scantily feathered, the upper tail-coverts barred with white, and not more than seven broad blackish bars on the tail-feathers. Below, both species are similar, but *N. forbesi* does not have the under wing-coverts barred as in *N. squamipila*.

The nearest ally of *N. forbesi* is undoubtedly *Ninox hantu* of Wallace from Bourou; but this bird, though agreeing in the paler and more cinnamon-rufous colours of the plumage, is distinguished by its scantily feathered toes, which have only a few hair-like bristles. Both species have the under wing-coverts uniform like the breast; but *N. hantu* is a much darker bird, with broader bars on the tail-feathers and no white on the wing-coverts; the under surface also is more uniform cinnamon-rufous, and does not show the white bars which distinguish *N. forbesi*.

Mr. Forbes informs us that he only met with this species on one occasion in Timor Laut, and this was during an excursion to the mainland on the 9th of August, 1882. While conducting a palaver with the natives to obtain permission to shoot near the village of Loetoe, his native hunters managed to procure a pair of this Owl while they were awaiting the result of Mr. Forbes's negotiations with the villagers. The birds were sitting in a thick bushy tree at no great height from the ground.

The following is a description of the typical specimen figured by Dr. Sclater:—

Adult male (type of species). General colour above reddish brown, rather more rufous on the upper tail-coverts; scapulars barred with white or yellowish buff, with narrow cross bars of dusky brown; wing-coverts like the back, the greater series barred with dusky brown and fulvous or white; bastard wing like the other coverts; primary-coverts nearly uniform dark brown, with slightly indicated reddish-brown cross bars; quills reddish brown barred with blackish, the interspaces being paler and either fulvous or whitish, the inner secondaries less barred; tail-feathers light rufous-brown, with eleven bars of dusky blackish on the centre ones, sixteen on the outer feathers, which show whitish interspaces for more than half of the length of the outer web; head and hind neck more dingy rufous-brown than the back, with the colour of which it is in slight contrast; base of forehead and lores white, extending above the fore part of the eye and having black shaft-lines; feathers below the eye and ear-coverts dingy reddish brown, like the head; base of cheeks and base of chin white; throat and breast tawny rufous, the latter slightly mottled with paler cross bars of fulvous or narrower ones of dusky; abdomen and flanks barred broadly with white and more narrowly with dusky brown, the latter with a conterminous line of tawny buff; thighs and under tail-coverts tawny rufous, the latter barred with dusky and with broader bands of yellowish white; under wing-coverts and axillaries tawny rufous, the edge of the wing white, the greater series of coverts and the quills below paler and more yellowish buff, more reddish brown towards the end of the quills, which are barred across with blackish brown; " bill pale corneous; feet pale yellow, covered with bristly hairs, soles of feet nearly orange; iris rich golden " (*H. O. Forbes*). Total length 11·5 inches, culmen 1·05, wing 7·4, tail 4·5, tarsus 1·1.

The female bird, which was brought over by Mr. Forbes on his return from Timor Laut (too late for us to figure on the Plate, which had unfortunately been printed off before Mr. Forbes's arrival), only differs from the male in being paler and in having the breast barred with pale cinnamon like the abdomen, but not so broadly.

The Plate is drawn from the type specimen, which Dr. Sclater was kind enough to lend us, and portrays the adult male, of about the full size, in two positions.

[R. B. S.]

NINOX DIMORPHA.

NINOX DIMORPHA.

Salvadori's Hawk-Owl.

Athene dimorpha, Salvad. Ann. Mus. Civic. Genov. vi. p. 308 (1874).
Ninox dimorpha, Sharpe, Ibis, 1875, p. 258.—Id. Cat. Birds in Brit. Mus. ii. p. 175 (1875).—Salvad. Ann. Mus. Civic. Genov. x. p. 118 (1877), xii. p. 40 (1878).—Ramsay, Proc. Linn. Soc. N. S. Wales, iii. p. 248 (1879).—Salvad. Ibis, 1879, p. 319.—Id. Orn. Papuasia e delle Molucche, i. p. 83 (1880).
? *Athene*, sp., Ramsay, Proc. Linn. Soc. N. S. Wales, i. p. 388 (1876), ii. p. 19 (1877).

WHEN we wrote the second volume of the 'Catalogue of Birds' this species was unknown to us, and we were indebted to the kindness of Count Salvadori for a sketch of the type specimen, from which we gathered that it might belong to the genus *Glaucidium*, or that it might even be the type of a new genus. The genus *Glaucidium*, however, is unknown in the Australian region, and the discovery of a species in Papuasia would have been somewhat surprising; but an examination of a specimen of *Ninox dimorpha* proves that all these speculations were wrong, and that the bird is a true *Ninox*, as might have been expected from the locality.

Its position in the last-named genus is very easily defined, for it can be recognized at once by its streaked breast, spotted hind neck, and banded wing-coverts, the latter resembling the back. It was originally discovered near Sorong, in North-western New Guinea, by Signor D'Albertis, and Mr. Ramsay has received an adult and a nestling bird from the vicinity of Port Moresby; of the identity of the latter specimen there may be some doubt, but that the species occurs also in South-eastern New Guinea is unquestionable, as Mr. H. O. Forbes has sent a specimen from the Sogeri district of the Astrolabe Mountains, of which we give a detailed description:—

Adult. General colour above dark brown, regularly banded across with light rufous, whiter on the scapulars, which have also large white patches externally; wing-coverts like the back, the rufous bars obscure on the lesser coverts, but especially distinct on the greater series, which resemble the quills; bastard-wing and primary-coverts blackish, with a few obscure rufescent bands; quills blackish, banded with rufous externally, the bands more ashy on the inner webs and at the ends of the feathers; upper tail-coverts like the back; tail-feathers blackish, barred with ashy rufous, the bands twelve in number on the centre feathers, as well as the outer ones, on which, however, they are not strictly conterminous on both webs; crown of head blackish brown, streaked with tawny rufous, with which colour the feathers are edged, the nape and hind neck being pale tawny rufous with large blackish-brown mesial spots; lores white, with some hair-like black plumes; eyebrows and base of forehead white, streaked with black; ear-coverts ashy grey, streaked with black; cheeks, feathers below the eye, and chin white with narrow black shaft-lines, the hinder cheeks tinged with tawny buff and more broadly streaked with blackish; sides of neck like the hind neck; throat, chest, and under surface of body pale tawny buff, streaked with black; the lower breast and abdomen rather whiter and more boldly streaked, sides of body and flanks buffy white; thighs and under tail-coverts white, the latter streaked with black; under wing-coverts and axillaries pale tawny buff, streaked with black; quills below blackish brown, profusely banded with ashy brown or fulvous: "bill lead-colour; toes yellow; iris yellow" (*D'Albertis*). Total length 11·5 inches, culmen 1·0, wing 8·1, tail 5·5, tarsus 1·3.

The figure in the Plate represents an adult bird, and is drawn from the specimen obtained by Mr. Forbes and described above.

[R. B. S.]

EPIMACHUS ELLIOTI, Ward.

EPIMACHUS ELLIOTI, *Ward.*

Elliot's Promerops.

Epimachus ellioti, Ward, Proc. Zool. Soc. 1873, p. 743.—Elliot, Monogr. Parad. pl. 20.—Beccari, Annali Mus. Civic. Genov. vii. p. 710.—Salvad. op. cit. ix. p. 190.—Sharpe, Cat. Birds Brit. Mus. iii. p. 163.

I AM fortunate in possessing the unique type of this splendid bird, which has been placed in the genus *Epimachus* by Mr. Edwin Ward, the original describer, and allowed to remain there by Mr. Elliot and Mr. Bowdler Sharpe, both of whom have recently monographed the Birds of Paradise. I retain it in the genus *Epimachus*, but with some hesitation; for it differs considerably in its structure from *E. speciosus*, the only other representative of the genus. I would point out the difference existing in the sharply ending tail-feathers of the present bird, and still more the very different shape of the beautiful flank-plumes. In *E. speciosus* they form gracefully falling plumes illuminated by a subterminal metallic band; but in *E. ellioti* there are three series of plumes on the flanks, alike in colour, but differing in size. Whether these differences are generic will remain for subsequent writers to decide, when perfect specimens of both sexes of *Epimachus ellioti* reach Europe. The style of coloration of the present bird is also very distinct, *E. speciosus* showing none of the beautiful velvety texture which strikes the observer at the first glance on beholding *E. ellioti*.

At present we are in ignorance as to the habitat of the latter bird. A single skin came into the possession of Mr. Ward, by whom it was described; and it afterwards passed into my hands, after being figured in Mr. Elliot's Monograph of the Paradise-birds. Dr. Beccari, though he tried hard to discover the species during his expedition to the Papuan Islands, did not succeed in discovering its home; but he believes that it will ultimately be found to inhabit the mountains of Waigiou. One may be almost certain that the same locality which possesses this brilliant species will also produce other interesting birds new to science.

The following is Mr. Elliot's description of the typical specimen:—

"Top of head rich amethyst; occiput and sides of neck also amethyst, changing in certain lights to a rich light greenish gloss; back, wings, upper tail-coverts, and tail brilliant violet-purple; the wings and the tail also marbled with a dark amethyst hue, like watered silk, changing according to the light; throat and upper portion of breast deep maroon colour, with purple reflections; a narrow reddish purple band crosses the lower part of the breast; sides of the breast, flanks, and rest of underparts dark green, the flank-feathers much elongated, and stretching beyond the wings; beneath the shoulder of the wing spring two rows of plumes, which are greenish at the base, graduating into deep purple, and terminating in a brilliant metallic blue, very much narrower on the upper row than the lower one. The plumage of the entire bird is very velvety in texture, and, with the exception of the metallic parts, appears black in ordinary lights; bill black, rich orange-yellow at the gape."

The principal figure in the Plate represents the species of nearly the size of life, and is drawn from the type specimen in my collection.

EPIMACHUS SPECIOSUS.

EPIMACHUS SPECIOSUS.

Great Promerops.

Le Grand Promérops de la Nouvelle-Guinée, Sonn. Voy. N. Guin. p. 163, pl. 101.
Le Promerops brun de la Nouvelle-Guinée, id. tom. cit. p. 164, pl. 100.
Grand Promerops à paremens frisés, Buff. H. N. Ois. vi. p. 472.
Promerops de la Nouvelle-Guinée, Buff. Pl. Enl. vi. pls. 638, 639.
Upupa speciosa, Bodd. Tabl. Pl. Enl. p. 39.
—— *striata*, Bodd. tom. cit. p. 39.
New-Guinea Brown Promerops, Lath. Gen. Syn. i. pt. 2, p. 694.
Grand Promerops, Lath. tom. cit. p. 695.
Upupa fusca, Gm. S. N. i. p. 468.
—— *magna*, Gm. tom. cit. p. 468.
Le Promérops rayé, Audeb. et Vieill. Ois. Dor. i. pl. 7.
Le Promerops à large parure, Levaill. H. N. Promér. et Guêp. pls. 13, 15.
Promerops striata, Shaw, Gen. Zool. viii. p. 144.
—— *superbus*, Shaw, tom. cit. p. 145.
Falcinellus superbus, Vieill. N. Dict. d'Hist. Nat. xxviii. p. 166.
—— *magnificus*, Vieill. tom. cit. p. 167.
Epimachus magnus, Cuvier, Règne Anim. i. p. 407.
—— *superbus*, Steph. Gen. Zool. xiv. p. 77.—Wagler, Syst. Av. *Epimachus*, sp. 1.—Less. Traité, p. 321, Atlas, pl. 73. fig. 1.—Rosenb. J. f. O. 1864, p. 123.
Cinnamolegus papuensis, Less. Ois. Parad. Syn. p. 32.—Id. H. N. pls. 39, 40.
Epimachus speciosus, Gray, Gen. B. i. p. 94.—Schl. Mus. P.-B. *Coraces*, p. 94.—Elliot, Monogr. Parad. pl. xix.—Salvad. Ann. Mus. Civic. Genov. vii. p. 785, ix. p. 190.—Sharpe, Cat. B. iii. p. 162 (1877).
—— *magnus*, Bp. Consp. i. p. 411.—Wall. Ibis, 1861, p. 287.—Id. P. Z. S. 1862, p. 160.—Id. Malay Arch. ii. p. 255.—Schl. J. f. O. 1861, p. 386.—Id. N. T. D. i. p. 332.
—— *maximus*, Gray, P. Z. S. 1861, p. 433.—Id. Hand-l. B. i. p. 105.—Beccari, Ann. Mus. Civic. Genov. vii. p. 710.—Id. Ibis, 1876, p. 249.

In spite of the long list of synonyms with which this species has been burdened by naturalists, the actual information respecting its habits is almost wanting; neither can I pretend to give a long account of the bird, simply for the reason that there is nothing to tell. I cannot weary my readers with a dissertation on the various incidents through which this fine Bird of Paradise has reached the very complicated synonymy which has marked its scientific history. Suffice it to say that, owing to our meagre knowledge of the bird in a natural state, the males and the females have generally been taken for separate species; and although imperfect skins have been sent to Europe in some numbers for the last hundred years, we have had to wait until quite recently for the gladdening of our eyes by the receipt of the perfect bird.

It is at once the largest and the most remarkable, if not the most beautiful, of the thin-billed Birds of Paradise, which comprise the Rifle-birds, the Twelve-wired *Seleucides*, and the lately discovered Sickle-billed *Drepanornis*. Only two species of *Epimachus* are known—the subject of the present article, and *E. ellioti*; the latter is still represented by the single type specimen in my collection, the habitat of which, though supposed to be the island of Waigiou, is not yet known for certain.

Mr. Wallace did not meet with the present species during his explorations in Papuasia. He says, "This splendid bird inhabits the mountains of New Guinea, in the same district with the Superb (*Lophorina atra*) and the Six-shafted (*Parotia sexpennis*) Paradise-birds, and, I was informed, is sometimes found in the ranges near the coast. I was several times assured by different natives that this bird makes its nest in a hole underground, or under rocks, always choosing a place with two apertures, so that it may enter at one and go out at the other. This is very unlike what we should suppose to be the habits of the bird; but it is not easy to conceive how the story originated if it is not true; and all travellers know that native accounts of the habits of animals, however strange they may seem, almost invariably turn out to be correct."

The following note appears in Dr. Beccari's Ornithological Letter:—

"The *Epimachi* have been separated from the other birds of Paradise; but I think this is paradoxical. The form and the length of the beak of *Epimachus maximus* is most variable; the young males and females are found with the beak only half the length of that of the adult males and females. This fact made me think at first that I had found the female of *E. ellioti*; but I was mistaken. An *Epimachus* seems to be found at Waigiou, and will probably be *E. ellioti*; but I was not able to return there as I had intended. *Epimachus maximus* and *Astrapia gularis* are only found on the highest and most difficult peaks of Mount Arfak, nearly always above 6000 feet elevation. Specimens in dark plumage are common enough; but those which have attained perfect plumage are rare, perhaps because they take some years to acquire it. Both of them live on the fruits of certain Pandanaceæ, and especially on those of the *Freycinetiæ*, which are epiphytous on the trunks of trees. The irides of the large *Epimachus* are dark brick-red."

The descriptions are taken from Mr. Bowdler Sharpe's 'Catalogue of Birds.'

Adult male. Above velvety black, with metallic feathers of coppery green on the head, middle of the back, and rump; lores and feathers on the side of the head metallic like the crown; entire under surface of body velvety black, with a purplish brown gloss on the sides of the body; on each side of the breast springs a tuft of sickle-shaped plumes in the shape of a fan, velvety black, tipped with a broad band of steel-blue, before which is a narrow subterminal band of purplish blue; flank-feathers long and drooping, the outer ones broadly tipped with metallic bronzy-green, before which is a double subterminal band of velvety black and purplish blue; wings velvety black, with a gloss of steel-blue; tail-feathers black, all but the three outermost feathers washed with steel-blue, the two centre ones entirely of this colour; bill and legs black.

Total length 26 inches, culmen 2·85, wing 7·2, tail 16·7.

Female. Upper part of head brownish red; rest of upper parts olive-brown, becoming slightly rufous on the rump and upper tail-coverts; secondaries reddish brown, edged with rufous; primaries dark brown, edge of outer web rufous; cheeks, throat, and upper part of breast brownish black; underparts white, narrowly barred with black; tail light brown, with a rufous tinge; bill long and slender, much curved, and, with the feet and tarsi, jet-black.

The figures in the Plate, which represent a male of about two thirds the natural size, and a male and a female very much reduced, are taken from a superb pair of skins in my own collection.

DREPANORNIS CERVINICAUDA.

DREPANORNIS CERVINICAUDA.

Bennett's Bird of Paradise.

Drepanornis albertisii (nec Sclater), Sharpe, Journ. Linn. Soc., Zool. xvi. p. 445 (1882).—Salvad. Orn. Papuasia, etc. iii. App. p. 552 (1882).
Drepanornis d'albertisii, Ramsay, Proc. Linn. Soc. N. S. W. iv. p. 469 (1880).—Id. op. cit.viii. p. 28 (1883).

For some time it was suspected both by Mr. Ramsay and by ourselves that the *Drepanornis* from South-eastern New Guinea was a different species from that of the Arfak Mountains, as the tail was always so much paler than in the north-western bird. Unfortunately only female specimens were at first obtained by the collectors in the Astrolabe Mountains, and it was only quite recently that Mr. Goldie succeeded in procuring the males.

Dr. Bennett of Sydney, who has always proved himself a true friend to science, became the possessor of specimens from the Astrolabe range in South-eastern New Guinea, and the British Museum is indebted to his liberality for the beautiful skins which now adorn that collection. They were previously submitted to Dr. Sclater, who exhibited them at a meeting of the Zoological Society on the 4th of December, 1883, and gave to them the name of *cervinicauda*.

Mr. Ramsay has likewise procured a series for the Sydney Museum, and has had the additional good fortune to obtain the nest and egg of this new Bird of Paradise, which he describes as follows:—

"The nest is a thin, rather flat structure, built between a horizontal bough in a fork of a thin branch; it has a slight depression about one inch deep, a network of wire rootlets are stretched across the fork, and the nest proper built on them; it is composed of wiry grasses of a light reddish-brown colour, the platform being of black wiry roots.

"The egg is in length 1·37, by 1 inch in breadth; it is of a light dull cream-colour, with a reddish tinge, spotted all over with oblong dashes of reddish brown and light purplish grey, closer on the thick end."

It is unnecessary to give a complete description of this species. It may be said to be exactly like *Drepanornis albertisi* from North-western New Guinea, but distinguished by its much paler rump and tail.

The type specimens are figured in the Plate, a representation of the male and female being given of about the natural size.

[R. B. S.]

DREPANORNIS ALBERTISI, *Sclater*.

DREPANORNIS ALBERTISI, Sclater.

D'Albertis's Bird of Paradise.

Drepanornis albertisi, Sclater, P. Z. S. 1873, p. 558, pl. 47.—D'Albertis, *t. c.* p. 558.—Elliot, Monogr. Paradiseidæ, pl. 21.
Epimachus wilhelminæ, Meyer, J. f. O. 1873, p. 404.—Id. Ibis, 1874, p. 303.

This remarkable new form of Paradise-bird was one of the most interesting discoveries made by Signor d'Albertis during his recent explorations in the island of New Guinea. He found it at Mount Arfak; and almost simultaneously, Dr. Meyer discovered it in the same locality. We are also informed by Dr. Sclater that previous to this, Baron von Rosenberg had seen a female in the collection of Mr. Van Duivenbode, at Ternate, as long ago as April 1871 (*cf.* Ibis, 1874, p. 187).

The following are the notes given by Signor d'Albertis with regard to its habits :—

"This will probably prove to be a new bird, both generically and specifically. It is a very rare bird, and many of the natives did not know it; but others called it 'Quarna.' The peculiarity of this species consists in the formation of the bill, head, and softness of the plumage. At first it does not appear to have the beauty peculiar to other birds of this class; but when observed more closely, in a strong light, the plumage is seen to be rich and brilliant : the feathers rising from the base of the beak are of a metallic green, and reddish copper-colour; the feathers of the breast, when smooth, are of a violet grey, and when raised form a semicircle round the body, reflecting a rich golden colour; other violet-grey feathers arise from the flanks, which are edged by a rich metallic violet tint; and when the plumage is entirely expanded the bird appears as if it had formed two semicircles round itself, and is very handsome.

"The tail- and wing-feathers are yellowish; underneath they are of a darker shade. The head is barely covered with small round feathers, which are rather deficient at the back of the ear. The shoulders are tobacco-colour; and under the throat black, blending into olive. The breast is violet-grey, banded by a line of olive, the rest white. The beak is black, eyes chestnut, and the feet of a dark leaden colour.

"This species is met with in the vicinity of Mount Arfak. Its food is not known, nothing having been found in the stomachs of those prepared except clean water."

The following complete descriptions are taken from Mr. Elliot's 'Monograph of the Paradiseidæ : '—

"*Male.*—Head covered with short, rather stiff, light-brown feathers, tipped with deep purple. Two spots of metallic-blue feathers between the eyes and bill, projecting above the eyes like horns; a spot of bare skin behind the eyes apparently red. Neck and back rufous brown. Primaries blackish brown, edged with light-rufous feathers on the outer webs. Secondaries light rufous brown on outer web, black on the inner, edged with very light reddish brown. The three innermost secondaries light reddish brown on both webs. Upper tail-coverts and tail bright reddish brown. Chin and throat metallic deep purple, black in certain lights. Breast covered with long feathers, grey, with rich purple reflections, and edged on the lower part with dull green, crossing the body in a narrow bar. From either side, near the shoulder of the wing, spring two tufts of feathers that extend beyond the breast-shield, of an intense metallic fiery red, tipped with purple. These when not elevated are altogether hidden by the outer feathers, which are uniform purple like the breast. From the flanks, just above the termination of the breast-shields, on either side project two long tufts of plumes, which extend to the end of the under tail-coverts, of the same colour as the breast, brownish grey, each feather tipped with very brilliant deep purple. The abdomen and under tail-coverts pure white, the former streaked with purplish grey on the upper portion. Bill very long, slender, and much curved, black. Feet and tarsi dull lead-colour.

"*Female.*—Head chestnut brown. Back and wings rufous brown. Primaries and secondaries blackish brown on inner web, outer web brown. Upper tail-coverts and tail light red. Chin and throat blackish brown, each feather with a central streak of light brown. Breast light brown, irregularly barred with dark brown. Flanks and lower parts of body yellowish brown, indistinctly barred with dark brown, except in the centre of the abdomen, which is light reddish white. Thighs reddish, barred with brown. Under tail-coverts pale reddish. Iris chestnut. Bill long, curved, and slender like that of the male, black. Feet and tarsi lead-colour."

For the opportunity of figuring the typical specimens of this bird, I am indebted to the courtesy of Signor d'Albertis. The birds are represented about the natural size.

DREPANORNIS BRUIJNII, Oust.

DREPANORNIS BRUIJNII, Oustalet.

Bruijn's Bird of Paradise.

Drepanornis bruijnii, Oustalet, Bull. Assoc. Scient. de France, 1880, p. 172.—Salvad. Orn. Papuasia e delle Molucche, ii. p. 553 (1881).—Guillemard, Proc. Zool. Soc. 1885, p. 649.

It is much to be regretted that we have been unable to procure a fully adult bird for our illustration of the present species, for to all appearances the specimen which has been lent to us by Dr. Guillemard for the purposes of this work is immature. At the same time it is somewhat singular that all the specimens so far procured by Mr. Bruijn's hunters in North-western New Guinea have been similar to the bird here figured.

Dr. Guillemard, who obtained two specimens during the cruise of the 'Marchesa' with Mr. Kettlewell, gives the following account of his getting them:—"While in Ternate Mr. Bruijn showed me the skins of two birds of the genus *Drepanornis* obtained by his hunters on the north coast of New Guinea a little to the eastward of the mouths of the Amberbaki River. One was marked 'female,' the other 'male'; but both were destitute of any brilliant colouring whatsoever. . . . . . Mr. Bruijn informed me that his hunters had obtained seven or eight examples of this species, but that, though of different sexes, they were all of the same sober colouring. Judging from the habits of others of the *Paradiseidæ*, notably in the case of *P. rubra*, where the immature males and females appear to live in districts quite apart from the adult male at certain seasons of the year, and from the fact that in this group of birds the males are all of brilliant colouring, we can safely predict that the adult male of this species has yet to be discovered, and that it will probably show a development of subalar plumes closely resembling that of *D. albertisi*."

It is no doubt true that when the fully-plumaged male becomes known considerable resemblance to the same sex of *D. albertisi* will be discovered, and a more accurate comparison of the two species will then be possible; but there can be no doubt that *D. bruijnii* is a well-marked species, even when founded on the immature bird. The size of the bill alone is sufficient to distinguish it, and the distribution of the bare patches on the face is also different; but the chief characters will no doubt be discovered when skins of the adult male are sent to Europe.

The following is a description of the specimen kindly lent to us by Dr. Guillemard :—

General colour above brown, with a slight tinge of olive; wing-coverts like the back, the outer median and the greater coverts washed externally with dull fawn-colour; bastard-wing and primary-coverts dusky brown, the latter shaded with fawn near the base; quills dusky brown, externally pale olive-brown, the secondaries washed with fawn-colour on the outer web; upper tail-coverts dull fawn-colour, washed with brown in the centre; tail-feathers clear fawn-colour; crown of head blackish, the feathers being of a velvety texture; the hind neck also shaded with blackish; sides of face bare; lores and a line of feathers from the gape along the side of the face blackish, the cheeks whity brown, black anteriorly, followed by a broad malar line of black; throat and under surface of body pale fawn-buff, regularly barred with blackish, the throat and fore neck more dusky and the cross bars smaller and more indistinct; the abdomen clearer buff and the bars wider and more distinct; sides of body and flanks like the abdomen; thighs and under tail-coverts also fawn-buff, barred with blackish; under wing-coverts and axillaries paler fawn-buff than the breast and indistinctly barred; quills below dusky, fawn-buff along the inner edge. Total length 12 inches, culmen 2·7, wing 5·6, tail 4·3, tarsus 1·25.

The figure in the Plate represents the bird of the size of life and is taken from Dr. Guillemard's specimen mentioned above.

[R. B. S.]

GRASPEDOPHORA MAGNIFICA.

CRASPEDOPHORA MAGNIFICA.

New-Guinea Rifle-bird.

Le Proméfil, Levaill. Ois. de Parad. p. 36, pl. 16 (1806).—Less. H. N. Ois. Parad. pl. 29 (1835).
Falcinellus magnificus, Vieill. N. Dict. d'Hist. Nat. xxviii. p. 167, pl. G 80. no. 3 (1818).
Epimachus splendidus, Steph. Gen. Zool. xiv. p. 77 (1826).
Epimachus magnificus, Wagler, Syst. Av., *Epimachus*, sp. 10 (1827).—Cuvier, Règne Anim. 1829, p. 440.—Less. Cent. Zool. p. 22, pls. 4, 5 (1830).—Id. Ois. Parad. Syn. p. 27 (1835).—Id. H. N. Ois. Parad. p. 218, pls. 32–34 (1835).—Bp. Consp. i. p. 412 (1850).—Gray, P. Z. S. 1859, p. 155.—Schl. J. f. O. 1861, p. 386; id. Mus. P. B., *Coraces*, p. 96 (1867).
Craspedophora magnifica, Gray, List Gen. B. p. 15 (1841).—Reichenb. Handb. *Scansoriæ*, p. 330, Taf. lcxi. figs. 4089–91 (1850).—Wall. P. Z. S. 1862, p. 160.—Rosenb. J. f. O. 1864, p. 128.—Salvad. Ann. Mus. Civic. Genov. ix. p. 191 (1874).
Epimachus paradiseus, Gray, Gen. B. ii. pl. xxxii. (nec Swains.).
Ptilornis magnificus, Gray, Handl. B. i. p. 105 (1869).
Ptiloris magnificus, Elliot, P. Z. S. 1871, p. 583.—Id. Mon. Parad. pl. xxiii. (1876).—Salvad. Ann. Mus. Civic. Genov. vii. p. 785 (1874).
Ptilorhis superbus, Beccari, Ann. Mus. Civic. Genov. vii. p. 173 (1875).—Sclater, Ibis, 1876, p. 252.
Ptilorhis magnifica, Sharpe, Cat. B. Brit. Mus. iii. p. 158 (1877).

SEVERAL years ago I described and figured a species of Rifle-bird from North-eastern Australia as *Ptilorhis magnifica*; and for a long time it was supposed by ornithologists that one species was common to New Guinea and the Cape-York peninsula. Mr. Elliot, however, in his work on the Birds of Paradise, pointed out certain differences between these two forms, which appear to justify their specific separation, and adopted for the Cape-York species the MS. name of *Ptilornis alberti*, proposed by the late Mr. G. R. Gray, after a study of the specimens in the British Museum. Mr. Bowdler Sharpe also concurs in the specific separation of these two Rifle-birds; and he points out that in the male of *P. magnifica* the breast becomes purple below the double pectoral band, and has not the oily-green lustre which distinguishes the same sex of *P. alberti*. I must also mention that the metallic lustre on the pectoral shields of these birds is of a different hue, although perhaps the greatest difference between the two species is exhibited by the female birds. Thus, the female of the New-Guinea Rifle-bird is entirely rufous on the upper surface, and has the head of the same colour as the back, whereas in the female of Prince Albert's Rifle-bird the head is ashy brown. As is the case with so many of the birds of New Guinea, we know nothing of the habits of the Rifle-bird inhabiting that region; but we may well suppose that they do not differ from those described by me in my work on the Birds of Australia. Dr. Beccari states that the eggs of the New-Guinea Rifle-bird have been discovered by one of Mr. Bruijn's hunters " in the branches of a tree called at Ternate 'Kaju tjapilong,' which is the *Calophyllum inophyllum*." He adds, " At present I have not the eggs before me; so I will write about them more fully another time, when I have been able to examine the man who found them."

The following descriptions are taken from Mr. Bowdler Sharpe's Catalogue of Birds.

"*Male.* Top of head and occiput, centre of throat, and entire upper part of breast shining bluish green, purple in certain lights; entire upper parts deep velvety black, with rich dark purple reflections; primaries black, with green reflections; a narrow line of green, red in some lights, beneath the metallic of the breast; flanks and abdomen purple; side plumes also purple, basal half and filamentary ends black; two centre tail-feathers shining green; remainder velvety black, with green reflections on their outer webs; bill, feet, and legs stout, black.

"*Female.* Above cinnamon-rufous, the wings and tail entirely of the same colour as the back, the inner webs browner; over the eye a narrow streak of white; lores and sides of face dusky brown, the former washed with rufous; the ear-coverts minutely streaked with rufous along the shafts of the feathers; cheeks white, the feathers somewhat scaly in appearance; a malar streak of dark brown on each side of the throat;

throat white, slightly mottled with minute dusky cross markings; rest of under surface of body dull white, very numerously and thickly barred across with dusky blackish. Total length 12 inches, culmen 1·95, wing 6·6, tail 4·45, tarsus 1·65.

"*Young male.* Similar to the adult female, but of a deeper rufous, the head and neck rather dingier than the back; a tolerably well-defined white eyebrow, the feathers edged with brown; lores and ear-coverts dusky chocolate-brown, with very few ochraceous shaft-streaks; under surface of body dirty white, very thickly barred across with black; the flank-plumes elongated, but barred exactly like the breast; from the base of the lower mandible a malar streak of dusky black, continued down the sides of the throat onto the sides of the chest. Total length 14·5 inches, culmen 2·3, wing 6·95, tail 4·9, tarsus 1·65."

It will be seen that I have not followed the nomenclature employed by me in the Supplement to the 'Birds of Australia.' The present bird and *P. alberti* ought to be placed in a separate genus from that which contains *P. paradisea* and *P. victoriæ*; and I therefore propose to adopt Mr. G. R. Gray's generic name of *Craspedophora* for the above two birds.

The figures represent a male and a female, about the natural size, drawn from specimens in my collection.

SELEUCIDES NIGRICANS.

SELEUCIDES NIGRICANS.

Twelve-wired Bird of Paradise.

Le Manucode à douze filets, Audeb. et Vieill. Ois. Dor. ii. p. 29, pl. 13.
Le Nébuleux, Levaill. Ois. de Parad. i. pls. 16, 17.
Le Promerops multifil, Levaill. H. N. Promer. et Guêp. pl. 17.
Paradisea nigricans, Shaw, Gen. Zool. vii. pt. 2, p. 489 (1809).
—————— *alba*, Blumenb. Abbild. nat. Gegenst. pl. 96.—Schleg. J. f. O. 1861, p. 386.
—————— *resplendescens*, Vieill. Nouv. Dict. xxviii. p. 165.—Id. Galerie Ois. p. 107, pl. 185.
Epimachus albus, Temm. Man. d'Orn. i. p. lxxxvi.—Wagl. Syst. Av. 1827, *Epimachus*, sp. 9.—Gray, Gen. B. ii.
    p. 94.—Id. P. Z. S. 1858, p. 190.—Id. List B. New Guinea, pp. 21, 55.—Id. P. Z. S. 1861, p. 433.—
    Wallace, P. Z. S. 1862, p. 160.—Schleg. Mus. P.-B., *Coraces*, p. 95.—Id. Nederl. Tijdschr. Dierk. iv.
    p. 49.—Gray, Hand-l. B. i. p. 105.
Twelve-wired Paradise Bird, Lath. Gen. Hist. iii. p. 199, pl. 48.
Seleucides acanthylis, Less. H. N. Ois. Parad. pls. 36–38.—Id. Syn. p. 29.
Nematophora alba, Gray, List Gen. B. i. p. 12.
Seleucides alba, Gray, List Gen. B., Addenda, p. 1.—Bp. Consp. i. p. 412.—Cab. Mus. Hein. i. p. 215.—Reichenb.
    Handb. Spec. Orn. Scansoriæ, p. 331, taf. 612, figs. 4092, 4093.—Wallace, Malay Archip. ii. p. 250.—
    Elliot, Monogr. Parad. pl. xxii.—Salvad. Ann. Mus. Civ. Genova, vii. p. 785.—Beccari, *t. c.* p. 713.—
    Scl. P. Z. S. 1876, p. 252.
Ptiloris nebulosus, Licht. Nomencl. p. 10.
Seleucides resplendens, Rosenb. Nat. Tijdschr. Nederl. Ind. xxv. p. 238.—Id. J. f. O. 1864, p. 123.
Epimachus resplendens, Rosenb. Reist. naar Geelvinkb. pp. 101, 116.
Seleucides ignota, Salvad. Ann. Mus. Civ. Genova, viii. p. 403; ix. p. 191; x. p. 154.—D'Albert. & Salvad.
    op. cit. xiv. p. 107.
—————— *niger*, Sharpe, Cat. B. iii. p. 159.
Epimachus resplendescens, Rosenb. Malay Arch. p. 552.
Seleucides nigricans, Salvad. Orn. della Papuasia &c. p. 561 (1881).

THE list of names given above shows that this species of Bird of Paradise has been known to writers for a long period. Most of the synonymy I have derived from Mr. Bowdler Sharpe's 'Catalogue of Birds,' and from the more complete list of works given by Count Salvadori in his recently published book on the birds of New Guinea. When I state that I have, by no means exhausted the synonymy of the species as set down by Count Salvadori, it may readily be imagined that the number of books in which reference is made to the species is very large indeed. I have not, however, full space for such lengthened synonymy in the present work, and must refer the reader to the above-mentioned volumes for further quotations. The Twelve-wired Bird of Paradise is the only representative of the genus *Seleucides*, which belongs to the slender-billed section of the Paradiseidæ. It is remarkable for its elongated flank-feathers, which are of a fine yellow colour, and have six shafts produced into thread-like plumes, whence the bird has received its ordinary English name. Unfortunately the beautiful yellow colour on the flanks fades away after death, and becomes white, when the bird loses much of its original beauty.

With regard to the nomenclature of the bird, I have come to the conclusion that the first name, which ought to be employed, is that of *nigricans* of Shaw, as has been set forth by Count Salvadori. As far as we know at present, it is entirely confined to New Guinea, over the whole of which great island it appears to be distributed. It is abundant on the Fly river, to judge by the large series obtained by Signor D'Albertis during his residence in Southern New Guinea. He found it living solitary, and frequently resting on the dead branch of a tree, uttering its note (which sounded like *Có-có-có*) in the early morning at the rising of the sun; during the day it was silent. Mr. Wallace, in his 'Malay Archipelago,' gives the following account of the species:—"The *Seleucides alba* is found in the island of Salwatty, and in the north-western parts of New Guinea, where it frequents flowering trees, especially sago-palms and pandani, sucking the flowers, round and beneath which its unusually large and powerful feet enable it to cling. Its motions are very rapid. It seldom rests more than a few moments on one tree, after which it flies straight off, and with great swiftness, to another. It has a loud shrill cry, to be heard a long way, consisting of 'cáh, cáh,' repeated five or six times in a descending scale; and at the last note it generally flies away. The males are quite solitary in their habits, although, perhaps, they assemble at certain times like the true Paradise-birds. All the specimens shot and opened by my assistant Mr. Allen, who obtained this fine bird during his last voyage to New Guinea, had nothing in their stomachs but a brown sweet liquid, probably the nectar of the flowers on which they had been feeding. They certainly, however, eat both fruit and insects; for a specimen, which I saw alive on board a Dutch steamer, ate cockroaches and paya fruit voraciously. This bird had the curious habit of resting at noon with the bill pointing vertically upwards. It died on the passage to Batavia; and I secured the body and formed a skeleton, which shows indisputably that it is really a Bird of Paradise. The tongue is very long and extensible, but flat, and a little fibrous at the end, exactly like the true Paradiseas.

"In the island of Salwatty the natives search in the forests till they find the sleeping-place of this bird, which they know by seeing its dung upon the ground. It is generally in a low bushy tree. At night they climb up the tree, and either shoot the birds with blunt arrows, or even catch them alive with a cloth. In New Guinea they are caught by placing snares on the trees frequented by them, in the same way as the Red Paradise-birds are caught in Waigiou."

Only on one occasion has the present species been known to have been brought alive to Europe, a single example having been procured by Signor G. E. Serruti in New Guinea, and presented by him to the late King of Italy. It survived, however, only a few months in Europe.

I take the following descriptions from Mr. Sharpe's 'Catalogue of Birds':—

Adult male. General colour above velvety black, with a strong gloss of oil-green when viewed from the light, with coppery bronze reflections; scapulars and wing-coverts resembling the back; greater coverts and secondaries fiery purple, the primaries black, with an external gloss of violet; tail fiery purple; head all round of a velvety texture, coppery purple above, oily green on the sides of the face and throat; fore neck and chest velvety black, forming a shield, somewhat shaded with oily green in the centre, the lateral plumes all tipped with bright metallic emerald-green, forming a fringe; rest of the under surface of body buffy yellow, the plumes of the flanks elongated and silky, and furnished with six thread-like shafts, produced to a great length, and curved backwards on the body; under wing-coverts black; bill black. Total length 12 inches, culmen 2·7, wing 6·45, tail 3·15, tarsus 1·75; threads reaching 10·2 inches beyond the flank-feathers.

Adult female.—General colour above bright chestnut-red; back of the neck and sides of the same black; the feathers of the mantle also mottled with black, the bases of the feathers being of this colour; crown of head and nape velvety black, with a purplish gloss when seen away from the light; wing-coverts and secondaries chestnut-red, like the back, the primaries black, chestnut on their outer webs; tail uniform chestnut; space around and behind the eye bare, as also a spot on the auricular region; ear-coverts black; sides of face and throat greyish white, faintly mottled with dusky bars of blackish; rest of under surface of body buffy brown, washed here and there with pale rufous, the whole transversely barred with somewhat irregular cross lines of blackish brown, broader on the fore neck and breast, and more faintly indicated on the abdomen, and especially on the long flank-feathers and under tail-coverts; under wing-coverts bright chestnut, with dusky blackish cross bars. Total length 12·5 inches, culmen 2·55, wing 6·5, tail 4·3, tarsus 1·7.

Young male.—At first resembles the adult female. A specimen collected by Mr. Wallace is in perfect plumage as regards its head, mantle, and breast, the rest of the body being in the chestnut plumage of the female, the tail being still entirely chestnut. At the same time the beautiful purple colour is being put on the wings by a gradual change of feather, and not by a moult; half the inner secondaries are chestnut, but more or less mottled with black, the purple colour appearing very plainly on the inner webs.

The first Plate represents the male bird, of the natural size; and I have thought it necessary to give a second illustration of this species, in order to show some of the changes of plumage. The second Plate represents a female and a young male in its first plumage, together with another bird, of the same sex, commencing to put on his adult livery.

SELEUCIDES NIGRICANS, ♀ and ♂ juv.

PARADIGALLA CARUNCULATA.

PARADIGALLA CARUNCULATA, Lesson.

Wattled Bird of Paradise.

Paradigalla carunculata, Less. Ois. Parad. p. 242 (1835); id. Rev. Zool. 1840, p. 1; Bp. Consp. i. p. 414 (1850); Sclater, P. Z. S. 1857, p. 6; Wall. P. Z. S. 1862, p. 160; id. Malay Arch. ii. p. 257 (1869); Elliot, Monogr. Parad. pl. xvii. (1873); Salvadori, Ann. Mus. Civic. Genov. vii. p. 784 (1875); Beccari, t. c. p. 711 (1875); Sclater, Ibis, 1876, p. 250; Salvadori, Ann. Mus. Civic. Genov. ix. p. 190 (1877); Sharpe, Catalogue of Birds, iii. p. 165 (1877).

Astrapia carunculata, Eydoux et Souleyet, Voy. Bonite, p. 83, pl. 4 (1841); Gray, Gen. B. ii. p. 326 (1846); Schlegel, J. f. O. 1861, p. 386; Rosenb. J. f. O. 1864, p. 131; Gray, Handl. B. ii. p. 17 (1870).

The extreme rarity of the present species in European collections may be imagined from the fact that Mr. Elliot, when writing his monograph of the birds of Paradise five years ago, could only cite two specimens as existing in the museums of the world. One of these was the original specimen procured by MM. Eydoux and Souleyet during the voyage of the 'Bonite,' and still preserved in the Paris Museum; and the second example was contained in the rich collection of the Philadelphia Academy. Since that time, however, perfect specimens have been obtained by the European travellers who have visited and explored the Arfak Mountains in North-western New Guinea.

It appears, indeed, to be somewhat rare even in this part of the great Papuan island; for out of five hundred and thirty-two specimens of Paradise-birds forwarded to Italy by Dr. Beccari and Mr. Bruijn, only fifteen belonged to the present species. The more recent explorers, MM. Laglaize and Raffray, have also met with the bird; and I possess in my own collection a fine pair procured by M. Laglaize in the Arfak Mountains.

Very little has been recorded concerning the habits of the Wattled Bird of Paradise, as the original discoverers did not themselves meet with the species in a living state; and the first person who has given us any account of the bird is Dr. Beccari, who has done so much to make us acquainted with the economy of the Paradiseidæ. In his Ornithological Letter, he writes:—"As to *Paradigalla carunculata*, I shot one from my hut, whilst it was eating the small fleshy fruits of an *Urtica*. It likes to sit on the tops of dead and leafless trees, like the *Mino dumonti*. The finest ornament of this bird are the wattles, which in the dried skin lose all their beauty. The upper ones, which are attached one on each side of the forehead, are yellow; those at the base of the lower mandible are blue, and have a small patch of orange-red beneath. The Arfaks call the *Paradigalla* 'Happoa.'"

As is the case with so many of the Birds of Paradise, the genus *Paradigalla* contains but one single species; and indeed it is impossible to find any one which is nearly allied to it. Its somewhat elongated tail places it close to *Astrapia*, which it also resembles in not possessing any of the wiry shafts which adorn the tail-feathers of most of the forms of Paradiseidæ. A glance at the Plate of *Astrapia nigra* will show how entirely different it is even from that, its nearest ally.

The following description is taken from Mr. Sharpe's 'Catalogue of Birds':—

Adult male. General colour velvety black above and below, a little browner on the under surface; wings and tail black, the inner secondaries with a purplish gloss under certain lights; head glossed with metallic steel-green; forehead, lores, and base of lower mandible bare; over each nostril a small tuft of black feathers; on each side of the base of the bill an erect wattled skin; round the eye a ring of black plumes; space below and behind the eye bare; bill and legs black. Total length 11·2 inches, culmen 0·55, wing 6·15, tail 4·85, tarsus 1·9.

Adult female. Similar to the male, but smaller.

The figures in the accompanying Plate represent the pair of birds in my own collection, obtained by M. Laglaize in the Arfak Mountains. For the opportunity of figuring the wattles as they appear in a state of nature, I am indebted to my friend Mr. D. G. Elliot, who sent me a sketch of these parts coloured from the recently killed bird by M. Raffray.

ASTRAPIA NIGRA.

ASTRAPIA NIGRA.

Gorget Paradise-bird.

Gorget Paradise Bird, Lath. Gen. Syn. i. p. 478, pl. 20 (1782).
Paradisea nigra, Gm. S. N. i. p. 401 (1788, ex Lath.).
Paradisea gularis, Lath. Ind. Orn. ii. p. 196 (1790).—Shaw, Gen. Zool. vii. pp. 69, 70 (1809).
Le Hausse-col doré, Aud. et Vieill. Ois. Dor. ii. p. 22, pls. 8, 9 (1802).
La Pie de Paradis, ou l'Incomparable, Levaill. Hist. Nat. Ois. Parad. i. pls. 20, 22 (1806).
Astrapia gularis, Vieill. N. Dict. d'Hist. Nat. iii. p. 37 (1816).—Id. Gal. Ois. i. p. 169, pl. 107 (1825) Less. Traité d'Orn. p. 338 (1831).—Id. Ois. Parad. Syn. p. 18 (1835).—Id. Hist. Nat. p. 106, pls. 21-23.—Schleg. J. f. O. 1861, p. 386.—Beccari, Ann. Mus. Civ. Genov. vii. p. 711.—Sclater, Ibis, 1876, p. 24 9.
Astrapia nigra, Steph. Gen. Zool. xiv. p. 75 (1820).—Gray, Gen. B. ii. p. 263 (1846).—Bp. Consp. i. p. 414 (1850).—Gray, P. Z. S. 1851, p. 436.—Wallace, P. Z. S. 1862, p. 154.—Finsch, Neu-Guinea, p. 173 (1865).—Wallace, Malay Archip. ii. p. 257 (1869).—Elliot, Monogr. Parad. pl. ix. (1873).—Salvad Ann. Mus. Civ. Genov. ix. p. 190 (1876).—Sharpe, Cat. B. Brit. Mus. iii. p. 165 (1877).
Epimachus niger, Schlegel, Mus. P.-B., Coraces, p. 94 (1867).

VARIOUS authors have endeavoured to accomplish the difficult task of classifying and defining the limits of the family *Paradiseidæ*; and I can only think of two other groups which present the same obstacles to classification, namely:—the Hornbills, where the shape of the casque is considered by some naturalists to be of generic, by others of specific importance only; and the Malkoha Cuckoos (*Phœnicophainæ*), where the shape and structure of the nostril varies so much as to induce some naturalists to place each species in a separate genus, whilst by others the form of the nostril is reckoned to be only a specific character. Mr. D. G. Elliot placed the Bower-birds along with the *Paradiseidæ*; but Mr. Sharpe has not included them, and I learn from him that they will be placed near the Thrushes in his arrangement of the class Aves. Professor Schlegel, again, places the present bird in the genus *Epimachus* near *E. speciosus*; but I think that its stout bill, so different from the sickle-shaped slender bills of the *Epimachi*, shows that the place of *Astrapia* is near to the true Paradise-birds, though its long tail is very different in form from that of the other stout-billed genera. I must say, however, that it seems to me impossible to keep such widely different forms of birds under one genus; and I cannot help thinking that a family which includes such widely different forms of birds as *Astrapia*, *Lophorina*, and the true *Paradiseæ*, may yet prove (strange as this may appear to most persons) to be the proper recipient for the Lyre-birds (*Menura*) of Australia.

As in the case of the other Birds of Paradise, little is known of the habits of this grand bird; but Dr. Beccari, who has shot the species in the Arfak Mountains, has given the following short account of its capture:—" *Epimachus maximus* and *Astrapia gularis* are only found on the highest and most difficult peaks of Mount Arfak, nearly always above 6000 feet elevation. Specimens in dark plumage are common enough; but those which have attained perfect plumage are rare, perhaps because they take some years to acquire it. Both of them live on the fruits of certain Pandanaceæ, and especially on those of the *Freycinetiæ*, which are epiphytous on the trunks of trees. The irides of the large *Epimachus* are dark brick-red, those of the *Astrapia* almost black; the neck-feathers of the latter are erectable, and expand into a magnificent collar round the head. The first day I went out at Atam, on June 23, I got both these species (two specimens of each), besides one *Drepanornis albertisi*, three *Paradigallæ*, one *Parotia*, and several other wonderful kinds of birds. It was a memorable day, because I ascended one of the peaks, and was surprised to find myself surrounded by four or five species of *Vaccinium* and *Rhododendron*, I also found an Umbellifer (a *Drymis*) and various other plants common to the mountains of Java, and there were also some mosses a foot and a half in height."

The habitat of this Bird of Paradise is the north-west of New Guinea, though it is said, on native report, to be an inhabitant of the Island of Jobi.

The following description is taken from Mr. Sharpe's Catalogue of Birds.—"*Adult male.* General colour above velvety black, with a purplish gloss; the wings black externally, glossed with purple; tail-feathers black, with wavy lines of dusky black under certain lights, the two centre feathers very long and glossed with rich purple; feathers of the head black, dense and velvety in texture, with a steel-blue gloss; from the nape a shield of golden green feathers springs; and the feathers of the hind neck are very long and tipped with the same golden green; from each side of the nuchal shield spreads a ruff of velvety steel-black plumes; the feathers of the throat also are steel-black, with a band of brilliant golden copper, which extends from behind the eye down the sides of the neck, and encircles the throat; rest of under surface of body rich velvety grass-green, the lateral plumes of the breast tipped with burnished emerald-green; the sides of the body and under wing- and tail-coverts dusky black; bill and legs black; 'iris almost black' (*Beccari*). Total length 28 inches, culmen 1·6, wing 8·8, tail 7, centre feathers 18."

The figures in the Plate are drawn from specimens in my own collection, and represent a male about the size of life, with reduced figures of a male and female in the distance.

LOPHORHINA SUPERBA.

LOPHORHINA SUPERBA.

Superb Bird of Paradise.

L'oiseau de Paradis de la Nouvelle Guinée, dit le Superbe, Brisson, Orn. iii. p. 169 (1760).—D'Aubent. Planches Enluminées, iii. pl. 632 (1774).
Oiseau de Paradis à gorge violette, Sonn. Voy. Nouv. Guinée, p. 157, pl. 96 (1776).
Paradisea superba, Pennant, in Forster, Ind. Zool. p. 40 (1781).—Scopoli, Del. Faun. et Flor. Insubr. ii. p. 88 (1783).—Shaw, Gen. Zool. vii. p. 494, pls. 63-65 (1809).—Id. & Nodder, Nat. Misc. xxiv. pl. 1021 (1813).—Wagler, Syst. Av. *Paradisea*, sp. 5 (1827).—Wallace, Ibis, 1859, p. 111.
Superb Bird of Paradise, Lath. Gen. Syn. vol. i. part 2, p. 479 (1782).
Paradisea atra, Bodd. Tabl. Pl. Enl. D'Aubent. p. 38 (1783).
Le Superbe, Audeb. et Vieill. Ois. Dor. ii. pl. vii. (1802).—Levaill. Hist. Nat. Ois. Parad. i. pls. 14, 15 (1806).
Paradisea furcata, Bechst. Kurze Uebers. p. 132 (1811).
Lophorina superba, Vieill. N. Dict. d'Hist. Nat. xviii. p. 184 (1817).—Id. Gal. Ois. i. p. 149, pl. xcviii. (1825).—Less. Traité, p. 337 (1831).—Id. Ois. Parad. Syn. p. 12 (1835).—Id. Hist. Nat. Ois. Parad. pls. 13, 14 (1835).—Bonap. Consp. Gen. Av. p. 414 (1850).—Wall. Ibis, 1861, p. 287.—Salvad. Ann. Mus. Civic. Genov. ix. p. 190 (1876).—Sharpe, Cat. Birds, iii. p. 179 (1877).
Epimachus ater, Schl. Mus. Pays-Bas, Coraces, p. 96, note (1867).
Lophorina atra, Wallace, Malay Arch. ii. p. 249 (1869).—Elliot, Monogr. Parad. pl. xi. (1873).—Salvad. Ann. Mus. Civic. Genov. vii. p. 783 (1875).—Beccari, tom. cit. p. 712 (1875).—Sclater, Ibis, 1876, p. 251.

To any one studying the Paradise-birds it soon becomes evident that there are several natural groups comprising this interesting family. First of all there are the long-billed *Epimachi* and Rifle-birds, then the larger species, with their enormously long flank-plumes of red or yellow, and lastly the smaller and more fantastic kinds with decorated mantles and tails. Amongst the latter there is great diversity of form: whether it takes the shape of a bare head, as in *Schlegelia*, or of an elongated racket to the centre tail-feathers, as in *Cicinnurus*, or of an elaborated breast-shield, as in *Diphyllodes*, we find that there are scarcely any two which are alike in ornamentation. Take, for example, the subject of the present article. It stands apart from all the others in the extraordinary mantle, which it is able to elevate behind its head,—and also in its remarkable breast-shield, unlike that of any known species. And this strikes me as being one of the most curious phenomena connected with Papuan ornithology—that there should be all these isolated genera of Paradisiidæ, many represented by a single species, and each so different. There seems to be no connecting link between the genera—*Parotia* standing alone with its six racketed plumes on the head, *Semioptera* with its streamlets on the wing, *Schlegelia* with its bare head, and so on, not one of these peculiar forms graduating into another.

The subject of our article is a native of the north of New Guinea, and it is still one of the rarest of the Paradisiidæ, specimens of it being still scarce in collections in this country.

Dr. Beccari does not give much information about the present species in his "Ornithological Letter" from North-western New Guinea. He merely observes:—"*Lophorina atra* is rather rarer than *Parotia*; but I must tell you that the abundance of fruit-eating birds in a given locality depends principally on the season at which certain kinds of fruit are ripe; therefore a species may be common in a place one month, and become rare or completely disappear in the next, when the season of the fruit on which it lives has passed."

The female of the Superb Bird of Paradise is similar in general appearance to that of *Parotia*, but is of course a smaller bird. I take the following descriptions from Mr. Sharpe's 'Catalogue of Birds':—

"*Adult male*. General colour above velvety black, somewhat glossed with bronzy purple; mantle produced into an elevated shield, composed of velvety black plumes, glossed under certain lights with bronze; wing-coverts velvety black, rather more distinctly glossed with purple than the back; quills and tail-feathers deep black, glossed with steel-blue; lores and nasal plumes forming an elevated crest of purplish black feathers; crown of head, nape, and hind neck spangled with metallic steel-coloured feathers, each of which has a sub-terminal bar of purple; sides of face, sides of neck, and entire throat deep coppery bronze; on the fore neck and breast a pectoral shield of bright metallic green plumes, most of which have a narrow edging of copper; remainder of under surface purplish black. Total length 9 inches, culmen 1·15, wing 4·55, tail 3·6.

"*Adult female.* Above deep chocolate-brown, the feathers of the top and sides of the head blackish brown; over the eye a few white-spotted plumes; wing-coverts and quills blackish brown, externally reddish; tail brown, externally dull rufous brown; throat white, all the feathers being black tipped with white; rest of the under surface buffy white, inclining to rufous on the flanks and under tail-coverts, the whole under surface barred across with dull brown; under wing-coverts rufous, barred across with brown. Total length 8·8 inches, culmen 1·05, wing 5·1, tail 4, tarsus 1·3."

The principal figure, drawn in a state of excitement, is of the natural size, with a reduced male and female in the distance.

LOPHORINA MINOR, Ramsay.

LOPHORHINA MINOR, Ramsay.

Lesser Superb Bird of Paradise.

Lophorhina superba minor, Ramsay, Proc. Linn. Soc. N. S. Wales, x. p. 242 (1885).
Lophorhina minor, Finsch u. Meyer, Zeitschr. ges. Orn. ii. p. 376, pl. xvii. (1885).—Meyer, op. cit. iii. p. 181, cum fig. (1886).

As Dr. Meyer has shown, the form of the head-shield in this species is different from that of *Lophorhina superba*, and would be quite sufficient to distinguish it, without the additional characters of the nasal plumes and the smaller dimensions, both of which serve to separate the south-eastern bird from its north-western representative in New Guinea.

In the single adult male of *L. minor* that has come under our notice, the nasal plumes are so disarranged that we cannot make out their form for certain, but they appear to be erect instead of spreading out in a small fan, as in *L. superba*. They are so figured by Dr. Madarász in the 'Zeitschrift' above quoted, and these plumes may be erroneously figured in our Plate of the present bird.

Lophorhina minor is so far only known from the Astrolabe Mountains in South-eastern New Guinea. The person who discovered it appears to have been Mr. Hunstein, one of the most energetic explorers in that region. He seems to have sent a considerable series of specimens from the Horseshoe range, as the British Museum was able to secure a pair of adults and a young male out of the duplicates. Mr. Forbes has more recently procured the species in the Sogeri district of the Astrolabe range; but unfortunately all his specimens were out of plumage, having been collected during the rainy season.

The differences between the males of *L. minor* and *L. superba* have been alluded to above, and they are apparent on a comparison of the Plates.

The female appears to differ from the female of *L. superba* in being olive-brown instead of chestnut on the back, and in having the wings bay externally instead of chestnut. The light eyebrow is also more prolonged and the underparts paler buff.

The figures in the Plate, which represent a male and female of the natural size, have been drawn from a pair of birds in the British Museum, collected by Mr. Hunstein.

[R. B. S.]

DIPHYLLODES RESPUBLICA, *Bonap.*

DIPHYLLODES RESPUBLICA.

Bare-headed Bird of Paradise.

Lophorina respublica, Bp. C. R. 1850, p. 131.

Diphyllodes respublica, Bp. Consp. i. p. 413 (1850).—Sclater, P. Z. S. 1857, p. 6.—Rosenb. J. f. O. 1864, p. 130.—Elliot, Monogr. Parad. pl. 14. (1873).

Paradisea wilsoni, Cassin, Journ. Acad. N. Sci. Philad. ii. p. 133, pl. 15. (1850).—Gray, P. Z. S. 1861, p. 436.—Sclater, P. Z. S. 1865, p. 465.—Schl. N. T. D. iii. p. 249 (1866).—Id. Mus. P.-B. *Coraces*, p. 87 (1867).—Gray, Hand-l. B. ii. p. 16 (1870).

Diphyllodes wilsoni, Wallace, P. Z. S. 1862, p. 160.—Newton, Ibis, 1865, p. 343.—Wall. Malay Arch. ii. p. 248 (1869).

Paradisea calva, Schl. N. T. D. ii. p. 1 (1865).

Schlegelia calva, Bernst. N. T. D. iii. p. 4, pl. 7 (1866).

THIS very beautiful Bird of Paradise was simultaneously described in the year 1850 by Prince Bonaparte in Europe and by the late Mr. Cassin in Philadelphia, but apparently in each case from an imperfect skin. Certainly the type of *P. wilsoni* in the Philadelphia Museum has not got its proper head, although all the rest of the body seems to be quite perfect; and as Prince Bonaparte does not mention the head, which, if it had been attached to the skin, could not have failed to attract his attention, we may suppose that, as in Mr. Cassin's specimen, the head of some other bird had been attached. Indeed the bare cranium is one of the chief peculiarities of the species—so much so that Dr. Bernstein instituted a new genus (*Schlegelia*) for it; but inasmuch as the allied species of *Diphyllodes*, if not absolutely bare, have the cranium clothed only with short stubby plumes, I have not deemed the characters sufficient to warrant a generic separation; this is also Mr. Elliot's conclusion.

To Dr. Bernstein, however, belongs the credit of discovering the home of this fine species. He found it in the island of Waigiou; and in the original note published by him, he thus characterized his proposed *Schlegelia calva*:—" Of the same size and general form as *Paradisea speciosa* and *P. wilsoni*; but the upper part of the head, from the forehead even to the nape, covered with bare skin, broken only by some transverse rows of little plumes. This bare skin is in the male of a very brilliant cobalt blue, in the female of a dirty blue, varied with red and with grey. The rows of little plumes, of which we have just spoken, answer almost to the sutures of the skull in young individuals. The other parts of the head and the chin are black; the posterior portion of the neck and the mantle are straw-yellow; the remainder of the back is of a fine red like that which adorns the plumage of *P. regia*; the fore neck and pectoral shield are of a beautiful dark green with metallic reflections; the breast and belly are black. The distribution of the colours in the female calls to mind those of the Wryneck (*Jynx torquilla*), especially on the lower parts."

In a further communication Dr. Bernstein states that the young male exactly resembles the female, but shows the velvety black plumes on the throat and lower part of the cheeks which are seen in the adult male. In the third volume of the 'Nederlandsch Tijdschrift' the species is fully described by him, and he writes in conclusion:—" This species, being distinguished from all the other known kinds of Paradise-birds by its crown and occiput being for the most part bare, I consider myself entitled to regard it as representing a new genus. This genus is allied, on account of its two centre tail-feathers being elongated and spirally twisted, to the genus *Diphyllodes* of Lesson, by the side of which it is convenient to range it.

"This bird is found in the island of Waigiou; but it inhabits the parts of the country more or less in the interior, and is much more rare than *Paradisea rubra*, which is moreover met with in the adjacent island of Gemien." Afterwards, however, Dr. Bernstein procured the species in the island of Batanta; and ten specimens of his collecting in these two localities are in the Leiden Museum.

The figures in the Plate are of the natural size.

DIPHYLLODES GULIELMI, III, Meyer.

DIPHYLLODES GULIELMI III., Meyer.

Waigiou Bird of Paradise.

Diphyllodes Gulielmi III., Meyer, P. Z. S. 1875, p. 31.
Paradisea (Diph.) Gulielmi III., Van Musschenbroek, Zool. Garten, 1875, p. 29.
Diphyllodes Gulielmi III. (v. Mussch. in lit.), Meyer, Mittheilungen a. d. kgl. zool. Mus. Dresden, 1875, pp. 4–8, pl. 1.

The present bird was discovered, in 1874, in the mountains of eastern Waigiou; and a short description was sent by Mr. S. C. T. Van Musschenbroek, the Dutch resident at Ternate, who is well known for his exertions in the cause of science. The original communication was made through Baron von Rosenberg, who takes the opportunity of remarking that it is also most likely to be found on the island of Batanta, as in general respects the avifaunas of the two islands are identical. Of course there is the probability of the above suggestion ultimately turning out correct; but it is a bold suggestion for a traveller of Von Rosenberg's experience to have made. The same paper expresses his opinion that the splendid new *Paradisea Raggiana* is a manufactured species, a supposition equally untrustworthy! Dr. Meyer about the same time communicated a note on the species to the Zoological Society, which I reproduce, as it so well expresses the differences between it and its allies.

"Notwithstanding there are more points of resemblance between the new species and *Diphyllodes speciosa* and *D. respublica* than between it and *Cicinnurus regius*, some features immediately remind one of the latter, *e. g.* the red colour and structure of the glossy feathers of the upper parts, the webless tail-shafts, at least at the beginning, and chiefly the similar fan, formed by elongated, broad and metallic-green-edged feathers, on the sides of the breast, not quite but nearly of the same colour and size as in *Cicinnurus regius*. This fan has been hitherto known to exist only in the latter species, except as regards the homologous organization of the large *Epimachus speciosus*.

"*Diphyllodes Gulielmi III.* has nearly the same-shaped crest, inserted on the neck, as *Diphyllodes speciosa*; but the crest seems to be somewhat smaller, and the colour of it differs from that of the latter species, as well as all other colours of the upper surface of the two birds, whereas the form and colour of the bill appear to be very similar in the two species. The new bird shows nothing of the broad line of iridescent feathers extending down from the chin over the breast, and nothing of the light-brown feathers of the shoulders and of the neck occurring in *Diphyllodes speciosa*.

"*Diphyllodes Gulielmi III.* reminds one of *D. respublica* in the shape of the green velvet feathers of the breast, and especially in the shape of the two elongated tail-shafts, with the single difference that these are webbed in *Diphyllodes respublica* from the beginning, and that they do not project so far as in *Diphyllodes Gulielmi III.*; but the breadths of the webs agree. The colour of the web is between that of the button of *Cicinnurus regius*, which is more of a green, and that of the web of *Diphyllodes speciosa*, which is more of a blue.

"From this short comparison with the allied species, it follows that *Diphyllodes Gulielmi III.* is a new species with very characteristic features, not to be confounded for a moment with any other hitherto known. These characteristic features are chiefly the shape of the elongated tail-shafts, with their web, the reddish colours of nearly all the upper parts, the violet-coloured belly, and the fan-like tufts at the sides of the breast."

I here insert a description of the female, with which I have been kindly favoured by Dr. Meyer:—

Whole upperside olive-brown; chin, throat, breast, belly, abdomen and under wing-coverts light yellow with fine brown stripes; each feather has several such light and dark markings, the lines getting smaller towards the upper part of the body; upperside of the wings brownish, secondaries and tertiaries margined yellow on the outer webs; underside silver-grey, basal portions of the inner webs cream-coloured; upper part of shafts brown, lower part whitish, underside of tail changing into grey, outer web lighter, lower parts of shafts white, upper part brown.

I owe the opportunity of figuring the present bird to the kindness of M. A. Bouvier, who lent the male specimen in his possession to Mr. Sharpe to bring from Paris for the purpose of this work. The example is intended for the Warsaw Museum.

Two males are represented in the Plate, of the natural size.

DIPHYLLODES SPECIOSA.

DIPHYLLODES SPECIOSA.

Magnificent Bird of Paradise.

Oiseau de Paradis de la Nouvelle Guinée, dit Le Magnifique, Montb. Pl. Enl. p. 194, pl. 631.—Sonn. Voy. Nouv. Guin. p 163, pl. 98.—Levaill. H. N. Ois. de Paradis, i. pls. 9, 10.
Magnificent Paradise-bird, Lath. Gen. Syn. ii. p. 477, pl. 19.
Paradisea speciosa, Bodd. Tabl. Pl. Enl. p. 38.—Gray, Gen. B. p. 323.—Schl. Mus. P. B. *Coraces*, p. 86.—Id. N. T. D. iv. p. 171.—Gray, Hand-l. B. ii. p. 16.
——— *magnifica*, Gm. S. N. i. p. 401.—Shaw, Gen. Zool. vii. p. 492, pl. 62.—Audeb. & Vieill. Ois. Dor. p. 15, pl. 4.—Wagler, Syst. Av., *Paradisea*, sp. 4.—Less. Traité, p. 338.—Wall. P. Z. S. 1862, p. 160.
——— *cirrhata*, Lath. Ind. Orn. i. p. 195.
Diphyllodes seleucides, Less. Ois. Parad. Syn. p. 16, pls. 19, 20.
——— *magnifica*, Gray, List Gen. B. 1841, p. 53.—Bp. Consp. i. p. 413.—Wall. Ibis, 1861, p. 287.—Rosenb. J. f. O. 1864, p. 130.
——— *speciosa*, Wallace, Malay Archipelago, ii. p. 247.—Elliot, Monogr. Parad. pl. 12.—Scl. P. Z. S. 1873, p. 697.—Wagner, J. f. O. 1873, p. 11, pl. 1, map 5.

Who was the original discoverer of this Bird of Paradise, seems to be a difficult question to decide; but it was probably first introduced to the scientific world by Sonnerat, who figures it in his 'Voyage à la Nouvelle Guinée.' Although Sonnerat doubtless procured specimens during his journey, and was thus the first to bring them to Europe, Montbeillard's work, in which a figure is also given, bears date two years before the volume of the first-mentioned author. There can be no doubt, however, that of these two early figures, one is a direct copy of the other; and from internal evidence in other parts of the 'Planches Enluminées,' I have every reason to believe that it was by no means an unfrequent practice for the authors of this standard work to make up plates from the figures and engravings of other works and then to colour them from the descriptions. Thus it is possible that Montbeillard's plate was made up from Sonnerat's already published figure, and that the date on the volume is the year in which the work was finished, and not of the actual part in which the plate of "Le Magnifique" appeared. Certain it is, however, that the species was very little known to the older authors; for after the works above mentioned, all the writers for years copied Montbeillard's plate into their books; and it is equally certain that, until the last ten years, none but skins of native preparation existed in the museums of the world. Recently, owing to the enterprise of the Dutch naturalists, a good series of perfect skins have reached Europe, and good examples are to be found in the British Museum and in my own collection. In Salwatti, according to Baron von Rosenberg, it cannot be very rare, judging from the number of skins collected; and it is found in the mountains both of this portion of New Guinea and of Mysol. In the former place it is called by the natives *Sabelo*; and in Mysol its name is *Arung-arung*. The late Dr. Bernstein procured numerous examples at Sorong, on the coast of New Guinea, facing Salwatti; and Von Rosenberg likewise records that it is an inhabitant of the eastern coast of the great island, where it is found both near the coast and in the interior. D'Albertis met with it in Atam; and during his last voyage Von Rosenberg discovered it to be an inhabitant of the island of Jobi.

Nothing is known of the habits of this beautiful bird, as it was not met with in a state of nature by Mr. Wallace. In his work on the Malay Archipelago is to be found the following note, from which it will be seen how recent has been the acquisition of properly prepared specimens by European naturalists:—

"From what we know of the habits of allied species, we may be sure that the greatly developed plumage of this bird is erected and displayed in some remarkable manner. The mass of feathers on the under surface are probably expanded into a hemisphere, while the beautiful yellow is no doubt elevated so as to give the bird a very different appearance from that which it presents in the dried and flattened skins of the natives, through which alone it is at present known. The feet appear to be dark blue."

The Plate represents two males of the size of life. The female is a little less than life-size.

DIPHYLLODES CHRYSOPTERA, Gould.

DIPHYLLODES CHRYSOPTERA, *Gould*.

Golden-winged Bird of Paradise.

Diphyllodes chrysoptera, Gould, MS., *undè*
―――――― *speciosa*, var. *chrysoptera*, Elliot, Monogr. Parad. pl. 13.

I HAVE for a considerable time possessed specimens of the present bird, which I lent to Mr. Elliot for figuring in his monograph of the family. He does not, however, consider it more than a "variety" of the ordinary species, as will be seen from the account which I extract from his work:—

"The only difference perceptible between these specimens and those of the well-known species with which I have compared them is that the wings are of a golden colour. In size they are equal. I do not consider that there are sufficient grounds shown for naturalists to regard these birds as belonging to a species distinct from the *D. speciosa*; and I have given a representation of them merely to exhibit a form of variation to which perhaps *D. speciosa* is subject in certain localities. The exact place from which they come is not known. . . . . The bird may be described as the same as *D. speciosa* with yellow secondaries."

I have reproduced Mr. Elliot's remarks because I wish my readers to know his exact opinion respecting the birds, especially as I am compelled to differ from him entirely in his belief that the present bird is nothing but a variety of *D. speciosa*. I possess two individuals of *D. chrysoptera*, both males, and agreeing with each other; and considering that only during the past year a new and beautiful species of this very genus has been discovered in the well-known island of Waigiou, I see nothing extraordinary in another island being the habitat of this, to me, very well-characterized bird. I consider it perfectly distinct from *D. speciosa*; and I am certain that before long its proper habitat will be brought to light.

The birds are represented of the natural size.

CICINNURUS REGIUS.

CICINNURUS REGIUS.

King Bird of Paradise.

The supposed King of the greater Birds of Paradise, Edwards, Birds, iii. pl. 3 (1750).
Le Petit Oiseau de Paradis, Briss. Orn. ii. p. 136 (1760).
Paradisea Regia, Linn. S. N. i. p. 166 (1766).—Shaw, Gen. Zool. vii. pt. ii. p. 497, pl. 67 (1809).—Less. Voy. Coq. i. p. 658, Atlas, pl. 26 (1826).—Wagler, Syst. Av. *Paradisea*, sp. 7 (1829).—Gray, Gen. B. ii. p. 323 (1847).—Schl. Hand-l. Dierk. i. p. 332, Atlas, pl. 4. fig. 46 (1857).—Gray, Hand-l. B. ii. p. 16 (1870).—Schl. M. P.-B. *Coraces*, p. 88 (1867).—Wall. Malay Arch. ii. p. 132 (1869).—Schl. N. T. D. iv. pp. 17, 49 (1873).—Wagner, Zool. Gart. 1873, p. 10.
Le Manucode, Buff. Hist. Nat. Ois. iii. p. 163, pl. 13.—Id. Pl. Enl. iii. pl. 496 (1774).—Vieill. Ois. Dor. (Oiseaux de Paradis) ii. p. 16, pl. 5 (1802).—Levaill. Ois. Parad. i. pls. 7, 8 (1806).
Le Roi des Oiseaux de Paradis, Sonn. Voy. N. Guin. i. p. 156, pl. 95 (1776).
King Paradise-bird, Lath. Gen. Syn. i. p. 475 (1782).
Cicinnurus regius, Vieill. Gal. Ois. i. p. 146, pl. 96 (1825).—Bp. Consp. i. p. 413 (1850).—Elliot, Monogr. Parad. pl. 16 (1873).—Salvad. & d'Albert. Ann. Mus. Civ. Genoa, 1875, p. 832.
Cicinnurus spiniturnix, Less. Ois. Parad. Synopsis, p. 14 (1835).—Id. Hist. N. Ois. Parad. p. 182, pls. 16, 17, 18 (1835).

ALTHOUGH one of the smallest of the Paradise-birds, the present species yields to none in the beauty of its plumage or the elegance of its form; while its wire-like caudal plumes are just as remarkable in structure as any of the fantastic decorations which adorn the larger kinds. Its range is, for a Bird of Paradise, rather extended. It seems to be found all over New Guinea, as it has been met with in the Bays of Lobo and Triton and, more recently, in the south-eastern part of the island, at Mount Epa, by Signor d'Albertis. A large number of specimens were collected in Salwatti by the Dutch travellers Bernstein and Von Rosenberg, as well as on the opposite coast of New Guinea, at Sorong &c. Von Rosenberg procured it in the island of Jobie, and also in the islands of Woxam and Wonoumbai of the Aru group. It is from the last-named islands and from Mysol that the largest number of specimens have come to this country, from the collections of Mr. Wallace and, more recently, of Mr. Cockerell. Notwithstanding the fact that specimens from all the above localities have been pronounced identical, I am in possession of facts which induce me to believe in the existence of, at least, two species of King Birds of Paradise.

As regards the habits of the *Cicinnurus* in Aru, it is impossible to do better than to quote the admirable account given by Mr. Wallace in his 'Malay Archipelago.' He says, "The first two or three days of our stay here were very wet, and I obtained but few insects or birds; but at length, when I was beginning to despair, my boy Baderoon returned one day with a specimen which repaid me for months of delay and expectation. It was a small bird, a little less than the Thrush. Merely in arrangement of colours and texture of plumage, this little bird was a gem of the first water; yet these comprised only half its strange beauty. Springing from each side of the breast, and ordinarily lying concealed under the wings, were little tufts of greyish feathers about two inches long and each terminated by a broad band of intense emerald-green. These plumes can be raised at the will of the bird, and spread out into a pair of elegant fans when the wings are elevated. But this is not the only ornament. The two middle feathers of the tail are in the form of slender wires about five inches long and which diverge in a beautiful curve. Almost half an inch of the end of this wire is webbed on the outer side only, and coloured of a fine metallic green; and being curved spirally inwards, they form a pair of elegant glittering buttons, hanging five inches below the body, and at the same distance apart. These two ornaments, the breast-fans and the spiral-tipped tail-wires, are altogether unique, not occurring on any other species of the eight thousand different birds that are known to exist upon the earth, and, combined with the most exquisite beauty of plumage, render this one of the most perfectly lovely of the many lovely productions of nature. My transports of admiration and delight quite amused my Aru hosts, who saw nothing more in 'Burong raja' than we do in the Robin or Goldfinch. Thus one of my objects in coming to the far east was accomplished. I had obtained a specimen of the King Bird of Paradise, which had been described by Linnæus from skins preserved in a mutilated state by the natives. I knew how few Europeans had ever beheld the perfect little organism I now gazed upon, and how very imperfectly it was still known in Europe. The emotions excited in the mind of a naturalist who has long desired to see the actual thing which he has hitherto known only by description, drawing, or badly preserved external covering, especially when that thing is of surpassing rarity and beauty, require the poetic faculty fully to express them. After the first

King bird was obtained, I went with my men into the forest; and we were not only rewarded with another in equally perfect plumage, but I was enabled to see a little of the habits of both it and the larger species. It frequents the lower trees of the less dense forests, and is very active, flying strongly with a whirring sound, and constantly hopping or flying from branch to branch. It eats hard stone-bearing fruits as large as a gooseberry, and often flutters its wings after the manner of the South-American Manakins, at which times it elevates and expands the beautiful fans with which its breast is adorned. The natives of Aru call it 'Goby-goby."

The figures in the Plate represent a pair of male birds from Aru, of the size of life, with a reduced figure of the female. According to notes made by Dr. Bernstein from the living bird, the male in breeding-plumage has the bill of a pale reddish yellow; feet clear cobalt blue; iris brownish yellow, tinged with grey. The female has the bill brownish black, passing to clear greenish brown near the angle of the gape; the gullet yellowish green, rather blackish; iris brownish grey; feet of a clear blue.

PAROTIA SEXPENNIS.

PAROTIA SEXPENNIS.

Six-plumed Bird of Paradise.

Le Sifilet de la Nouvelle Guinée, Montb. Pl. Enl. 111, pl. 633.
L' Oiseau de Paradis à gorge dorée, Sonnerat, Voy. N. Guin. p. 158, pl. 97.
Gold-breasted Paradise-bird, Lath. Gen. Syn. ii. p. 481.
Paradisea penicillata, Scop. Del. Faun. et Flor. Insubr. ii. p. 88 (ex Sonn.).
———— *aurea*, Gm. S. N. i. p. 402 (ex Lath.).
———— *sexsetacea*, Lath. Ind. Orn. i. p. 196.—Shaw, Gen. Zool. vii. pt. 2, p. 496, pl. 66.—Wagler, Syst. Av. Paradisea, sp. 6.
Le Sifilet, Levaill. H. N. Ois. Parad. i. pls. 12, 13.—Audeb. et Vieill. Ois. Dor., Parad. pl. 6.
Parotia sexsetacea, Vieill. Gal. des Ois. ii. p. 148, pl. 97.—Less. Ois. Parad. Syn. p. 10, pls. 11 bis, 12.
———— *aurea*, Gray, List Gen. B. p. 39 (1840).
Paradisea sexpennis, Gray, P. Z. S. 1861, p. 436.—Schl. Mus. P.-B. *Coraces*, p. 92.—Gray, Hand-l. B. ii. p. 16.—Schl. N. T. D. iv. p. 50.
Parotia sexpennis, Wall. Malay Archip. ii. p. 251.—Elliot, Monogr. Parad. pl. 10.

It is evident, from the recent researches of travellers, that we are at last beginning to solve the ornithological mysteries of New Guinea; and if no other proof existed, the history of the present bird would furnish us with one. It is now nearly one hundred years ago since the Six-plumed Bird of Paradise was figured by Montbeillard; and until about three or four years ago we knew of this beautiful species nothing but the fact that a few specimens, of native preparation, existed in some of the great collections of Europe. Even its exact *habitat* was unknown; but the correctness of the supposition that it came from New Guinea has been amply proved by the investigations of recent travellers. The first person who procured perfect specimens of the bird was Baron von Rosenberg. He discovered the species in the mountains of the northern part of New Guinea; and after him, Signor d'Albertis obtained it in the Arfak mountains, whence also Dr. Meyer brought to Europe numerous examples. I am enabled to give an illustration of the bird from a pair of his beautiful specimens, which are now in my collection.

The female was known to Lesson, by whom it was figured in his 'Histoire Naturelle des Oiseaux de Paradis' (pl. 12); and at the same time he gives a curious illustration of a male in imperfect plumage, retaining still some of the immature brown feathers on the breast. A better illustration of the female bird is given in Mr. Elliot's work. The latter gentleman has figured the male with the white feathers protruding far over the bill; and I recently saw a specimen with the feathers thus recurved, in the possession of Professor S. L. Steere, of Michigan University. On his way through London, back to America, he very kindly brought me the bird to examine; and although all the specimens which I have seen had these white feathers reflected backwards so as to form a broad frontal band as represented in my plate, it was evident that the position of the frontal plumes in Mr. Steere's specimen was quite natural; and there can be no doubt that when alive the bird raises and depresses these feathers at will.

The following remarks are from the pen of Signor d'Albertis:—

"Although this species has been known many years, it is not yet accurately understood, having only been described from birds in a mutilated condition. My observations have been made in the natural haunts of these elegant birds, from numerous specimens both living and dead. These birds are found in the north of New Guinea. I met with them about thirty miles from the coast, at an elevation of 3600 feet above the level of the sea, near Mount Arfak. I have never found the adult male in company with females or young birds, but always in the thickest parts of the forest; the females and young birds are generally found in a much lower zone. This Paradise-bird is very noisy, uttering a note like '*guaad-guaad*.' It feeds upon various kinds of fruits, more especially a species of fig which is very plentiful in the mountain-ranges; at other times I have observed it feeding on a small kind of nutmeg. To clean its rich plumage this bird is accustomed, when the ground is dry, to scrape (similarly to a gallinaceous bird) around places clear of all grass and leaves, and to roll over and over again in the dust produced by the clearing, at the same time crying out, extending and contracting its plumage, elevating the brilliant silvery crest on the upper part of its head, and also the six remarkable plumes from which it derives the specific name of *sexpennis*. On seeing its eccentric movements at this time, and hearing its cries, one would consider it to be engaged in a fight with some imaginary enemy. This bird is named 'Caran-a' by the natives. I have also a skeleton of a young male of this species, which, although not in a perfect state, may no doubt be interesting as showing the form of the cranium, on which there is an admirable muscular structure which enables the bird to elevate the feathers of the head. The feathers at the nape of the neck exhibit, when the rays of light strike upon them, a rich and brilliant metallic hue. The eyes are of a light blue, with a circle of a pale yellowish green colour."

The figures in the Plate are of the natural size.

PAROTIA LAWESI, Ramsay.

PAROTIA LAWESI, *Ramsay.*

Lawes's Bird of Paradise.

Parotia lawesi, Ramsay, Proc. Linn. Soc. N.S. Wales, x. p. 243 (1885).—Finsch u. Meyer, Zeitschr. ges. Orn. ii. p. 375, pl. xvi. (1885).

THIS species represents the Six-plumed Bird of Paradise (*P. sexpennis*) of North-western New Guinea, in the Astrolabe Mountains, in the south-eastern portion of that great island. It was first met with by Mr. Hunstein in the Horseshoe range at a height of 7000 feet, and Mr. H. O. Forbes has also come across the species in the Sogeri district of the Astrolabe range.

As might be expected, the southern bird is closely allied to its north-western representative, but it is, nevertheless, quite a distinct species. The pectoral shield, when examined in the same light as that of *P. sexpennis*, is much more fiery metallic-golden, and shows less of the green shade which is seen in all specimens of *P. sexpennis*. The metallic band on the nape is also differently composed, the general effect being purplish or steel-blue, whereas the centre of this band in *P. sexpennis* is green. It is true that a slight greenish tinge can be discovered at the base of the metallic plumes in *P. lawesi*, but it is of the faintest possible character. The silvery patch on the crown is differently disposed in the two species, being fan-shaped in *P. sexpennis* and not descending to the base of the bill, all the frontal plumes being velvety black with a brownish gloss, exactly like the rest of the head. The white patch in *P. lawesi* comes to an obtuse point above the nostrils, and is differently shaped to that of *P. sexpennis*.

The female of *P. lawesi* is also distinct from the female of *P. sexpennis*, being rufous underneath with black bars, while the upper surface is also of a more chestnut tint.

The young male at first resembles the adult female.

The measurements of the pair of *P. lawesi* in the British Museum are as follows:—

Male. Total length 13 inches, culmen 1·15, wing 6·15, tail 5·0, tarsus 2·15.
Female. „ „ 9·5 „ „ 1·10, „ 6·10, „ 4·1, „ 2·0.

The figures in the Plate have been drawn from a pair of birds in the British Museum, collected in the Horseshoe range by Mr. C. Hunstein. The adult male and female are represented of the full size.

[R. B. S.]

PARADISEA DECORA, Salv. et Godm.

PARADISEA DECORA, Salv. & Godm.

Grey-chested Bird of Paradise.

Paradisea decora, Salvin and Godman, Ibis, 1883, pp. 131, 202, pl. viii.
Paradisea susannæ, Ramsay, Proc. Linn. Soc. N. S. Wales, viii. p. 21 (1883).

The discovery of a new Bird of Paradise must always be a matter of interest to naturalists, and especially when the species proves to be of so fine a character as the present bird. The home of *Paradisea decora* is Fergusson Island, in the D'Entrecasteaux group, whence comes the beautiful *Manucodia comrii*, also figured by us in the 'Birds of New Guinea.'

We owe the discovery of this beautiful bird to Mr. A. Goldie, to whose energies science has been indebted for many years. He has given the following account of its capture:—

"The Birds of Paradise were shot on Fergusson Island, one of the D'Entrecasteaux group, in the mountains, at a considerable elevation above the sea, the first specimen obtained having been secured at the lowest point. The plumed males and the younger individuals were generally seen three or four together. Once heard, their call was unmistakable, being very like that of *Paradisea raggiana*; but the plumed and wired birds, after giving that call a few times, added to it a peculiar shrill whistle. Their motions whilst calling were identical with those of *P. raggiana*; but, so far as we were able to observe, they had no particular tree for dancing in. The females were found alone.

"We neither saw nor heard *P. raggiana* on these islands; and the new bird is not found on the mainland. On showing it to the natives of Chad's Bay and China Straits along with a specimen of *Paradisea raggiana*, they, in both cases, made us to understand that the latter is found in their country, whilst the former is not; but two or three of them in China Straits who had traded to the D'Entrecasteaux Islands made signs that the new bird was to be found there."

The nearest ally of the present species is *P. raggiana*, and, like that bird, it has red flank-plumes; but it differs in its yellow back and grey chest, the latter extending up to the green throat, and not being separated from it by a yellow collar.

The description of the species, as given by Messrs. Salvin and Godman, is so complete that we transcribe it entire. We may add that the typical series, described by them, is now in the British Museum.

"Fergusson Island was so named by Capt. Moresby, who calculated that the mountain called Kilkerran (the highest near the northern coast) reaches an altitude of 6000 feet above the sea. Two other large islands, Goodenough Island and Normandy Island, lie close to Fergusson Island, and form the chief islands of the group.

"*Paradisea decora*, as we have proposed to call this Bird of Paradise, combines the characters of some of the previously known species. The side-plumes are like those of *P. sanguinea*, each feather having its barbs towards the end wide apart and destitute of barbules. These feathers are similarly formed in *P. sanguinea*; but in the present bird the barbs are even wider apart. The 'wires' of the tail are like those of *P. apoda*, *P. minor*, and *P. raggiana*, the feathers having a simple stem on which the atrophied barbs become more and more evanescent till they disappear at the middle of the feather to reappear again at its extremity. In *P. sanguinea* the stem is broad and flattened.

"The side-plumes of *P. decora* are very peculiar, inasmuch as a number of the anterior plumes are quite short, with the barbs of each feather much lengthened towards the end; the distal ends of these feathers are deep rich vinous red, and appear as if the pigment which colours the elongated plumes were concentrated in these shorter ones.

"The breast of *P. decora* is of a soft vinaceous lilac, and in this respect differs from that of all its congeners. The throat is velvet-green, showing two shades, owing to the feathers nearer the chin reflecting the light at a different angle. This darker-looking patch is larger in *P. decora* than in the other species. The back, except the narrow green forehead, is straw-coloured, like that of *P. minor*.

"Mr. Goldie's series contains males in all stages of development. The youngest are like the females, but with the throat green; the breast is of a ruddy tint, vermiculated with dusky marks on each feather. In the first plumage the two central rectrices are narrow and elongated, but with barbs &c. as in the perfect feather; they project beyond the rest of the tail-feathers, the length of the projection varying. In some males (perhaps young birds, perhaps birds out of nuptial plumage) these feathers are much more elongated and the length of the barbs of the middle of each feather is much reduced; but these lengthen again so as to form a small spatule.

"The moult to the nuptial plumage proceeds in various ways: sometimes the lilac feathers of the breast are the first to appear; in others the wiry rectrices are the first to take the place of their predecessors. In some cases these latter are fully grown before the ornamental side-plumes make their appearance. In others, again, they grow contemporaneously with these plumes."

The figures in the Plate represent the adult male and female of about the natural size, and have been drawn from the typical specimens in the British Museum.

[R. B. S.]

PARADISEA PAPUANA.

PARADISEA PAPUANA.

Papuan Bird of Paradise.

Paradisea papuana, Bechst. Kurze Uebers. p. 131 (1811).—Gray, Gen. B. ii. p. 323 (1847).—Bonap. Consp. Gen. Av. i. p. 413 (1850).—Gray, P. Z. S. 1859, p. 157.—Wallace, Ibis, 1859, p. 111; 1861, p. 287.—Schl. J. f. O. 1861, p. 385.—Sclater, P. Z. S. 1862, p. 123.—Wall. tom. cit. p. 160.—Rosenb. J. f. O. 1864, p. 129.—Schl. Nederl. Tijdschr. Dierk. iv. pp. 17, 49.
Paradisea bartletti, Goodwin, P. Z. S. 1860, p. 244 (the young male).

The specific name usually applied to this gorgeous bird is *minor*; I must therefore state my reasons for not adopting that title in the present instance. To call it *minor*, when it is the largest, is unphilosophical. I consider that there are two species or races confounded under one name; the examples brought home by Wallace from New Guinea, of which I have four or five, are by no means so fine or large as those from Mysol and other localities. I therefore propose to restrict the name *papuana* to the larger bird, and to adopt that of *minor* for the smaller. The term *papuana* is also applicable to the specimens sent from Jobi, as Count Salvadori states that the individuals from that island are larger, and the male birds have the plumes longer and much more finely developed, while the dull chestnut colour of the head in the females and young males is darker, occupying the whole of the head, and being abruptly separated from the yellow colour of the neck.

The present species has been on two occasions sent alive to this country, and has lived in the Zoological Gardens. The first specimens were brought by Mr. Wallace from Singapore in April 1862; and one of these lived till the 25th of December 1863, the other till the 28th of March 1864, in the Gardens. The second couple were obtained from the young French traveller, M. Léon Laglaize, who brought four living specimens with him on his return from his expedition to New Guinea. All these four specimens lived through the winter in the Jardin des Plantes at Paris; and two of them subsequently passed into the Zoological Society's collection in the Regent's Park, where one of them still survives.

Mr. Wallace gives the following account of his obtaining these specimens and their subsequent journey:—
"When I returned home in 1862, I was so fortunate as to find two adult males of this species in Singapore; and as they seemed healthy and fed voraciously on rice, bananas, and cockroaches, I determined on giving the very high price asked for them (£100), and to bring them to England by the overland route under my own care. On my way home I stayed a week at Bombay, to break the journey and to lay in a fresh stock of bananas for my birds. I had great difficulty, however, in supplying them with insect food; for in the Peninsular and Oriental steamers cockroaches were scarce, and it was only by setting traps in the store-rooms, and by hunting an hour every night in the forecastle, that I could secure a few dozen of these creatures, scarcely enough for a single meal. At Malta, where I stayed a fortnight, I got plenty of cockroaches from a bakehouse; and when I left I took with me several biscuit-tins full, as provision for the voyage home. We came through the Mediterranean in March with a very cold wind; and the only place on board the mail steamer where their large cage could be accommodated was exposed to a strong current of air down a hatchway, which stood open day and night; yet the birds never seemed to feel the cold. During the night-journey from Marseilles to Paris it was a sharp frost; yet they arrived in London in perfect health, and lived in the Zoological Gardens for one and two years respectively, often displaying their beautiful plumes to the admiration of the spectators. It is evident, therefore, that the Paradise-birds are very hardy, and require air and exercise rather than heat; and I feel sure that if a good-sized conservatory could be devoted to them, or if they could be turned loose in the tropical department of the Crystal Palace, or the Great Palm-house at Kew, they would live in this country for many years."

The following account of these two birds when in the Gardens, was written by Mr. Bartlett for publication in my friend Mr. Elliot's 'Monograph of the Paradiseidæ:'—"When the two birds of Paradise first arrrived at the Gardens in April 1862, their plumes were quite short, only about five inches long. The birds had moulted; and the new feathers were growing in a thick bunch on each side below their wings. They appeared in good health, and were active and lively. I soon ascertained how fond they were of meal-worms and other insects; and they fed freely upon fruit, boiled rice, &c.;

a little cooked flesh was also acceptable to them. Their mode of hopping about from perch to perch and clinging to the bars or wires of the cage reminded one of a Jay or Jackdaw. They were fond of a bath, and were very careful in dressing and drying their fine plumes. These were about two months in growing to their full perfection; and it was a charming sight to see them when in full plumage. When uttering their loud call the body was bent forward, the wings spread open and raised up, frequently over their heads, meeting the plumes, which were spread in the most graceful manner, every feather vibrating in a way that almost dazzled the sight. During this display the bird would become greatly excited, and sometimes turn almost under the perch or branch, the head and neck being bent so low down. At this period we found that they would not agree, but attacked each other; and we were therefore obliged to keep them separated by a wire division. They hopped about like Jays or Jackdaws, never ran like Starlings or Magpies, and, when on the ground, raised the points of the plumes so that they should not touch the earth. They soon became very tame, and would take food from the hand; and the sight of a meal-worm would bring them down from their perch immediately. The moult was extremely rapid, the fine plumes being thrown off in a few days; and these appeared to grow all at the same time in a bunch. It is therefore certain that these birds, after they attain the adult plumage, lose it only during the annual moult, like the Peacock and many other richly ornamented birds."

I have been constrained to give an additional Plate to exhibit the young male and the female of the natural size, which I could not do on the same Plate with the adult male in full display of the plumes.

The young male has been regarded as a different bird, and named *P. bartletti*, as will be seen on reference to the synonymy.

The accompanying illustrations represent an adult male in the nuptial plumage, and a young male and a female on the second Plate, all of the natural size.

PARADISEA PAPUANA.
(Young Male and Female)

PARADISEA APODA, *Linn.*

PARADISEA APODA, *Linn.*

The Greater Bird of Paradise.

The Greater Bird of Paradise, Edwards, Birds, iii. pl. 110.
L'Oiseau de Paradis, Brisson, Orn. ii. p. 130, pl. xiii.
Paradisea apoda, Linn. Syst. Nat. i. p. 166.—Wagler, Syst. Av., *Paradisea*, sp. 1.—Bonap. Consp. Gen. Av. i. p. 412.—Gray, P. Z. S. 1861, p. 436.—Wallace, Ibis, 1859, p. 111; 1861, p. 289.—Schlegel, Mus. Pays-Bas, Coraces, p. 78.—Wallace, Malay Archip. ii. p. 238.—Gray, Handl. B. ii. p. 16.—Elliot, Monogr. Parad. pl. i.—Salvad. Ann. Mus. Civic. Genov. ix. p. 191.—Sharpe, Cat. B. iii. p. 167.
Paradisea major, Shaw, Gen. Zool. vii. p. 480, pl. 58.—Less. Ois. de Paradis, Synopsis, p. 6.—Id. Hist. Nat. p. 155, pl. 6.
Paradisea apoda, var. *wallaciana*, Gray, P. Z. S. 1858, p. 181.

I HAVE in my collection some skins of this splendid bird which would justify the specific name of *apoda* bestowed upon it by Linnæus; and I suspect that nearly every one of the specimens in public collections mounted before the last twenty years would be found to have other birds' feet attached to them instead of their own proper appendages. The reason is that, until very recently, all the skins of the Great Bird of Paradise which reached Europe were in a mutilated condition—generally without feet, and often without wings; and I well remember the admiration which was roused by the arrival of Mr. Wallace's beautiful perfect examples, and the interest which the exhibition of Mr. Bartlett's mounted specimen excited, when it was first exhibited in the British Museum. Now that complete skins are the rule, and badly prepared ones the exception, it is interesting to glance at the past history of the species, and to peruse the accounts of the first describers of its remarkable plumage. "When the earliest European voyagers," writes Mr. Wallace, in his 'Malay Archipelago,' "reached the Moluccas in search of cloves and nutmegs, which were then rare and precious spices, they were presented with the dried skins of birds so strange and beautiful as to excite the admiration even of those wealth-seeking rovers. The Malay traders gave them the name of 'Manuk dewater' (or God's birds); and the Portuguese, finding that they had no feet or wings, and not being able to learn any thing authentic about them, called them 'Passaros de Sol' (or Birds of the Sun); while the learned Dutchmen, who wrote in Latin, called them 'Avis paradiseus' (or Paradise-bird). John van Linschoten gives these names in 1598, and tells us that no one has seen these birds alive; for they live in the air, always turning towards the sun, and never lighting on the earth till they die; for they have neither feet nor wings, as, he adds, may be seen by the birds carried to India, and sometimes to Holland; but being very costly they are rarely seen in Europe. More than a hundred years later Mr. William Funnel, who accompanied Dampier, and wrote an account of the voyage, saw specimens at Amboyna, and was told that they came to Banda to eat nutmegs, which intoxicated them, and made them fall down senseless, when they were killed by ants. Down to 1760, when Linnæus named the largest species *Paradisea apoda* (the footless Paradise-bird), no perfect specimen had been seen in Europe, and absolutely nothing was known about them. And even now, a hundred years later, most books state that they migrate annually to Ternate, Banda, and Amboyna, whereas the fact is that they are as completely unknown in these islands in a wild state as they are in England." I may remark that Edwards had probably a complete specimen in 1750, as he mentions the figures in the older authors, such as Willughby, and remarks, "As none of these were satisfactory to me, I have given this figure and description of a *perfect bird*, which may more than answer the purposes of so many;" and again:—"It hath legs and feet of a moderate proportion and strength for its bigness, shaped much like those of Pyes or Jays, of a dark brown colour, armed with claws of middling strength." The fact remains, however, that the vast majority of skins received in Europe before Mr. Wallace's expedition, were mutilated and footless. He writes:—"The native mode of preserving them is to cut off the wings and feet, and then skin the body up to the beak, taking out the skulls. A stout stick is then run up through the specimen, coming out at the mouth. Round this some leaves are stuffed, and the whole is wrapped up in a palm-spathe and dried in the smoky hut. By this plan the head, which is really large, is shrunk up almost to nothing, the body is much reduced and shortened, and the greatest prominence is given to the flowing plumage. Some of these native skins are very clean, and often have wings and feet left on; others are dreadfully stained with smoke; and all give a most erroneous idea of the proportions of the living bird." The following notes on the habits are also given by Mr. Wallace:—"The Great Bird of Paradise is very active and vigorous, and seems to be in constant motion all day long. It is very abundant, small flocks of females and young males being constantly met with; and though the full-plumaged birds are less plentiful, their loud cries, which are heard daily, show that they also are very numerous. Their note is 'Wauk-wauk-wauk-wok-wok-wok,' and is so loud and shrill as to be heard a great distance, and to form the most prominent and characteristic animal-sound in the Aru Islands. The mode of nidification is unknown; but the natives told me that the nest was made of leaves placed on an ants' nest, or on some projecting limb of a very lofty tree, and believe that it contains only one young bird. The egg is quite unknown; and the natives declared they had never seen it; and a very high reward offered for one by a Dutch official did not meet with success. They moult about January or February; and in May, when they are in full plumage, the males assemble early in the morning to exhibit themselves. This habit enables the natives to obtain specimens with comparative ease. As soon as they find that the birds have fixed upon a tree on which to assemble, they build a little shelter of palm leaves in a convenient place among the branches; and the hunter ensconces himself in it before daylight, armed with his bow and a number of arrows terminating in a round knob. A boy waits at the foot of the tree; and when the birds come at sunrise, and a sufficient number have assembled, and have begun to dance, the hunter shoots with his blunt arrow so strongly as to stun the bird, which drops down, and is secured and killed by the boy without its plumage being injured by a drop of blood. The rest take no notice, and fall one after another till some of them take the alarm."

The *Paradisea apoda*, as far as we have any certain knowledge, is confined to the mainland of the Arru Islands, never being found in the smaller islands which surround the centre mass. It is certainly not found in any of the parts of New Guinea visited by the Malay and Bugis traders, nor in any of the other islands where Birds of Paradise are obtained. But this is by no means conclusive evidence; for it is only in certain localities that the natives prepare skins, and in other places the same birds may be abundant without ever becoming known. It is therefore quite possible that this species may inhabit the great southern mass of New Guinea, from which Arru has been separated; while its near ally (*P. papuana*) is confined to the north-western peninsula. I may remark that Mr. Wallace's prediction that this species would be found on the southern part of New Guinea has been verified by Signor D'Albertis, who recently showed me a fine skin obtained by himself on the Fly River far in the interior of S.E. New Guinea. This specimen was a trifle smaller, and brighter in colour, than Mr. Wallace's Arru specimens, of which I have a fine series.

The figures in the Plate, which I and Mr. Hart have drawn with extreme care, render any detailed description of the Great Bird of Paradise unnecessary. Like *P. papuana* it has the long flank-plumes yellow, and not red as in *P. raggiana* and *P. sanguinea*, and it also wants the yellow collar on the fore neck. The female of *P. papuana* is white on the breast, while in the same sex of *P. apoda* the breast is maroon brown like the back.

The Plate represents an adult male and female, of the size of life, drawn from Arru specimens in my own collection.

PARADISEA SANGUINEA, Shaw.

PARADISEA SANGUINEA, Shaw.

Red Bird of Paradise.

Paradisea sanguinea, Shaw, Gen. Zool. (1809), vol. vii. pt. 1, p. 487, pl. 59.
Paradisea rubra, Vieill. Gal. Ois. (1825), vol. i. p. 152, pl. 99.—Wall. Proc. Zool. Soc. (1862), p. 160.—Id. Ibis (1859), p. 111 (1861), p. 287.—Malay Archip. vol. ii. pp. 214, 221, 243.
Red Bird of Paradise, Lath. Gen. Hist. of Birds (1822), vol. iii. p. 186, sp. 4.
L'Oiseau de Paradis Rouge, Levaill. Hist. Nat. des Ois. Parad. (1806), vol. i. pl. 6.
Le Paradis Rouge, Vieill. Ois. dor. (1802), vol. ii. p. 14, pl. 3.
Paradisea sanguinea, Elliot, Mon. Paradiseidæ, pl. 5.

As I am unable to add any thing concerning the history of *Paradisea sanguinea* to what has been said so well both by Mr. Wallace and also by Mr. Elliot, in his 'Monograph of the Paradiseidæ,' I take the liberty of copying the remarks of the latter, who says :—

"This beautiful bird, remarkable for the rich red plumes that spring from its sides and afford so conspicuous a decoration, is found upon the island of Waigiou and the neighbouring ones of Ghemien and Batanta. The list of synonyms given above will serve to show that for a long time it has been known to, and quoted by, many authors; yet we were practically ignorant of its nature and mode of life until Mr. Wallace visited one of the islands where it has its home, and published his account of it in the work to which I have so often had occasion to allude. I will let Mr. Wallace tell his story in his own words regarding the capture of this beautiful species.

"'When I first arrived I was surprised at being told that there were no Paradise-birds at Muka, although there were plenty at Bessir, a place where the natives caught them and prepared the skins. I assured the people I had heard the cry of these birds close to the village; but they would not believe that I could know their cry. However, the very first time I went into the forest I not only heard but saw them, and was convinced there were plenty about; but they were very shy, and it was some time before we got any. My hunter first shot a female; and I one day got very close to a fine male. He was, as I expected, the rare red species, *Paradisea rubra*, which alone inhabits this island and is found nowhere else. He was quite low down, running along a bough searching for insects, almost like a Woodpecker; and the long black ribaud-like filaments in his tail hung down in the most graceful double curve imaginable. I covered him with my gun, and was going to use the barrel, which had a very small charge of powder and No. 8 shot, so as not to injure his plumage; but the gun missed fire, and he was off in an instant among the thickest jungle. Another day we saw no less than eight fine males at different times, and fired four times at them; but though other birds at the same distance almost always dropped, these all got away, and I began to think we were never to get this magnificent species. At length the fruit ripened on the fig-tree close to my house, and many birds came to feed on it; and one morning, as I was taking my coffee, a male Paradise-bird was seen to settle on its top. I seized my gun, ran under the tree, and, gazing up, could see it flying across from branch to branch, seizing a fruit here and another there; and then, before I could get a sufficient aim to shoot at such a height (for it was one of the loftiest trees of the tropics), it was away into the forest. They now visited the tree every morning; but they stayed so short a time, their motions were so rapid, and it was so difficult to see them, owing to the lower trees which impeded the view, that it was only after several days' watching, and one or two misses, that I brought down my bird—a male in the most magnificent plumage. . . . . I had only shot two Paradiseas on my tree when they ceased visiting it, either owing to the fruit becoming scarce, or that they were wise enough to know there was danger. We continued to hear and see them in the forest, but after a month had not succeeded in shooting any more; and as my chief object in visiting Waigiou was to get these birds, I determined to go to Bessir, where there are a number of Papuans who catch and preserve them. I hired a small outrigger boat for this journey, and left one of my men to guard my house and goods. . . . . My first business was to send for the men who were accustomed to catch the Birds of Paradise. Several came; and I showed them my hatchets, beads, knives, and handkerchiefs, and explained to them as well as I could by signs the price I would give for fresh-killed specimens. It is the universal custom to pay for every thing in advance; but only one man ventured to take goods to the value of two birds. The rest were suspicious, and wanted to see the result of the first bargain with the strange white man, the only one who had ever come to their island. After three days my man brought me the first bird—a very fine specimen, and alive, but tied up in a small bag, and consequently its tail- and wing-feathers were very much crushed and injured. I tried to explain to him, and to others that came with him, that I wanted them as perfect as possible, and that they should either kill them or keep them on a perch with a string to their leg. As they were now apparently satisfied that all was fair, and that I had no ulterior designs upon them, six others took away goods, some for one bird, some for more, and one for as many as six. They said they had to go a long way for them, and that they would come back as soon as they caught any. At intervals of a few days or a week some of them would return, bringing me one or more birds; but though they did not bring any more in bags, there was not much improvement in their condition. As they caught them a long way off in the forest, they would scarcely ever come with one, but would tie it by the legs to a stick, and put it in their house till they caught another. The poor creature would make violent efforts to escape, would get among the ashes, or hang suspended by the leg till the limb was swollen or half-putrefied, and sometimes die of starvation and worry. One had its beautiful head all defiled by pitch from a dammar torch; another had been so long dead that its stomach was turning green. Luckily, however, the skin and plumage of these birds is so firm and strong that they bear washing and cleaning better than almost any other sort; and I was generally able to clean them so well that they did not perceptibly differ from those I had shot myself. Some few were brought me the same day they were caught; and I had an opportunity of examining them in all their beauty and vivacity. As soon as I found they were generally brought alive, I set one of my men to make a large bamboo cage, with troughs for food and water, hoping to be able to keep some of them. I got the natives to bring me branches of a fruit they were very fond of; and I was pleased to find they ate it greedily, and would also take any number of live grasshoppers I gave them, stripping off the legs and wings, and then swallowing them. They drank plenty of water, and were in constant motion, jumping about the cage from perch to perch, clinging to the top and sides, and rarely resting a moment the first day till nightfall. The second day they were always less active, although they would eat as freely as before; and on the morning of the third day they were almost always found dead at the bottom of the cage, without any apparent cause. Some of them ate boiled rice, as well as fruits and insects; but, after trying many in succession, not one out of ten lived more than three days. The second or third day they would be dull, and in several cases they were seized with convulsions and fell off the perch, dying a few hours afterwards. I tried immature as well as full-plumaged birds, but with no better success, and at length gave it up as a hopeless task, and confined my attention to preserving specimens in as good a condition as possible.

"'The Red Birds of Paradise are not shot with blunt arrows, as in the Aru Islands and some parts of New Guinea, but are snared in a very ingenious manner. A large climbing Arum bears a red reticulated fruit, of which the birds are very fond. The hunters fasten this fruit on a stout forked stick, and provide themselves with a fine but strong cord. They then seek out some tree in the forest on which these birds are accustomed to perch, and, climbing up it, fasten the stick to a branch, and arrange the cord in a noose so ingeniously that, when the bird comes to eat the fruit, its legs are caught; and by pulling the end of the cord, which hangs down to the ground, it comes free from the branch and brings down the bird. Sometimes, when food is abundant elsewhere, the hunter sits from morning till night under his tree, with the cord in his hand, and even for two or three whole days in succession, without even getting a bite; while, on the other hand, if very lucky, he may get two or three birds in a day. There are only eight or ten men in Bessir who practise this art, which is unknown anywhere else in the island.'"

Male.—Fore part of the head, chin, cheeks, and throat rich metallic grass-green, appearing black upon the chin. Over each eye the feathers are raised, forming two short tufts. Back of head orange-yellow. All the feathers of the head are short, velvety, and closely pressed together, and project over the bill, above and below, hiding the nostrils. Upper part of back, scapulars, shoulders, upper part of breast and rump orange-yellow. Wings, tail, back, and entire underparts dark chestnut-brown, darkest on the breast, where it is almost a blackish brown. From each side beneath the wings springs a mass of deep red plumes, which glisten like glass, as is seen in the upper part of the plumage of *Cicinnurus regius*, becoming white towards the ends on both webs and shafts, the former widely separated and hair-like. From the lower part of the back fall two very long and webless shafts, black and twisted, which descend on either side of the tail in graceful double curves, and, like the side plumes, constitute very conspicuous appendages.

Female.—Fore part of head, chin, cheeks, and throat very dark chestnut-brown. Back of head, and upper part of breast yellow. Upper part of back or mantle dark ochre-yellow. Entire rest of plumage, including wings and tail, dark brownish chestnut. Bill horn-colour. Feet and tarsi black.

Hab. Waigiou (*Wallace*); islands of Ghemien and Batanta (*Bernstein*).

PARADISEA RAGGIANA, Sclater.

PARADISEA RAGGIANA, Sclater.

Marquis de Raggi's Bird of Paradise.

Paradisea raggiana, Sclater, P. Z. S. 1873, pp. 559, 697.—Elliot, Monogr. Paradis. pl. 3 (1873).—Salvad. Ann. Mus. Civ. Genova, vii. p. 829 (1875).—D'Albertis, P. Z. S. 1875, p. 532.

TAKING it all in all, I consider this the grandest Paradise-bird that has been discovered since Wallace sent home the startling and singular *Semioptera*. That so large and splendid a species should have remained totally unknown up to the period of Signor d'Albertis's visit to the southern parts of New Guinea serves to show the probability that other treasures remain for future explorers in this great unknown land. When the rumour first spread abroad that a second red-plumed Bird of Paradise existed, suspicions arose that these two birds were merely varieties of *P. rubra*, than which there could be nothing more fallacious, since it not only differs from that species, but belongs to another section of the Paradiseidæ, of which *P. papuana* and *P. apoda* are examples.

That the *Paradisea raggiana* holds a place in the south of New Guinea, as the *P. papuana* does in the north, is very evident. It is also quite certain that it is a bird equally common in all the parts yet visited by Signor d'Albertis, Captain Moresby, and other travellers; naturalists therefore need not fear that our collections will long remain without examples. In a note from Signor d'Albertis, addressed to the Secretary of the Zoological Society of London, published in the 'Proceedings' for 1875, is perhaps the best description of this bird yet received. He says:—"I was rather fortunate in my excursion into the mountains; for I found *Paradisea raggiana*, and obtained some beautiful specimens in full dress. In its voice, movements, and attitudes it perfectly resembles the other species of the genus. It feeds on fruit; and I could find no trace of insects in the seven individuals which I prepared. It inhabits the dense forest, and is generally found near the ravines—perhaps because the trees on the fruit of which it feeds prefer the neighbourhood of water. The female is always smaller in size than the male; and I find this sex less abundant, because, as I believe, it is the season of incubation. The female is more like the same sex of *P. apoda* than that of *P. papuana*. The young male is like the female, but often recognizable by having distinguishable traces of the yellow collar which in the old male divides the green of the throat from the breast-feathers. The irides are of a rather bright yellow, and the feet lead-colour with a reddish tinge. The long flank-feathers in individuals recently killed have a very bright tint, which they lose in a few days—even in a few hours. The two middle tail-feathers are filiform, as those of *P. apoda* and *P. papuana*, and in no stage of development resemble those of *P. rubra*. These two feathers are not so long as in *P. apoda*, and about equal to those of *P. papuana*. Like its sister species, *P. raggiana* is an inquisitive bird, and often approaches from branch to branch within a few yards of the hunter, and remains motionless for some seconds to observe its pursuer, stretching out its neck, flapping its wings, and emitting a peculiar cry, upon the sound of which other individuals come forward to join it. When one is wounded and cries out, many others come forward as if to protect it, and approach quite near, descending to the lowest boughs. The adult males frequent the tops of the highest trees, as Mr. Wallace observed in the other species, and as I also remarked in my former expedition. As regards the nidification I have as yet obtained no information." Specimens of all ages and both sexes were procured. Would that this enterprising traveller could set our minds at rest by letting us know something of the nidification, colouring of the eggs, &c.

The following description is from a perfect specimen of a male in the possession of Mr. Stone :—

Total length from bill to tail 14 inches, wing 7, tail 6, tarsi 2¼.

Bill pale blue; feet lead-colour with a pinky tint; forehead and throat beautiful green, almost metallic; crown, sides, and neck fine orange-yellow; this fine yellow also extends across the throat, separating the green from the rich maroon of the chest (which is very intense); all the upper and under surface reddish brown, with a distinct mark of yellow on the shoulders; long flank-feathers splendid blood-red, becoming brown towards the ends, the extremities of which are dirty white; the two long filamentous feathers narrow, and terminating in a spatula like that of *P. papuana* in form. Count Salvadori describes the female as being "smaller than the male; of a vinaceous chestnut-colour, paler below; the sinciput, sides of head, fore part of neck, and breast deep chestnut; occiput and hinder neck, as well as the sides of the latter, yellowish. The entire green colour of the forehead and throat is wanting, as well as the pale yellowish straw-coloured ring under the throat, the yellow band across the wing, and the ornaments of the tail and flanks.

The figures in the accompanying Plate are both males, that in the foreground being the size of life.

MANUCODIA COMRII, *Slater*.

MANUCODIA COMRII, Sclater.

Curl-crested Manucode.

Manucodia comrii, Sclater, Proc. Zool. Soc. 1876, p. 459, pl. xlii., et P. Z. S. 1877, p. 43.

THE discovery of this interesting bird in the southern parts of New Guinea confirms the opinion so generally offered by naturalists, that this highly gifted country may yet have many stores left for our contemplation and study; indeed, as time runs on, it is constantly yielding fruits to those who may enter its precincts, let it be on the north or the south. Since the time of Wallace's visit the Dutch explorer Von Rosenberg, Dr. Meyer, Hr. Bruijn, and the Italians D'Albertis and Beccari have each immortalized their names by their discoveries in Papuan ornithology. But that of the fine bird under consideration is due to a medical officer of Her Majesty's Service; and the species itself may be regarded as one of the most valuable of the novelties lately transmitted to us, particularly so from its adding another member to the Paradiseidæ. We find in the 'Proceedings' of the Zoological Society, as above quoted, the following note by Dr. Sclater:—
" Dr. Comrie has placed in my hands, for determination, some bird-skins collected by him while serving as medical officer in H.M.S. 'Basilisk' during its recent survey of the south-east coast of New Guinea under the command of Captain Moresby. The collection contains thirteen specimens, belonging to eleven species, of which one is quite new to science, and two others are only known from single specimens."

"This *Manucodia*," Dr. Sclater continues, "may be regarded as by far the finest and largest species of the genus yet discovered. It is immediately distinguishable from *M. chalybeia* and *M. atra* by its much larger size and longer bill, which is deeply sulcated at the nostrils. The characteristic curling of the feathers is extended to a greater degree, and pervades the whole of the head and neck; the feathers of the abdomen are black at the base, broadly margined with purple. Dr. Comrie obtained a single specimen of this fine bird in May 1874 in Huan Gulf. It was shot flying amongst the trees in the scrubby forest, about a quarter of a mile from the coast."

In a subsequent communication to the Society Mr. Sclater called attention to the original and unique specimen of his *Manucodia comrii* (P. Z. S. 1876, p. 459), now belonging to the collection of the Marquis of Tweeddale, the President of the Society, which since it was described and figured, had undergone a most efficient "remake" in Mr. Bartlett's able hands.

The curly feathers of the head were now much more apparent and better developed than shown even in the woodcut given with the original description, and formed standing ridges over each eye. The tail was not flat, as would appear from the figure (pl. xlii.), but "boat-shaped" as in some of the American Grakles (*Quiscalus*), *i. e.* with the median tail-feathers elevated above the lateral. The two middle tail-feathers were very peculiar in construction, being shorter by three quarters of an inch than the next pair, and having the inner webs twisted round over the outer, so as to show their under surfaces.

The highly singular characteristic of the turnover feathers is also seen in the Green Manucode (*M. chalybea*), where the frizzly plumes, particularly those of the face and neck, are even more singular. Such eccentricities are really not to be accounted for, as we cannot conceive they can be for any useful purpose.

I must not conclude this paper without returning my thanks to the Marquis of Tweeddale and Dr. Sclater for the loan of the specimen from which the accompanying drawing was taken. The principal figure is of the size of life.

MANUCODIA CHALYBEA, Bodd.

MANUCODIA CHALYBEA, *Bodd.*

Green Manucode.

Blue-green Paradise-bird, Lath. Gen. Syn. vol. ii. p. 482. sp. 7 (1782).
Manucodia chalybea, Bodd. Tabl. Plan. Enl. d'Hist. Nat. de Daubent. (1783).—Elliot, Mon. of the Paradiseidæ, pl. vi. (1873).
Paradisea viridis, Gmel. Syst. Nat. vol. i. pt. i. p. 402. sp. 8 (1788).
Paradisea chalybea, Lath. Ind. Ornith. vol. i. p. 197. sp. 10 (1790).
Le Chalibé, Vieill. Ois. Dor. vol. ii. p. 24, pl. 10 (1802).—Levaill. Ois. de Parad. p. 64, pl. 23 (1806).
Cracticus chalybeus, Vieill. Nouv. Dict. d'Hist. Nat. vol. v. p. 355 (1816).
Phonygama chalybæa, Less. Trait. Ornith. p. 344. sp. 1 (1831).
Phonygama viridis, Gray, Gen. Birds, vol. ii. p. 303. sp. 1 (1849).—Bon. Consp. Gen. Av. p. 368 (1850).— Rosenb. Journ. für Orn. 1864, p. 122.
Manucodia viridis, Gray, Proc. Zool. Soc. 1861, p. 436.
Chalybea viridis, Schleg. Tijdsch. Dierk. pt. v. p. 49.

A GLANCE at the above list of synonyms will show that this species has been known to naturalists for nearly a hundred years. Formerly all the skins arrived without either feet or wings, like many other New-Guinea birds; latterly, however, perfect specimens have come to hand, those appendages not having been removed. That this species is a true Bird of Paradise all must now admit, whatever opinions have been entertained to the contrary. We have now at least five species of the genus *Manucodia*, and are living in hope that we may not long be kept in the dark as to their habits and economy, which are likely to be characterized by some peculiarities.

The following is taken from Mr. Elliot's 'Monograph of the Paradiseidæ:'—

"The Green Manucode resembles in some respects its relative the *M. atra*, but may be distinguished in nearly all stages of plumage by the frizzled state of the feathers upon the head and neck; in some specimens, indeed, these peculiarly formed feathers encroach upon the upper part of the back and breast; and they frequently possess variegated colours of different metallic hues.

"Feathers of the head short, closely pressed together, elongated over the eyes into diminutive tufts, black, with the tips a bright metallic blue; throat, neck, and upper part of breast light green, the feathers on the breast appearing as though the tips were frizzled. Back and wings bright metallic purplish blue, very glossy upon the secondaries; primaries rich brown; underparts similar to the back, but the blue not so rich. Tail purplish blue like the wings. Bill black; feet black."

A still more perfect description taken from Mr. Sharpe's Catalogue of the Birds in the British Museum, iii. p. 182, is as follows:—

"*Adult Male.*—Head purple, the feathers compressed and close-set; the nape slightly washed with steel-greenish, as also the hinder neck and mantle; back rich purple, the feathers of the interscapular region rather recurved; wings and tail rich purple, the inner webs of the feathers blackish, the outer wing-coverts somewhat shaded with steel-black; sides of the face and neck deep green, the feathers compressed and velvety like those of the crown; those of the chin, throat, and fore neck extending onto the sides of the neck, crinkled and curled and of an oily-green colour; the rest of the under surface deep purple, the feathers being tipped with this colour, less broadly on the vent and under tail-coverts, a few of the abdominal plumes with a slight greenish reflection; under wing-coverts black, the outer edge of the wing washed with green; bill and legs black. Total length 14·5 inches, culmen 1·65, wing 6·85, tail 5·9, tarsus 1·55."

"*Hab.* New Guinea. Regions near the coasts and mountain-ranges of the interior (*Rosenberg*)."

The figure on the Plate is of the size of life.

PHONYGAMA PURPUREO-VIOLACEA, Meyer.

PHONYGAMA PURPUREOVIOLACEA, Meyer.

Purple-and-Violet Manucode.

Phonygama purpureoviolacea, Meyer, in Madarász, Zeitsch. ges. Orn. ii. p. 375, taf. xv. (1885).

The species of *Phonygama* from South-eastern New Guinea are difficult to determine, as the changes to which the metallic colours are subject under the influences of abrasion or wearing of the feathers are at present indistinctly understood. The species from North-western New Guinea is *Phonygama keraudreni*, and we have ourselves described from Southern New Guinea two species of the genus, *P. hunsteini* and *P. jamesii*. With these we have compared a series of *P. purpureoviolacea* procured by Mr. H. O. Forbes in the Astrolabe Mountains, as well as an example obtained by Mr. Hunstein himself in the Horseshoe range.

P. hunsteini is much larger than any of the Astrolabe specimens; its colour is a dull purple with scarcely any gloss, and the colour of the head and crest-feathers is metallic oily green, of diminished lustre. It will probably be found that *P. hunsteini* is an inhabitant of one of the islands off the coast, and not of New Guinea itself. No information, beyond that it had come, like other birds in the collection, from East Cape in South-eastern New Guinea, was given with the type specimen of *P. hunsteini*; but it is quite possible that the real habitat is Normanby Island, where Mr. Hunstein also collected. At any rate the species appears distinct from *P. jamesii* and *P. purpureoviolacea*, of which it could only be a worn and bleached individual, and even then the larger size is not accounted for.

The series before us at the present moment leaves very little doubt that the *Phonygama* recently described by Dr. Meyer, and figured by us in the accompanying Plate, is distinct from *P. keraudreni* and *P. hunsteini*; but it is apparently the same as *Phonygama jamesii*, a species described by us in 1877 from Aleya, in South-eastern New Guinea. The chief difference between these two species is, that *P. purpureoviolacea* is more purple above and steel-blue below, and *P. jamesii* is metallic green above and steel-green below. But between these extremes of colour every transition is found in the series now before us; and it should be noted that the type specimen of *P. jamesii* is moulting, and that the old feathers of the wing are very dull purple, while the new ones are bright purplish blue externally. In fine, without asserting dogmatically that *P. jamesii* and *P. pupureoviolacea* are the same, we have very little doubt in our own minds that they are, and that the steel-blue and green shades become gradually faded into purple or purplish blue.

The figures in the Plate represent an adult bird in two positions, drawn from a specimen procured by Mr. Hunstein in the Horseshoe range, and now in the British Museum.

[R. B. S.]

LYCOCORAX OBIENSIS, Bernst.

LYCOCORAX OBIENSIS, Bernst.

Obi Paradise-Crow.

Lycocorax obiensis, Bernst. Journ. für Orn. 1864, p. 410.—Id. Nederl. Tijdschr. Dierk. ii. p. 350 (1865).—Schl. op. cit. iii. p. 192 (1866).—Id. Mus. Pays-Bas, Coraces, p. 132 (1867).—Gray, Hand-list of Birds, ii. p. 17, no. 6263 (1870).—Sharpe, Cat. Birds in Brit. Mus. iii. p. 185 (1877).—Salvad. Ann. Mus. Civic. Genov. xvi. p. 199 (1880).—Id. Orn. Papuasia e delle Molucche, ii. p. 495 (1881).—Guillemard, Proc. Zool. Soc. 1885, p. 573.

This species appears to be confined to the Obi group of islands in the Moluccas, where it replaces *Lycocorax pyrrhopterus* of Batchian and Gilolo, and *L. morotensis* of Morotai or Morty Island. It is distinguished from both by the greenish wash on the upper parts, and it has the quills blacker than in *L. pyrrhopterus*. Count Salvadori and ourselves both regarded the white on the base of the quills as a distinctive character of *L. morotensis*, but Dr. Guillemard, who has recently visited the Obi Islands and obtained five specimens of the present species, states that all his series, excepting one bird, had a white mark on the primaries. The exception was in the case of a female bird, which was duller in colour than the males and had the primaries buff.

Dr. Bernstein, the discoverer of the species, procured it in Obi Major and Obi Lattoo, but he states that, like *L. morotensis*, it is a difficult bird to procure, as it frequents the thick forest. Its note is described by Dr. Bernstein as "whunk."

The following description of an adult bird is copied from the British Museum 'Catalogue of Birds,' and is taken from a specimen in that institution:—

"General colour above and below of a dull rifle-green, somewhat glistening; tail black, the feathers slightly washed with green on the outer web; quills blackish brown, the least wing-coverts edged with dull green like the scapulars, the rest of the coverts and secondaries slightly washed with green on the outer web, the primaries much paler brown; bill and feet black. Total length 13·5 inches, culmen 1·95, wing 7·75, tail 6·75, tarsus 1·9."

Dr. Guillemard says that the iris is crimson, but that in the female bird referred to above it was brown.

The figure in the Plate represents an adult bird of about the size of life, and is drawn from a specimen kindly lent to us by Dr. Guillemard.

[R. B. S.]

ÆLURŒDUS STONII, *Sharpe.*

ÆLURŒDUS STONII, Sharpe.

Stone's Cat Bird.

Ælurœdus stonii, Sharpe, Nature, Aug. 17, 1876, p. 339.—Salvad. Ann. Mus. Civic. Genova, ix. p. 193 (1876). —Sharpe, Proc. Linn. Soc. xiii. p. 495 (1877).—Ramsay, Proc. Linn. Soc. New S. Wales, iii. p. 268 (1879); iv. p. 97 (1879).—Salvad. Orn. della Papuasia &c. p. 678 (1881).

Count Salvadori, in his great work on the birds of New Guinea, mentions his having examined the type of this species in the British Museum; but he considers that it is by no means a very well-marked form, and may yet prove to be identical with *Æ. buccoides*. This was at first my own idea; and although I had the drawing prepared some years ago, I never ventured to publish it, but relied on the arrival of more specimens to decide the validity of the species. I have since seen several examples from South-eastern New Guinea, and I have found all the characters noted by Mr. Sharpe to be fully borne out in all the birds examined by me; so that now I have no alternative but to recognize Stone's Cat Bird as a well-founded species of the genus *Ælurœdus*. It is closely allied to *Æ. buccoides* of North-western New Guinea (which bird, by the way, has been said by Count Salvadori to have been obtained on the Fly river also), but is distinguished by the small spots on the under surface and the darker head. It represents *Æ. buccoides* in South-eastern New Guinea.

Mr. E. P. Ramsay records the capture of a few specimens by Mr. A. Goldie, about fifteen miles inland from Port Moresby. They were found in dense scrubs, feeding on fruit and berries.

The following description has been copied from Mr. Sharpe's original account of the species:—

Adult.—General colour above bright green, some of the feathers tinged with blue; wings green, like the back, the inner webs dusky brown, the primaries externally washed with yellow, the secondaries tipped with the latter colour; tail green, blackish on the inner webs of the outermost rectrices, which are tipped with white; head dark brown, slightly washed with olive; hind neck yellowish buff, mottled with black centres to the feathers, those adjoining the mantle spotted with green; sides of face and throat pure white, with a few tiny spots of black on the ear-coverts, larger on the sides of the neck; rest of under surface of body ochraceous buff, the fore neck and chest minutely spotted with green, the flanks also with a few tiny spots of the latter colour; under wing-coverts yellowish buff, the edge of the wing washed with green. Total length 9·3 inches, culmen 1·15, wing 5·05, tail 3·5, tarsus 1·55.

The figure in the Plate, which represents the species of about the natural size, has been drawn from the typical specimen kindly lent to me by Mr. Bowdler Sharpe.

AILURŒDUS MACULOSUS, Ramsay.

AILURŒDUS MACULOSUS, Ramsay.

Queensland Cat-bird.

Ælurœdus maculosus, Ramsay, P. Z. S. 1874, p. 601.

THIS addition to the group of Cat-birds (*Ailurœdus*) is of very great interest to me, as I have for years suspected that more species of the genus would be discovered, and the affinities of Mr. Ramsay's bird afford further evidence of the close zoological relations existing between the avifauna of Australia and that of New Guinea and the Aru Islands. Instead of being related to the ordinary species (*Ailurœdus Smithii* vel *crassirostris*), the Queensland species comes nearer to the black-eared Cat-bird of the Aru Islands (*Ailurœdus melanotis*), from which it is distinguished by its smaller size, dark-green coloration, and thickly mottled under surface. From the common Cat-bird it is at once to be told by its black ear-coverts and by other characters which are well set forth by Mr. Ramsay in the description quoted below. Although our knowledge of its habits is at present meagre, more details than usual in the case of novelties are given by the above-named naturalist. It will probably be found that, like *A. Smithii*, the present species does not build a bower as do *Ptilonorhynchus* and *Chlamydera*.

The following is Mr. Ramsay's description :—

"The whole of the head, including the chin, feathers at the base of the lower mandible, and ear-coverts, black, having a spot of white tinged with olive-green at the end of each feather; the spots on the back of the head and neck become more distinctly tinged with olive-green as they approach the shoulders, where they are lost in the brown-green margin of the feathers, which on their under surface are of a bluish green tint.

"A narrow line down the centre of each feather on the top of the head and neck black; line over the eye and just in front of it white tinged with olive-green; lower part of the ear-coverts black; upper part immediately behind the eye centred with white and tinged slightly with olive-green; sides of the lower part of the neck olive-green, each feather indistinctly marked with a whity-brown spot.

"Whole of the under surface olive-green, being brightest on the flanks, each feather on the chest centred with a conspicuous and somewhat heart-shaped blotch of white, which on the abdomen and flanks becomes lanceolate in form, more acute on those feathers between the flanks and round the vent, which are nearly white, having a brownish crescentic mark near the margin; the spots on the abdomen are almost obsolete. Under tail-coverts olive-green, centred and broadly margined with white tinged with olive-green; the basal portion of all the feathers brown; under surface of the tail brown, with a tinge of bluish green in certain lights; all but the two centre tail-feathers tipped with white on both surfaces; the under surface of the wings, inner webs of primaries, and secondaries dark brown, becoming almost white on the margins of the inner webs. The inner webs of primaries and secondaries, and the basal part of the inner webs of the tertiaries, blackish brown on the upper surface; the margin of the wings and under wing-coverts white, with a somewhat crescent-shaped mark of dark brown, margined with green, in the centre; the basal portion of the feathers dark brown. The narrow outer web of the primaries above bluish green; all the upper surface from the back of neck and mantle bright grass-green; on the tertiaries to approximate secondaries a small spot of white at the tips of the outer webs only; some of the lesser wing-coverts have also a similar spot in certain specimens; and a few of the feathers between the shoulders have a bluish green tinge; two centre tail-feathers and the outer webs of all except the first feather on either side bright grass-green above; the lower portion of the outer web of the first tail-feather on either side tinged with green; the inner webs of all the tail-feathers except the two centre ones blackish brown above, tinged with green near the shafts of the feathers, but becoming blackish brown near the tips, which end in a white marginal spot more largely developed on the inner webs and of greater extent on the outer feathers, diminishing into a narrow white line on those next the centre tail-feathers. Bill light horn-colour.

"Total length 10 to 11 inches; bill from forehead 1, from eye 1·9 to 2, from nostril 0·6, from angle of the mouth 1·45, width at base 0·5; wing from flexure 5·5; tail 4·3; tarsus 1·6 to 1·8, of a dark bluish horn-colour. Claws light brown.

"This new species is at least one quarter less in size than the New-South-Wales bird, *A. Smithii*, and is only found on the East-Coast ranges, about Cardwell, at Rockingham Bay. Several specimens were obtained, but unfortunately so late in the season that the plumage was much worn and discoloured.

"Its note resembles that of *A. Smithii*, but is not so distinct a cry, and less bat-like, clearer and more of a whistle. They assemble in small flocks from ten to twenty in number, and frequent the palms and native fruit-trees in company with *Ptilonopus superbus* and *Carpophaga assimilis*."

The figure is of the natural size.

AILURŒDUS MELANOTIS.

AILURŒDUS MELANOTIS.

Black-cheeked Cat-bird.

Ptilonorhynchus melanotis, Gray, P. Z. S. 1858, p. 181.—Id. Cat. Mamm. & Birds N. Guin. p. 37.—Von Rosenb. J. f. O. 1864, p. 122.—Schl. Mus. P. B. Coraces, p. 118.—Gray, Handl. B. i. p. 294.—Schl. N. T. D. iv. p. 51.
Æluroedus melanotis, Elliot, Monogr. Paradiseidæ, pl. 35.

THE discovery of this genuine and unmistakable Cat-bird in the Papuan Islands, forms part of the evidence that the latter are Australian, and not Indian, in their geographical affinity.

The *Ailurœdus melanotis* of the Aru Islands is the largest of the five species of this genus now known to us: the nearest ally is the *A. arfakianus* of the mainland of New Guinea, and by some may be considered the same species; Mr. Meyer, however, has separated them; and I have followed him and given figures of both. Whatever they may ultimately prove to be, it will be seen that the mainland bird differs materially both in the colour and form of its markings. The newly discovered species at Rockingham Bay, in Queensland, is also nearly allied, but is at once separated by its much smaller size, and less developed markings on the tips of the secondaries, a feature so prominent in the Aru bird as to remind us very forcibly of the round marks so abundantly dispersed over the whole upper surface in *Chlamydodera maculata*. These marks, both in form and colour, are not the usual tippings of such feathers, but are rounder and more sharply defined; in colour they will be found deeply tinted with yellowish grey, affording a strong contrast to the pure white ends of the tail-feathers. These facts in my mind tend to confirm the alliance between the members of the genus *Ailurœdus* and *Chlamydodera*, while the length and greater development of the mantle in most of the species shows an alliance to many of the Birds of Paradise. Mr. Wallace did not fail to obtain specimens of the present bird during his visit to the Aru Islands; neither was Mr. Cockerell less active in this respect: from these two sources our cabinets are now well supplied. The Dutch travellers, Baron von Rosenberg and Mr. Hoedt, have also forwarded a large series to the Leiden Museum, from the islands of Wokan, Trangan, and Maykor, all belonging to the Aru group. Nothing whatever has been written about its habits.

Male.—Bill fleshy white; crown of the head, nape, and mantle black, with longitudinal spots of buffy brown occupying the centre of each feather; lores grey; ears black, separated from the back of the head by a narrow line of buffy white; throat greyish white, mottled with dark brown; all the undersurface dull greenish yellow, each feather having darker tips and a hair-like stripe of white down the centre; the breast generally darker than the belly and flanks, the latter being strongly suffused with green; all the upper surface lively grass-green; the tips of all the secondaries greyish white, all the tail-feathers tipped with pure white, the outermost ones largely, while in the two centre ones the white marks are almost obsolete; legs and toes bluish black.

Total length 12½ inches, bill 1⅜, wing 6½, tail 4¾, tarsus 2.

The sexes are much alike in colour; but the female is decidedly the smallest.

AILURŒDUS ARFAKIANUS, Meyer.

AILURŒDUS ARFAKIANUS.

Arfak Cat-bird.

Ælurœdus melanotis, Sclater, P. Z. S. 1873, p. 697 (nec Gray).
―――― *arfakianus*, Meyer, Sitz. Akad. Wien, lxix. p. 82.

It will be seen, on reference to the synonymy, that Dr. Sclater considered the Cat-bird collected by Signor d'Albertis in the Arfak mountains to be the same as *Ailurœdus melanotis* of the Arru Islands; consequently in his list of New-Guinea birds, published in the 'Proceedings,' it is given under that name. Dr. Meyer, on the other hand, has separated them; but, although I have followed Meyer rather than Sclater, I am not altogether satisfied as to the correctness of this view, and fear it must remain for future observers to determine this point. I may, however, state that the Arfak bird is much more strongly coloured, the spots being blacker and more sparkling; to which it must be added that it is of smaller size. I regret that nothing is known of the habits of the bird, or even of its distribution in New Guinea. It will not be long, however, before the ornithological secrets of this *terra incognita* will be laid bare; and we may expect that, when the interior of this great island is explored, not only will the ranges of many well-known species be extended, but new forms even of Bower-birds will be discovered. Each new traveller adds some striking novelty or other to the ornithological series; and there is probably no country in the world which will yield such a rich harvest of unknown species as New Guinea.

I must refer my readers to the accompanying Plate for a correct idea of the present species, as Signor d'Albertis's hurried departure for Italy with his collections gave me no time to describe the bird in detail.

The figure is of the natural size.

AILUROEDUS BUCCOIDES.

AILURŒDUS BUCCOIDES.

Barbet-like Cat-bird.

Kitta buccoides, Temm. Pl. Col. 575.—Von Rosenb. J. f. O. 1864, p. 122.
Cissa buccoides, Gray, Gen. B. iii., App. p. 14.
Ptilonorhynchus buccoides, Gray, P. Z. S. 1858, p. 194.—Id. Cat. Mamm. & B. N. Guin. p. 37.—Schl. Mus. P. B., Coraces, p. 118.—Gray, Hand-l. B. i. p. 294.—Schl. N. T. D. iv. p. 49.
Æluroedus buccoides, Elliot, Monogr. Parad. pl. 36.—Sclater, P. Z. S. 1873, p. 697.

Although this bird, aptly named *buccoides* by Temminck on account of its resemblance to the Barbets of the genus *Megalæma*, has been known for so many years, and specimens not unfrequently turn up in collections from New Guinea, there is nothing recorded of its habits or economy. It is confined to the islands of New Guinea and Waigiou, the Leiden Museum possessing a considerable number of specimens collected in these localities by the late Dr. Bernstein. The original specimen was procured at Triton Bay, on the west coast of New Guinea; and other localities mentioned by Professor Schlegel are Sorong, the north coast of Salwatti, and the island of Batanta.

In its general green coloration this bird resembles the other species of the genus; but it may be readily distinguished by the more decided spotting of the breast and by the very distinct whitish streaks which are seen on the nape of the neck; besides which it is by far the smallest species of the whole genus, a group of birds intimately allied to the Bower-birds (*Chlamydodera* and *Ptilonorhynchus*); but up to the present moment we have no information as to whether the members of the genus *Ailurœdus* ever construct a bower.

Upper surface of body, the wings included, grass-green, the secondaries slightly tipped with yellow; head olive-brown washed with green; all the feathers of the neck pale yellow, with large terminal spots of black, giving a streaked appearance to this part; a great many of the interscapulary plumes barred with yellow, some spotted with black; tail green; sides of face and throat buffy white, the former thickly spotted with black; rest of under surface ochraceous buff washed with green, with a black spot to each of the feathers, disappearing on the under tail-coverts; the spots greenish on the lower abdomen; under wing-coverts salmon-buff, the outermost with brown-mottled bases; bill yellowish white; tarsi and feet lead-colour. Total length 10 inches; culmen 1·0. wing 4·9, tail 3·6, tarsus 1·4.

Apparently there is no difference in the sexes; and the figures in the accompanying Plate are as nearly life-size as possible.

AILUROEDUS MELANOCEPHALUS, Ramsay.

ÆLURŒDUS MELANOCEPHALUS, *Ramsay*.

Black-naped Cat-bird.

Æluroedus melanocephalus, Ramsay, Proc. Linn. Soc. N. S. Wales, viii. p. 25 (1883).—Finsch u. Meyer, Zeitschr. ges. Orn. ii. p. 394 (1885).

WE have already alluded in the present work to the differences between *Æluroedus melanotis*, from the Aru Islands, and *Æ. arfakianus*, from Mount Arfak; but at the time when we wrote we were unable to convince ourselves that those two species were specifically distinct. We now consider that we were in error, since we have seen additional examples of both, and now we have a third representative species in *Æ. melanocephalus* from South-eastern New Guinea. Of this latter bird we have seen several specimens collected by Mr. Goldie and Mr. H. O. Forbes in the Astrolabe Mountains, and by Mr. Hunstein in the Horseshoe range of the Owen Stanley Mountains.

The present species differs from *Æ. arfakianus* by its black lores and chin and in the uniform character of the breast and abdomen.

We subjoin a description of an adult male collected by Mr. Hunstein on the Owen Stanley Mountains:—

Adult male. General colour above grass-green, the upper tail-coverts slightly washed with lighter green; the upper mantle varied with ovate spots of ochreous buff in the centre of the feathers; wing-coverts like the back, the median and greater coverts and the bastard-wing faintly tipped with ashy ochreous buff; primary-coverts and quills externally green like the back, the primaries washed with bluish on the outer web, the secondaries tipped with ochreous white, less distinct on the primaries; tail-feathers dark green on the outer web, black internally, all the feathers tipped with white, increasing in extent towards the outer ones; crown of head black, with ovate spots of ochreous buff, smaller on the forehead and nape, the latter being almost entirely black; hind neck ochreous buff, the feathers margined with black; lores black, surmounted by a line of ochreous-buff-spotted feathers; feathers round eye and ear-coverts black, with a line of buff-spotted feathers below the eye; behind the ear-coverts a line of whitish down the sides of the neck; fore part of cheeks black, as well as the chin; throat and sides of neck ochreous buff, mottled with black edges to the feathers; fore neck and remainder of under surface of body rufescent ochre, with greenish edges on the feathers of the chest, the breast and abdomen more uniform; all the feathers with more or less distinct white shaft-lines; sides of body and flanks like the breast, and washed with greenish; thighs dull greenish; under tail-coverts like the abdomen, with white shaft-lines; under wing-coverts and axillaries ashy, tipped with whitish; quills below dusky, ashy along the inner edge. Total length 11·5 inches, culmen 1·3, wing 5·7, tail 4·6, tarsus 1·7.

The figure in the Plate represents an adult male of the natural size, drawn from a specimen collected by Mr. H. O. Forbes.

[R. B. S.]

SCENOPÆUS DENTIROSTRIS, Ramsay.

SCENOPŒUS DENTIROSTRIS, *Ramsay*.

Toothed-billed Bower-bird.

Scenopœus dentirostris, Ramsay, P. Z. S. 1875, p. 591.—Id. Proc. Linn. Soc. New South Wales, ii. p. 188 (1878).

ONLY the naturalist who has devoted his life to the pursuit of a particular study, and has spent long years in the determination to render his writings on the various interesting subjects he has so long been treating as complete as lies in his power, can imagine the feelings of satisfaction with which I contemplate my work on the Birds of Australia. The time which I spent in working out the birds of that country is one of the most pleasing recollections of a life spent in the pursuit of natural history; and I cannot but congratulate myself on the fact that the additions to the avifauna of that continent have not been very numerous since I published my last volume on its ornithology. At the same time it would be vain to suppose that the field is exhausted, when such extraordinary new forms as that which is figured on the opposite Plate continue to be discovered, constituting, as it does, the type, not only of a new species, but of an entirely new genus. No one who looks for one moment at the extraordinary doubly toothed bill of the present bird can doubt the propriety of Mr. Ramsay's instituting a separate genus for so anomalous a bird.

Nothing is at present known of the habits of this species, or if it really builds a bower; but Mr. Ramsay observes that, like the Catbirds, it clears a large space under the brushwood some nine or ten feet in diameter, and ornaments the cleared parts with tufts of gaily tinted leaves and young shoots. The only specimens yet procured were shot with a rifle by Inspector Johnstone of Cardwell; and the habitat is given as the Bellenden-Ker range and the dense brushes clothing the steep sides of "Sea-view range," on the north-east coast of Queensland.

The following is the description by Mr. Ramsay :—

"The whole of the upper surface, wings, and tail rich olive-brown, the inner webs of the primaries and secondaries blackish brown, their margins near the base buffy white; under surface of the shoulders yellowish buff, with remains of broken bars of blackish brown on the smaller feathers; the under wing-coverts yellowish buff, with cross bars of dull brown; under primary-coverts buff, crossed more distinctly with dull brown; under surface of primaries and secondaries dark ashy brown, the basal half of the inner margin buff tinged with a faint wash of light rufous; flanks olive-buff; abdomen buff; under tail-coverts olive-buff, each feather barred with two or more lanceolate marks of dull olive-brown, under surface of the tail dull brown; throat, neck below, chest, and the rest of the under surface buffy white, each feather margined with olive-brown, which becomes lighter and less distinct on the lower parts, and almost obsolete on the flanks and abdomen; on the throat and chest the margins are almost black, and tinged with yellowish olive on the sides of the neck and chest, and the buff central portion of a deeper tint; the under surface has the appearance of being broadly streaked with lanceolate marks of buff, which become more and more indistinct as they approach the under tail-coverts, becoming obsolete on the abdomen.

"Total length 11 inches; wing 5·7; tail 4; tibia 2·2; tarsus 1·2; hind toe 0·6, its claw 0·4, its width 0·25; inner toe 0·65, its claw 0·3; middle toe 0·9, its claw 0·35; outer toe 0·7, its claw 0·3; width of the sole of the foot 0·35; bill from gape 1·2, from forehead 1·1, from the nostril 0·6, height at nostril 0·6, width at nostril 0·5, culmen 1·1; upper mandible black, lighter at the tip; lower mandible blackish brown; gape yellow; legs and feet black, claws brown."

It is now five years since I received information of the discovery of this bird from Mr. Coxen, then Honorary Curator of the Museum at Brisbane, with a promise that he would get the loan of the mutilated skin and forward the same to me for the purpose of figuring in the 'Birds of New Guinea.' Almost immediately after the receipt of Mr. Coxen's letter, I read the sad news of his nearly sudden decease, a source of the deepest regret both to myself as a relative, and all persons who had the pleasure of knowing him. Besides the skin, the head of a second example was forwarded, which greatly added to the interest of the subject. We now anxiously await more examples.

CHLAMYDODERA ORIENTALIS, *Gould.*
Queensland Bower-bird.

Chlamydodera nuchalis (part.), Ramsay, Proc. Zool. Soc. 1868, p. 385.—Id. Proc. Linnean Soc. of New S. Wales, ii. p. 188.
Chlamydodera orientalis, Gould, Ann. & Mag. of Nat. Hist. ser. 5, vol. iv. p. 74.

In the 'Annals of Natural History' for July 1879 I described a new species of Bower-bird, and made some remarks on the *Chlamydoderæ*, proposing the name of *Chlamydodera orientalis* for the new species. I have now the pleasure of giving figures which will enable the reader to distinguish the western Bower-bird from the eastern. Putting aside the very distinct Cat bird (*Æluræus*), Regent bird (*Sericulus*), the Satin Bower-bird (*Ptilonorhynchus*), each of which is easily recognizable by structural characters as well as a particular style of plumage, we come to the members of the genus *Chlamydodera*, containing the "spotted" Bower-birds, the most characteristic species of which are *C. maculata, C. guttata,* and *C. occipitalis*; these have each the plumage on the upper surface thickly ocellated, as well as beautiful lilac frills on the back of the neck. *Chlamydodera cerviniventris* has not the beautiful frilled nape of the above-named species. In *C. orientalis* the tendency to a spotted plumage becomes less marked, and still less so in the larger *C. nuchalis*, in this respect leading off, as it were, towards the uniformly coloured species of New Guinea (the *Amblyornis inornatus*, or Garden Bower-bird, figured in the present work).

Mr. Ramsay has recorded a large Bower-bird from Queensland, but considered it similar to the great bird from Western Australia, *C. nuchalis*, which has been for many years called by that name.

In size the present bird is somewhat less than *C. nuchalis*, and differs from that species in the light edgings to the feathers of the head and upper surface, and especially on the wings, all these parts having somewhat of a banded appearance, which is not so visible as in the species from Western Australia. This species was lately brought to England by Mr. Waller, and is now, as well as its bower, in the fine collection at Liverpool. Both species are represented in the British Museum, as well as in my own collection, where the specimens can be consulted by any student who wishes to verify the differences between them.

The figure in the accompanying Plate is about the size of life, and is drawn from a specimen in my collection.

CHLAMYDERA OCCIPITALIS, Gould.

CHLAMYDODERA OCCIPITALIS, Gould.

Large-frilled Bower-bird.

Chlamydodera occipitalis, Gould, Annals and Mag. Nat. Hist. 4th series, vol. xvi. p. 429 (1875).

In describing the present species in 1875, I wrote as follows:—"Of all the acts performed by birds, that of building themselves beautiful bowers, variously decorated with shells, bleached bones, glittering stones, and gaudily coloured feathers, must ever be ranked amongst the most interesting traits in connexion with ornithology. At present the only known country in which these playing-places or halls of assembly are constructed is Australia. Whoever may have the good fortune to lift up the curtain which separates New Guinea from other countries, may probably find others." Not four years have yet elapsed since those words were penned; and they already appear to have been prophetic. In that short space of time how much has been done to unravel the mysteries of the avifauna of New Guinea, resulting in no small addition to our knowledge of the Bower-birds! The south-eastern part of that great island has been ransacked by English, Italians, and Dutch; and these collectors have made us acquainted with the birds which are met with in that newly explored country. Then, again, in the north-western portion of the island, we owe to the energies of the distinguished Italian traveller, Dr. Beccari, the discovery of *Amblyornis inornata*, whose capabilities in bower-building and decoration seem to eclipse those of any of the other species.

The male of the present bird (*Chlamydodera occipitalis*) is one of the novelties I have but lately received. It is true I had in my collection a female from some unknown locality, which always seemed to me to differ from those of all the other species I had seen; but I hesitated to describe it, and it was not till a year or so previous to 1875 that the receipt of the fine male bird which is figured in the accompanying Plate enabled me to publish its specific characters. That its true habitat was Northern Queensland I was pretty certain; and in confirmation of this opinion, Mr. Janson, a highly respectable person and dealer in natural-history specimens, from whom I purchased the skin, kindly writes to me as follows:—

"Dear Sir,—On reference to my books I find that you purchased the Bower-bird of me on the 20th of January, 1872; and, referring to my old letters, I see that it formed part of a collection made by Mr. Jardine at Port Albany, North Australia.

"Yours truly and respectfully,
"Edward W. Janson."

If it should be proved that the Cape-York district is its true locality, it will then be the representative of *C. guttata* of the north-west, and *C. maculata* of the south. The principal distinguishing characteristic, in addition to its somewhat larger size, consists in the extreme beauty of its occipital patch, which is nearly twice as large as in the species mentioned, and is even of a more brilliant lilac colour, particularly if the frill be turned up and seen from beneath.

I have drawn in the accompanying Plate a male, about the size of life. The decorative bower forming part of the illustration is taken from a photograph of some unknown species sent to me by the late Mr. Coxen, of Brisbane; it may, or may not, be that of the present bird.

AMBLYORNIS INORNATA.

AMBLYORNIS INORNATA.

Gardener Bower-bird.

Ptilorhynchus inornatus, Schleg. Tijdsch. Dierk. vol. iv. p. 51 (1871).
Amblyornis inornata, Elliot, Ibis, 1872, p. 114.—Id. Monogr. Paradis. pl. xxxviii. (1873).—Sclater, P. Z. S. 1873, p. 697.—Salvad. Ann. Mus. Civ. Genov. vii. p. 780 (1875).—Id. op, cit. ix. p. 193 (1876-77).—Beccari, tom. cit. p. 352, tav. viii.—Salvad. op. cit. x. p. 151 (1877).

It was not until Baron Von Rosenberg penetrated into the interior of North-western New Guinea that the present species became known to science. It was first described by Professor Schlegel, who received specimens direct from the Baron, under the name of *Ptilorhynchus inornatus*. Shortly afterwards Mr. Elliot, when preparing his Monograph of the Paradiseidæ, pointed out that, in addition to the similarity of the sexes, the plumes which hide the nostril in *Ptilorhynchus* were wanting in the new species; and he therefore separated it under the name of *Amblyornis*.

The Malays call the bird "Gardener," from the floral decorations he gives to his bower and the lawn before it.

The following abstract of the article published by Dr. Beccari appeared in the 'Gardeners' Chronicle' for March 16th, 1878, from which I transcribe it:—

"The *Amblyornis inornata*, or, as I propose to name it, the Bird gardener, is a Bird of Paradise of the dimensions of a Turtledove. The specific name '*inornata*' well suggests its very simple dress. It has none of the ornaments common to the members of its family, its feathers being of several shades of brown, and showing no sexual differences.

"It was shot some years ago by the hunters of Mynheer von Rosenberg. The first descriptions of its powers of building (the constructions were called 'nests') were given by the hunters of Mynheer Bruijn. They endeavoured to bring one of the nests to Ternate; but it was found impossible to do this, both by reason of its great size and the difficulty of transporting it.

"I have fortunately been able to examine these constructions at the remote places where they are erected. On June 20, 1875, I left Andai for Hatam, on Mount Arfak. I had been forced to stay a day at Warmendi to give rest to my porters. At this time only five men were with me; some were suffering from fever, and the remaining porters declined to proceed. We had been on our way since early morning; and at 1 o'clock we intended to proceed to the village of Hatam, the end of our journey.

"We were on a projecting spur of Mount Arfak. The virgin forest was very beautiful. Scarcely a ray of sunshine penetrated the branches. The ground was almost destitute of vegetation. A little track way proved that the inhabitants were at no great distance. A limpid fountain had evidently been frequented. I found here a new *Balanophora*, like a small orange or a small fungus. I was distracted by the songs and the screams of new birds; and every turn in the path showed me something new and surprising. I had just killed a small new marsupial (*Phascologale dorsalis*, Pet. and Doria) that balanced itself on the stem of a great tree like a squirrel; and turning round, I suddenly stood before the most remarkable specimen of the industry of an animal. It was a hut or bower close to a small meadow enamelled with flowers. The whole was on a diminutive scale. I immediately recognized the famous nests described by the hunters of Bruijn. I did not suspect, however, then that they had any thing to do with the constructions of the *Chlamydoderæ*. After well observing the whole, I gave strict orders to my hunters not to destroy the little building. That, however, was an unnecessary caution, since the Papuans take great care never to disturb these nests or bowers, even if they are in their way. The birds had evidently enjoyed the greatest quiet until we happened, unfortunately for them, to come near them. We had reached the height of about 4800 feet; and after half an hour's walk we were at our journey's end.

"*The Nest.*—I had now full employment in the preparation of my treasure; and I gave orders to my people not to shoot many of the birds. The nest I had seen first was the nearest one to my halting-place. One morning I took colours, brushes, pencils, and gun, and went to the spot. I there made the sketch which I now publish (fig. 56, p. 333). While I was there neither host nor hostess were at home. I could not wait for them. My hunters saw them entering and going out, when they watched their movements to shoot them. I could not ascertain whether this bower was occupied by one pair or by several pairs of birds, or whether the sexes were in equal or unequal numbers—whether the male alone was the builder, or whether the wife assisted in the construction. I believe, however, that such a nest lasts for several seasons.

"The *Amblyornis* selects a flat even place around the trunk of a small tree that is as thick and as high as a walking-stick of middle size. It begins by constructing at the base of the tree a kind of cone, chiefly of moss, of the size of a man's hand. The trunk of the tree becomes the central pillar; and the whole building is supported by it. On the top of the central pillar twigs are then methodically placed in a radiating manner, resting on the ground, leaving an aperture for the entrance. Thus is obtained a conical and very regular hut. When the work is complete many other branches are placed transversely in various ways, to make the whole quite firm and impermeable. A circular gallery is left between the walls and the central cone. The whole is nearly 3 feet in diameter. All the stems used by the *Amblyornis* are the thin stems of an orchid (*Dendrobium*), an epiphyte forming large tufts on the mossy branches of great trees, easily bent like straw, and generally about 20 inches long. The stalks had the leaves, which are small and straight, still fresh and living on them—which leads me to conclude that this plant was selected by the bird to prevent rotting and mould in the building, since it keeps alive for a long time, as is so often the case with epiphytical orchids.

"The refined sense of the bird is not satisfied with building a hut. It is wonderful to find that the bird has the same ideas as a man; that is to say, what pleases the one gratifies the other. The passion for flowers and gardens is a sign of good taste and refinement. I discovered, however, that the inhabitants of Arfak did not follow the example of the *Amblyornis*. Their houses are quite inaccessible from dirt.

"*The Garden.*—Now let me describe the garden of the *Amblyornis*. Before the cottage there is a meadow of moss. This is brought to the spot and kept free from grass, stones, or any thing which would offend the eye. On this green turf flowers and fruits of pretty colour are placed so as to form an elegant little garden. The greater part of the decoration is collected round the entrance to the nest; and it would appear that the husband offers there his daily gifts to his wife. The objects are very various, but always of vivid colour. There were some fruits of a Garcinia like a small-sized apple. Others were the fruits of Gardenias of a deep yellow colour in the interior. I saw also small rosy fruits, probably of a Scitamineous plant, and beautiful rosy flowers of a splendid new Vaccinium (*Agapetes amblyornithis*). There were also fungi and mottled insects placed on the turf. As soon as the objects are faded they are moved to the back of the hut.

"The good taste of the *Amblyornis* is not only proved by the nice home it builds. It is a clever bird, called by the inhabitants 'Buruk Gurea' (master bird), since it imitates the songs and screamings of numerous birds so well that it brought my hunters to despair, who were but too often misled by the bird. Another name of the bird is 'Tukan Robon,' which means a gardener."

The general colour of the present species is dark brown, rather more rufous on the head and upper back; the wings also rufous brown, the primaries darker brown; tail dark brown; entire under surface dark buff. The sexes, according to Professor Schlegel, are alike in plumage.

The figures in the Plate represent a pair of birds of the present species, the larger one being a little under life-size. They are drawn from specimens in my own collection. The representation of the garden and cabin is adapted from the woodcut given in the 'Gardeners' Chronicle,' which is, in turn, derived from Dr. Beccari's original illustration.

AMBLYORNIS SUBALARIS, Sharpe.

AMBLYORNIS SUBALARIS, Sharpe.

Orange-crested Bower-bird.

Amblyornis subalaris, Sharpe, Journ. Linn. Soc., Zool. xvii. p. 408 (1884).—Finsch u. Meyer, Zeitschr. ges. Orn. ii. p. 390, pl. xxii. (1885).

MR. GOLDIE first discovered this species in the Astrolabe Mountains, in South-eastern New Guinea, and his specimen remained with us for a long time in the British Museum before we ventured to describe it as new. It was, however, so evidently distinct from the *Amblyornis inornata* of North-western New Guinea that we described it at last, and events have proved that we were correct in judging from the female bird alone.

The male, which is recognized at a glance by its splendid orange crest, was first found by Mr. C. Hunstein in the Horseshoe range of the Owen Stanley Mountains, and a capital figure of it is given by Dr. von Madarász in the 'Zeitschrift' (*l. c.*). We were somewhat surprised to find that the male of this dull-looking species turned out to be such a fine bird, and it seems reasonable to doubt whether we really yet know the full-plumaged male of *A. inornata*. It is true that the nest and breeding-habits of the latter species have been described by Dr. Beccari, and the sexes are supposed to be alike in colour; but it is just possible that a crest is donned during the nesting-season by the male.

Adult male. General colour above uniform dark olive-brown, rather more olive on the back, rump, and upper tail-coverts; wing-coverts like the back, bastard-wing, primary-coverts, and quills olive-brown externally, internally dark brown; tail-feathers dark brown, washed with olive-brown externally; crown of head with an immense crest of orange, the lateral and frontal feathers edged and tipped with blackish brown; base of forehead dusky olive-brown; hind neck lighter olive-brown; lores ashy, sides of face, eyebrow, and ear-coverts dark olive-brown; cheeks and entire under surface of body light olive-brown, streaked down the centre of the feathers with ochreous buff, the sides of body and flanks rather browner; thighs dusky brown; under tail-coverts fulvous, with ochreous-buff centres to the feathers, the long ones edged with dark brown; under wing-coverts and axillaries orange-buff or tawny; quills below dusky, ochreous along the inner edge. Total length 8·3 inches, culmen 1·0, wing 5·0, tail 3·4, tarsus 1·3.

Adult female. Differs from the male in having no orange crest, the head being like the back. Total length 8·3 inches, culmen 0·9, wing 4·8, tail 3·3, tarsus 1·4.

Mr. Forbes has sent specimens of both sexes, killed in the rainy season. The whole of the colours are paler and more olive, and the ochreous tints of the under surface are much paler, especially on the under wing-coverts. The male is only distinguished from the female at this season of the year by the greater amount of clear ochreous on the underparts.

The figures in the Plate represent an adult male and female, drawn from a pair procured by Mr. Hunstein in the Horseshoe range.

[R. B. S.]

XANTHOMELUS AUREUS.

XANTHOMELUS AUREUS.

Golden Bird of Paradise.

Golden Bird of Paradise, Edwards, Nat. Hist. Birds, iii. p. 112 (1750).
Oriolus aureus, Linn. Syst. Nat. i. p. 163 (1766).—Gray, Gen. of B. i. p. 232 (1845).—Id. Hand-list B. i. p. 293 (1869).
Paradisea aurea, Lath. Ind. Orn. ii. p. 195 (1790).
Le Paradis orangé, Audeb. et Vieill. Ois. Dor. ii. p. 26, pls. 11, 12 (1802).
Le Loriot de Paradis, Levaill. Hist. Nat. Ois. Parad. i. pls. 18, 19 (1806).
Paradisea aurantia, Shaw, Gen. Zool. vii. p. 499, pl. 68 (1809).
Lophorhina aurantia, Stephens, Gen. Zool. xiv. p. 76 (1826).
Sericulus aurantiacus, Less. Traité, p. 339 (1831).—Id. Ois. Parad. Syn. p. 20 (1835).—Id. H. N. Ois. Parad. p. 201, pls. 25, 25 *bis*, 25 *ter* (1835).
Sericulus aureus, Bonap. Consp. Gen. Av. i. p. 349 (1850).—Gray, P. Z. S. 1861, p. 435.—Wall. Ibis, 1861, p. 287.—Id. P. Z. S. 1862, p. 160.—Id. Malay Arch. ii. p. 257 (1869).—Schl. Mus. P.-B. Coraces, p. 98 (1867).—Sclater, Ibis, 1876, p. 248.
Xanthomelus aureus, Bonap. Comptes Rendus, xxxviii. p. 538 (1854).—Elliot, Ibis, 1872, p. 112.—Id. Monogr. Parad. pl. 15 (1873).—Salvad. Ann. Mus. Civic. Genov. vii. p. 783 (1875).—Id. *op. cit.* ix. p. 192 (1876).—Sharpe, Cat. Birds, iii. p. 186 (1877).—Salvad. Ann. Mus. Civic. Genov. x. p. 152 (1877).
Sericulus xanthogaster, Schlegel, Nederl. Tijdsch. Dierk. v. p. 50 (1871).—Salvad. Ibis, 1876, p. 267.—Id. Atti R. Accad. Torino, xi. p. 688 (1876).
Chlamydodera xanthogastra, Elliot, Ibis, 1872, p. 113.—Id. Monogr. Parad. pl. xxxiii. (1873).—Sclater, P. Z. S. 1873, p. 697.

It will be seen that this remarkable Bird of Paradise was described and figured by Edwards more than one hundred and twenty years ago; and yet it is only within the last five years that we have been able to get perfect examples for our European collections. It may readily be imagined that the dried skins prepared by the natives afforded us but a very faint idea of its beauty; and therefore it gives me great pleasure to acknowledge my obligation to my friend Count Salvadori, who, during his visit to this country, was so kind as to bring with him some lovely specimens for my use in the present work; and I must record *en passant* my appreciation of the zeal shown by his countrymen in the scientific explorations recently made by Italian naturalists in New Guinea. It must be remembered that the present species was one of the few Birds of Paradise which Mr. Wallace was unable to obtain in a perfect state; and it had also baffled the endeavours of the Dutch naturalists to obtain the bird in the adult plumage.

The Golden Bird of Paradise is an interesting species in every way. Ornithologists who have studied Mr. Elliot's 'Monograph of the Paradisiidæ' will remember that he has included in that family the Bower-birds, and on one of his plates he figured a species as belonging to the latter group under the name of *Chlamydodera xanthogastra*. This bird had been previously described as a *Sericulus* by Prof. Schlegel, from specimens sent to the Leiden Museum from North-western New Guinea by Baron von Rosenberg; so that Mr. Elliot was but following Dr. Schlegel in assigning it a place among the Bower-birds; and there was nothing in the habitat of the species to render it unlikely that it would prove to be a *Chlamydodera*, as the latter genus is found all over many parts of Australia, and one species at least, *C. cerviniventris*, is very common in South-eastern New Guinea. It was therefore with considerable surprise that ornithologists must have regarded a communication from Count Salvadori to 'The Ibis,' stating that he had come to the conclusion that *Sericulus xanthogaster* of Schlegel was nothing more than the young of *Xanthomelus aureus*. Having looked carefully into the matter myself, I have not the least doubt that Count Salvadori is perfectly right in his conclusions.

In his third volume of the 'Catalogue of Birds' Mr. Bowdler Sharpe has not admitted the Bower-birds into the family Paradisiidæ, in that respect differing from Mr. Elliot; but he places *Xanthomelus* among the true Paradise-birds, leading from them to the Orioles. A further knowledge of the habits of the species is very desirable and can alone determine whether it should be placed among the Bower-birds.

In the interesting letter of Dr. Beccari, translated in 'The Ibis' for 1876 (p. 248), we find the only published account of the habits of the Golden Bird of Paradise. He writes:—"*Sericulus aureus* I killed on the same fig-tree, near Atam, where D'Albertis obtained the greater part of his birds. It has much the same habits as a Bird of Paradise, lives on fruits, especially on figs; one does not find more than two or three individuals together, usually only one male and one female; the younger males and females are very different; the iris is clear straw-colour. It is a very lively and shy bird; when the male is killed the female and another, perhaps a young male, return again to their food on the same tree, and then are seen no more. Although it is found at an elevation of 3000 feet or more, it seems more abundant in the hills near the sea, but is always most difficult to find, because in each of the localities which it frequents there are only a few pairs. Its song, according to my hunters, has much resemblance to the 'zigolio' of the *Nectariniæ*, but rather more strong and sonorous. Only the crest of feathers on the head is erectile. The Arfaks call it *Komeida*."

During his stay in Atam Signor D'Albertis obtained three specimens of this species, one adult and two young birds. He gives the colour of the soft parts as follows:—"Bill black; feet lead-colour; iris chestnut-brown." These indications are probably those of the young birds, as the iris in the adult is said by Beccari to be 'straw-colour.'"

The accompanying Plate represents two males of the natural size, with a reduced figure of a female.

ORIOLUS DECIPIENS.

ORIOLUS DECIPIENS.
Deceptive Oriole.

Mimeta decipiens, Sclater, Proc. Zool. Soc. 1883, p. 199.

THE present species illustrates an example of mimicry in the class of Birds which it would be difficult to surpass for interest. Although amongst insects cases of mimicry are not unfrequent, they are comparatively rare amongst birds, and the Moluccan Orioles are known as perhaps the best illustration of this peculiar phenomenon in ornithology. In the island of Bourou there is a well-known Honey-eater (*Philemon moluccensis*) the plumage of which is almost exactly reproduced by the Oriole of the island (*Oriolus bourouensis*); and now in the Tenimber Islands we find the same thing taking place, the Honey-eater (*Philemon plumigenis*) being mimicked by the Oriole (*Oriolus decipiens*). By many ornithologists these Orioles are separated under the generic heading of *Mimeta*; but structurally they cannot be separated from the genus *Oriolus*, from which they differ only in their dull-coloured plumage.

The Deceptive Oriole is very closely allied to the species from Bourou, *O. bourouensis*; but the latter has the throat pale brown without any dusky spots; the rufous edgings to the wing-coverts are also more pronounced.

The following description is taken from the type specimen :—

Adult female. General colour above brown, lighter on the head and mantle, darker and more ashy brown on the lower back, rump, and upper tail-coverts; the head distinctly streaked with dark brown; the sides of the neck and hinder neck mottled with ashy grey, the feathers having narrow ashy grey edges; wing-coverts and quills deep brown, more ashy on the primaries, many of the coverts and quills slightly washed with pale rufous externally; tail-feathers light brown, with paler tips; an indistinct eyebrow of white, spotted with dark brown; ear-coverts uniform dark brown; cheeks whitish, spotted with dark brown, rather more marked on the malar line; centre of throat white, with a few dusky spots; remainder of under surface from the fore neck downwards ruddy ashy brown, a little darker and more ashy on the thighs and under tail-coverts; the chest slightly mottled with dark brown shaft-streaks; under wing-coverts, axillaries, and quill-lining rufous; " bill, legs, and feet black; iris dark brown " (*W. O. Forbes*). Total length 12 inches, culmen 1·45, wing 6·1, tail 5·0, tarsus 1·25.

The Plate represents this species of the full size, the figures being drawn from the typical examples collected by Mr. Forbes.

[R. B. S.]

CHÆTORHYNCHUS PAPUENSIS, Meyer.

CHÆTORHYNCHUS PAPUENSIS, Meyer.
Arfak Drongo.

Chætorhynchus papuensis, Meyer, Sitz. Akad. Wiss. Wien, lxix. p. 493 (1874).—Sharpe, Cat. B. Brit. Mus. iii. p. 242, pl. xiii. (1877).

This little species of King Crow or Drongo Shrike was discovered by Dr. Meyer during his visit to New Guinea, and, although not a showy bird, is one of great interest to the naturalist, who is ever on the watch for forms which evince unexpected affinities to others of a distant habitat. In the present instance we find, in the Arfak Mountains of North-western New Guinea, a genus of Drongo which, in size, general form, and especially in the scaly nature of its plumage, reminds us forcibly of the small genus *Chaptia* of India and Malayana. In one respect, however, *Chætorhynchus* differs not only from *Chaptia*, but from all other genera of the *Dicruridæ*; and that is, in the shoulder-spot of white which adorns the New-Guinea bird. The general coloration of the Drongos is black, relieved by a certain gloss or metallic spangles. An intermixture of white is of rare occurrence in the family; it occurs, however, in some of the Indian species of *Buchanga* (which have a little white on the lower parts), and is most conspicuous in *Dicrurus mirabilis* of the Philippines (which has a pure white breast and abdomen).

The Arfak Drongo has a white spot on the shoulders; but this is by no means a conspicuous feature, as may be gathered from the fact that when Dr. Meyer described the species he did not notice it, nor did Mr. Bowdler Sharpe, in his 'Catalogue of Birds,' nor did Mr. Keulemans when he drew the figure for the latter work. Indeed it was only when Dr. Meyer came to have one of the typical specimens mounted for the Dresden Museum that the shoulder-spot became suddenly visible; and its first discovery is due to him. I have therefore much pleasure in giving the first accurate figure of *Chætorhynchus*. I find that the white spot varies much in size, and is almost absent in some specimens. I have examined the typical examples obtained by Dr. Meyer, as well as others procured by M. Laglaize, one of which is in the collection of the British Museum.

The following description is given by Mr. Sharpe in the work above alluded to, and is taken from one of the original types:—

General colour black, with a steely gloss, the head more brightly glossed with steel-green, the feathers rounded and somewhat scale-like in appearance; least wing-coverts steel-black, the median, greater, and primary-coverts black, narrowly edged with steel-green; quills black, the secondaries margined with steel-green, the innermost glossed with the latter colour; tail black, slightly glossed with steel-green on the edges of the feathers; sides of face and under surface of body black, the breast glossed with steel reflections; under wing-coverts greyish black; bill and feet black. Total length 7 inches, culmen 0·7, wing 4·2, tail 3·5, tarsus 0·7.

The figures in the Plate represent the species of the natural size; they are drawn from the type specimens, lent to me by Dr. Meyer.

DICRANOSTREPTUS MEGARHYNCHUS.

DICRANOSTREPTUS MEGARHYNCHUS.

New-Ireland Drongo.

Edolius megarhynchus, Quoy et Gaimard, Voyage de l'Astrolabe, Zool. vol. i. p. 184. pl. 6.
Edolius intermedius, Lesson, Traité d'Orn. p. 380.
Dicrurus megarhynchus, Gray, Gen. B. i. p. 286.—Bonap. Consp. Gen. Av. i. p. 352.—Gray, Cat. Birds of New Guinea &c., p. 33.—Id. P. Z. S. 1861, p. 435.—Finsch, Neu-Guinea, p. 171.—Gray, Handl. B. i. p. 287.—Sclater, P. Z. S. 1869, p. 119.
Dicranostreptus mezorhynchus, Jerdon, Birds of India i. p. 430, note.
Dicranostreptus megarhynchus, Reichenb. Syst. Taf. lxxxviii. fig. 12.—Sclater, P. Z. S. 1877.—Sharpe, Catalogue of Birds, iii. p. 256.
Dissemurus megarhynchus, Tweeddale, Ibis, 1878, p. 79.

This large species of Drongo was described in 1830, by MM. Quoy and Gaimard, in their account of the Zoology of the Voyage of the 'Astrolabe.' Unfortunately, the habitat of the species was given as Dorey, New Guinea, a mistake which led to its being included in more than one list of New-Guinea birds. It has also been recorded from the Ké Islands by Mr. Gray; but this is doubtless owing to a misprint (as the Marquis of Tweeddale suggests), the locality being intended to apply to *Dicrurus megalornis*, which is confined to the last-named group of islands, and the locality of which is omitted in the proper column of places tabulated by Mr. Gray. In 1869, Dr. Sclater recorded it from the Solomon Islands; and more recently it has occurred in the collections made by Mr. George Brown in New Ireland. Specimens are also contained in the British Museum, which were presented by Captain Lambrick, R.N., who obtained them at Carteret Harbour, New Ireland; and the species will probably be found to be an inhabitant of that island, and of the Solomon group. Lord Tweeddale considers that it is a member of the genus *Dissemurus*, to which the Indian and Indo-Malayan Racket-tailed Drongos belong; and if this is the case, its appearance in a truly Papuan habitat is interesting and at the same time remarkable. The extraordinary length of the tail and the absence of a true racket, which is one of the characters of *Dissemurus*, taken in connexion with the locality inhabited by the species, induce me to consider it generically distinct, and I therefore follow Dr. Sclater and Mr. Sharpe in its nomenclature.

The following description is taken from the 'Catalogue of Birds' of the latter author:—

Above blue-black, with a very slight steel gloss on the back, rump, and upper tail-coverts; head and sides of neck metallic steel-purple, the feathers on the latter elongated and somewhat lanceolate; frontal plumes, lores, and sides of face dull purplish black; under surface of body dusky purplish black, with faintly indicated tips of glossy green on some of the feathers of the throat and chest; under wing-coverts purplish black like the breast; wings above dull glossy steel-green; quills purplish black, the secondaries externally dull steel-green, not so metallic as the coverts; tail-feathers purplish black, glossed with metallic steel-green on the outer edge of both webs, more distinctly and broadly on the centre ones, the outermost feather elongated and twisted in an inward curl at the tip; bill and feet black. Total length 20·5 inches, culmen 1·5, wing 7, tail 6, tarsus 1·05.

The specimen figured (of the natural size) in the Plate was sent by Mr. George Brown from New Ireland, and was kindly lent to me by Dr. Sclater.

RECTES LEUCORHYNCHUS, Gray.

RECTES LEUCORHYNCHUS, *Gray*.

White-billed Wood-Shrike.

Rectes leucorhynchus, Gray, Proc. Zool. Soc. 1861, p. 430.—Finsch, Neu-Guinea, p. 170 (1865).—Meyer, Sitz.
 k. Akad. Wissensch. Wien, lxix. p. 206 (1874).—Rosenb. Malay. Arch. p. 395 (1879).
Colluricincla leucorhyncha, Gray, Hand-list B. i. p. 386, no. 5838 (1869).
Pseudorectes leucorhynchus, Sharpe, Cat. Birds in Brit. Mus. iii. p. 288 (1877).
Rhectes leucorhynchus, Salvad. Ann. Mus. Civic. Genov. xv. p. 43 (1879).—Id. Orn. Papuasia e delle Molucche,
 ii. p. 206 (1881).

This Wood-Shrike is easily distinguished from all the other Papuan members of the group by its white bill and sombre coloration. It was originally described by the late Mr. George Gray, from Gagie Island, in the Moluccas; but the original specimens, now in the British Museum, were obtained by Mr. Wallace in Waigiou, and we do not understand Mr. Gray's authority for the occurrence of the species in Gagie. It has since been obtained by Dr. Beccari both in Waigiou and in Batanta. Nothing has yet been recorded concerning its habits.

The following description is copied from the British Museum 'Catalogue of Birds':—

"*Adult male.* Above rufous brown, the rump and the tail a little brighter, the crown and sides of the head slightly darker; wing-coverts like the back, but a little darker brown; quills dark brown, externally dark rufous brown; under surface of body deep rufous fawn-colour, the throat paler; under wing- and tail-coverts uniform with the breast; bill yellowish white. Total length 12 inches, culmen 1·15, wing 5·7, tail 5·2, tarsus 1·5."

The Plate represents two specimens of this bird, of the natural size, and is drawn from individuals in the British Museum.

[R. B. S.]

RECTES CERVINIVENTRIS, Gray.

RECTES CERVINIVENTRIS, Gray.

Fawn-breasted Wood-Shrike.

Rectes cerviniventris, Gray, Proc. Zool. Soc. 1861, p. 430.—Finsch, Neu-Guinea, p. 176 (1865).—Meyer, Sitz. k. Akad. Wissensch. Wien, lxix. p. 208 (1874).—Sharpe, Cat. Birds in Brit. Mus. iii. p. 286 (1877).—Rosenb. Malay. Arch. p. 395 (1879).
Rhectes cerviniventris, Salvad. Ann. Mus. Civic. Genov. xv. p. 44 (1879).—Id. Orn. Papuasia e delle Molucche, ii. p. 200 (1881).

As with *R. leucorhynchus*, so with the present bird, we cannot find Mr. Gray's authority for stating that it is found in the island of Gagie, as no specimen from this island exists in the British Museum, where the type specimens are deposited. The original specimen came from Waigiou, where Mr. Wallace discovered the species, and where it has since been met with by Mr. Bruijn, who has also found it in Batanta. The late Dr. Bernstein likewise procured it in the island of Ghemien. The habitat Gagie must therefore be entirely erroneous, though how the mistake arose we are unable to conjecture.

The following description is copied from the British Museum 'Catalogue of Birds':—

"*Adult male.* General colour above ashy olive, inclining to olive on the rump and upper tail-coverts; the head crested, ashy grey, this colour pervading the mantle and back; sides of face and sides of neck rather darker ashy; wings olive-brown, the inner webs of the feathers darker brown, the outer webs washed with reddish olive; tail-feathers brown, washed with olive on the edges of the feathers, the outer ones reddish brown, edged with deep fawn-colour on the outer web; throat and fore neck ashy, washed with ochre; rest of under surface of body chestnut fawn-colour, including the thighs, under wing- and tail-coverts; under surface of body dark ashy brown, the inner webs rufous fawn-colour. Total length 8·6 inches, culmen 1·05, wing 3·75, tail 3·5, tarsus 1·25."

The Plate represents an adult bird in two positions, of the size of life, the figures being drawn from a specimen in the British Museum.

[R. B. S.]

RECTES UROPYGIALIS, Gray.

RECTES UROPYGIALIS, Gray.

Rufous-and-Black Wood-Shrike.

Rectes uropygialis, G. R. Gray, Proc. Zool. Soc. 1861, pp. 430, 435.—Finsch, New Guinea, p. 170 (1865).—
    Meyer, Sitz. k. Akad. Wissensch. Wien, lxix. p. 208 (1874).—Sharpe, Cat. Birds Brit. Mus. iii.
    p. 285 (1877).—Rosenb. Malay Arch. p. 395 (1879).
Colluricincla uropygialis, Gray, Handl. B. i. p. 385, no. 5, 836 (1869).
Rectes uropygialis ceramensis, Meyer, Sitz. k. Akad. Wissensch. Wien, lxix. p. 208 (1874).—Salvad. Proc. Zool.
    Soc. 1878, p. 96 (note).
Rectes tibialis, Sharpe, Cat. Birds Brit. Mus. iii. p. 285 (1877).
Rhectes uropygialis, Salvad. Ann. Mus. Civ. Genova, xv. p. 42 (1879).—Id. Orn. della Papuasia, &c. p. 193
    (1881).

THERE are three species of the genus *Rectes* which are very closely allied to each other, and which are remarkable for their red and black plumage. These are *R. uropygialis*, from North-western New Guinea and Mysol, *R. dichrous*, spread over the greater part of New Guinea, and *R. aruensis* from the Aru Islands. Of these three species the present is perhaps the finest, being slightly the largest in size. It differs from *R. dichrous* in having the whole of the rump and upper tail-coverts black, whereas in the last-named bird these parts are chestnut like the rest of the back. *R. aruensis* resembles the present bird in having the rump and upper tail-coverts black; but, besides being rather smaller, it has the breast slightly paler and washed with black.

Nothing has been recorded concerning the habits of this species, or of those of any member of the genus *Rectes*; but we can easily suppose that they are not greatly different from those of the Australian *Colluriocincla*. Whereas *R. dichrous* appears to be spread over the greater part of the island of New Guinea, *R. uropygialis* takes its place in the western and north-western part of the island, and extends into Mysol. Mr. Sharpe was inclined at one time to separate the New-Guinea bird as *Rectes tibialis*; but he informs me that he is now quite convinced of its identity with the Mysol bird.

The following description is taken from Mr. Sharpe's Catalogue of Birds in the British Museum:—

Adult male. Head, which is strongly crested, and nape black, as also the entire sides of the face, sides of neck, throat, fore neck, and chest; hind neck, mantle, and upper back rich maroon-chestnut, rather more orange on the hind neck; lower back, rump, and upper tail-coverts black; wings and tail black; remainder of under surface of body, from the chest downwards, including the thighs, under wing- and tail-coverts, rich maroon-chestnut; bill and legs black.

Total length 11 inches, culmen 1·1, wing 4·7, tail 4·55, tarsus 1·36.

Adult female. Similar to the adult male, and quite as richly coloured. Total length 11 inches, culmen 1·25, wing 4·95, tail 4·75, tarsus 1·35.

The figures in the Plate represent an adult bird in two positions, of about the natural size, and are drawn from an example in my own collection.

RECTES JOBIENSIS, *Meyer.*

RECTES JOBIENSIS, Meyer.

Jobi-Island Wood-Shrike.

Rectes jobiensis, Meyer, Sitz. k. Akad. der Wiss. zu Wien, lxix. p. 205 (1874).—Sclater, Ibis 1874, p. 417.—
Salvad. Ann. Mus. Civ. Genova, vii. p. 173 (1875), viii. p. 40 (1876).—Sharpe, Cat. B. Brit. Mus.
iii. p. 287 (1877).—Gieb. Thes. Orn. iii. p. 412 (1877).
Rhectes jobiensis, Salvad. Ann. Mus. Civ. Genova, xv. p. 43 no. 12 (1879).—Id. Orn. Papuasia, &c. ii. p. 201
(1881).

THE Shrike-like genus *Rectes* consists of a small number of birds found in New Guinea and in the neighbouring islands. It is divided into two sections, consisting of those species which have a distinct cap of darker colour on the head, and those species which have no cap whatever; and the present bird will be seen at once to belong to the second section. Its nearest allies therefore are *Rectes cervineiventris* from Waigiou, and *R. cristatus* of New Guinea; but it is very distinct from both of these by reason of its foxy red colour both above and below.

The present bird was discovered by Dr. A. B. Meyer in the island of Jobi, which is situated in the Bay of Geelvink, in North-western New Guinea. He preserved several specimens of it; and it has also been met with there by Dr. Beccari and in the neighbouring little islet of Krudu.

Dr. Meyer says that the note of this bird is a "*chrrr*;" and its habits are doubtless the same as those of the rest of the genus, of which at present we unfortunately know but little.

The following description of the species I take from Mr. Bowdler Sharpe's Catalogue of Birds:—

Adult male. General colour above bright foxy red, paler on the head; wing-coverts like the back; quills dark brown, externally rufous, the inner secondaries entirely rufous; tail-feathers chestnut, the shafts rufous; lores, sides of face, and under surface of body deep rich tawny buff, darker on the throat and fore neck, and inclining to chestnut on the lower flanks and under tail-coverts; under wing-coverts bright tawny, like the breast; quills dusky brown below, rufous on the outer web and along the inner web; bill light yellowish horn-colour; feet blackish, claws pale. Total length 9·2 inches, culmen 11·5, wing 4·65, tail 4·35, tarsus 1·35.

Adult female. Similar in colouring to the male.
Total length 10·5 inches, wing 4·6, tail 4·3, tarsus 1·25.

Young male. Similar in colour to the adult, but rather duller; the bill shorter and duller brown.

The above descriptions were taken from the typical specimens lent to Mr. Sharpe by Dr. Meyer; and from one of these I have drawn the accompanying figure, which is of about the natural size.

RECTES ARUENSIS, Sharpe.

RECTES ARUENSIS, Sharpe.

Aru-Island Wood-Shrike.

Rectes dichrous, pt. (nec Bp.), Gray, Proc. Zool. Soc. 1858, pp. 173, 193.—Id. List Mamm. &c. New Guinea, pp. 33, 58 (1859).—Id. Proc. Zool. Soc. 1861, p. 435.
Rectes aruensis, Sharpe, Cat. Birds in Brit. Mus. iii. p. 285 (1877).—Salvad. Proc. Zool. Soc. 1878, p. 96.
Rhectes aruensis, Salvad. Ann. Mus. Civic. Genov. xv. p. 42 (1879).—Id. Rep. Voy. 'Challenger,' p. 79 (1881).
—Id. Ucc. Papuasia e delle Molucche, ii. p. 194 (1881).—Meyer, Zeitschr. ges. Orn. i. p. 284 (1884).
Rhectes analogus, Meyer, Zeitschr. ges. Orn. i. p. 284 (1884).

THE present species belongs to the capped section of the genus *Rectes*, and its nearest ally is *Rectes uropygialis* from New Guinea and Mysol. It is, however, a much smaller bird than the latter, which it represents in the Aru Islands, and is further distinguished by the black of the upper tail-coverts not extending over the rump, while on the under surface of the body the black descends down the centre of the breast and abdomen instead of ending abruptly on the breast.

The birds which we have described as the young have recently been considered by Dr. A. B. Meyer to belong to a distinct species, which he has named *Rectes analogus*. With all due respect to Dr. Meyer, who is a recognized authority on all matters relating to the zoology of Papuasia, we still think that there is only one species inhabiting the Aru Islands, and we consider *Rectes analogus* to be the same species as *R. aruensis*. The whole subject, however, of the colour of the sexes and young birds in this genus is so imperfectly understood, in our opinion, that we are free to admit that Dr. Meyer's second species from the Aru Islands possesses quite as good claims to specific distinction as some of those allowed by naturalists to inhabit New Guinea.

The following are the descriptions of specimens in the British Museum, published by us in the 'Catalogue of Birds' (*l. c.*):—

Adult male. Above very bright chestnut, the upper tail-coverts black; head crested, black all round, with the sides of face, fore neck, and chest also entirely black; rest of under surface of body deep tawny buff, the whole of the breast shaded with black; under wing-coverts deep ochreous, those near the edge of the wings black; wings and tail deep black; bill and legs black. Total length 10 inches, culmen 1·2, wing 4·3, tail 4·1, tarsus 1·35.

Young. Paler ochre-rufous below, not mixed with black on the breast; the cap and black throat almost as strongly defined as in the adult, but both are much washed with brown; rump dingy ochreous brown, washed with chestnut, the tail-coverts inclining to blackish; quills and tail blackish, the outer webs of the quills dingy ochreous brown; tail-feathers brown, slightly washed with rufous, and inclining to black near the base and upwards along the centre of the feathers. Total length 10 inches, culmen 1·2, wing 4·45, tail 4·3, tarsus 1·3.

The figures in the Plate represent the type of *Rectes aruensis* and the type of *R. analogus*, the latter having been kindly lent to us by Dr. Meyer.

[R. B. S.]

THE
BIRDS OF NEW GUINEA

AND THE

ADJACENT PAPUAN ISLANDS,

INCLUDING MANY

NEW SPECIES RECENTLY DISCOVERED

IN

AUSTRALIA.

BY

JOHN GOULD, F.R.S.

COMPLETED AFTER THE AUTHOR'S DEATH

BY

R. BOWDLER SHARPE, F.L.S. &c.,

ZOOLOGICAL DEPARTMENT, BRITISH MUSEUM.

VOLUME II.

LONDON:
HENRY SOTHERAN & CO., 36 PICCADILLY.
1875–1888.

[All rights reserved.]

PRINTED BY TAYLOR AND FRANCIS,
RED LION COURT, FLEET STREET.

CONTENTS.

VOLUME II.

Plate			Part	Date
1.	Artamides unimodus	Slaty-grey Cuckoo-Shrike	XVI.	1884.
2.	„ temmincki	Blue Cuckoo-Shrike	XI.	1880.
3.	Campephaga strenua	Blue-grey Campephaga	II.	1876.
4.	Graucalus pusillus	Ramsay's Cuckoo-Shrike	XVII.	1884.
5.	„ axillaris	Bruijn's Cuckoo-Shrike	XIV.	1883.
6.	„ maforensis	Mafoor-Island Cuckoo-Shrike	XIII.	1882.
7.	Edoliisoma montanum	Mount Arfak Cuckoo-Shrike	XIII.	1882.
8.	„ poliopsis	Grey-faced Cuckoo-Shrike	XXII.	1886.
9.	Lalage mœsta	Black-browed Caterpillar-catcher	XV.	1883.
10.	Micrœca assimilis	Western Micrœca	XI.	1880.
11.	Gerygone dorsalis	Rufous-backed Gerygone	XVI.	1884.
12.	Pseudogerygone notata	White-spotted Flycatcher	XXI.	1886.
13.	„ chrysogastra	Yellow-bellied Flycatcher	XXII.	1886.
14.	„ cinereiceps	Grey-headed Flycatcher	XXII.	1886.
15.	Heteromyias cinereifrons	Ashy-fronted Flycatcher	X.	1879.
16.	Monachella muelleriana	Chat-like Flycatcher	XIII.	1882.
17.	Pœcilodryas bimaculata	Black-and-White Flycatcher	XVI.	1884.
18.	„ albifacies	Southern White-faced Flycatcher	XIII.	1882.
19.	„ placens	Yellow-banded Robin	X.	1879.
20.	Hypothymis rowleyi	Rowley's Blue Flycatcher	XIII.	1882.
21.	Todopsis cyanocephala	New-Guinea Todopsis	VIII.	1878.
22.	„ bonapartii	Bonaparte's Todopsis	VIII.	1878.
23.	„ wallacii	Wallace's Todopsis	VIII.	1878.
24.	„ grayi	Gray's Todopsis	VIII.	1878.
25.	Malurus alboscapulatus	Pied Malurus	IV.	1877.
26.	Rhipidura rubrofrontata	Rufous-fronted Fantail Flycatcher	XXI.	1886.
27.	„ leucothorax	White-breasted Fantail Flycatcher	XVIII.	1884.
28.	„ cockerelli	Cockerell's Fantail Flycatcher	XVII.	1884.
29.	„ opistherythra	Larat Fantail Flycatcher	XVI.	1884.
30.	„ hamadryas	Rufous-backed Fantail Flycatcher	XV.	1883.
31.	„ fuscorufa	Timor-Laut Fantail Flycatcher	XV.	1883.
32.	„ dryas	Wood-Fantail	II.	1876.
33.	„ hyperythra	Rufous-breasted Fantail Flycatcher	XXII.	1886.
34.	Myiagra fulviventris	Buff-bellied Flycatcher	XIX.	1885.
35.	„ cervinicauda	Fawn-tailed Flycatcher	XVIII.	1884.
36.	„ ferrocyanea	Purple-backed Flycatcher	XVII.	1884.
37.	Machærirhynchus albifrons	White-fronted Flycatcher	IV.	1877.
38.	„ nigripectus	Black-breasted Flycatcher	IV.	1877.
39.	Heteranax mundus	Forbes's Pied Flycatcher	XVI.	1884.
40.	Arses telescophthalmus	Frilled-necked Flycatcher	X.	1879.
41.	„ batantæ	Large Frilled-necked Flycatcher	X.	1879.
42.	„ aruensis	Little Frilled-necked Flycatcher	IX.	1879.
43.	„ insularis	Orange-collared Flycatcher	IX.	1879.
44.	Piezorhynchus brodiei	Brodie's Flycatcher	XVIII.	1884.
45.	„ browni	Brown's Flycatcher	XVIII.	1884.
46.	„ richardsi	Richards's Flycatcher	XVII.	1884.
47.	„ castus	White-crowned Flycatcher	XVI.	1884.
48.	„ vidua	White-backed Pied Flycatcher	XVI.	1884.
49.	„ squamulatus	Scaly-necked Pied Flycatcher	XVI.	1884.
50.	„ medius	Coppinger's Flycatcher	XXV.	1888.
51.	„ axillaris	White-tufted Flycatcher	XXV.	1888.
52.	Monarcha periophthalmicus	Black-spectacled Flycatcher	XIV.	1883.
53.	„ kordensis	Mysore Yellow Flycatcher	V.	1877.
54.	„ melanonota	Papuan Yellow Flycatcher	V.	1877.
55.	Peltops blainvillii	Broad-billed Flycatcher	I.	1875.
56.	Pomarea rufocastanea	Rufous-and-chestnut Flycatcher	XVIII.	1884.
57.	„ castaneiventris	Chestnut-bellied Flycatcher	XVIII.	1884.
58.	„ ugiensis	Ugi-Island Flycatcher	XVII.	1884.

ARTAMIDES UNIMODUS.

ARTAMIDES UNIMODUS.
Slaty-grey Cuckoo-Shrike.

Graucalus unimodus, Sclater, Proc. Zool. Soc. 1883, pp. 55, 198.

This fine Cuckoo-Shrike is very nearly allied to *Artamides cæruleigriseus* of the Aru Islands, like which species it has buff-coloured axillaries and under wing-coverts. Excepting in this respect, however, the two birds are quite different; for *A. cæruleigriseus* is blue-grey instead of slaty grey, and the male has only the lores, feathers in front of the eye, the base of the chin, and the base of the cheeks black, showing none of the black on the throat and fore neck which distinguishes *A. unimodus*.

The females of the two species are also distinct; for besides the blue-grey colour of *A. cæruleigriseus*, the latter has the lores grey, and no black on the chin or cheeks, whereas in *A. unimodus* the lores and the base of the chin and base of the cheeks are black.

The female bird was shot by Mr. H. O. Forbes in the island of Larat in the Tenimber group on the 4th of August 1882, and the male was obtained in Loetoer, on the mainland of Timor Laut, in September of the same year.

Mr. Forbes informs us that he found this species frequenting the mangroves along the shores on both occasions when he procured it.

The following descriptions are taken from the type specimens kindly lent to us by Dr. Sclater:—

Adult male (type of species). General colour above uniform slaty grey; lesser and median wing-coverts like the back; greater coverts, bastard wing, primary-coverts, and quills blackish, washed externally with ashy grey and edged with slaty grey like the back, lighter on the margins of the primaries; upper tail-coverts like the back, but crossed with dusky bars under certain lights and with more or less of a subterminal shade of black; tail-feathers black, barred across with dusky under certain lights, the feathers edged with ashy grey round the tips; forehead slightly shaded with darker slate-colour than the crown; nasal plumes, lores, feathers round the eye, ear-coverts, sides of hinder crown, and entire throat and centre of fore neck greenish black, the lower part of this latter washed with slaty grey; sides of neck and rest of under surface slaty grey, darker on the under tail-coverts; thighs black; under wing-coverts pale fawn-colour or buff with slaty-grey centres to the feathers, all the lower greater coverts and those near the edge of the wings blackish slate-colour; quills dusky below, ashy grey on their inner face; "bill, legs, and feet black; iris dark brown" (*H. O. Forbes*). Total length 13·8 inches, culmen 1·25, wing 7·65, tail 6·5, tarsus 1·15.

Adult female. Differs from the male in wanting the black on the throat and fore neck; the lores and feathers in front of the eye, a spot at the base of the cheeks, and another smaller mark at the base of the chin black; axillaries and under wing-coverts more wholly ochreous buff, less distinctly washed with slaty grey in the centres; thighs grey like the rest of the under surface; "iris black" (*H. O. Forbes*). Total length 13·6 inches, culmen 1·3, wing 6·2, tail 5·9, tarsus 1·15.

The Plate represents the adult male and female described above, of about the natural size. They are now in the British Museum.

[R. B. S.]

ARTAMIDES TEMMINCKI.

ARTAMIDES TEMMINCKI.
Blue Cuckoo-Shrike.

Ceblepyris temmincki, S. Müller, Ver. Natuurl. Gesch. Land- en Volkenk. p. 191.
Campephaga temmincki, Gray, Gen. of Birds, i. p. 283.—Finsch, Neu-Guinea, p. 172.—Gray, Hand-list of Birds, i. p. 337, no. 5081.
Graucalus temmincki, Bonap. Consp. Gen. Av. i. p. 354.—Hartl. Journ. für Ornith. 1864, p. 446.—Walden, Trans. Zool. Soc. viii. pp. 68, 113, pl. xii.—Meyer, Ibis, 1879, p. 129.
Artamides temmincki, Sharpe, Brit. Mus. Cat. Birds, iv. p. 15.—Id. Abhandl. Mus. Dresden, Abth. iii. p. 363.

I HAVE already figured several *Graucali* or Cuckoo-Shrikes in my work on the Birds of Australia, and have hitherto been content to keep them in the genus *Graucalus*; but Mr. Sharpe, who has recently classified the family of the *Campephagidæ*, considers that there are really four genera in which the Australian Cuckoo-Shrikes ought to be placed, and he arranges the species figured in my work as follows. *Graucalus tenuirostris* should be placed in the genus *Edoliisoma*; *Campephaga karu*, *C. leucomela*, and *C. numeralis* in *Lalage*; so that only *Graucalus melanops*, *G. parvirostris*, *G. mentalis*, *G. hypoleucus*, and *G. swainsoni* remain in the genus *Graucalus*; *Pteropodocys phasianella* he admits to be generically distinct. As Count Salvadori, who has also studied these birds, agrees with Mr. Sharpe in many of his conclusions, I have deemed it best in the present work to adopt the arrangement of the last-named author. *Graucalus*, as a genus, appears to be widely distributed, as it occurs not only in Africa and Madagascar, but extends all over India and Ceylon, through the Burmese countries to Formosa, and down the Malayan peninsula, throughout the Moluccas, to Australia and Tasmania; and one species has even straggled to New Zealand. The members of the genus *Artamides*, on the other hand, are not Australian, though they occur in New Caledonia and thence extend through the New-Hebrides group, New Guinea, and the Moluccas to the Indo-Malayan islands, the Malayan peninsula, and the Andamans: they are remarkable for very much stronger and stouter bills than the remainder of the Cuckoo-Shrikes. Both the genera *Graucalus* and *Artamides* contain representatives which are barred below, and others which are more or less uniform in coloration.

The present species belongs to the uniform section of the genus *Artamides*, and is one of the most beautifully coloured not only of the genus to which it belongs, but also of the whole family of Cuckoo-Shrikes, which are, as a rule, remarkably plain-plumaged birds. Its home is the Island of Celebes; and here it appears only to inhabit the mountains. It was discovered by Forsten in the neighbourhood of Gorontalo; and Dr. Meyer procured four specimens near Kakas, in the mountains of the Minahassa (about 2000 feet high). He never saw it elsewhere; so that at present it is only known to inhabit the north-eastern promontory of Celebes.

The following description of the species is given by Mr. Sharpe in his 'Catalogue :'—

"*Adult.* General colour greyish azure, with more or less of a cobalt hue, especially on the wings and tail, the inner webs of the quills and tail-feathers being blackish; base of forehead, lores, and feathers in front of the eye blackish with a blue gloss; under surface of body azure-blue like the upper; under wing-coverts like the breast, the lower series and the underside of the quills ashy. Total length 12 inches, culmen 1, wing 5·95, tail 5·9, tarsus 0·95." A second specimen in the British Museum collected by Dr. Meyer had the tips to the bastard wing-feathers and the inner secondaries white, and is probably either a female or a younger bird.

My figure represents the species of the full size, and is drawn from a specimen in my own collection.

CAMPEPHAGA STRENUA, *Schl.*

Blue-grey Campephaga.

Campephaga strenua, Schl. N. T. D. iv. p. 45.—Sclater, P. Z. S. 1873, p. 697.—Meyer, Sitz. Wien, lxix. p. 10.

This remarkable species was discovered by Baron von Rosenberg during his last voyage. Two specimens were obtained—one in the island of Jobi, and the other in the interior of the northern peninsula of New Guinea. D'Albertis procured it in Atam; and Dr. Meyer shot altogether four specimens—three at Rubi, a place situated at the southern extremity of Geelvink Bay, and one on the Arfak mountains. These are apparently all the specimens at present in Europe; I am not aware of any other individuals existing in any collection.

Professor Schlegel makes the following observations on the species:—"The new species which we introduce under the name of *Campephaga strenua* is remarkable for its large size and excessively robust bill; it is, moreover, very easily recognized by its entirely blue-grey colour, broken only by the black lores, nasal plumes, fore part of chin, primaries, and tail-feathers, as well as by the clear isabelline rufous which pervades the axillaries and lower wing-coverts, with the exception of the greater series, which are white. It is to be remarked, moreover, that the median wing-coverts have the inner web black, whilst the black of the basal part of the primaries passes partially into whitish. Bill and feet black; iris blackish brown. Total length 6″ 6‴, tail 5″ 10‴, bill from front 15‴, breadth of bill at forehead 8‴, tarsus 13‴, middle toe 10‴."

Dr. Meyer obtained the female, which, he says, differs from the male in the absence of black on the wings and chin, these parts being bluish grey in the females like the rest of the body. A hen bird which he obtained in the Arfak mountains had the head and under surface of body brighter blue than the other examples obtained at Rubi.

My Plate represents the bird of the natural size, and was drawn from a specimen lent me by Signor d'Albertis.

GRAUCALUS PUSILLUS, Ramsay.

GRAUCALUS PUSILLUS, *Ramsay*.

Ramsay's Cuckoo-Shrike.

Graucalus, sp. nov., Ramsay, Nature, 1879, p. 125.
Graucalus pusillus, Ramsay, Proc. Linn. Soc. N. S. Wales, iv. p. 71 (1879).—Salvad. Ann. Mus. Civic. Genov. xv. p. 35 (1879).—Id. Ibis, 1880, p. 128.—Id. Orn. Papuasia, etc. ii. p. 140 (1881).
Graucalus dussumieri (?), Ramsay, Proc. Linn. Soc. N. S. W. iv. p. 71 (1879).
Graucalus salamonis, Ramsay, Proc. Linn. Soc. N. S. W. iv. p. 314 (1879).

THIS species is a small island race of *Graucalus axillaris*, the males of the two species being almost identical in colour. In the present bird, which comes from the Solomon Islands, the colour is a clearer slaty blue in both sexes and the under wing-coverts and axillaries are barred with black and white, and the bars are narrower.

In the females of *G. pusillus* the thighs are uniform blue-grey, whereas in *G. axillaris* they are barred with black and white like the rest of the under surface.

As far as is known at present, this Cuckoo-Shrike is found only in Guadalcanar, where it was discovered by Mr. Cockerell.

The following descriptions have been drawn up from the typical specimens:—

Adult male. General colour above clear blue-grey; wing-coverts like the back; large outer feather of bastard wing black; primary-coverts and quills black, externally blue-grey, more hoary on the outer webs of the primaries; tail-feathers black, the centre feathers ashy for two thirds of their length on the inner web, not quite so extended on the outer web, the outermost feather dull ashy at the tip; head clear blue-grey like the back; a narrow frontal line and lores black, occupying a distinct space in front of the eye and extending below the latter, as well as the upper and under edge of the eyelid; sides of face, ear-coverts, cheeks, and entire under surface of body clear blue-grey, with a tiny spot of black at the base of the chin and base of cheeks; under tail-coverts darker grey, with blackish shaft-stripes; axillaries barred, ashy grey on the inner web and on the base of the outer, the latter otherwise barred with black and ashy whitish; under wing-coverts slaty blue, with less distinct bars of black and ashy whitish; quills ashy blackish below, greyer on the inner web especially near the base, the inner web edged with white. Total length 9 inches, culmen 0·7, wing 4·95, tail 4·2, tarsus 0·8.

Adult female. Differs from the male in having the base of the forehead greyish black instead of deep black, the loral patch being also ashy blackish; the throat and chest are uniform blue-grey, but the rest of the under surface is barred with black and ashy white on a blue-grey ground; on each side of the lower back a conspicuous patch of silky white; thighs uniform blue-grey. Total length 8·5 inches, culmen 0·65, wing 4·95, tail 3·1, tarsus 0·8.

The Plate represents the male and female birds of the natural size, the figures being drawn from the type specimens which Mr. Ramsay has kindly lent to us.

[R. B. S.]

GRAUCALUS AXILLARIS, *Salvad.*

GRAUCALUS AXILLARIS, Salvad.

Bruijn's Cuckoo Shrike.

Graucalus axillaris, Salvad., Ann. Mus. Civ. Genova, vii. p. 925 (1875).—Sharpe, Mittheil. Zool. Mus. Dresden, i. p. 366 (1878).—Id. Cat. B. Brit. Mus. iv. p. 27 (1879).

THERE is not much to record respecting this interesting species of Cuckoo Shrike, which, at the time we write, has been too recently discovered for us to know any thing of its habits and economy.

Although the male is to all appearance uniform in coloration, it has the axillaries and under wing-coverts barred with black and white; and this at once distinguishes it from any of the allied species of *Graucalus*. The female is quite different, and belongs to the section of *G. lineatus* and *G. maforensis*.

Originally discovered by Mr. Bruijn at Mansema in the Arfak Mountains, it has lately been sent by Mr. Goldie from the Taburi district, at the back of the Astrolabe range, in South-eastern New Guinea. He gives the native name as *Shorara*.

The following descriptions, taken from the typical specimens, are reproduced from Mr. Sharpe's Catalogue of Birds :—

"*Adult male.* General colour above slaty grey, with a cast of lighter and more bluish grey; lesser wing-coverts like the back, the median and greater series rather darker than the back; quills black, externally edged with dark slaty grey; the secondaries outwardly entirely dark grey like the greater wing-coverts; tail black, with a slight ashy shade on the centre feathers; a narrow frontal line, lores, and feathers in front of the eye black; ear-coverts darker grey than the head; under surface of body slaty grey, the under wing-coverts and axillaries barred across with white; quills ashy-black below, grey on the inner web, which is edged internally with white. Total length 8·7 inches, culmen 0·75, wing 5·4, tail 4·4, tarsus 0·85 (*Mus. Civ. Genov.*).

"*Adult female.* Differs from the male in having the throat and fore neck alone slaty grey; the rest of the under surface barred with black and white, the black bars always the broadest, but especially so on the sides of the body and on the under tail-coverts. The general tone of the grey upper surface is darker, the quills being blackish and the secondaries narrowly edged with white like the primaries. Total length 9 inches, culmen 0·8, wing 5·65, tail 4·4, tarsus 0·85 (*Mus. Civ. Genov.*).

"*Young male.* Like the old female, but with some of the uniform grey breast-feathers of the adult male plumage appearing below; upper tail-coverts tipped with white, of which there are scarcely any traces in the adult female. Wing 5·4 inches; tail black, the middle feathers dark slaty grey with a black tip (*Mus. Civ. Genov.*)."

The pair of birds represented in the accompanying illustration are drawn from specimens collected by Mr. Goldie in South-eastern New Guinea, and now form part of the national collection. The birds are shown of about the size of life.

[R. B. S.]

GRAUCALUS MAFORENSIS.

GRAUCALUS MAFORENSIS.
Mafoor Island Cuckoo-Shrike.

Campephaga maforensis, Meyer, Sitz. Akad. Wien, lxix. p. 386 (1874).
Graucalus maforensis, Salvad. Ann. Mus. Civic. Genov. vii. p. 927 (1875).—Sharpe, Mitth. zool. Mus. Dresden,
    i. p. 365, pl. xxx. (1878).—Id. Brit. Mus. Cat. B. iv. p. 41 (1879).—Salvad. Ann. Mus. Civic. Genov.
    xv. p. 35 (1879).—Id. Orn. della Papuasia &c. p. 141 (1881).

This species was discovered by Dr. Meyer in the island of Mafoor in the Bay of Geelvink, and belongs to that section of the genus *Graucalus* which contains *G. lineatus* of Australia and *G. axillaris* from Mount Arfak. Mr. Sharpe states that the female is so like that of *G. axillaris* that it can scarcely be separated, but the male bird is very different, and is distinguished by having the breast and abdomen black, crossed by very narrow wavy lines of white, whereas in *G. axillaris* the male is perfectly uniform below. Comparing *G. maforensis* with *G. lineatus* of Australia, Count Salvadori says that the colour is of a more intense ashy blue, and the transverse striations on the abdomen are much less strongly defined.

The following descriptions of the type specimens are given by Mr. Sharpe in his 'Catalogue of Birds:'—

"*Adult male.* Above light bluish grey, the wing-coverts like the back; bastard-wing plumes black, narrowly edged with white; primary-coverts black on the inner webs, externally grey, rather lighter on the extreme margin; quills black, the primaries with a narrow edging of hoary grey, the secondaries like the back, blackish on the inner web; tail-feathers entirely black (only three feathers remaining); sides of face a little darker bluish grey than the head; a narrow frontal line, lores, feathers in front of the eyes and at base of cheeks black; throat and chest bluish grey; breast and remainder of under surface of body black, crossed with very fine narrow lines of greyish white; thighs bluish grey; under wing-coverts and axillaries black, barred with white like the breast, but the bars much broader; quills ashy grey below, narrowly edged with white along the inner web; 'iris yellow' (*Meyer*). Total length 8·8 inches, culmen 0·75, wing 4·7, tail 3·8, tarsus 0·8.

"*Adult female.* Differs from the male in having much broader white cross bars on the under surface, the black ones, however, being broader than the white ones; above, the colour is darker bluish grey than in the male, the black at the base of the forehead being more dusky and not so velvety black; the secondaries as well as the primaries have a narrow and nearly obsolete hoary grey edging, and are much more broadly margined with white on the inner web below; tail black, the outer feathers with a narrow white tip, the centre feathers dark grey for the greater part of their extent, black at the tips. Total length 8·8 inches, culmen 0·75, wing 4·85, tail 4, tarsus 0·8."

The male and female birds depicted in the Plate have been drawn from the typical examples, which were kindly lent to me by Dr. Meyer for the purpose.

[R. B. S.]

EDOLIISOMA MONTANUM.

EDOLIISOMA MONTANUM.

Mount Arfak Cuckoo-Shrike.

Campephaga montana, Meyer, Sitzb. k. Ak. Wiss. zu Wien, lxix. p. 386 (1874).
Edoliisoma montanum, Salvad. Ann. Mus. Civ. Gen. vii. p. 927 (1875).—Sharpe, Mitth. zool. Mus. Dresd. i.
    pp. 367, 369 (1878).—Id. Cat. B. iv. p. 46 (1879).—Salvad. Ann. Mus. Civ. Gen. xv. p. 35, n. 19
    (1879).—Id. Orn. Papuasia &c. ii. p. 147 (1881).

This species was discovered in the Arfak Mountains by Dr. Meyer, and has since been met with in the same locality by Dr. Beccari and Mr. Bruijn's hunters. It is one of the finest members of the genus *Edoliisoma*, the female being more brightly coloured than is usual in this genus, and not being barred below as so many of the hen birds are.

The male bird has the upper surface grey, and the lower parts entirely black; and the female is recognized by its perfectly black tail from all the other species of *Edoliisoma* with which it might be confounded.

The following description of the species is taken from Mr. Bowdler Sharpe's Catalogue:—

"*Adult male.* General colour above blue-grey, including the wing-coverts; greater coverts black, edged with grey; bastard wing, primary-coverts, and quills black, with a steel-green gloss on the margin, the inner secondaries broadly edged with grey; tail-feathers black, glossed with steel-green on the margins; a narrow frontal line, lores, feathers round the eye, sides of face, sides of neck, and entire under surface of body glossy black, the thighs ashy grey; under wing-coverts black; greater series of under wing-coverts and quills, below ashy grey. Total length 9·5 inches, culmen 0·8, wing 5·4, tail 4·3, tarsus 1.

"*Adult female.* Differs from the male in being of a lighter blue-grey both above and *below*; only the lores, feathers at base of forehead and in front of the eye, front half of eyelid, a spot at base of lower mandible, and chin black. Total length 9·5 inches, culmen 0·75, wing 5·05, tail 4·1, tarsus 1.

"The typical specimens lent to me by Dr. A. B. Meyer measure, in inches, as follows:—

		Total length.	Culmen.	Wing.	Tail.	Tarsus.
a. ♂.	Arfak Mountains.	9·2	0·75	5·3	4·3	1
b. ♀.	,, ,,	8·8	0·8	5·0	4·1	1

"The female is not fully adult, and has the inner secondaries and the outer tail-feathers tipped with white; the under tail-coverts are rufous, barred with dark slaty-grey, from which it may be inferred that the young bird is entirely marked in this manner on the under surface."

The figures in the Plate represent a pair of birds of about the size of life. They are drawn from the typical specimens kindly lent to me by Dr. Meyer.

[R. B. S.]

EDOLIISOMA POLIOPSE, Sharpe.

EDOLIISOMA POLIOPSE, Sharpe.

Grey-faced Cuckoo-Shrike.

Edoliisoma poliopsa (err. typ.), Sharpe, Journ. Linn. Soc., Zool. xvi. pp. 318, 433 (1882).

The original specimens of this Cuckoo-Shrike were sent by Mr. Goldie from the Morocco district in the Astrolabe Mountains, S.E. New Guinea, where its native name is said to be "Nagioa." Mr. Forbes has found it in the Sogeri district of the same range of mountains. Two specimens were sent by Mr. Goldie, both of them evidently females; but notwithstanding this fact, we described the species as new without any hesitation.

Many of the species of the genus *Edoliisoma* can scarcely be told apart, if the males only are examined, whereas the females are very distinct and easily recognizable. Thus it was that we felt certain that the male, when discovered, would be found scarcely to differ from the same sex of *E. schisticeps*, but there would be no mistake about the females; and now that Mr. Forbes has discovered the male there can be no doubt on the point. The validity of the species rests therefore on the female bird, which may be distinguished from the hen of *E. schisticeps* by the slaty-grey colour of the chin, ear-coverts, and fore part of the cheeks.

The following is a description of both sexes:—

Adult male. General colour above slaty grey; lesser wing-coverts like the back; median and greater coverts clearer and more French-grey; bastard-wing, primary-coverts, and quills black, the secondaries externally French-grey like the greater wing-coverts; upper tail-coverts like the back; centre tail-feathers grey, with a broad black band at the end; the remainder black, with a broad grey tip at the outer ones; crown of head, hind neck, and mantle a shade darker slaty grey than the lower back and rump; lores and nasal plumes black; feathers round eye, ear-coverts, and cheeks blackish, with a slight greyish shade; chin blackish; throat and under surface of body leaden grey, clearer and more slaty grey on the lower breast and abdomen, as well as on the sides of body and flanks; thighs and under tail-coverts, under wing-coverts and axillaries pale slaty grey; quills below blackish; the inner webs ashy whitish towards their bases. Total length 8 inches, culmen 0·75, wing 4·4, tail 3·25, tarsus 0·9.

Adult female. Different from the male. General colour above dark chestnut, more dusky on the mantle and upper back, where the feathers are obscurely dark-shafted; the scapulars like the mantle; the lower back, rump, and upper tail-coverts lighter and more maroon-brown; two centre tail-feathers chestnut, with a subterminal mark of blackish; remainder of tail-feathers black, tipped with chestnut, increasing in extent towards the outermost, which is also chestnut along the outer web; wing-coverts chestnut; bastard-wing and primary-coverts black; quills black, externally chestnut, broader on the secondaries, the innermost of which are entirely chestnut; entire head and nape, as well as the side of the face and ear-coverts, slaty grey, blackish on the lores and on extreme base of forehead and below the eye; the ear-coverts also blackish; fore part of cheeks and chin ashy grey; hinder cheeks and throat chestnut, barred with grey; remainder of under surface rich chestnut, becoming paler towards the flanks and under tail-coverts; under wing-coverts like the breast, as also the axillaries; quills black below, rufous along the inner web. Total length 6·8 inches, culmen 0·8, wing 4·2, tail 3·4, tarsus 0·85.

The Plate represents both sexes of the size of life, the figures being drawn from specimens procured by Mr. H. O. Forbes in the Astrolabe range.

[R. B. S.]

LALAGE MŒSTA, *Sclater*.

LALAGE MŒSTA, Sclater.

Black-browed Caterpillar-catcher.

Lalage mœsta, Sclater, Proc. Zool. Soc. 1883, p. 55.

THE present species is one of the novelties discovered by Mr. H. O. Forbes during his expedition to Timor laut. It is intermediate in coloration between *Lalage timoriensis* of Timor, and *L. pacifica* and *L. terat*, having the tail of the first-named bird, but the wing-coverts patterned as in the last-named species: from both of these, however, it may be distinguished by its pure white rump-band, and by the absence of a white eyebrow. From *L. timoriensis* it may be told by the lesser wing-coverts being greenish-black like the back; but, like that species, it has only two of the outer tail-feathers tipped with white. In *L. timoriensis*, also, there is a very distinct superciliary streak of white, whereas in the present bird the white is confined to a rather broad streak above the lores.

The following is a description of the typical specimens, which have been kindly lent to me by my friend Dr. Sclater. They will be ultimately deposited in the British Museum.

Adult male. General colour above glossy greenish black, the feathers of the lower back with ashy grey bases; rump white; the upper tail-coverts steel-black, tipped with white; wing-coverts like the back, the median and greater series with white spots at the tip, forming a double wing-bar, the white endings to the median series being much the broadest; bastard wing and primary-coverts uniform black; quills black, the secondaries edged with white on the outer web, this white, however, not extending to the base of the feathers, though it reaches to the tip of the outer web of the inner secondaries; tail-feathers black with a steely gloss on the margins, the two outer feathers with broad white ends, feathers in front of the eye black, surmounted by a white loral streak from the base of the nostrils to above the fore part of the eye; upper and lower margin of eyelid white; ear-coverts white, black on the upper margin; cheeks, sides of the neck, and entire under surface of the body, as well as the thighs, under tail-coverts, under wing-coverts and axillaries, pure white; quills black below, but with about half of the inner web white, forming a large white patch on the under surface of the quills. Total length 7 inches, culmen 0·55, wing 3·85, tail 3·0, tarsus 0·9.

The bird described is apparently an adult male. A female sent by Mr. Forbes differs in being less glossy black above, and in having the rump-band more ashy white, the blackish subterminal markings, which are concealed in the male, being ashy and more conspicuous in the female. The throat and chest are slightly tinged with buff, and the lower series of under wing-coverts are white with blackish tips. These last two characters may be the remains of immaturity. Total length 7 inches, culmen 0·55, wing 3·7, tail 3·0, tarsus 0·95.

The figures in the Plate represent a pair of birds of the natural size.

[R. B. S.]

MICROECA ASSIMILIS, Gould

MICRŒCA ASSIMILIS, Gould.

Western Micrœca.

Micrœca assimilis, Gould, Proc. Zool. Soc. 1840, p. 172.—Id. B. Austr. Intr. p. xl (1848).—Bonap. Consp. Gen.
  Av. i. p. 321 (1850).—Reichenbach, Vög. Neuholl. p. 287 (1850).—Cab. Mus. Hein. Th. i. p. 52 (1850).
  —Gould, Handbook B. Austr. i. p. 260 (1856).—Ramsay, Proc. Linn. Soc. New South Wales, ii. part
  2, p. 182 (1878).—Sharpe, Cat. B. Brit. Mus. iv. p. 124 (1879).
Myiagra assimilis, Gray, Gen. of Birds, i. p. 261.
Muscicapa assimilis, Gray, Hand-list of Birds, i. p. 324, no. 4856 (1869).

THE genus *Micrœca* was instituted by me in 1840 for the reception of the present bird and the *M. fascinans* of Latham, also an Australian species. Since that time only two or three species have been discovered in the Papuan Islands and North-eastern Australia; so that in *Micrœca* we have a thoroughly Australian genus of Flycatchers, representing the true Flycatchers of Europe and Asia. It may be noticed that the two best-known Australian species, *Micrœca fascinans* and *M. assimilis*, in their sober brown coloration are not unlike the Common Flycatcher (*Butalis grisola*) of England.

The present species very closely resembles *M. fascinans* of New South Wales, of which it is the representative in Western Australia; but it is smaller, and, instead of having the outer tail-feathers pure white, has only the tip of the inner web and the outer edge of the external tail-feather white.

The species is described in full by Mr. Sharpe, whose words I here transcribe:—

"*Adult male.* General colour above earthy brown, the upper tail-coverts darker sepia-brown; wing-coverts brown; quills brown, narrowly edged with lighter brown, the secondaries with dull white; four centre tail-feathers dark brown, the next two on each side tipped with white on the inner web, the white tip gradually increasing in size towards the outermost, which has also the outer web white; a narrow frontal line of dull white drawn backwards over the fore part of the eye; in front of the eye a dusky spot; ear-coverts brown with a slight dash of rufous; cheeks, throat, abdomen and under tail-coverts white, the chest slightly shaded with light brown, the sides of the body more distinctly light earthy brown; axillaries pale rusty brown; under wing-coverts whity brown, with dusky bases; bill and feet bluish brown; iris reddish brown. Total length 4·5 inches, culmen 0·45, wing 3·35, tail 2·25, tarsus 0·55."

The female is similar to the male.

Mr. Sharpe also notices a specimen from North-western Australia, collected by Mr. Elsey, in which there are small white spots at the tips of the primary-coverts. These he believes to be a sign of immaturity.

All the specimens which have as yet come under my notice have been from Western Australia; but Mr. Ramsay also gives the Gulf of Carpentaria as a habitat for the species. I have never seen it from anywhere but Western Australia.

The figures in the Plate represent this species of the natural size, and are drawn from skins in my own collection. There is no difference in the sexes.

GERYGONE DORSALIS, *Sclater*.

GERYGONE DORSALIS, Sclater.

Rufous-backed Gerygone.

Gerygone dorsalis, Sclater, Proc. Zool. Soc. 1883, p. 199.

THIS is one of the most distinct species of the genus *Gerygone*. The rufous colour of the back, so markedly in contrast with the grey head, is a character approached by none of the other species of the genus. Mr. Forbes, who discovered this new bird, sent a large series from Moloe Island in the Tenimber group:—

The following is a description of the typical specimen kindly lent to me by Dr. Sclater:—

Adult male. General colour rufous or bay; the lesser and median wing-coverts like the back; greater series, bastard wing, and primary-coverts, as well as the quills, dusky brown, edged with rufous like the back; tail-feathers light brown, margined with rufous, all but the centre feathers with a blackish shade before the tips, which are pale brown; near the end of the inner webs a white spot, which increases in extent towards the outermost feather; head dull ashy brown, contrasting with the back; a spot of dusky white on the lores; ear-coverts light ashy brown as well as the sides of the neck; feathers below the eye, cheeks, and under surface of body white; the sides of the breast and flanks light rufous or bay; thighs also light rufous; under tail-coverts buffy whitish; under wing-coverts and axillaries yellowish white; quills dusky below, yellowish white along the inner web; " bill, legs, and feet black; iris black " (*H. O. Forbes*). Total length 4 inches, culmen 0·55, wing 2·1, tail 1·5, tarsus 0·8.

Adult female. Resembling the male in colour, but with the head a trifle duller. Wing 2·15 inches, tarsus 0·85.

The Plate represents a male and female of this species, of the natural size.

[R. B. S.]

PSEUDOGERYGONE NOTATA.

PSEUDOGERYGONE NOTATA.

White-spotted Flycatcher.

Gerygone chrysogaster, pt., Gray, Proc. Zool. Soc. 1861, p. 434.
Gerygone neglecta, pt., Wallace, Proc. Zool. Soc. 1865, p. 475.—Salvad. Ann. Mus. Civic. Genov. vii. p. 957 (1875).
Gerygone notata, Salvad. Ann. Mus. Civic. Genov. xii. p. 344 (1878), xiv. p. 504 (1879).—Id. Orn. Papuasia e delle Molucche, ii. p. 99 (1881).
Pseudogerygone notata, Sharpe, Notes from the Leyden Museum, i. p. 29 (1878).—Id. Cat. Birds in Brit. Mus. iv. p. 227 (1879).
Leptotodus tenuis, Meyer in Madarász, Zeitschr. ges. Orn. ii. p. 197, pl. ix. fig. 2 (1884).

THE members of the genus *Pseudogerygone* have been separated by us from true *Gerygone* on account of the different proportions of the quills, the second primary being equal to the secondaries in the former, while it is considerably longer than the secondaries in the genus *Gerygone*. Whether these differences are sufficient to separate the two genera is a matter for ornithologists to consider; but there can be no question as to the convenience of dividing the great genus *Gerygone* into two sections, when we have as good characters for separation as those mentioned above.

The type of the present species was discovered by Dr. Beccari on the river Wa Samson, in North-western New Guinea, and was described by Count Salvadori. A specimen had been collected by Mr. A. R. Wallace in the island of Mysol many years previously, but had been referred either to *G. chrysogaster* or *G. neglecta* by previous writers. The Leyden Museum contains specimens from Salwati, obtained by Dr. Bernstein; and quite recently Dr. A. B. Meyer has received a specimen from Amberbaki in North-western New Guinea, which he described as belonging to a new genus and species. He very kindly sent us the type specimen for examination, and we found by comparison that it was identical with the specimens of *Pseudogerygone notata* in the British Museum.

The following is a copy of the description given by us in the British Museum Catalogue of Birds:—

"General colour above dark greenish olive, rather more rufescent on the upper tail-coverts; least wing-coverts like the back; greater and median wing-coverts dark brown, edged and tipped with yellow; quills dark brown, externally edged with the same olive as the back; tail-feathers brown, with olive margins; lores buffy whitish; ear-coverts olive, with lighter shaft-streaks; throat and breast white, slightly tinged with yellow; the abdomen, thighs, and under tail-coverts olive-yellow; under wing-coverts white, as also the axillaries, washed with yellow, especially on the edge of the wing; quills dusky brown below, buffy white along the inner edge of the quills. Total length 4·2 inches, culmen 0·45, wing 1·95, tail 1·55, tarsus 0·6."

The figures in the Plate are drawn from the type specimen of *Leptotodus tenuis*, which was kindly lent to us by Dr. Meyer. They represent an adult bird in two positions.

[R. B. S.]

PSEUDOGERYGONE CHRYSOGASTRA.

PSEUDOGERYGONE CHRYSOGASTRA.

Yellow-bellied Flycatcher.

Gerygone chrysogaster, Gray, Proc. Zool. Soc. 1858, pp. 171, 191.—Id. Cat. Mamm. etc. New Guinea, pp. 25, 56 (1859).—Rosenb. J. f. O. 1864, p. 122.—Finsch, Neu-Guinea, p. 166 (1865).—Meyer, Sitz. k. Akad. Wien, lxx. p. 118 (1874).—Salvad. & D'Albert. Ann. Mus. Civic. Genov. vii. p. 820 (1875).—Salvad. tom. cit. p. 956 (1875).—Id. op. cit. ix. p. 26 (1876), xiv. p. 503 (1879).—Id. & D'Albert. t. c. p. 63 (1879).—Salvad. Orn. Papuasia e delle Molucche, i. p. 97 (1881).
Acanthiza chrysogaster, Gray, Hand-list Birds, i. p. 219, no. 3131 (1869).
Gerygone xanthogaster (lapsu), Salvad. Ann. Mus. Civic. Genov. vii. p. 958 (1875).
Gerygone chrysogastra, Sharpe, Journ. Linn. Soc., Zool. xiii. p. 495 (1878).
Pseudogerygone chrysogastra, Sharpe, Notes from the Leyden Mus. i. p. 29 (1878).—Id. Cat. Birds in Brit. Mus. iv. p. 226 (1879).

The late Mr. George Robert Gray described the present species from the Aru Islands, where it was discovered by Mr. Wallace; Mr. Gray also included specimens from Mysol and Waigiou, but the birds from the two last-named islands are really distinct species, and have been separated as *P. notata*, Salvad., and *P. neglecta*, Wallace.

Besides the Aru Islands, where Dr. Beccari has also met with the species, it has been found in South-eastern New Guinea as well as in the Island of Jobi, in the Bay of Geelvink. Count Salvadori, however, notices some slight differences in the Jobi specimen, which he thinks may indicate a distinct species.

The present bird is one of many Aru species which are also found to inhabit South-eastern New Guinea, and no differences can be detected between specimens from these two localities. D'Albertis met with it on the Fly River, and again at Mount Epa and Naiabui. Mr. H. O. Forbes has also recently sent a specimen from the Sogeri district, in the Astrolabe Mountains. Count Salvadori also believes that *Gerygone inconspicua* of Ramsay (Proc. Linn. Soc. N. S. Wales, iii. p. 116) is identical with the present species; and this seems likely enough, the only point in which the description disagrees with that of *P. chrysogastra* being in the presence of "an oblique blackish spot from in front to under the eyes across the gape."

Adult male. General colour olive-brown, the upper tail-coverts more rusty brown; wing-coverts like the back, the greater series darker brown, narrowly edged with olive-brown like the back; quills dark brown, externally edged with olive; tail dark brown, edged with rusty brown like the upper tail-coverts; lores and eyelid dull white; ear-coverts and sides of face light ashy brown; cheeks, throat, and breast white, the remainder of the under surface sulphur-yellow, the thighs browner; under wing-coverts pale yellow, with white bases: "bill and feet dusky" (*Wallace*); "iris black" (*D'Albertis*): "bill black; legs and feet faint purplish brown; iris brown, with a ring of pale orange outside" (*H. O. Forbes*). Total length 4·7 inches, culmen 0·45, wing 2·2, tail 1·8, tarsus 0·65.

Adult female. Similar to the male in colour. "Bill black; feet reddish white" (*Wallace*). Total length 4·2 inches, culmen 0·45, wing 2, tail 1·7, tarsus 0·65.

The Plate represents an adult bird in two positions, and has been drawn from a specimen procured in the Astrolabe Mountains by Mr. H. O. Forbes.

[R. B. S.]

PSEUDOGERYGONE CINEREICEPS.

PSEUDOGERYGONE CINEREICEPS, Sharpe.

Grey-headed Flycatcher.

Pseudogerygone cinereiceps, Sharpe, Nature, 1886, p. 340.

THERE is not much to say about this little Flycatcher, which belongs to an Australian group which have much the appearance of the Willow-Warblers of more northern latitudes. The present species was discovered by Mr. H. O. Forbes in the Sogeri district of the Astrolabe Mountains, at a height of 1750 feet above the sea. Its nearest ally appears to be *P. flavilateralis* of New Caledonia, which is yellow-sided like *P. cinereiceps*, but is a larger bird and has a great deal of white on the tail.

The following description is taken from the typical specimen :—

Adult female. General colour above yellowish green, rather more olive on the upper tail-coverts ; lesser and median coverts like the back ; greater coverts, bastard-wing, primary-coverts, and quills dusky brown, externally edged with yellowish green, brighter on the margin of the quills, the innermost secondaries also washed with greenish ; tail-feathers ashy brown, edged with yellowish green, with a very distinct subterminal bar of black before the tip, which is ashy brown, with a tiny spot of white at the end of the inner web, scarcely visible ; crown of head ashy with a faint tinge of green ; lores white, extending above the fore part of the eye ; feathers round eye, ear-coverts, and cheeks ashy brown ; throat and under surface of body white, the sides of the upper breast ashy brown ; lower breast and abdomen purer white ; sides of body and flanks pale sulphur-yellow ; thighs ashy brown ; under tail-coverts white, washed slightly with yellow ; under wing-coverts and axillaries white, edged with yellow ; quills below dusky, white along the inner edge : " bill black ; legs and feet lavender ; iris rich lake " (*H. O. Forbes*). Total length 3·5 inches, culmen 0·4, wing 1·9, tail 1·25, tarsus 0·65.

The Plate gives an illustration of an adult bird in two positions, drawn from the specimen described above.

[R. B. S.]

HETEROMYIAS CINEREIFRONS.

HETEROMYIAS CINEREIFRONS.

Ashy-fronted Flycatcher.

Pœcilodryas? cinereifrons, Ramsay, P. Z. S. 1875, p. 588.—Id. Proc. Linn. Soc. N. S. W. ii. p. 182.
Heteromyias cinereifrons, Sharpe, Cat. B. Brit. Mus. iv. p. 239 (1879).

This interesting bird was described by Mr. E. Pierson Ramsay from a specimen shot at Dalrymple's Gap, near Cardwell, Queensland, and was placed by him, with doubt, in the genus *Pœcilodryas*, from which it has lately been removed by Mr. Sharpe, who has made it the type of a new genus, which he calls *Heteromyias*. It is allowed by the latter gentleman to stand very close to *Pœcilodryas*, a genus instituted by myself in 1865, and of which the type is *P. cerviniventris*, figured by me in the 'Birds of Australia;' but the bill is differently shaped, being higher at the nostrils than it is broad; and this peculiarity allies it very closely to the genus *Metabolus*, whose single species, *M. rugensis*, is apparently confined to the Caroline Islands. Beyond these few remarks respecting the scientific history of the present species I am unable to add any thing, its habits and economy being at present entirely unknown. In fact, at the precise moment when I write, there exists but a single complete specimen of the bird, viz. the type in the Australian Museum at Sydney, a full description of which, as given by Mr. Sharpe in his Catalogue, I transcribe:—

"*Adult.* General colour above rusty brown, the head and nape dark ashy grey, shaded slightly with brown except on the forehead, which accordingly looks lighter grey; over the eye a broad streak of light French grey, extending to the sides of the nape; feathers in front of the eye dusky greyish black; round the eye a ring of feathers, blackish where it joins the loral spot before, and the ear-coverts behind; fore part of cheeks and feathers just below the eye white; ear-coverts rusty brown, blackish just under the eye; a chin-spot and feathers at base of lower mandible greyish black; throat and centre of breast and abdomen white; chest and fore neck, as well as the sides of the breast, light grey; the sides of the body and under tail-coverts tawny buff; thighs grey; under wing-coverts white, the axillaries tipped with tawny; greater series dark ashy, forming a patch on the under wing-coverts; quills ashy brown below, white at the base of both webs; upper wing-coverts ashy grey, the median and greater series dark brown slightly shaded with rusty brown; bastard wing externally ashy, internally dark brown; primary-coverts uniform dark brown; quills dark brown, paler towards the tips, edged externally with rusty brown, the inner secondaries entirely of the latter colour and resembling the back, all the quills but the latter white at base, forming a bar across the wing; upper tail-coverts rufous, the tail-feathers brown, washed on the edges with rusty brown, inclining to rufous near their bases, the outer feathers narrowly tipped with white on the inner web. Total length 6·3 inches, culmen 0·7, wing 3·85, tail 2·7, tarsus 1·2."

I have availed myself of Mr. Ramsay's permission to figure the type of this species before its return to Sydney; and in the accompanying Plate I have given two illustrations of the bird (in two positions), both of the natural size.

MONACHELLA MUELLERIANA.

MONACHELLA MUELLERIANA.

Chat-like Flycatcher.

Muscicapa mulleriana, Schlegel (nec *Muscicapa mulleri*, Temm.), Ned. Tijdschr. Dierk. iv. p. 40 (1871).
Monachella saxicolina, Salvad. Ann. Mus. Civ. Gen. vi. p. 63 (1874).—Beccari, ibid. vii. p. 709 (1875).—D'Albertis, Sydney Mail, 1877, p. 248.—Id. Ann. Mus. Civ. Gen. x. p. 11 (1877).—Salvad. op. cit. x. p. 11 (note), p. 133 (1877).—D'Alb. & Salvad. op. cit. xiv. p. 59 (1879).—Salvad. op. cit. p. 501 (1879).—Id. Orn. Papuasia &c. p. 83 (1881.)
Monachella mulleriana, Salvad. Ann. Mus. Civ. Gen. vi. p. 308 (1874, note).—Sharpe, Cat. B. Brit. Mus. iv. p. 240 (1879).
Micrœca albofrontata, Ramsay, Proc. Linn. Soc. New S. Wales, iii. p. 304 (1879), iv. pp. 90, 98 (1879)—Salvad. Ibis, 1859, p. 323.

I AM sorry that I cannot follow my friend Count Salvadori in calling this bird *Monachella saxicolina* (although his reasons are worthy of some consideration), as he has only preferred to use that name to avoid the confusion that might take place between *Muscicapa muelleriana* of Schlegel and *Muscicapa muelleri* of Temminck. I do not think, however, that there is really much chance of this confusion, as the latter bird is a Flycatcher belonging to the genus *Erythromyias*, and is an inhabitant of Sumatra and Borneo. Professor Schlegel's name having been published three years before that of Count Salvadori, it has an undoubted claim to priority. In other respects the name of *saxicolina* is extremely well chosen, as indicating the habits of the bird; and Signor D'Albertis states that when he first saw the species on the torrents of the Arfak Mountains he really thought that it was a true *Saxicola*.

Dr. Beccari also says that in the above-named locality he found the species abundant, but only in the streams near the sea. During his expedition up the Fly river, D'Albertis met with the species along the banks, and relates how, when the water was low, it perched on small rocks at the side of the river, and was continually in motion flying after insects. Mr. Goldie also states that during his recent expedition to the Astrolabe Mountains he found the species, in company with *Grallina bruijnii*, flying about creeks and hopping amongst the stones.

As far as we know, the present bird is exclusively confined to New Guinea. A specimen was procured by Solomon Müller in Lobo Bay as long ago as 1828, though it does not seem to have been described till 1871. It also inhabits the Arfak Mountains, and has been met with at Karons by M. Laglaize. In the southern part of the island D'Albertis met with it on the Fly river; and Mr. Goldie procured it on the Goldie river inland from Port Moresby. He has more recently met with it in the Morocco district, at the back of the Astrolabe range of mountains in South-eastern New Guinea: here it is called *Iada*.

The following description is taken from Mr. Sharpe's 'Catalogue of Birds:'—

"*Adult male.* General colour above light French grey, paler on the lower back, the rump and upper tail-coverts white; wings and tail dark brown; crown of head and nape dark brown, as also the feathers above the eye and the upper edge of the eyelid, the brown narrowing on the forehead to the base of the bill; lores and feathers over the front of the eye pure white; between the eye and the base of the bill a triangular patch of blackish feathers; sides of face and ear-coverts, as well as entire under surface of body, creamy white; under wing-coverts dark brown; 'bill and feet black' (*D'Albertis*). Total length 5·5 inches, culmen 0·6, wing 3·8, tail 2·4, tarsus 0·65.

"*Adult female.* Similar to the male. Total length 5·3 inches, culmen 0·6, wing 3·55, tail 2·25, tarsus 0·65."

Count Salvadori says that the young bird has the head and the wing-coverts blackish brown spotted with white, the back dull whitish varied with dusky, and the tips of the tail-feathers white.

The figures in the Plate represent a pair of birds of the size of life, and are drawn from a specimen collected by Mr Goldie, and now in the British Museum.

[R. B. S.]

PŒCILODRYAS BIMACULATA.

PŒCILODRYAS BIMACULATA.

Black-and-White Flycatcher.

Myiolestes? *bimaculata*, Salvad. Ann. Mus. Civic. Genov. vi. p. 84.
Pachycephala? bimaculata, Salvad. Ann. Mus. Civic. Genov. vii. p. 935 (1875), x. p. 142 (1877).
Pœcilodryas bimaculata, Sharpe, Notes Leyden Mus. i. p. 25 (1878).—Id. Cat. Birds Brit. Mus. iv. p. 244 (1879).
—Salvad. Ann. Mus. Civic. Genov. xiv. p. 502 (1879).—Id. Orn. Papuasia, etc. ii. p. 85 (1882).
Pœcilodryas sylvia, Ramsay, Trans. Linn. Soc. N. S. Wales, viii. p. 5 (1883).

The genus *Pœcilodryas* consists of two groups, or sections, characterized by the colour of the abdomen; in one section the abdomen is white, in the other yellow. The white-bellied group may be further subdivided into those which have the throat white and those which have a black throat. The present species belongs to the latter section, which now contains three species—*P. bimaculata*, *P. æthiops*, and *P. albinotata*, the latter being distinguished by its bluish-grey upper surface.

P. bimaculata and *P. æthiops* have the upper surface black, with the rump and upper and under tail-coverts white; but they may easily be distinguished from each other, *P. bimaculata* having the abdomen white, a long white patch on the sides of the fore neck and chest, and the inner wing-coverts black; whereas in *P. æthiops* the abdomen is black, the inner wing-coverts are white, forming a shoulder-patch, and there is no white on the sides of the fore neck.

The present species was discovered in north-western New Guinea by Signor D'Albertis, and has been met with in the same locality by Dr. Beccari and M. Laglaize. It has been more recently obtained in the Astrolabe Mountains by Mr. Goldie, and a specimen from that range of mountains has been lent to me by Mr. Ramsay.

The following description is copied from the 'British Museum Catalogue of Birds':—

"*Adult male.* Above velvety black; upper tail-coverts white, forming a band across; wings and tail black; sides of face, sides of neck, throat, and breast black, as well as the flanks and thighs; abdomen, vent, and under tail-coverts white; on each side of the chest a broad line of pure white feathers running from the sides of the lower throat to the sides of the upper breast; under wing-coverts black; quills ashy black below. Total length 5·1 inches, culmen 0·6, wing 3·3, tail 2, tarsus 0·85."

The figures in the Plate, which represent an adult bird of the natural size, are drawn from the Astrolabe-Mountain specimen lent to us by Mr. Ramsay, the type of his *P. sylvia*. [R. B. S.]

POECILODRYAS ALBIFACIES, Sharpe.

PŒCILODRYAS ALBIFACIES, *Sharpe*.

Southern White-faced Flycatcher.

Pœcilodryas albifacies, Sharpe, Journ. Linn. Soc. (Zool.) vol. xvi. p. 318 (1882).

ALTHOUGH closely allied to *Pœcilodryas leucops* of North-western New Guinea, Mr. Goldie sent so many specimens from the Astrolabe Mountains, all of which presented the same characters, that there can be little doubt of the two species being distinct. The present bird is very similar to *P. leucops*, but has the whole of the region round the eye white. In *P. leucops* the feathers in front of, below, and round the eye are black, leaving only a large loral spot of white, the facial appearance of *P. albifacies* being therefore quite different when the two birds are compared.

The native name is said by Mr. Goldie to be "Iddimattamatta;" he also states that the legs are yellow; and this is all we know concerning the species beyond the fact that it was procured by the above-named collector in the Choqueri district at the back of the Astrolabe Mountains.

The following description I copy from Mr. Sharpe's essay on Mr. Goldie's collection, published in the Linnean Society's Journal.

"*Adult.* General colour above olive-green, with a concealed spot of silky white on the sides of the rump; lesser and median wing-coverts like the back; primary-coverts and greater series dusky brown, edged with olive-green, the latter slightly tinged with rufous brown near the tips; quills dusky brown, externally olive, a little more yellow in colour than the back; tail-feathers light brown, edged with olive, and having a small tip of ashy white at the end of the inner web; forehead blackish, extending over the eye; top of head dark slaty grey with blackish shaft-streaks to the feathers, which are also very faintly tinged with olive; lores, feathers in front of the eye impinging on the forehead, feathers above and around the eye, as well as the space below the eye, pure white; ear-coverts slaty black; cheeks and chin white faintly washed with yellow; throat and under surface of body bright yellow, the sides of the breast and flanks olive-greenish, a tinge of which is also in the centre of the breast; axillaries bright yellow; under wing-coverts white washed with yellow, with a dusky patch near the edge of the wing, which is also yellow; quills ashy-brown below, whitish along the edge of the inner web. Total length 4·5 inches, culmen 0·55, wing 2·85, tail 1·9, tarsus 0·8."

The figures in the Plate represent a pair of birds of the size of life, drawn from the typical specimens in the British Museum. [R. B. S.]

POECILODRYAS PLACENS.

POECILODRYAS PLACENS.

Yellow-banded Robin.

Eopsaltria placens, Ramsay, Proceedings of the Linnean Society of New South Wales, iii.
Pœcilodryas flavicincta, Sharpe, Annals and Mag. of Nat. Hist. 5th series, vol. iii. p. 313.
Pœcilodryas placens, Sharpe, Journ. Linn. Soc., Zool. xiv. p. 630.

My first acquaintance with this brightly coloured Robin was in the month of March 1879, when five specimens were sent to this country by Mr. Kendal Broadbent, whose recent researches in South-eastern New Guinea have earned him a reputation as one of the best collectors in Australia. The collection has been described by Mr. Bowdler Sharpe in the 'Journal of the Linnean Society of London;' but a diagnosis of the present species was published by him in the April number of the 'Annals.' Scarcely, however, had Mr. Sharpe's description been published and become beyond recall, when a paper of Mr. Ramsay's was received in this country, containing an account of the collections made in south-eastern New Guinea by Messrs. Goldie and Broadbent. This paper purports to have been read as long ago as the 30th of September 1878; and at any rate the description of Mr. Ramsay's *Eopsaltria placens* must have been published long before that of Mr. Sharpe's *Pœcilodryas flavicincta*. The former gentleman remarks on the structural peculiarities of the species as showing a likeness to the genus *Leucophantes* of Sclater; and that genus, Mr. Sharpe has just shown us in the fourth volume of his 'Catalogue of Birds,' must be considered a synonym of my genus *Pœcilodryas*.

It is much to be regretted that the specimens sent by Mr. Broadbent were sold in London with an assurance that they had been sent direct to England, whereas it now turns out that a portion of the collection had also been sent to Sydney. Hence Mr. Sharpe and Mr. Ramsay were both led to describe the new species independently of each other; and thus a bird coming from such a recently explored field as S.E. New Guinea is introduced to the notice of ornithologists with two synonyms affixed to it in the twinkling of an eye. The skins forwarded to Sydney are marked by Mr. Broadbent as having come from the mountain-scrub of Goldie's River.

The following is a translation of Mr. Sharpe's original description of *P. flavicincta*.

Adult. General colour above yellowish green; the wing-coverts and quills dusky black, edged with the green colour of the back; tail-feathers dusky brown, externally edged with green, and having a small white tip; crown and nape dark ashy grey; chin, fore part of cheeks, and ear-coverts uniform with the head, the latter rather blacker; hinder part of cheeks, lower part of throat, and jugular region bright yellow, as also the sides of the head, forming a broad collar across the throat; fore neck and upper breast yellowish green; rest of under surface of body very bright yellow; under wing-coverts and axillaries white washed with bright yellow. Total length 5·3 inches, culmen 0·7, wing 3·65, tail 2·2, tarsus 0·9.

One of the figures in the accompanying Plate represents a specimen in my own collection, the other that in the British Museum.

HYPOTHYMIS ROWLEYI.

HYPOTHYMIS ROWLEYI.

Rowley's Blue Flycatcher.

Zeocephus rowleyi, Meyer, in Dawson Rowley's Orn. Misc. iii. p. 163 (1878).
Hypothymis rowleyi, Sharpe, Brit. Mus. Cat. Birds, iv. p. 278 (1879).

The late Marquis of Tweeddale, in his well-known memoir on the Birds of Celebes, has certainly proved that, while it possesses a large number of peculiar forms, the island of Celebes is rather Indian in the affinities of its avifauna than Austro-Malayan; but inasmuch as there is also a strong Moluccan element perceptible in the birds of the island, I propose still to include them in the present work, as the completion of my 'Birds of Asia' prevents me from figuring them in that work, to which perhaps they would more properly belong.

If, as Dr. Meyer first suggested, the present species had turned out to be a true *Zeocephus*, it would have been a fact of the highest interest, as the latter genus is hitherto known only from the Philippine archipelago. It is scarcely less interesting, however, to find that it belongs to the strictly Indian genus *Hypothymis*, which thus gains a more extended range. I am glad that the specific name chosen by Dr. Meyer will perpetuate the memory of such an ardent naturalist and true patron of science as the late Mr. George Dawson Rowley, whose untimely death was a real loss to ornithology.

Hypothymis rowleyi is nearly allied to *H. puella* from Celebes and the Sula Islands, and, like that species, has no black collar or nape-spot, and has no black on the forehead or chin. It differs from *H. puella*, however, in having the under surface of a light silvery blue, while the colour of the back is a greyish cobalt: in the Celebean species the upper and under surface are alike in their shade of blue. The type specimen in the Dresden Museum still remains unique; it was procured by Dr. Meyer's collectors in Great Sangi Island, the avifauna of which, as far as we know, is Celebean in its character.

I take the accompanying description of the type from Mr. Sharpe's ' Catalogue of Birds : '—

"*Adult*. General colour above greyish cobalt-blue, more grey on the rump; lesser and median wing-coverts like the back, the greater series and the quills light bluish grey, edged externally with the same greyish cobalt as the back of the quills, internally dusky blackish; tail greyish blue, dusky blackish on the inner webs; sides of face more dusky greyish blue than the head; cheeks and under surface of body light silvery bluish, darker on the throat, and more dusky blue on the sides of the breast; under wing-coverts and axillaries like the breast. Total length 6·5 inches, culmen 0·65, wing 3·65, tail 3·6, tarsus 0·95."

I am indebted to Dr. Meyer for the loan of the type specimen, from which the accompanying life-size figures have been drawn.

TODOPSIS CYANOCEPHALA, (Q & G.)

TODOPSIS CYANOCEPHALA, Q. & G.

New-Guinea Todopsis.

Todus cyanocephalus, Quoy et Gaimard, Voy. de l'Astrolabe, Zool. p. 227, pl. v. fig. 4.
Philentoma cyanocephala, Jacq. et Pucher. Voy. Pôle Sud, Zool. iii. p. 79, pl. xx. fig. 2.
Tchitrea? cæruleocephala, Gray, Gen. B. i. p. 260.
Todopsis cæruleocephala, Bp. Comptes Rendus, xxxviii. p. 652.
Todopsis cyanocephala, Gray, P. Z. S. 1858, p. 177; 1859, p. 156; 1861, p. 434.—Id. Cat. Mamm. & Birds New Guinea, p. 27.—Finsch, Neu-Guinea, p. 168.—Sclater, P. Z. S. 1873, p. 696.—Meyer, Sitz. k. Akad. Wien, lxix. pp. 74, 78.—Salvad. Ann. Mus. Civic. Genov. x. p. 148.
Tchitrea cyanocephala, Gray, Hand-l. B. i. p. 334, no. 5031.

ONE of the greatest difficulties I have had to contend with in the present work has been the utter want of any notes on the habits of most of the birds which it becomes my duty to figure. In the work on the Birds of Australia I had my own personal observations to record, the result of two years' acquaintance with the birds in their native wilds; in the Birds of Asia there is generally a chance of finding some field-notes among the papers of Mr. Hume or some of his excellent coadjutors in India; while, of course, in writing the 'Birds of Great Britain,' there was always a plethora of works to consult, which rendered it rather a matter of selection than of copying. In the present work, however, the case is quite different. Many of the species figured are from obscure or little-known localities, penetrated by European naturalists for the first time, where no leisure for studying the habits of the birds shot was obtainable; or the collection has been made by trained native hunters, whose only object is to shoot and preserve specimens, and from whom, of course, no information on the economy of the birds can be expected. The above remarks have been called forth by the fact that all my attempts to gain the slightest clue to the habits of *Todopsis* have failed, neither the early nor the recent travellers in New Guinea having given us any information on the subject. This is the more to be regretted, as I find myself at variance with several ornithologists as to the position of these little birds. From my knowledge of the Australian Wrens of the genus *Malurus*, I cannot help the conviction that, notwithstanding their long broadened bills and plentiful bristles, the members of the genus *Todopsis* are Warblers and not Flycatchers, and should be placed in the vicinity of the above-named genus. Count Salvadori arrived at this conclusion quite independently; and I am glad to have his support in this opinion. On the other hand, Mr. Sharpe considers them to be Flycatchers, and regards this (the usual) view of their position as strictly correct. I am indebted to my friend Mr. Sharpe for the opportunity of seeing his MSS. on these birds, and for permission to copy the careful descriptions of this and the succeeding species from the fourth volume of the 'Catalogue of Birds,' as follows:—

Adult male. Crown of head bright turquoise-blue, extending in a narrow band down the nape and hind neck; lores, and a narrow frontal band, feathers above the eye, sides of face and of neck, and hinder neck velvety black; middle of the back and scapulars bright cobalt, as also the lesser wing-coverts; greater and primary coverts black, externally edged with purple; quills black, the secondaries externally bright cobalt, the innermost purplish blue; entire back from below the mantle velvety black, with a slight purple gloss; upper tail-coverts deep purple; tail-feathers black, dull blue on the outer webs, the two centre feathers entirely shaded blue; cheeks and entire under surface of body dark purple; under wing-coverts black.

Total length 6 inches, culmen 0·65, wing 2·35, tail 2·4, tarsus 1·0. Signor D'Albertis describes the bill, feet, and iris as black.

Adult female. General colour above chestnut-brown, the wing-coverts like the back; quills dark brown, externally edged with chestnut; crown of head bright turquoise-blue, extending in a band down the nape; a narrow frontal line, lores, sides of face, and sides of the hinder crown purplish black; cheeks and throat dull cobalt; lower throat, breast, and centre of the body dull creamy white, the lower throat laterally cobalt; sides of body, flanks, thighs, and under tail-coverts chestnut, glossed with lilac on the sides of the upper breast; under wing-coverts chestnut, the outer lower coverts buffy white, as also the edge of the wing; lower surface of quills dark brown, edged with rufous along the inner web; tail light indigo, waved across with dusky lines under certain lights, all the feathers rather broadly tipped with white.

Total length 5·2 inches, culmen 0·65, wing 2·25, tail 2·35, tarsus 0·95.

The present species seems to be confined to North-western New Guinea. Wallace procured it at Dorey, D'Albertis at Ramoi; and it was obtained in Salwatti by Mr. Bruijn's hunters.

The figures in the Plate represent a male and a female, of the natural size, drawn from specimens lent to me from the Dresden Museum under the direction of Dr. Meyer.

TODOPSIS BONAPARTII, Gray.

TODOPSIS BONAPARTII, *Gray*.

Bonaparte's Todopsis.

Todopsis cyanocephala, Gray, P. Z. S. 1858, p. 177, pl. cxxxiv (nec Quoy et Gaimard).
Todopsis bonapartii, Gray, P. Z. S. 1859, p. 156, 1861, p. 434.—Finsch, Neu-Guinea, p. 168.—Meyer, Sitz. k. Akad. Wien, lxix. pp. 78, 80.—Sharpe, Journ. Linn. Soc. xiii. pp. 316, 498.
Todopsis, sp., Salvad. Ann. Mus. Civic. Genov. ix. p. 25.
Tchitrea bonapartii, Gray, Hand-l. B. i. p. 334, no. 5032.

The present species very closely resembles the preceding one, and was mistaken for it at first by the late Mr. George Robert Gray in 1858. In the course of the following year, however, the receipt of the true *Todopsis cyanocephala* from New Guinea showed him that the species from the Aru Islands was a different one; and he named it thereupon after Prince Bonaparte, the original proposer of the genus. Since that date very little information has been added to our knowledge of the Aru bird, until in 1876 it was discovered on the mainland of South-eastern New Guinea by the late Dr. James, who found it there, about eight miles east of Yule Island, "inhabiting clumps of trees and shrubs in the midst of scrub." This solitary note is, I believe, all that has ever been published respecting the habits of any species of *Todopsis*. Signor D'Albertis met with it in the same part of New Guinea, at Naiabui. Count Salvadori, in recording the latter specimen, thought that the bird from South-eastern New Guinea might be *T. mysoriensis* of Meyer —a bird which certainly very much resembles *T. bonapartii*, but differs (so Mr. Sharpe tells me) in having the upper back black and only the mantle ultramarine, whereas the latter colour is more extended in *T. bonapartii*, occupying both the mantle and upper back.

The following descriptions are from Mr. Sharpe's 'Catalogue of Birds:'—

Adult male. Crown of head bright cobalt of an enamelled texture, running in a narrower line down the nape; a narrow frontal line, lores, feathers above and below the eye, cheeks, ear-coverts, sides of neck and hinder neck, the latter washed with purple, mantle, scapulars, and lesser wing-coverts purplish cobalt; the greater series of coverts and the inner secondaries black, externally edged with purple; primaries black, with scarcely any purple edgings; back velvety black, glossed with purple, the upper tail-coverts of the last-named colour; tail-feathers purplish black, inclining to duller black on the inner webs; entire under surface of body deep purple, much brighter on the breast and flanks; under wing-coverts black glossed with purple. Total length 5·8 inches, culmen 0·65, wing 2·3, tail 2·5, tarsus 0·95.

Adult female. Crown of head cobalt, extending in a rather broad band down the nape; a narrow frontal line, lores, sides of face including a narrow eyebrow, and the sides of the hinder crown and nape purplish black; upper surface of body maroon-chestnut, as also the scapulars and wing-coverts, some of the outermost coverts of the thumb spotted with lilac; quills dark brown, externally edged with rufous like the back, the secondaries tipped with pale rufous; tail-feathers dull indigo, obscurely waved under certain lights, and broadly tipped with white; entire throat purplish blue, descending onto the sides of the chest; centre of of the fore neck, chest, and middle of the body white, the sides of the body light maroon-chestnut, including the thighs and under tail-coverts; the sides of the upper breast distinctly glossed with lilac; under wing-coverts very light rufous, paler on the lower series, a spot on the edge of the wing cobalt-blue; "bill black; feet dusky olive; iris dark" (*Wallace*, MS.). Total length 5·9 inches, culmen 0·65, wing 2·25, tail 2·65, tarsus 0·9.

In the Plate I have figured a male of the natural size, with the head of a male, and in the centre is a bird which I take to be a young one, all from the Aru Islands. I was at first inclined to believe that it might be an old female exhibiting differences from the same sex of *T. cyanocephala* and wanting the blue throat. As, however, the British Museum contains blue-throated females of *T. bonapartii* from the Aru islands, I believe now that it must be only an immature bird, which has not gained the blue throat of the adult.

TODOPSIS WALLACII, Gray.

TODOPSIS WALLACII, Gray.
Wallace's Todopsis.

Todopsis wallacei, Gray, P. Z. S. 1861, pp. 429, 434, pl. xliii. fig. 2.—Finsch, Neu-Guinea, p. 168.—Meyer, Sitz. k. Akad. Wien, lxix. p. 81.
Tchitrea wallacei, Gray, Handl. B. i. p. 334, no. 5033.

This pretty little species is rather different in structure of bill from the other members of the genus. The bill is longer, rather more curved; and the bristles are longer and more numerous, reaching to the end of the bill. In colours also it is very distinct; but I do not feel disposed to found a new generic title at present on these characters.

There is, however, another point in connexion with this bird to which I must call attention; and this has reference to the small *Todopsis* which comes from the Aru Islands, and of which I have a single specimen in my collection. I have compared this with the typical examples from Mysol in the British Museum, and I notice that the Aru bird has a black streak under the eye and the spots on the head are in the form of rounded tips of blue to the feathers of the crown, quite different from the lanceolate tips which appear in the Mysol species. If the receipt of future specimens should confirm my impression that the Aru bird is distinct, I propose the name of *Todopsis coronata* for it.

Wallace's *Todopsis* differs from all the others in its small size, exceedingly delicate legs and toes, and the whitish colouring of the entire undersurface, and in having the tail tipped with white, which is only found in the females of the other members of the genus. Nothing has been recorded of the habits of this species, which is described as follows by the late Mr. George Robert Gray:—

"Top of the head black, with the tips of the feathers light blue, and the shaft bluish white; back rufous, wings and tail dark brown; wing-coverts tipped with white; round the eyes, lores, ear-coverts, and beneath the body white; quills narrowly margined with white; the tips of the outer tail-feathers also white.

"Total length 4 inches 7 lines, bill from gape 8 lines, wings 2 inches.

"The young bird is rufous-white on the throat; the bill is black tipped with yellowish white, differs from that of the typical *Todopsis* in being longer and somewhat curved and in having the bristles as long as the bill."

The three figures in the Plate are of the size of life, the one in the foreground representing the *T. coronata* from Aru, the centre bird being drawn from one of the typical specimens in the British Museum.

TODOPSIS GRAYI, Wall.

TODOPSIS GRAYI, *Wall.*

Gray's Todopsis.

Todopsis grayi, Wallace, P. Z. S. 1862, p. 166.—Finsch, Neu-Guinea, p. 168.—Beccari, Ann. Mus. Civic. Genov. vii. p. 709.—Meyer, Sitz. k. Akad. Wien, lxix. pp. 81, 212.
Tchitrea grayi, Gray, Hand-list of Birds, i. p. 334, no. 5034.

This is one of the most distinct species of *Todopsis* yet discovered, being totally unlike any of the other members of the genus, its silvery grey or glaucous coloration being quite sufficient to distinguish it at a glance. It was discovered by Mr. Wallace in the mountains of Sorong in North-western New Guinea; and it still remains one of the rarest species of the genus, as I am only aware of one other instance of its capture, viz. the specimen mentioned by Beccari as having been obtained by Mr. Bruijn's hunters in the Arfak mountains. It is probable, however, that the *Todopsis sericyanea* of Rosenberg, and the *Myiagra glauca* of Schlegel may also belong to the present species; but at present I have not been able to determine this for certain.

It is much to be regretted that Dr. Beccari never met with this bird himself, as from his pen we should have undoubtedly received some information as to its habits, of which I can at present record absolutely nothing. The following description of the type is from Mr. Sharpe's 'Catalogue of Birds:'—

Adult. General colour above dull blue, brighter and more cobalt on the lower back and rump; wing-coverts brown, slightly washed externally with olive, and tipped with dull verditer blue; quills brown, broadly washed externally with olivaceous brown; tail-feathers brown, externally bluish, slightly washed with olivaceous on the margins and tipped with white; head dull greyish blue, inclining to ashy verditer above the eyes and on the sides of the crown: feathers in front of the eye blackish, as also a broad streak above the ear-coverts; round the eye a circlet of verditer feathers; ear-coverts verditer blue, the feathers rather elongated, and streaked with brighter cobalt; cheeks and throat silvery whitish, with a bluish gloss, the rest of the under surface pale ashy blue, washed with a cobalt shade; under wing-coverts like the breast; thighs olivaceous brown.

Total length 5·7 inches, culmen 0·7, wing 2·55, tail 2·5, tarsus 1·0.

The figures in the Plate, which represent the species in two different attitudes, are taken from the type specimen in the British Museum.

MALURUS ALBOSCAPULATUS, *Meyer*.

MALURUS ALBOSCAPULATUS, Meyer.

Pied Malurus.

Malurus alboscapulatus, Meyer, Sitzungsberichte der k.-k. Akademie der Wissenschaften zu Wien, Bd. lxix. p. 496 (1874).—T. Salvadori, Ann. del Mus. Civ. di Genova, vol. vii. 1875, p. 778.

THE discovery of this little bird in New Guinea is a welcome addition to the Australian genus *Malurus*. I say Australian genus; for it is in that country where all the other known species are found, and over which they are generally dispersed. They are divisible into several little sections to which generic terms might be given—the blue-crowned bird of Tasmania forming part of a group which differs from the variegated and more gorgeous species of the mainland, the delicate white-winged birds inhabiting the interior, the red-backed frequenting the great grass-beds of the plains, being as many natural divisions. The nests of all the species are dome-shaped; and many of the kinds are foster-parents of the little Bronzy Cuckoo *Chalcites lucidus* &c. The two principal figures in the accompanying Plate are copied from a bird in Dr. Meyer's Arfak collection, while the other is from a specimen collected on the south coast. I have taken considerable trouble to satisfy myself that the birds received from these distant localities are really identical; and I may state that size, and size alone, is the only difference that exists between them, the southern bird being by about one sixth the smallest in all admeasurements. Until I have an opportunity of seeing more specimens than I have done, I shall regard the two birds, although so widely distributed, as one and the same. Of M. Salvadori's *naimii* I have not seen a specimen, and am therefore unable to state if it is a female or young male or a different bird from the one under consideration.

A few words will sufficiently describe the *Malurus alboscapulatus*. Body and tail shining velvety black; wings brownish, on each shoulder a large glowing white spot; bill, feet, and tarsi black.

Hab. Arfak Mountains, New Guinea.

The figures in the accompanying Plate are of the natural size.

RHIPIDURA RUBROFRONTATA, Ramsay.

RHIPIDURA RUBROFRONTATA, *Ramsay.*

Rufous-fronted Fantail Flycatcher.

Rhipidura rubrofrontata, Ramsay, Proc. Linn. Soc. N. S. Wales, iv. p. 82 (1879).—Layard, Ibis, 1880, p. 293.—
Ramsay, Proc. Linn. Soc. N. S. Wales, vi. p. 178 (1881).—Salvad. Orn. Papuasia e delle Molucche,
ii. p. 68 (1881), iii. p. 533 (1882).
Rhipidura rufofronta, Ramsay, Nature, xx. p. 125 (1879).—Salvad. Ann. Mus. Civic. Genov. xiv. p. 508 (1879).
Rhipidura rufofrontata, Salvad. Ibis, 1880, pp. 127, 129.

THE present species belongs to the group of Fantailed Flycatchers which have the lower part of the back cinnamon-rufous, contrasting in colour with the mantle and upper back. It is very closely allied to the Australian *Rhipidura rufifrons*, but is a smaller bird and differs, moreover, in the following characters, viz. :— The tips of the tail-feathers are pure white and not ashy white, and the under tail-coverts are deep rufous instead of pale cinnamon-buff; the black band of the fore neck is narrower, and the ear-coverts are blacker. The rufous base on the tail is more restricted in *R. rubrofrontata*, where it is hidden by the coverts, whereas in *R. rufifrons* it occupies nearly half the tail-feathers.

It would thus appear that *R. rubrofrontata* is the representative of *R. rufifrons* in the Solomon Islands, where it was discovered by Mr. Cockerell in the Island of Guadalcanar. Mr. Ramsay at one time considered it to be identical with Canon Tristram's *R. russata* from S. Christoval; but as he lent us the type specimen of the present bird, we were enabled to compare the two species, and can affirm that they are quite distinct.

The following description is taken from the typical specimen lent to us by Mr. E. P. Ramsay :—

Adult male. General colour above brown, the lower back, rump, and upper tail-coverts rufous; wing-coverts like the back, the outer ones slightly tinged with rufous; quills dark brown, edged with paler brown, tinged with rufous on the inner secondaries; tail-feathers blackish brown, broadly tipped with white, and having the base of the outer web rufous, this being almost entirely concealed by the rufous upper tail-coverts; crown of head brown like the back; forehead and feathers above the eye orange-rufous, as well as the upper edge of the eyelid, the hinder frontal plumes tipped with tiny brown spots like the rest of the crown; lores, feathers below the eyes, and ear-coverts blackish brown, with a white mark on the lower eyelid; cheeks and throat white, followed by a black band across the lower throat; the feathers of the fore neck scaly, black with white edges; sides of the fore neck and chest dark ashy, the sides of the breast brown, with a rufous tinge; centre of abdomen white; thighs and under tail-coverts orange-rufous; axillaries and under wing-coverts brown washed with rufous, the lower ones white; quills dusky below, ashy whitish along the inner web. Total length 5·2 inches, culmen 0·5, wing 2·6, tail 2·9, tarsus 0·7.

In the Plate two figures are given, of the natural size, drawn from the type specimen which Mr. Ramsay lent to us.

[R. B. S.]

RHIPIDURA LEUCOTHORAX, Salvad.

RHIPIDURA LEUCOTHORAX, Salvad.

White-breasted Fantail Flycatcher.

Rhipidura leucothorax, Salvad. Ann. Mus. Civic. Genov. vi. p. 311 (1874).—Id. & D'Albert. op. cit. vii. p. 820 (1875).—Salvad. op. cit. ix. p. 25 (1876), x. p. 134 (1876).—Oustalet, Bull. Soc. Philom. 1877, p. 6.—Sharpe, Cat. B. Brit. Mus. iv. p. 327 (1879).—D'Albert. & Salvad. Ann. Mus. Civic. Genov. xiv. p. 60 (1879).—Salvad. t. c. p. 498 (1879).—Id. Orn. Papuasia, etc. ii. p. 58 (1881).
Rhipidura episcopalis, Ramsay, Proc. Linn. Soc. N. S. W. ii. p. 371 (1878).—Salvad. Ibis, 1879, p. 323.

This species belongs to the section of the genus *Rhipidura* with blackish under tail-coverts and rounded white spots on the chest. The large white breast-patch distinguishes it from all its near allies.

In North-western New Guinea the present bird has been found in Hatam and at Mariati by Bruijn, and D'Albertis procured specimens in Southern New Guinea on the Fly River, near Hall Bay and at Naiabui. Mr. Ramsay has also described a bird as *R. episcopalis* from Goldie River, in the interior of South-eastern New Guinea, which is certainly the same as *R. leucothorax*. It should be noticed, however, that the single example in the British Museum, determined by Count Salvadori himself to be *R. leucothorax*, differs from his description in having the chin white, and may belong to another species, unless it is the young bird.

The following is a description of the type specimen of *Rhipidura episcopalis*, which has been lent to us by Mr. Ramsay :—

Adult male. General colour above earthy brown, the head a little more dusky than the back and blacker on the forehead; scapulars like the back; wing-coverts black, spotted with white at their ends; primary-coverts blackish; quills dark sooty brown; upper tail-coverts black; tail-feathers black, tipped with white, increasing in extent toward the outermost; lores and base of forehead, feathers above and below the eye, and ear-coverts black, extending on to the sides of the neck; above the eye a band of white separated from the crown by a narrow line of black; hinder part of cheeks white, widening out on the sides of the throat and neck, and forming a large white patch; fore part of cheeks and throat black, widening out on the fore neck and chest, and having rounded white spots on the latter; remainder of breast white, the adjoining chest-feathers black, tipped with white, or white with black edges; sides of body and flanks pale ashy brown; vent and under tail-coverts black, the latter with white tips; thighs black; axillaries white with black bases; under wing-coverts black, tipped with white; quills dusky below, pale brown along the edge of the inner web. Total length 6·8 inches, culmen 0·65, wing 3·05, tail 3·5, tarsus 0·85.

The Plate has been drawn from the specimen above described, which is represented in two positions, of the size of life.

[R. B. S.]

RHIPIDURA COCKERELLI, Ramsay.

RHIPIDURA COCKERELLI.

Cockerell's Fantail Flycatcher.

Sauloprocta cockerelli, Ramsay, Proc. Linn. Soc. New South Wales, iv. p. 81 (1879).—Salvad. Ann. Mus. Civic. Genova, xiv. p. 508.—Id. Ibis, 1880, p. 129.—Id. Orn. Papuasia, etc. ii. p. 53 (1881).—Id. *op. cit.* iii., App. p. 531 (1882).
Rhipidura cockerelli, Ramsay, Proc. Linn. Soc. New South Wales, vi. p. 181 (1881).—Tristr. Ibis, 1882, p. 142.—Ramsay, Ibis, 1882, p. 473.

This species was described by Mr. Ramsay from a specimen obtained in Guadalcanar, in the Solomon group, by Mr. Cockerell. In the absence of specimens Count Salvadori was inclined to think that it might be the same species as *R. tricolor*, an identification which was rather warmly resented by Mr. Ramsay, who certainly might be supposed to know thoroughly so familiar a species as *R. tricolor* of Australia. On examining the type specimen, which Mr. Ramsay has kindly lent to us, we find that it is not of the *Sauloprocta* type at all, but is, as Count Salvadori has more recently suggested, a true *Rhipidura* as regards coloration. As, however, we cannot admit that *Sauloprocta* is generically distinct from *Rhipidura*, it follows that its position in the latter genus would be between *R. perlata* and *R. tricolor*, in the same black-throated section as the last-named bird. It may, however, be distinguished by its smaller size, by the ovate drops of white on the fore neck and chest, and by the broad white margins to the inner secondaries.

The following description is taken from the type specimen lent to us by Mr. Ramsay:—

Adult male (type of species). General colour above brownish black, the long feathers of the rump tipped with white and forming a tolerably complete band; lesser wing-coverts like the back; median and greater coverts blackish brown, narrowly edged with black, the latter with a small spot of white at the ends; bastard wing, primary-coverts, and quills blackish brown, the inner secondaries broadly margined with white, extending round the ends of the innermost; upper tail-coverts and tail-feathers blackish brown; head black, deeper than the back; over the eye a spot of silvery white; cheeks, throat, and chest black, the cheeks slightly varied with a few white tips to the feathers, the lower throat, fore neck, and chest variegated with large ovate subterminal drops of white on each feather; breast, abdomen, and under tail-coverts white, with a few blackish edgings to the breast-feathers; thighs black; axillaries blackish brown tipped with white; under wing-coverts black, with white spots at the ends of the feathers; quills dusky blackish below, browner along the edge of the inner web. Total length 6·3 inches, culmen 0·6, wing 3·35, tail 3·4, tarsus 0·6.

The Plate represents the adult male in two positions, drawn from the typical example. The figures are of the natural size.

[R. B. S.]

RHIPIDURA OPISTHERYTHRA, *Sclater.*

RHIPIDURA OPISTHERYTHRA, Sclater.
Larat Fantail Flycatcher.

Rhipidura opistherythra, Sclater, Proc. Zool. Soc. 1883, p. 197.

The present species belongs to the Australian group of Rufous-backed Fantail Flycatchers; it is very nearly allied to *Rhipidura rufa*, but is much larger and has a white spot on the lores. The throat is also white instead of being dull ashy, and the tail likewise seems different, as *R. rufa* is described as having broad tips of pale rufous to the tail-feathers, which cannot be said to be the case with the present species.

Mr. H. O. Forbes discovered this Flycatcher on the islands of Larat and Maroe in the Tenimber group. The latter island is distant about twenty miles to the north-west of the mainland of the northern island, which the natives call Yamdena, and which is known to Europeans as Timor Laut.

The following description is taken from the typical specimen:—

Adult male. General colour above dusky brown, gradually shading off into rufous on the back, and deepening into ferruginous on the lower back, rump, and upper tail-coverts; lesser wing-coverts like the back, slightly washed with ferruginous; median and greater coverts, bastard wing, primary-coverts, and quills dusky brown, narrowly edged with deep ferruginous; tail-feathers dusky brown, externally ferruginous, the centre feathers broadly tipped with the latter colour, the others becoming reddish brown towards their ends, this colour increasing in extent towards the outermost feather; lores dull white, forming a rather conspicuous spot; no eyebrow; ear-coverts dusky brown, with pale shaft-lines; cheeks and throat white; fore neck and chest pale ochraceous buff, deepening into pale reddish buff on the rest of the under surface; thighs and under tail-coverts more ferruginous; under wing-coverts and axillaries like the breast; quills below dusky, inner edges ashy rufous; "upper mandible sooty brown, lower one sooty brown at tip, pale flesh-colour at the base; legs and feet lavender-pink; iris dark brown" (*H. O. Forbes*). Total length 6·8 inches, culmen 0·6, wing 2·8, tail 4, tarsus 1.

The type specimen, for the loan of which we are indebted to the kindness of Dr. Sclater, is represented in the accompanying Plate of the natural size and in two positions.

[R. B. S.]

RHIPIDURA HAMADRYAS, Sclater.

RHIPIDURA HAMADRYAS, Sclater.

Rufous-backed Fantail Flycatcher.

Rhipidura hamadryas, Sclater, Proc. Zool. Soc. 1883. p. 54.

This species belongs to the section of the genus *Rhipidura* with the rump and upper tail-coverts rufous or cinnamon-coloured. Its nearest allies would appear to be *R. semicollaris* from Eastern Timor and *R. dryas* from Northern Australia; but from both of these it is distinguished by the mantle being rufous like the rest of the back, and by the ashy tips to the tail-feathers.

The only specimens known were procured in Larat by Mr. H. O. Forbes during his first expedition to the Tenimber group; and I am indebted to Dr. Sclater for lending me the type specimen from which the accompanying description has been drawn up.

Adult female. General colour above dark ferruginous or cinnamon-rufous; lesser wing-coverts brown, with a slight wash of rufous; median and greater coverts, bastard-wing, primary-coverts and quills dusky brown, narrowly margined with ashy rufous, a little clearer rufous on the inner greater coverts and the secondaries; upper tail-coverts like the back; tail-feathers ashy brown, all but the centre ones with a broad tip of greyish white to the inner web, before which is a shade of black, the outer feathers also shaded with ashy whitish at the end of both webs; the bases of all the tail-feathers edged with ferruginous; forehead bright ferruginous, extending backwards to above the eye; remainder of the crown and hind neck dusky brown; lores and feathers below the eye blackish; ear-coverts sooty brown; cheeks and throat white, with a large spot of black on the lower part of the latter; remainder of under surface creamy buff; the sides of the upper breast ashy olive; sides of body and flanks a little deeper fulvous, as also the thighs; under tail-coverts fulvous white; under wing-coverts and axillaries ashy fulvous, the latter with whitish edgings; quills below dusky; inner edge of quills ashy; "legs and feet black; iris dark brown or black" (*H. O. Forbes*). Total length 5·6 inches, culmen 0·5, wing 2·4, tail 3, tarsus 0·8.

The Plate represents an adult female of this species, of the natural size, in two positions.

[R. B. S.]

RHIPIDURA FUSCORUFA, Sclater.

RHIPIDURA FUSCO-RUFA, *Sclater*.

Timor-Laut Fantail Flycatcher.

Rhipidura fusco-rufa, Sclater, Proc. Zool. Soc. 1883, p. 197, pl. xxvii.

This Flycatcher appears to us, after a comparison with the "Keys to the species" of *Rhipidura* given in the British-Museum 'Catalogue of Birds,' and in Count Salvadori's 'Ornitologia della Papuasia,' to belong to a totally new section of the genus. Its rufous outer tail-feathers and a peculiar distribution of rufous and brown coloration about the bird render it quite distinct from any other known species. In the character of the tail it approaches somewhat the female of *R. brachyrhyncha*, Schlegel, from New Guinea; but there the resemblance ends, as the coloration of the rest of the plumage is totally different. Mr. Forbes sent a large series of this species from Larat, Moloe, and Loetoe; and it would appear to be common in the Tenimber Islands which he visited.

The following is a description of the type specimen:—

Adult male. General colour above dull chocolate-brown, the head and hind neck duller and more sooty brown; wing-coverts dusky brown, the least and median series tipped and the greater series edged with pale ferruginous; bastard wing, primary-coverts, and quills dusky black; the secondaries edged with pale ferruginous, more broadly on the inner ones; the four centre tail-feathers entirely dusky blackish, the next pair blackish, excepting for a shade of pale ferruginous towards the end of the feather, the three outermost entirely pale ferruginous with yellowish-white shafts; sides of face, lores, and ear-coverts dusky blackish; the fore part of the cheeks white, and a half-concealed white spot above the eye; throat and fore neck white, the sides of the latter ashy streaked with white; remainder of the under surface very pale ferruginous, deepening in tint on the thighs, vent, and under tail-coverts; under wing-coverts and axillaries richer ferruginous; quills dusky below, ashy rufous along the edge of the inner webs; "bill, legs, and feet black; iris dark brown" (*H. O. Forbes*). Total length 6·75 inches, culmen 0·75, wing 3·6, tail 3·5, tarsus 0·75.

The Plate represents an adult male of this species, in two positions, of the natural size.

[R. B. S.]

RHIPIDURA DRYAS, *Gould*.

RHIPIDURA DRYAS, Gould.

Wood-Fantail.

Rhipidura dryas, Gould, Birds of Australia, 8vo edition, vol. i. page 242.

THE members of the genus of Flycatchers called *Rhipidura* are so universally dispersed over Australia, New Guinea, the Philippines, and India that it would be difficult to name either of these countries where one or other of the members are not to be found. To enter into the specific characters of all the known species would be out of the question in a work like the 'Birds of Australia;' this can only be properly and effectively done by the monographist; suffice it to say that the present bird, notwithstanding what I have stated in my 'Introduction to the Birds of Australia,' is very distinct from any other species I have ever met with; and a comparison of the accompanying Plate with figures of its allies will at once convince ornithologists that this is the case.

The following is extracted from my 'Handbook to the Birds of Australia,' at the page quoted above.

"This bird differs from *R. rufifrons* in being of a smaller size, in its dark-grey tail-feathers being more largely tipped with white, and merely fringed with rufous at the base only, in the breast being white, crossed by a distinct band of black, and devoid of the dark spotted markings seen on the chest of its ally.

Total length $5\frac{3}{4}$ inches, wing $2\frac{5}{8}$, tail $3\frac{1}{4}$, tarsus $\frac{5}{8}$.

R. dryas inhabits the north-western portion of Australia, where it appears to be as common as *R. rufifrons* is in the south-eastern. I had several specimens, all of which bore a general resemblance to each other.

The majority of these little birds dwell in the utmost recesses of the forest; there they fan out their large tails, and make their displays without ever being seen, unless man in his wanderings and investigations should intrude upon their privacy. In the capture of their insect prey these tiny-billed gnat-catchers exhibit themselves in many graceful attitudes, sometimes running along the branches of trees, or over large stones, or restlessly darting here and there after *Aphidii* and other minute flies. When they are by nature prompted to breed, they construct without exception the neatest and most charming of nests, the grassy materials with which the walls are formed being woven together with the webs of the most minute spiders. In this frail structure, about the size of an egg-cup, two beautifully speckled eggs are deposited.

There is no outward difference between the sexes.

The Plate represents two individuals of the size of life.

RHIPIDURA HYPERYTHRA, Gray.

RHIPIDURA HYPERYTHRA, Gray.

Rufous-breasted Fantail Flycatcher.

Rhipidura rufiventris (nec Vieill.), Verh. Land- en Volkenk. p. 185 (1839–44).—Gray, Gen. B. i. p. 259 (1846).—Bp. Consp. i. p. 323 (1850).—Sclater, Journ. Linn. Soc. ii. p. 162 (1858).—Gray, Proc. Zool. Soc. 1858, pp. 176, 192.—Id. Cat. Mamm. etc. New Guinea, pp. 28, 57 (1859).—Id. Proc. Zool. Soc. 1861, p. 434.—Rosenb. J. f. O. 1864, p. 119.—Finsch, Neu-Guinea, p. 169 (1865).

Rhipidura hyperythra, Gray, Proc. Zool. Soc. 1858, pp. 176, 192.—Id. Cat. Mamm. etc. New Guinea, pp. 28, 57 (1859).—Id. Proc. Zool. Soc. 1861, p. 434.—Rosenb. J. f. O. 1864, p. 119.—Finsch, Neu-Guinea, p. 169 (1865).—Gray, Hand-list Birds, i. p. 331, no. 4977 (1869).—Sclater, Proc. Zool. Soc. 1873, p. 696.—Sharpe, Cat. Birds in Brit. Mus. iv. p. 338 (1879).—Salvad. Ann. Mus. Civic. Genov. xiv. p. 499 (1879).—Sharpe, Journ. Linn. Soc., Zool. xvi. p. 431 (1882).—Salvad. Orn. Papuasia e delle Molucche, ii. p. 65 (1881), iii. App. p. 532 (1882).

Rhipidura, sp., Gray, Hand-list Birds, i. p. 231, no. 4976 (1869).

Rhipidura muelleri, Meyer, Sitz. k. Akad. Wien, lxx. p. 502 (1874).—Salvad. Ann. Mus. Civic. Genov. x. p. 135 (1877).

Rhipidura castaneothorax, Ramsay, Proc. Linn. Soc. N. S. Wales, iii. p. 270 (1879), iv. p. 98 (1879).—Salvad. Ibis, 1879, p. 323.

This is a very distinct species of Fantail Flycatcher, and is easily recognized by its style of colour on the underparts, the chin and cheeks being white, the throat black, and the breast rufous.

It was first described by S. Müller from Lobo Bay, in New Guinea, but he gave it the name of *Rhipidura rufiventris*, which already belonged to the species from Timor. Count Salvadori states that he could not find the type of Müller's species in the Leiden Museum, and it was apparently exchanged away to the late Mr. Gould, as his collection contained a specimen of this bird marked as from Lobo Bay, which is now in the British Museum, having been purchased with the rest of the Gould Collection.

Mr. Wallace subsequently discovered the species in the Aru Islands, and we cannot perceive any difference between examples from the last-named locality and from New Guinea. It seems to inhabit the whole of the latter island, as Dr. Meyer found it at Rubi, and Signor D'Albertis on the Arfak Mountains. It was afterwards obtained by Mr. Broadbent on the Goldie River, about 40 miles inland from Port Moresby, and was named by Mr. Ramsay *R. castaneothorax*. We have, however, examined specimens from this part of New Guinea, and can affirm that they are the same as the Aru Island and Lobo birds. Mr. Goldie has forwarded a specimen from the Morocco district in the Astrolabe Mountains, where he says it is called by the natives "Urobiagga." Mr. H. O. Forbes has likewise sent a pair from the Sogeri district in the same range of mountains, where he obtained them at an altitude of 2000 feet.

The following is a description of the type specimen of *R. hyperythra* in the British Museum:—

General colour above slaty grey, a little darker on the crown and sides of the head; the lores and feathers round the eye blackish; over the eye a distinct white streak, and a second narrow line of white above the upper edge of the ear-coverts less distinctly indicated; cheeks, chin, and moustache white; centre of throat black, widening out upon the lower throat; rest of under surface of body orange-rufous, slightly paler towards the lower abdomen and vent; thighs slaty grey; under wing-coverts pale orange-buff; quills ashy below, whitish along the inner webs; wings above resembling the back, the greater series and secondaries brown, washed externally with grey; the primaries entirely brown; median and greater coverts tipped with buff or whitish spots; tail slaty black, the three outer feathers tipped with white, with an obscure subterminal band of dull brown: "bill black, the lower mandible yellow; feet dusky olive" (*Wallace*). Total length 6·4 inches, culmen 0·5, wing 2·8, tail 3·2, tarsus 0·65.

The male bird collected by Mr. Forbes has the head blacker than in the female, but otherwise the sexes are alike in colour.

The figures in the Plate represent a male and female of the natural size, drawn from the pair of specimens obtained by Mr. Forbes in the Astrolabe Mountains.

[R. B. S.]

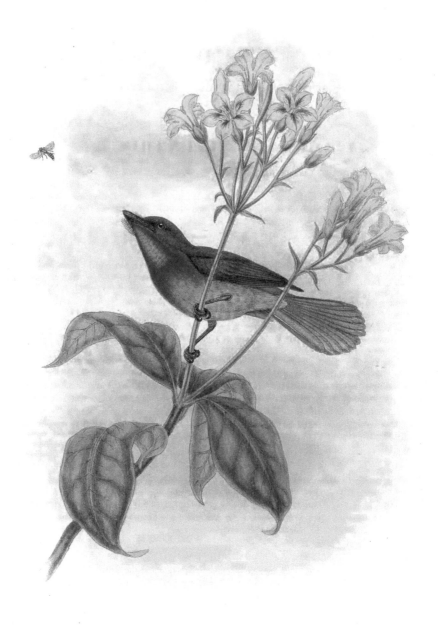

MYIAGRA FULVIVENTRIS, Sclater.

MYIAGRA FULVIVENTRIS, *Sclater*.

Buff-bellied Flycatcher.

Myiagra fulviventris, Sclater, Proc. Zool. Soc. 1883, pp. 54, 200.—Meyer, Zeitschr. gesammte Orn. i. p. 194 (1884).

This is one of the most distinct of all the species of grey-backed Flycatchers of the genus *Myiagra*, as it can only be confounded with two other species, viz. *M. erythrops*, which has a rufous forehead, and with the female of *M. galeata*. The latter, however, has the upper parts much duller grey, and the rufous of the under surface is paler and almost uniform, the abdomen being only a shade lighter than the throat and breast ; but in *M. fulviventris*, as may be seen from the Plate, the throat and breast are of a fine rich orange-rufous, contrasting with the paler colour of the abdomen and sides of the body.

This species was discovered by Mr. H. O. Forbes in the Tenimber Islands, having been met with by him in Larat as well as in Loetoer.

The following descriptions are taken from the typical specimens in the British Museum :—

Adult male. General colour above leaden grey, with a slight greenish gloss ; wing-coverts like the back ; bastard-wing slightly more dusky ashy ; primary-coverts and quills dusky, externally edged with ashy grey, rather paler on the primaries ; tail-feathers pale slaty grey, with lighter ashy margins, and barred indistinctly with dusky under certain lights, and the outer feathers slightly tipped with fulvous ; lores dull ashy ; ear-coverts leaden grey like the head ; cheeks, throat, and breast deep rich orange-rufous, the abdomen, sides of body, and under tail-coverts pale fawn-buff ; thighs more ashy ; axillaries and under wing-coverts a little deeper rufous-buff ; quills dusky below, ashy along the edge of the inner web : " bill black ; legs and feet black ; the soles yellow ; iris dark brown " (*H. O. Forbes*). Total length 5·5 inches, culmen 0·55, wing 2·8, tail 2·8, tarsus 0·75.

Adult female. Very similar to the male, but rather paler, especially on the under surface of the body : " bill lavender " (*H. O. Forbes*). Total length 5 inches, culmen 0·6, wing 2·55, tail 2·4, tarsus 0·75.

Another male bird collected by Mr. Forbes in Loetoer has the bill marked as lavender-coloured, like the female above described, which is from Larat. The bird from which the description of the adult male is taken is also from Larat, and is the one which is figured in the Plate.

[R. B. S.]

MYIAGRA CERVINICAUDA, *Tristr.*

MYIAGRA CERVINICAUDA, Tristr.

Fawn-tailed Flycatcher.

Myiagra cervinicauda, Tristram, Ibis, 1879, p. 439.—Salvad. Ibis, 1880, p. 130.—Tristr. tom. cit. p. 246.—Salvad. Orn. Papuasia, etc. ii. p. 79 (1881).—Ramsay, Proc. Linn. Soc. N. S. W. vi. p. 726 (1881).—Tristr. Ibis, 1882, pp. 137, 142.—Salvad. Ann. Mus. Civic. Genov. xviii. p. 423 (1882).—Id. Orn. Papuasia, etc. iii. p. 533 (1882).—Ramsay, Proc. Linn. Soc. N. S. W. vii. p. 24 (1882).

MANY of the *Myiagræ*, or Broad-billed Flycatchers, seem to be more easily recognized by the females than by the male birds. Such is certainly the case with the present species; for the male is scarcely to be distinguished from the same sex of *M. melanura* of the New Hebrides, whereas the hens of the two species are easily separable, the clear grey head and ear-coverts, the fawn-coloured abdomen and under tail-coverts, as well as the fawn-coloured tail-feathers, at once distinguishing *M. cervinicauda*. As might be expected, *M. ferrocyanea* is also very like the present species; but the male is distinguished by its purplish upper surface, and the female by the white underparts.

The present species is doubtless peculiar to the Solomon group of islands, having been met with in Rendova by Capt. Richards and by Mr. Morton in Ugi.

The following is a description of a pair of birds lent to us by Mr. Ramsay:—

Adult male. General colour above dull bottle-green, becoming greyer on the lower back and rump; upper tail-coverts like the back; wing-coverts like the back; bastard wing, primary-coverts, and quills blackish, externally edged with the same colour as the back, rather greyer on the secondaries; tail-feathers greenish black, with a greyish shade on the edges; lores and feathers below the eye velvety black; sides of face, ear-coverts, cheeks, throat, and chest greenish black; breast and remainder of under surface, including the thighs and under tail-coverts, white; axillaries and under wing-coverts white, the edge of the wing mottled with greenish black; quills blackish below, white along the edge of the inner web. Total length 5·7 inches, culmen 0·55, wing 2·65, tail 2·3, tarsus 0·6.

Adult female. General colour above rufous-brown, the upper tail-coverts fawn-coloured and contrasting with the back; wing-coverts blackish, edged with rufous-brown like the back; bastard wing, primary-coverts, and quills blackish, with scarcely any rufous margin, except on the secondaries; two centre tail-feathers entirely brown; the next two brown, with a small fawn-coloured tip; the next brown along the inner web, fawn-coloured at the tip and along the outer web, the two outermost entirely pale fawn-colour; head and nape French grey, contrasting with the back; a line across the base of the forehead, lores, and eyelid pale tawny buff; ear-coverts French grey like the crown; cheeks, throat, and breast deeper tawny, the abdomen and thighs lighter and more tawny brown, deeper again on the sides of the body and the under tail-coverts; axillaries and under wing-coverts like the breast. Total length 5·2 inches, culmen 0·65, wing 2·45, tail 2·3, tarsus 0·65.

The figures in the Plate are drawn from the pair of birds above described, and represent an adult male and female of the natural size.

[R. B. S.]

MYIAGRA FERROCYANEA, Ramsay.

MYIAGRA FERROCYANEA, Ramsay.

Purple-backed Flycatcher.

Myiagra ferrocyanea, Ramsay, Proc. Linn. Soc. N. S. W. iv. p. 78 (1879).—Salvad. Ann. Mus. Civ. Genov. xiv. p. 508 (1879).—Id. Ibis, 1880, p. 129.—Id. Orn. Papuasia, etc. ii. p. 79 (1881), iii. p. 533 (1882).
Myiagra pallida, Ramsay, Proc. Linn. Soc. N. S. W. iv. p. 78 (1879).—Salvad. Ann. Mus. Civ. Genov. xiv. p. 508 (1879).—Id. Ibis, 1880, p. 129.—Id. Orn. Papuasia, etc. ii. p. 79 (1881).

On comparing the male of the present species with its allies, it will be found at once to be very distinct by reason of its purple back. The female was described by Mr. Ramsay as a distinct species (*M. pallida*), although he appears to have entertained doubts at the time whether it might not be the hen bird of a black-throated male. We have no hesitation in adopting the latter view, although, according to some observers, it would be quite possible to have two species, as dissimilar as the sexes of the present bird, living side by side in the same island. Where a black-throated male occurs, however, along with a red-throated bird, we believe that the latter will be found to be the female, and that it is only in certain islands that both sexes are red- or white-throated. The nearest approach to the coloration of the females of *M. ferrocyanea* occurs in *M. cervinicauda* of Tristram; but here the rufous on the chin and under surface of body at once distinguishes them.

The following descriptions are taken from the original specimens lent to us by Mr. Ramsay:—

Adult male (type of *Myiagra ferrocyanea*, Ramsay). General colour above deep purplish, the head, which is much crested, being more of a deep steel-blue colour; wing-coverts purple like the back; bastard wing, primary-coverts, and quills blackish brown, narrowly edged with blue-black; upper tail-coverts blue-black, not so purple as the back; tail-feathers blue-black; lores velvety black; sides of face, ear-coverts, cheeks, sides of neck, throat, and fore neck deep purple, inclining to steel-blue on the throat; remainder of under surface from the chest downwards and including the thighs, under tail-coverts, axillaries, and under wing-coverts pure white; quills dusky below, white along the edge of the inner web. Total length 5·3 inches, culmen 0·55, wing 2·75, tail 2·5, tarsus 0·65.

Adult female (type of *Myiagra pallida*, Ramsay). General colour above pale reddish brown, with an ashy shade on the scapulars and lower back, as well as the lesser wing-coverts; median and greater coverts, bastard wing, primary-coverts, and quills dusky brown, externally edged with light rufous, the margins nearly obsolete on the bastard-wing and primary-coverts; upper tail-coverts light rufous; tail-feathers pale brown, externally edged with pale rufous, the outer feathers almost entirely pale rufous, dusky brown towards the tips and along the outer webs; crown of head, hind neck, and mantle French grey; lores and eyelid whitish; feathers below the eye and ear-coverts French grey; cheeks and entire under surface of body white, faintly shaded with fulvous on the fore neck; lower flanks, thighs, and under tail-coverts washed with pale rufous; axillaries and under wing-coverts white; quills dusky brown below, pale rufescent along the edge of the inner web. Total length 5·3 inches, culmen 0·55, wing 2·55, tail 2·45, tarsus 0·55.

The figures in the Plate represent the two sexes of the size of life, and are drawn from the types of *M. ferrocyanea* and *M. pallida* respectively.

[R. B. S.]

MACHÆRIRHYNCHUS ALBIFRONS, G.R.Gray.

MACHÆRIRHYNCHUS ALBIFRONS, *G. R. Gray*.

White-fronted Flycatcher.

Machærirhynchus albifrons, G. R. Gray, Proc. Zool. Soc. 1861, p. 429.

"This bird," says Mr. G. R. Gray, " is, in many respects, like the *M. xanthogenys*; but it is at once distinguished from that species by the front and streak over the eyes being white, and by the bill being rather narrower and slightly sharper in front.

"The young bird is of a yellowish olivaceous, with the front and eyebrows pale rufous; throat and breast white; the latter is waved with fuscous; beneath the rest of the body yellow, and olivaceous on the sides; wings and tail fuscous, with the coverts of the former margined with white, while the quills are margined with yellow."

In size and general appearance this species approaches the Australian bird to which I have given the name of *flaviventer*; but on comparing the drawings of these two birds the most casual observer may discern the difference that exists between them. It will be necessary for me to state that my figures were taken from Mysol specimens collected by Mr. Wallace.

I have now figured in the present work three very distinct species of this singular group of Flycatchers; a fourth is said to exist in Mr. Gray's *M. xanthogenys*, a plate of which will appear when I get good Aru specimens for illustration.

Total length 4" 11''', bill from gape 9½''', wing 2" 4'''.

Hab. Waigiou and Mysol.

MACHÆRIRHYNCHUS NIGRIPECTUS, Schl.

MACHÆRIRHYNCHUS NIGRIPECTUS, Schlegel.

Black-breasted Flycatcher.

Machærirhynchus nigripectus, Schlegel, Obs. Zool. v., Ned. Tijdschr. voor de Dierk. iv. p. 43 (1873).—George Dawson Rowley, Proc. Zool. Soc. May 1876, p. 414.
Machærorhynchus nigripectus, T. Salvadori, Ann. del Mus. Civ. di St. Nat. di Genova, vii. p. 768, 1875.—Atti della Reale Acad. delle Scienze di Torino, February 1875, vol. x. p. 378.

Of this very rare species I have received two specimens, from which the accompanying drawing was taken, one very kindly lent to me by Geo. Dawson Rowley, Esq., of Brighton, the other from Dr. Meyer, of Dresden, through the hands of Mr. Gerrard. Both these, I believe, are from the northern part of New Guinea. I regret to be unable to state any particulars as to the habits and disposition of this bird; but its peculiarly constructed bill would naturally lead me to infer that aphides and very soft-winged gnats constitute a great portion of its food.

Face hoary, eye surmounted by a stripe of yellow; throat, ear-coverts, neck, and under surface bright yellow; on the chest a somewhat lengthened tuft of feathers, which are black at their bases and yellow at their tips, giving the appearance of a dark patch on this part of the under surface; crown of head and upper surface generally very dark grey tinged with olive; tail-feathers black, all but the four centre ones margined and tipped with white; wings blackish, the lesser and greater wing-coverts and tertiaries tipped with white; bill and legs black. In size this bird about equals *M. xanthogenys* and *M. albifrons*.

Total length 4½ inches, wing 3⅜, tail 2½, tarsi ⅝, bill ¼.

The figures in the accompanying Plate, and which I suppose to represent the male and female, are of the size of life.

Hab. Arfak Mountains, New Guinea.

HETERANAX MUNDUS.

HETERANAX MUNDUS.

Forbes's Pied Flycatcher.

Monarcha mundus, Sclater, Proc. Zool. Soc. 1883, p. 54, pl. xii. fig. 2.

IN describing this species, Dr. Sclater seems to have had some suspicion that it was not a typical *Monarcha*, for he figures the bill alongside that of *M. castus*, in order to show the difference between them. On the arrival of the typical specimens in the British Museum, we at once compared them with those species which seemed to be their natural allies in the genus *Piezorhynchus*, and we found that the Timor-Laut bird differed in the structure of its bill from all of them. It is closely allied to the Australian genus *Sizura*; but the latter has the bill flattened, although it is very narrow; whereas in *M. mundus* the bill is not only narrow, but is strongly compressed, so that it is higher than broad at the nostrils. Under these circumstances, we have felt compelled to propose the new generic term of *Heteranax* ($\xi\tau\epsilon\rho\sigma\varsigma$ = alter, and $\check{\alpha}\nu\alpha\xi$ = rex).

Mr. Forbes informs us that he shot this species near the village of Waitidal, on the island of Larat; it was found not far from the coast.

The following description is taken from the original specimens :—

Adult male. General colour above iron-grey, with a band of silky white plumes across the rump; wing-coverts glossy blue-black, including the bastard wing and primary-coverts; quills black, externally edged with iron-grey, broader on the secondaries; upper tail-coverts and four central tail-feathers blue-black, the next pair with a white spot near the tip, the latter fringed with black, the other feathers broadly ending in white, increasing in extent towards the outermost, where the white occupies the terminal half of the feather; forehead, lores, a narrow superciliary line, ear-coverts, and feathers round the eye velvety black; feathers below the eye, cheeks, sides of face, and sides of neck pure white; base of cheeks and centre of throat black; sides of the throat and rest of the under surface pure white, with a slight wash of grey on the flanks; thighs black; under wing-coverts and axillaries pure white; edge of wing black; quills dusky blackish below, ashy along the edge of the inner web. Total length 6·4 inches, culmen 0·7, wing 3·25, tail 2·8, tarsus 0·85.

Adult female. Like the male, but with the black not extending so far down the throat. Wing 3·1 inches.

Dr. Sclater having very kindly lent us the typical pair of specimens, before their deposition in the British Museum, we have been enabled to give a figure of them both. They are represented in the Plate, of the size of life.

[R. B. S.]

ARSES TELESCOPHTHALMUS.

ARSES TELESCOPHTHALMUS.

Frilled-necked Flycatcher.

Muscicapa telescophthalma, Garnot, Voy. Coquille, i. pt. 2, p. 593, pl. 19. fig. 1 (1826).
Muscicapa enado, Less. Voy. Coquille, i. pt. 2, p. 643, pl. 15. fig. 2 (1826).—Sclater, Proc. Linn. Soc. 1858, p. 181.
Arses telescophthalmus, Lesson, Traité, p. 387 (1831).—Bp. Consp. i. p. 326 (1850).—Salvad. Ann. Mus. Civ. Genov. x. p. 132 (1877).—Sharpe, Cat. B. Brit. Mus. iv. p. 409 (1879).
Monarcha telescophthalma, Swains. Classif. B. ii. p. 257 (1837).—Gray, Gen. B. i. p. 260 (1846).—Id. P. Z. S. 1858, p. 177, 1859, p. 156.—Id. Cat. B. N. Guinea, p. 30 (1859).—Id. P. Z. S. 1861, p. 435.—Finsch, Neu-Guinea, p. 169 (1865).—Gray, Handl. B. i. p. 320 (1869).
Monacha telescophthalma, Swains. Nat. Libr. Flycatchers, p. 140 (1837).

ALL the species of the genus *Arses* are remarkable for a frill or tippet round the back part of the neck, and for a naked skin of blue which surrounds the eye; the former, it appears, the bird has the power of erecting. That the appearance which I have given to the present species in the accompanying Plate is no exaggeration may be believed from the fact that Signor D'Albertis brought me a specimen, killed by himself, dried in an erect position, in order to show the way in which these birds are capable of elevating this frill. Mr. Broadbent had also preserved some of his specimens with an evident view to show this peculiarity. The present species has been known the longest, having been described in 1826. It appears to be confined to Northern New Guinea and to Mysol, whence numerous specimens have been sent to England, and where a large series also appears to have rewarded the exertions of the Dutch and Italian travellers. It is to be recognized from the allied species by its larger black chin-spot. The female may be told at a glance from the same sex of *A. aruensis* by its white lores, and from the females of *A. batantæ* and *A. insularis* by its deep orange-chestnut back, which renders the orange collar round the hind neck only a little darker than the rest of the upper surface. The head is blackish or very dark grey; and the colour of the tail, which is rufous-brown edged with chestnut, is also a distinguishing peculiarity of the species. For the following descriptions I am indebted to Mr. Sharpe's Catalogue:—

"*Adult male.* General colour glossy steel-black; the scapulars and lower mantle-feathers white at their ends, where they adjoin the lower back, which, with the rump, is white, the bases to the feathers being grey; wings entirely black; upper tail-coverts and tail black; crown of head, sides of face and ear-coverts, fore part of cheeks, chin, and upper throat black, the plumes of the head of a velvety texture, and with a slight steel gloss; hinder part of cheeks, sides of neck, and a collar round the hind neck, as well as entire under surface of body pure white, including the under wing-coverts and axillaries; thighs black, as also the edge of the wing; 'bill pearly grey; feet dull lead-colour or ashy; iris black; fleshy wattle round the eye sky-blue' (*D'Albertis*).

"Total length 6·4 inches, culmen 0·6, wing 3·15, tail 2·85, tarsus 0·7.

"*Adult female.* General colour orange-brown, brighter on the hind neck; wings dusky brown, all the coverts and quills externally orange-brown or rufous; crown of head dull slate-colour, as well as the feathers round and below the eye; in front of the latter a large spot of dull white; sides of face, sides of neck, throat, and breast orange; remainder of under surface pure white, the flanks washed with reddish brown; thighs entirely of the latter colour; under wing-coverts whitish, washed slightly with orange; quills dull brown below, rufous along the inner webs; 'bill dusky; feet ashy; iris black' (*D'Albertis*).

"Total length 6·3 inches, culmen 0·6, wing 3·2, tail 2·9, tarsus 0·65."

Fine specimens of this bird are in my own collection; it will also be found in the British Museum, and Leyden, Dresden, and many other collections both on the Continent and in America.

ARSES BATANTÆ, Sharpe

ARSES BATANTÆ, *Sharpe.*

Large Frilled-necked Flycatcher.

Arses batantæ, Sharpe, Notes Leyden Mus. i. No. 5, p. 20 (1879).—Id. Cat. B. Brit. Mus. iv. p. 411 (1879).

THIS species is the largest of the genus *Arses* yet discovered; and it is remarkable that a yellow bill and eye-wattle, instead of blue ones, are sometimes seen in adult males. The female is also very different from that sex of the allied species in its coloration, having the back brownish orange, with the wings of the same colour, the innermost secondaries being exactly like the back, as are also the rump and upper tail-coverts; the tail also is plain orange-chestnut, with a faint shade of brown at the end; the head is dark grey, and the lores white.

The islands inhabited by the present bird are Waigiou and Batanta, whence, Mr. Sharpe informs me, the Leyden Museum has a considerable series collected by the late Dr. Bernstein. I transcribe the following detailed description of the species from Mr. Sharpe's 'Catalogue of the Flycatchers:'—

"*Adult male.* Similar to *A. telescophthalmus*, but rather larger, and sometimes having the wattle round the eye and bill yellow. Total length 7 inches, culmen 0·7, wing 3·4, tail 3, tarsus 0·75.

"*Adult female.* General colour above light rufous, the wing-coverts and inner secondaries like the back; primary-coverts and primaries dusky brown, externally edged with light rufous, the secondaries more broadly, the inner ones being almost entirely orange-rufous, with a shade of dusky brown on the inner web; tail-feathers pale rufous, the centre feathers dusky brown, with pale rufous edges and shafts; crown of the head dark grey, as also the feathers round the eye, which are slightly mottled, with white bases; loral spot dull white; ear-coverts, sides of face, and throat light orange-buff; remainder of the under surface white, the sides of the breast washed with orange; the tibial plumes and under wing-coverts light orange-buff, the edge of the wing deeper orange; axillaries buffy white; quills ashy brown below, rufescent along the inner web. Total length 7 inches, culmen 0·65, wing 3·3, tail 3, tarsus 0·8."

The figures in the Plate are drawn from a pair of birds collected by Mr. Wallace in the island of Waigiou, and now in my own cabinet.

ARSES ARUENSIS, Sharpe.

ARSES ARUENSIS, Sharpe.

Little Frilled-necked Flycatcher.

Arses telescophthalmus, Salvad. & D'Albert. (nec Garn.), Ann. Mus. Civ. Gen. vii. p. 819 (1875).—Salvad. op. cit. ix. p. 24 (1876).—D'Albert. op. cit. x. p. 19 (1877).—Ramsay, Proc. Linn. Soc. N. S. W. i. p. 391 (1877).—Sharpe, Journ. Linn. Soc. xiii. pp. 316, 497 (1878).
Arses aruensis, Sharpe, Notes Leyden Mus. i. No. 5, p. 21 (1879).—Id. Cat. B. Brit. Mus. iv. p. 410 (1879).—Salvad. Ann. Mus. Civ. Genov. xiv. p. 59 (1879).

At the time that I made my drawings of the different species of *Arses*, which was towards the end of the year 1878, I had occasion to examine closely the series of these birds in my collection; and I came to the conclusion that five species of the genus could be recognized. On communicating my impressions to Count Salvadori, I received a note from him to the effect that his opinion entirely coincided with mine, and that, in the MS. of his forthcoming work on the Birds of New Guinea, he had conferred new titles upon two of the species, which were not previously named. Mr. Bowdler Sharpe, who was at that time absent from England on a visit to Leyden, whither he had gone for the purpose of examining the collection of Flycatchers in the Museum, brought me back word that he had become convinced from a study of the splendid series of *Arses* in that institution that there were two species still undescribed, and that he had bestowed the names of *A. batantæ* and *A. aruensis* upon them, and had left the descriptions in Professor Schlegel's hands for publication. A full account of the two birds appeared in the 'Notes from the Leyden Museum' in January of the present year; and on receiving notice, Count Salvadori very properly suppressed the MS. names which he had given to the species. Mr. Sharpe's title of *aruensis* can scarcely be called well chosen, as the bird is by no means confined to the Aru Islands, but extends apparently along the entire coast of Southern New Guinea. Signor D'Albertis met with it on the Fly River; and it has been procured near Port Moresby by Mr. O. C. Stone and Mr. Kendal Broadbent. The present bird is distinguished by the smaller size of the male and the less-extended black spot on the throat. The female is to be recognized by its brown back and tail, contrasting strongly with the orange collar round the hind neck; the head also is jet-black, and the loral spot orange. The following descriptions are extracted from Mr. Sharpe's 'Catalogue,' the synonymy of which has been here corrected so as to include the references to the bird's occurrence in South-eastern New Guinea, which that author forgot by some accident to place under the heading of the present species.

"*Adult male.* Crown of head, ear-coverts, and sides of face, fore part of cheeks, chin, and extreme upper edge of throat blue-black; round the hind neck a broad white collar, joining the sides of the neck, which, with the hinder cheeks and the rest of the under surface of the body, are pure white, excepting the tibial plumes, which are black; mantle, scapulars, and upper back blue-black; wing-coverts also blue-black, the terminal half of the inner greater coverts and the ends of the scapulars white; primary-coverts and quills black, edged with blue-black; lower back and rump white; upper tail-coverts blue-black; tail-feathers black, washed with blue-black on their margins; under wing-coverts and axillaries white; quills blackish below, ashy along the inner edge of the primaries, white on the secondaries. Total length 6·5 inches, culmen 0·6, wing 3·1, tail 2·9, tarsus 0·7.

"*Adult female.* General colour above dusky orange-brown, the tail also dull brown, with dusky-rufous edges to the feathers; least wing-coverts like the back, the median and greater series dusky brown with orange-brown edges, and tips of lighter orange; primary-coverts and quills brown with dull orange-rufous margins, broader on the secondaries; crown of head, feathers below the eye, and ear-coverts glossy black; a loral spot of orange; round the hind neck a collar of bright orange rufous, contrasting strongly with the back; cheeks, throat, and breast bright orange-rufous, deepening almost to chestnut on the latter; remainder of under surface white, the flanks washed with ashy, the sides of the upper breast orange-rufous; thighs dusky brown; under wing-coverts and axillaries white, the edge of the wing orange; quills dull brown, rufescent along the edge of the inner web. Total length 6 inches, culmen 0·55, wing 2·95, tail 2·75, tarsus 0·7."

Signor D'Albertis says that the bill and feet are dull ashy, the eyes black, and that the bird feeds on insects. The figures in the Plate represent a pair of birds of the size of life, and are taken from specimens in my own collection. They were collected by the late Dr. James in South-eastern New Guinea; and according to the latter gentleman the soft parts are as follows:—"eyes very dark brown surrounded by a disk of blue; bill pale blue at base, lighter at tip; feet and tarsi dark leaden colour."

ARSES INSULARIS, Meyer

ARSES INSULARIS.

Orange-collared Flycatcher.

Monarcha insularis, Meyer, Sitzungsb. k. Akad. Wien, lxix. p. 395.
Arses insularis, Sclater, P. Z. S. 1878, p. 579.—Sharpe, Notes from Leyden Mus. i. no. 5, p. 20.

This beautiful species of *Arses* was discovered by Dr. Meyer in the island of Jobi, in the Bay of Geelvink, North-west New Guinea, and would appear to be by no means uncommon in that locality, as he collected a considerable number of specimens. It was also met with by the Italian traveller Beccari in the same island. More recently it has been obtained during the voyage of H.M.S. 'Challenger,' at Humboldt Bay in New Guinea, where the ship touched for half an hour on the 23rd of February, 1875. Mr. Bowdler Sharpe, in the first volume of 'Notes from the Leyden Museum,' has described two new species of *Arses*, and has given a list of the species now known to belong to this genus. They are:—
A. telescophthalmus (Garnot), from New Guinea and Mysol; *A. batantæ* (Sharpe), from the islands of Waigiou and Batanta; *A. aruensis* (Sharpe), from the Aru Islands and South-eastern New Guinea; *A. kaupi* (Gould), from North-eastern Australia; and *A. insularis* (Meyer), from North-eastern Australia. All these species, which are known to me personally, seem to be well founded; but perhaps the handsomest of all is the subject of the present article, the male of which may be distinguished from all the other species of the genus *Arses* by the orange collar round the hind neck and the light orange or Naples-yellow colour of the lower throat and breast. The female is more closely allied to those of the other kinds of *Arses*, but has the back of an olive-brown, which contrasts strongly with the orange collar round the hind neck. It has a grey head like the females of *A. telescophthalmus* and *A. batantæ*, and also a white loral spot. These last characters distinguish it from the hen of *A. aruensis*, which has a black head and an orange loral spot.

The following descriptions of *A. insularis* are taken from Mr. Sharpe's 'Catalogue of Birds':—

"*Adult male.* General colour above glossy blue-black, the feathers of the crown of a velvety and somewhat scaly nature; feathers of lower mantle tipped with white where they adjoin the scapulars, which, as well as the lower back, are also white with black bases; rump ashy grey, some of the feathers white at the ends; upper tail-coverts and tail jet-black; wings entirely black; lores, feathers, round the eye, ear-coverts, and chin black; cheeks, throat, breast, and sides of neck pale orange, extending in a collar round the hind neck; rest of under surface, as well as the under wing-coverts and axillaries, pure white; small coverts along the edge of the wing black; thighs black. Total length 6·4 inches, culmen 0·65, wing 3·25, tail 2·75, tarsus 0·7."

The following note of the soft parts is from the 'Proceedings' as above cited:—"Eyes large; ring surrounding the eye large, and of a sky-blue; bill and legs of a darker blue or violet."

The figures in the accompanying Plate represent the two sexes, kindly lent to me by Dr. Mëyer, of the size of life.

PIEZORHYNCHUS BRODIEI, Ramsay.

PIEZORHYNCHUS BRODIEI.

Brodie's Flycatcher.

Monarcha brodiei, Ramsay, Proc. Linn. Soc. N. S. W. iv. p. 80 (1878).—Salvad. Ibis, 1880, p. 129.—Id. Orn. Papuasia, etc. ii. p. 26 (1880).
Monarcha barbata, Ramsay, Nature, xx. p. 125 (1879).
Monarcha barbatus, Salvad. Ann. Mus. Civic. Genov. iv. p. 507 (1879).

This species belongs to a group of Flycatchers of the genus *Piezorhynchus*, which seems to be peculiar to the Solomon Islands, for its members do not agree with any of the Papuan or Moluccan forms of the genus. In the white under surface, black throat, and black ear-coverts the present bird approaches *P. guttulatus*, which it also resembles in having white tips to the outer tail-feathers; but its entirely black upper surface distinguishes it at once both from *P. guttulatus* and *P. morotensis*, which are grey above.

The habitat of the present species is the Solomon Archipelago, where it was found by Mr. Cockerell in Guadalcanar and Lango.

We describe the typical examples as follows:—

Adult male. General colour above blue-black, the feathers of the head rather velvety in texture, and scale-like on the fore part of the crown; scapulars black; least wing-coverts black, tipped with white where they adjoin the median series, which, with the greater coverts, are pure white with concealed black bases; bastard wing, primary-coverts, and quills black; tail-feathers black, the three outer feathers tipped with white; lores, feathers below the eye, fore part of cheeks, and ear-coverts blue-black; a large patch of white extending from the hinder cheeks along the sides of the neck and united to the breast; throat black, the feathers long, rounded, and scale-like; rest of under surface of body from the lower throat downwards pure white; thighs black, edged with white; axillaries and under wing-coverts white, with a black patch near the edge of the wing; quills blackish below, ashy along the edge of the inner web. Total length 5·7 inches, culmen 0·5, wing 2·95, tail 2·6, tarsus 0·75.

Adult female. Different from the male. General colour above brown, with faintly indicated dusky centres to the feathers of the forehead and crown, which are denser and somewhat scaly in appearance; lesser wing-coverts like the back; the median series white at base and pale fulvous brown at the ends, forming an indistinct wing-bar; greater series dark brown, edged externally with reddish brown; bastard wing and primary-coverts dark brown; quills dusky brown, externally of the same colour as the back; upper tail-coverts and tail-feathers black, the three outer ones broadly tipped with white, increasing in extent towards the outermost; lores dull whitish, obscured with dull brown tips to the feathers; eyelid dusky slate-colour; ear-coverts dusky grey; cheeks and sides of neck pale orange-rufous; throat scaly in appearance, as in the male, and of a dusky grey colour washed with fulvous brown; fore neck and sides of breast pale orange-rufous, extending on to the flanks; centre of breast, abdomen, and under tail-coverts white, with a slight tinge of orange-buff; thighs grey; under wing-coverts and axillaries light orange-rufous; quills dusky below, whitish along the edge of the inner web. Total length 5·5 inches, culmen 0·6, wing 3·0, tail 2·7, tarsus 0·75.

The typical specimens are represented in the Plate, of the size of life, together with a young male in intermediate plumage; and we have to thank Mr. E. P. Ramsay for the opportunity of describing and figuring these interesting specimens.

[R. B. S.]

PIEZORHYNCHUS BROWNI, Ramsay.

PIEZORHYNCHUS BROWNI, *Ramsay*.

Brown's Flycatcher.

Monarcha (Piezorhynchus) browni, Ramsay, Proc. Zool. Soc. 1882, p. 711.—Salvad. Orn. Papuasia, etc. iii. App. p. 531 (1882).

Though closely resembling *P. brodiei* in general appearance, this species may be distinguished by the longer bill, the greater extent of white on the outer tail-feathers, and especially by the much greater extent of black on the throat, which reaches to the chest. The consequence of this is that the white patch on the hinder cheeks is entirely shut off from the white breast, whereas in *P. brodiei* the two join.

The only specimen of *P. browni* that we have seen is the typical example from Marrabo, in the Solomon Islands, of which we give a description. We are indebted for the loan of this example to the kindness of our friend Mr. E. P. Ramsay.

Adult male. General colour above blue-black; lesser wing-coverts like the back, those adjoining the greater series broadly tipped with white; median wing-coverts white, the outer ones black at the base; greater coverts white, forming with the others a large wing-patch; bastard wing, primary-coverts, and quills black; tail-feathers black, all but the two centre feathers tipped with white, increasing in extent greatly towards the outermost; lores, feathers below the eye, and ear-coverts black; fore part of cheeks also black, the hinder cheeks white, extending in a large patch on to the sides of the neck; throat blue-black, the feathers scaly in appearance; fore neck and chest also black, united by the black sides of the neck and the mantle; breast, abdomen, and under tail-coverts white, the feathers adjoining the black throat having black bases; thighs black; under wing-coverts and axillaries white, with black bases to the feathers; coverts near the edge of the wing black; quills black below. Total length 7 inches, culmen 0·7, wing 3·45, tail 3·3, tarsus 0·85.

The Plate represents the adult male above described, in two positions, of the size of life.

[R. B. S.]

PIEZORHYNCHUS RICHARDSI, Ramsay.

W. Hart del et lith. Mintern Bros. imp.

PIEZORHYNCHUS RICHARDSII, *Ramsay*.

Richards's Flycatcher.

Piezorhynchus richardsii, Ramsay, Proc. Linn. Soc. N. S. W. vi. p. 177 (1881).
Pomarea richardsii, Tristram, Ibis, 1882, pp. 136, 142.—Salvad. Ann. Mus. Civic. Genova, xviii. p. 422 (1882).
Monarcha richardsii, Salvad. Orn. Papuasia, etc. iii., App. p. 529 (1882).

AFTER careful examination we incline to consider this species a true *Piezorhynchus*, as Mr. Ramsay has also determined it to be, and not a *Pomarea*, as Canon Tristram calls it, although in its style of coloration it very much resembles *P. castaneiventris*. It is distinguished from all the members of the genera *Piezorhynchus* and *Pomarea* by the remarkable white patch on the hinder crown and nape, which serves to characterize the species at once from all the other Flycatchers of the same group.

The following is the description of an adult male shot in Rendova Island, in the Solomon group, by Lieut. Richards, R.N., who also met with it in the island of Ugi:—

Adult male. General colour above velvety black, with slight rufous edges to some of the upper tail-coverts; wing-coverts like the back; greater coverts, bastard wing, primary-coverts, and quills dull black; tail-feathers black; lores, nasal plumes, forehead, and sinciput black, as well as the fore and hind part of the eyelid; upper and lower part of eyelid as well as the vertex and entire nape pure white, extending on to the sides of the neck behind the ear-coverts; ear-coverts, sides of face, cheeks, throat, and chest glossy black; feathers of lower chest black, tipped with chestnut like the rest of the under surface of the body, which is entirely chestnut; thighs black; axillaries and under wing-coverts black, tipped with pale chestnut, with a patch of black near the edge of the wing; quills blackish below, ashy along the edge of the inner web; "bill horn-colour; feet black; iris black" (*Richards*). Total length 5·5 inches, culmen 0·7, wing 3·0, tail 2·55, tarsus 0·65.

The specimen from which the above description was taken has been lent to us by Canon Tristram, in whose collection it now remains. The Plate represents the adult male of the natural size, and has been drawn from the same individual.

[R. B. S.]

PIEZORHYNCHUS CASTUS.

PIEZORHYNCHUS CASTUS.

White-crowned Flycatcher.

Monarcha castus, Sclater, Proc. Zool. Soc. 1883, p. 53, pl. xii. fig. 1.

This beautiful species of Flycatcher belongs to the pied section of the genus *Monarcha* (or of the genus *Piezorhynchus*, as set forth in the 'Catalogue of Birds'), which includes *M. verticalis*, *M. leucotis*, *M. loricatus*, and other well-known Moluccan and Papuan species. It would appear to resemble *M. verticalis* very closely, and, like that species, has the ear-coverts, sides of neck, and summit of crown white, while the distribution of white on the wing-coverts is the same; but it would seem to be at once distinguished by the lines of white spots on the throat, and by having the inner secondaries edged with white instead of their being entirely pure white as in *M. verticalis*.

Only a single specimen was contained in Mr. Forbes's collection, from which we gather that the species is rather rare in the part of the Tenimber Islands visited by him.

The following is a description of the original type:—

Adult male. General aspect above variegated; the back and mantle bluish black, somewhat mixed with grey on the lower back as it approaches the rump, the lower part of which, with the upper tail-coverts, is white; least wing-coverts black; entire median series and inner greater coverts white; outer greater coverts black tipped with white; bastard wing, primary-coverts, and quills black, the secondaries narrowly fringed with white near the tip of the outer web; the innermost secondaries broadly edged with white externally; centre tail-feathers black, the next with a small triangular tip of white, which increases in extent towards the outer feathers until the external one is white for more than the terminal half; forehead, lores, and a narrow superciliary line extending backwards and forming a broad nuchal collar bluish black; the crown of the head pure white encircled by the black aforesaid; feathers below the eye, cheeks, and throat black, which extends in a broad band and joins the black nuchal collar above mentioned, thus encircling the white ear-coverts, which form a conspicuous patch; the lower throat decorated with white tips to the feathers, forming a triple line; sides of neck white, extending backwards so as to form a more or less distinct collar separating the nuchal band from the mantle; remainder of under surface from the throat downwards pure white, excepting the hinder aspect of the thighs, which is black; the white of the under surface slightly sullied with a grey shade; under wing-coverts and axillaries white, excepting the edge of the wing and the adjacent coverts, which are black; quills blackish below, edged with white along the inner web; "bill lavender, tip black; legs and feet lavender; iris reddish brown" (*H. O. Forbes*). Total length 5·75 inches, culmen 0·55, wing 2·7, tail 2·7, tarsus 0·75.

The Plate represents the typical specimen of the present species, of the natural size, in two positions.

[R. B. S.]

PIEZORHYNCHUS VIDUA, *Tristr.*

PIEZORHYNCHUS VIDUA, Tristram.

White-backed Pied Flycatcher.

Piezorhynchus vidua, Tristram, Ibis, 1879, p. 439.—Salvad. Ibis, 1880, p. 130.—Tristram, Ibis, 1880, p. 246.
Piezorhynchus melanocephalus, Ramsay, Proc. Linn. Soc. N. S. W. iv. p. 468.
Monarcha vidua, Salvad. Orn. Papuasia, etc. ii. p. 24 (1881).

THE original specimen of this Flycatcher was obtained by Lieut. Richards, R.N., at Makira Harbour, San Cristoval, in the Solomon group of islands, on the 3rd of October 1878; and on the 21st of May 1879 the same gentleman procured a second specimen in the same locality, and this was named by Mr. Ramsay *Piezorhynchus melanocephalus*. Count Salvadori expressed at one time an opinion that the present bird was identical with Mr. Ramsay's *Monarcha brodiei*, but afterwards, in his great work on the birds of New Guinea, he came to the conclusion that they were really different species; and this proves to be the case, now that we have examined the typical specimens of both birds.

The present species belongs to a section of the genus *Piezorhynchus* which is very distinct from all the other groups of these Pied Flycatchers by reason of the white collar and white rump. The only species with which it could be confounded is *P. squamulatus*, which has similar white markings on the wings, but has the feathers of the fore neck edged with black, so as to present a scaled appearance.

The following is a description of the type specimen:—

Adult male. General colour above blue-black on the mantle, upper back and scapulars; lower back and rump white, as well as the upper tail-coverts; lesser wing-coverts black, with ovate spots of white, the median and greater coverts white, with rather broad black edgings; bastard wing, primary-coverts, and quills black, the inner secondaries with a broad mark of white towards the end of the outer web; tail-feathers black, the three outer feathers tipped with white, increasing in extent towards the outermost; head blue-black, separated from the mantle by a broad collar of white, which also occupies the sides of the neck; lores, feathers round the eye, cheeks, and ear-coverts black; throat also black; remainder of the under surface of body pure white; thighs black; under tail-coverts white, as also the under wing-coverts and axillaries; the coverts near the edge of the wing black; quills blackish below, with their extreme inner base white; "feet ash-colour; bill black; iris grey" (*Richards*). Total length 5·8 inches, culmen 0·65, wing 2·95, tail, 2·9, tarsus 0·8.

The figures in the Plate are drawn from the typical specimen, which has been very kindly lent to us by Canon Tristram.

[R. B. S.]

PIEZORHYNCHUS SQUAMULATUS, Tristr.

PIEZORHYNCHUS SQUAMULATUS, *Tristram*.

Scaly-necked Pied Flycatcher.

Piezorhynchus squamulatus, Tristram, Ibis, 1882, pp. 136, 142.—Ramsay, tom. cit. p. 472.
Monarcha squamulatus, Salvad. Ann. Mus. Civic. Genov. xviii. p. 423 (1882).—Id. Orn. Papuasia, etc. iii. App. p. 530 (1882).

The island of Ugi in the Solomon group is the home of this Flycatcher, which belongs to the same section of the genus *Piezorhynchus* as *P. vidua* of Tristram, to which species it is closely allied. Like that bird it has a white collar round the neck, and the lower back and rump also white; but it is easily recognized by the black and white mottlings on the fore neck, which are not present in the San Cristoval species.

The subject of the present article was discovered by Lieut. Richards, R.N., who has brought to our knowledge so many fine species of birds from the Solomon Archipelago. It appears to be confined to the island of Ugi, whence Mr. Ramsay also informs us that he has received several specimens.

The following is a description of the type specimen:—

Adult male. General colour above blue-black, the crown of the head separated from the mantle by a broad white collar; lower back, rump, and upper tail-coverts white; lesser wing-coverts like the back; median and greater coverts with large spots of white at the ends, margined with black in the median series, and extending to the edge of the feathers in the greater series, where the white is more largely developed, especially on the inner ones; bastard wing, primary-coverts, and quills black, the innermost secondaries with a triangular spot of white at the ends; tail-feathers black, with the two outer ones tipped with white (N.B. The tail is imperfect, and three white-tipped feathers may exist); lores, sides of face, ear-coverts, cheeks, and throat black; the fore neck white, all the feathers edged with black, producing a scaly appearance; remainder of under surface of body pure white; thighs black; under wing-coverts and axillaries white, the edge of the wing black, as well as the adjoining coverts; quills black below, white at extreme base of inner web; "bill and feet drab colour; iris light brown" (*Richards*). Total length 6 inches, culmen 0·5, wing 3·1, tail 2·8, tarsus 0·8.

We are again indebted to our friend Canon Tristram for the opportunity of figuring and describing this interesting species in the present work. The Plate represents the male bird of the natural size, and is drawn from the original specimen, which, as far as this country is concerned, still remains unique in Canon Tristram's collection.

[R. B. S.]

PIEZORHYNCHUS MEDIUS, Sharpe.

PIEZORHYNCHUS MEDIUS, Sharpe.

Coppinger's Flycatcher.

Piezorhynchus medius, Sharpe, Rep. Zool. Coll. Voy. H.M.S. 'Alert,' p. 14 (1884).

THE specimen from which the present species was characterized was obtained during the voyage of the surveying-ship 'Alert,' by Dr. Coppinger, the naturalist attached to the expedition. He procured a male at Port Molle, in Queensland, in the month of May 1888; and an examination of the individual in question induced us to reconsider the relations of the species of *Piezorhynchus*, to which it is allied. Writing in the year 1879, we had recognized four species of this particular group of Flycatchers, viz. *P. bernsteini* from the Island of Salwati, *P. nigrimentum* from Amboyna and Goram, *P. trivirgatus* from Timor, and *P. gouldi* from North-eastern Australia. The latter species had previously been united with *P. trivirgatus*, but was separated in 1860 by the late Mr. G. R. Gray; and in writing our account of the 'Alert' collections we acknowledged our error in uniting with it *P. albiventris* of Gould.

Dr. E. P. Ramsay, in his latest list (1888) of the Birds of Australia, gives the habitat of *P. gouldi* as from Cape York to the Wide-Bay district of Eastern Australia, as far as the Richmond and Clarence Rivers, to New South Wales. *P. albiventris* is said to occur only in the Gulf of Carpentaria and the Cape York district, probably extending to Rockingham Bay. He has apparently overlooked our description of *P. medius*, which is closely allied to *P. albiventris*, and, like that species, has the upper tail-coverts black; but differs from it in having the sides of the body orange-rufous instead of white. *P. gouldi* has the sides of the body orange-rufous as in *P. medius*, but has the upper tail-coverts grey.

No notes on the habits of *P. medius* have yet been recorded; but they are doubtless exactly the same as those of the allied Australian Flycatchers, described by Mr. Gould.

Dr. Coppinger describes the soft parts as follows:—"Iris black; bill light grey; legs and feet dark."

The figures in the Plate are taken from the typical specimen in the British Museum, and represent two male birds of the size of life.

[R. B. S.]

PIEZORHYNCHUS AXILLARIS.

PIEZORHYNCHUS AXILLARIS.

White-tufted Flycatcher.

Monarcha axillaris, Salvad. Ann. Mus. Civic. Genov. vii. p. 291 (1875), xiv. p. 495 (1879).—Id. Orn. Papuasia e delle Molucche, ii. p. 30 (1881).
Piezorhynchus axillaris, Sharpe, Cat. Birds in Brit. Mus. iv. p. 426 (1879).

This plain-coloured but elegant Flycatcher is easily distinguished from all the members of the genus *Piezorhynchus* by its pure white axillaries, which contrast strongly with the general black plumage, so that it forms a distinct section of the genus to which it belongs.

It is apparently a very local species; for, as far as is known at present, it is an inhabitant only of the Arfak Mountains in North-western New Guinea, where Mr. Bruijn's hunters procured an adult female in May and a young female in June, while the type specimen was obtained by Dr. Beccari in July, at Profi, in the Arfak Mountains, at a height of 3400 feet. No other travellers appear to have met with the species; but Mr. Bruijn sent a few specimens to the Leyden Museum, one of which Professor Schlegel allowed us to have for the British Museum.

The following descriptions are taken from the typical adult males and the young female specimen in the Genoa Museum. They were examined by us during Count Salvadori's visit to England, when he kindly allowed us to take descriptions of them for the purposes of the British Museum 'Catalogue of Birds':—

Adult male. General colour above and below glossy blue-black; wing-coverts blackish, edged with glossy blue-black, like the back; tail black; under wing-coverts and axillaries pure white, the latter forming a conspicuous patch on each side of the breast. Total length 5·7 inches, culmen 0·5, wing 3·2, tail 2·95, tarsus 0·75.

Female (not quite adult). Differs from the male in being dark slate-colour above and below, with a slight bluish gloss, the breast brown (apparently remains of young plumage); wing brownish black; tail black. Total length 6 inches, wing 3·1, tail 2·9, tarsus 0·7.

The figures in the Plate are of the size of life, and are drawn from the male specimen in the British Museum.

[R. B. S.]

MONARCHA PERIOPHTHALMICUS, Sharpe.

MONARCHA PERIOPHTHALMICUS, Sharpe.

Black-Spectacled Flycatcher.

Monarcha periophthalmicus, Sharpe, Journ. Linn. Soc. (Zool.) vol. xvi. p. 318 (1882).

ALL the members of the genus *Monarcha* are birds of very elegant coloration; and the present species is very delicately coloured. It belongs to the section of the genus with black wings and tail, to which *Monarcha canescens* and *M. frater* also appertain. It is, indeed, very closely allied to the last-named bird, the type of which is now in the Civic Museum of Genoa, but a full description of which will be found in the British-Museum 'Catalogue of Birds' by Mr. Sharpe, and also in Count Salvadori's 'Ornitologia della Papuasia.' In both these works the forehead and chin are spoken of as black, while the region round the eye is white. Although I have not been able to compare the type of *M. periophthalmicus* with that of *M. frater*, yet I have very little doubt as to its distinctness; for it has the fore part of the crown black, as well as the forehead, while the entire region of the eye is also black.

The following is a copy of the full description contributed by Mr. Sharpe to the 'Journal of the Linnean Society' in a paper read by him on the 6th of April, 1882:—

"General colour above pearly grey, a little darker on the upper tail-coverts, which have concealed black bases; lesser and inner median and greater coverts pearly grey like the back; bastard wing, primary-coverts, as well as the outer median and greater series and the quills, black, only the innermost secondaries externally pearly grey; tail-feathers black; forehead and sinciput, lores, fore part of cheeks, feathers below the eye and a broad ring round the eye black; chin and upper throat black; lower throat, fore neck, and chest, pearly grey, as also the sides of the neck; remainder of under surface of body, as well as the thighs and under tail-coverts and the axillaries and under wing-coverts, cinnamon-buff; quills blackish below. Total length 5·5 inches, culmen 0·75, wing 3·4, tail 2·75, tarsus 0·75."

The figure in the Plate is drawn from one of the typical specimens in the British Museum. It represents a male bird of the natural size.

[R. B. S.]

MONARCHA KORDENSIS, Meyer.

MONARCHA KORDENSIS, Meyer.

Mysore Yellow Flycatcher.

Monarcha kordensis, Meyer, Sitzungsberichte der k. Akad. Wissensch. Wien, Band lxix. p. 202 (1874).

In introducing this fine new bird to my readers, I cannot do better than quote Dr. Meyer's own words on the subject, as the differences between this bird and *M. melanonota* are very clearly expressed by him.

In the above-mentioned paper, published in the 'Proceedings' of the Vienna Academy, he writes as follows:—" I procured near the village of Kordo on the island of Mysore, in the north of Geelvink's Bay, eight specimens of a *Monarcha* which is closely related to *M. chrysomela* (i. e. *M. melanonota*, Scl.) of New Guinea, a species of which I also got nine specimens from different localities (Nappau, Passim, Andai, and Puta on the Arfak Mountains); but nevertheless the Kordo specimen is easily to be distinguished.

"*M. kordensis* is not yellow as is *M. chrysomela* auct., but is orange-coloured; and the head, which in *M. chrysomela* has an orange hue only, has a very distinct fiery tinge in the species from the neighbouring island. Further, only the upper back is black, and not the whole back, as in *M. chrysomela*; so that it might be described as follows—'upper parts orange, with a black patch on the uppper mantle.' Otherwise the colours are distributed as in *M. chrysomela*.

"The female of *M. kordensis* differs in a similar conspicuous way from the female of *M. chrysomela*, the underparts being orange, nearly as bright as in the male, and not brownish yellow as in *M. chrysomela*.

"The throat is whitish, the region of the throat deep orange; the head is darker orange, with a brownish hue, not olive-coloured as in *M. chrysomela*; upper parts olive-coloured with an orange gloss; wing-coverts with broad yellow tips. Total length 6·4 inches; culmen 0·5, wing 3·3, tail 2·75. A young male differs from the female only in having some feathers of the throat deep black."

But if *M. kordensis* differs from *M. melanonota*, as I freely confess it does, it is so closely allied to the true *M. chrysomela* from New Ireland that Dr. Cabanis and Reichenow considered the two species to be identical (*cf.* J. f. O. 1876, p. 320). Dr. Sclater, however, informs me that a comparison of the birds from these two localities induces him to consider them distinct, as *M. kordensis* has the black colour more circumscribed on the back and extending further down the throat; the black colour also narrowly surrounds the eye, which is not the case in *M. chrysomela*.

I an indebted to the kindness of my friend Dr. Meyer for the loan of one of the typical specimens of his *M. kordensis*, of which the following is a description :—

Adult male. Similar to *M. melanonota*, but distinguished at a glance by its more yellow back, the scapulars being yellow like the wing-coverts; the black patch in the middle of the back is confined to the mantle; the secondaries are much more broadly bordered with yellow, the latter colour being much richer and inclining to fiery orange on the head and neck; the white spot below the front of the eye very small.

Total length 5·9 inches; culmen 0·6, wing 2·9, tail 2·5, tarsus 0·75.

The figures in the Plate represent a pair of this species of the natural size.

MONARCHA MELANONOTA, Sclater.

MONARCHA MELANONOTA, *Sclater*.

Papuan Yellow Flycatcher.

Monarcha chrysomela auct, ex Novâ Guineâ.
Arses chrysomela, Gray, Handl. B. no. 4805 (1869, nec Garn.).

Dr. Sclater has brought to my notice the fact that this species, usually called *M. chrysomela* by authors, is not the true *M. chrysomela* of Garnot, which is from New Ireland. It will be seen that the present bird, which is an inhabitant of New Guinea and Mysol, is distinguished by the greater extent of the black on the back, and by other characters mentioned by Dr. Meyer. I have not had a sufficient series for comparison; but I question whether *M. melanonota* can really be separated specifically from *M. aruensis* of Salvadori (Ann. Mus. Civic. Genova, vi. p. 309), and in the case of their identity the latter name will have precedence.

Nothing has been recorded concerning the habits of this species, which, by reason of its brilliant coloration, is one of the most beautiful of all Flycatchers.

I add a full description of the sexes of this species, which has been figured from some of Dr. Meyer's specimens. He procured it on the mainland of New Guinea, at Nappau, Passim, Andai, and Puta. I have a specimen in my collection obtained by Mr. Hoedt at Waigaama in Mysol.

Adult male. Back, from the nape to the rump, purplish black; entire head and nape, sides of neck, and hinder part of ear-coverts golden yellow, deepening to orange on the crown; a narrow frontal line, lores, fore part of ear-coverts, and entire throat glossy purplish black; under the eye a spot of white feathers; round the eye a ring of purplish black plumes, yellow where they adjoin the yellow of the ear-coverts; least and median series of wing-coverts yellow; primary-coverts and greater series black, the innermost of the latter yellow, some of the outer greater coverts also slightly tipped with yellow; quills purplish black, the innermost secondaries broadly edged with yellow; rump and upper tail-coverts bright yellow; tail black, with slight indications of a tiny yellow tip; under surface of body, from the fore neck downwards, brilliant yellow; thighs black; under wing-coverts yellow, the lower ones whiter; quills blackish below, white along the inner webs.

Adult female. General colour above olive-green, rather yellower on the head; wing-coverts and quills brown, externally edged with the same olive-green as the back; tail-feathers light brown, washed with olive-green on both webs; lores dull whitish, tipped with dusky; in front of the lower part of the eye a white spot; sides of face olive-green like the back; entire under surface of body yellow, more dingy on the throat, the chin somewhat dusky; under wing-coverts pale yellow; quills light brown below, edged with fulvous brown along the inner webs; "bill bluish, the tip black; feet lead-colour; iris dark" (*Wallace*).

The principal figures in the accompanying Plate are of life-size.

PELTOPS BLAINVILLII.

PELTOPS BLAINVILLEI.

Broad-billed Flycatcher.

Eurylaimus Blainvillei, Garnot, Voy. Coquille, i. p. 595, pl. 19.—Finsch, Neu-Guinea, p. 160.
Peltops Blainvillii, Wagler.—Gray, Gen. B. i. p. 66.—Bp. Consp. i. p. 169.—Reich. Handb. Merop. p. 59, pl. 440. figs. 3211, 3212.—Scl. Journ. Linn. Soc. vol. ii. (1858) p. 160.—Gray, Cat. Mamm. &c. N. Guin. p. 19.—Id. P. Z. S. 1861, p. 433.—Wall. P. Z. S. 1862, p. 165.—Gray, Hand-l. B. i. p. 319.—Scl. Ibis, 1872, p. 177.—Id. P. Z. S. 1873, p. 696.
Erolla Blainvillei, Less. Traité, p. 260.
Platysomus Blainvillei, Swains. Classif. B. ii. p. 261.

This curious form of Flycatcher is only found in New Guinea and Mysol, Von Rosenberg having discovered it in the latter island. The original specimen was procured at Dorey in New Guinea; and Mr. Wallace also met with it in the north-western part of the same island, while Signor d'Albertis shot an example at Sorong. So rare has the species always been in collections, that a special examination of its structure has always been difficult; thus the bird has until lately been placed with the Broad-bills (Eurylæmidæ). Dr. Sclater was the first to recognize the true affinities of the genus; and I cannot do better than quote his observations on the subject:—

"The genus *Peltops*, containing the single species *P. Blainvillii* of New Guinea, has been usually referred to the Eurylæminæ, or Broad-bills, and the group thus formed united in the same family with the Rollers (Coraciadæ), the Todies (Todidæ), and the Motmots (Momotidæ), or, at all events, placed in their immediate neighbourhood. Several errors are, in my opinion, embraced in this classification.

"In the first place, *Peltops* has nothing whatever to do with the Eurylæmidæ, being a truly Muscicapine form allied to *Monarcha* and *Machærirhynchus*, as the most casual examination of its structure at once shows. The mistake, no doubt, comes from the somewhat exaggerated form of the bill in *Peltops*, and from its general coloration resembling that of *Cymbirhynchus*. The rarity of *Peltops* has prevented the error from being discovered. On examining the wing of *Peltops* it will be seen that the first primary is short or 'spurious' (as in all the true Oscines) when it exists at all. In *Cymbirhynchus* there are ten fully formed primaries. There is also a conspicuous difference in the size of the feet in the two forms, these organs being strong and thick in *Cymbirhynchus*, while they are feeble and weak in *Peltops*, as in other Muscicapidæ. The relegation of *Peltops* to the Muscicapidæ also removes an anomaly in geographical distribution, it being obviously strange that no otherwise exclusively Indo-Malayan type, such as the Eurylæmidæ, should have a single outlier in New Guinea."

General colour, both above and below, glossy black, including the wings and tail; ear-coverts and a patch on the interscapulary region white; rump, vent, upper and under tail-coverts crimson. Total length 7 inches, culmen 0·95, wing 3·9, tail 3·4, tarsus 0·6.

My Plate is drawn from Signor d'Albertis's Sorong specimen, which he kindly lent me when in London. The figure is life-size.

POMAREA RUFOCASTANEA.

POMAREA RUFOCASTANEA.

Rufous-and-Chestnut Flycatcher.

Monarcha rufocastanea, Ramsay, Proc. Linn. Soc. N. S. Wales, iv. pp. 79, 313 (1879).—Salvad. Ibis, 1880, p. 129.
Monarcha rufocastaneus, Salvad. Ann. Mus. Civic. Genov. xiv. p. 508 (1879).
Pomarea castaneiventris, Salvad. Orn. Papuasia, etc. ii. p. 11 (1881, pt.).

This species resembles very closely *Pomarea castaneiventris* of the Solomon group of islands, and to the latter bird it has been united by Count Salvadori. Our own idea was also that these two species were identical, judging from Mr. Ramsay's description; but the latter gentleman, during his visit to England as Commissioner for New South Wales to the International Fisheries Exhibition, brought over the types of his *Monarcha rufocastanea*, and we are enabled to say that they are not the same species.

On comparing males it is evident that *P. rufocastanea* is a smaller bird, is duller and more slaty black above and lighter chestnut below. The axillaries are entirely bay-coloured like the breast, whereas in *P. castaneiventris* the axillaries have distinct blackish bases and are rufous for not quite the terminal half.

The females of the two species differ much more than the males. That of *M. rufocastanea*, besides being a much paler bird, is distinguished at once by its light grey throat, this being black in *P. castaneiventris*.

The following are the descriptions of the typical specimens:—

Adult male (type of species). General colour above slaty black with a purplish-blue gloss, the hinder neck and mantle more ashy; lesser and median wing-coverts like the back; greater coverts, bastard wing, primary-coverts, and quills blackish brown, with slight remains of lighter brown edgings; tail-feathers blue-black; lores and crown of head like the back; ear-coverts also slaty black, surmounted by a faint indication of an iron-grey streak from above the fore part of the eye to about the end of the ear-coverts; feathers below the eye and cheeks deeper black, with a slight wash of ashy grey; sides of neck and entire throat dull cindery grey; fore neck and rest of underparts clear chestnut or bay, including the thighs and under tail-coverts; axillaries also bay-coloured like the breast; the under wing-coverts bay, except near the edge of the wing, where they are either dull ashy or are bay with ashy bases. Total length 5·3 inches, culmen 0·65, wing 2·75, tail 2·5, tarsus 0·7. (*Mus. Austr.*)

Adult female. Different from the male. Above slaty grey, rather lighter on the crown of the head; lesser and median wing-coverts like the back; greater coverts, bastard wing, primary-coverts, and quills brown, with narrow reddish-brown margins; tail-feathers dingy blackish brown; sides of crown paler grey than the top of the head, which is like the back; eyelid dusky slate-colour; feathers below the eye hoary whitish; ear-coverts, sides of face, and cheeks ashy grey, as also the sides of the neck, which are somewhat washed with rufous; throat slaty grey, the lower part and the fore neck washed with rufous, and gradually deepening into the rich bay of the breast and underparts; thighs ashy, washed with rufous; axillaries and under wing-coverts slaty grey, the former edged with rufous near the ends. Total length 5·5 inches, culmen 0·65, wing 3·0, tail 2·7, tarsus 0·7.

The two sexes are figured of the natural size, the birds being drawn from the typical examples, for the loan of which we are indebted to Mr. E. P. Ramsay.

[R. B. S.]

POMAREA CASTANEIVENTRIS.

POMAREA CASTANEIVENTRIS.

Chestnut-bellied Flycatcher.

Monarcha castaneiventris, Verr. Rev. et Mag. de Zool. 1858, p. 304.—Gray, Birds Trop. Isl. of the Pacific Ocean, p. 19 (1859).—Id. Hand-list of Birds, i. p. 320, no. 4793 (1869).—Salvad. Orn. Papuasia, etc. iii. App. p. 529 (1882).
Pomarea castaneiventris, Sharpe, Cat. Birds Brit. Mus. iv. p. 435, pl. xi. fig. 2 (1879).—Tristr. Ibis, 1879, p. 439.
Pomarea? castaneiventris, Salvad. Orn. Papuasia, etc. ii. p. 11 (1881).

Though described by M. Verreaux in 1858, the real home of the present species has only recently been ascertained for certain. For many years its habitat was recorded as "Oceania;" but so unlike was it to any Oceanic species of Flycatcher, that, in the 'Catalogue of Birds,' the latter locality was regarded as doubtful. Since that book was written, however, the bird has been met with in the Solomon Islands, thus fixing its habitat beyond a doubt. Capt. Richards procured it at Makira Harbour in San Cristoval, and Mr. Ramsay has received it from his collectors in the Solomon Islands. The latter gentleman has also described a nearly allied species as *P. rufocastanea*, which was considered by Count Salvadori to be identical with *P. castaneiventris*; but after comparing the two species, we consider them to be separable, and our readers will find the distinguishing characters noted under the heading of the former bird.

The following is a copy of the description of the adult male given in the 'Catalogue of Birds,' and drawn up from the type in the British Museum :—

"General colour above glossy black; wings black, the quills somewhat browner; tail black; sides of face, sides of neck, throat, and fore neck glossy black like the upper surface; remainder of under surface of body chestnut; under wing-coverts and axillaries chestnut, those near the edge of the wing black; quills ashy black below, whitish along the edge of the inner web. Total length 7 inches, culmen 0·7, wing 3·6, tail 3, tarsus 0·75."

The female differs from the male in being iron-grey instead of purplish black; the breast and abdomen are dark chestnut as in the male, but the throat is iron-grey.

The figures in the Plate are drawn from a pair of birds lent to us by Mr. E. P. Ramsay; they represent a male and female of the natural size.

[R. B. S.]

POMAREA UGIENSIS, Ramsay.

POMAREA UGIENSIS, *Ramsay*.

Ugi-Island Flycatcher.

Pomarea (Monarcha) ugiensis, Ramsay, Journ. Linn. Soc., Zool. xvi. p. 128 (1881); Reichenow & Schalow, J. f. O. 1882, p. 224.
Pomarea ugiensis, Tristram, Ibis, 1882, pp. 136-142; Salvad. Ann. Mus. Civ. Genov. xviii. p. 442 (1882).
Monarcha ugiensis, Salvad. Orn. Papuasia, etc. ii. p. 531 (1882).

This species was discovered by the Rev. George Brown in the island of Ugi, and Captain Richards afterwards met with it in the same island, to which it is probably confined. It is a very large species, and exceeds in size both *P. castaneiventris* of San Christoval and *P. nigra* of the Society and Marquesas Islands; its uniform glossy black plumage, which is peculiar to both sexes, is also a striking characteristic of the species. Count Salvadori would refer it to the genus *Monarcha*; but after carefully examining the specimens in the British Museum, we believe that its place is in the genus *Pomarea*, where it has been placed by Mr. Ramsay and Canon Tristram.

The following is a description of the typical specimen, which has been lent to us by Mr. Ramsay:—

Adult female (type of species). General colour above glossy blue-black; lesser and median wing-coverts black, edged with the same colour as the back; greater coverts, bastard wing, primary-coverts, and quills black, slightly and almost imperceptibly washed with blue-black on the outer web; tail-feathers glossy black, with dusky cross markings under certain lights; lores and base of forehead velvety black; sides of face and ear-coverts, cheeks and under surface of body glossy blue-black, duller black in the centre of the abdomen; "bill blue-black, whitish on the tips and margins of the mandibles; legs and feet black" (*Brown*). Total length 7 inches, culmen 0·8, wing 3·5, tail 3·15, tarsus 0·85.

In the Plate are given two representations of this species, of about the natural size, in different positions; the figures are drawn from the type specimen above described.

[R. B. S.]

THE
BIRDS OF NEW GUINEA

AND THE

ADJACENT PAPUAN ISLANDS,

INCLUDING MANY

NEW SPECIES RECENTLY DISCOVERED

IN

AUSTRALIA.

BY

JOHN GOULD, F.R.S.

COMPLETED AFTER THE AUTHOR'S DEATH

BY

R. BOWDLER SHARPE, F.L.S. &c.,
ZOOLOGICAL DEPARTMENT, BRITISH MUSEUM.

VOLUME III.

LONDON:
HENRY SOTHERAN & CO., 36 PICCADILLY.
1875–1888.

[All rights reserved.]

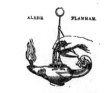

PRINTED BY TAYLOR AND FRANCIS,
RED LION COURT, FLEET STREET.

CONTENTS.

VOLUME III.

Plate			Part	Date.
1. Geocichla schistacea	Meyer's Ground-Thrush		XXIV.	1888.
2. Criniger chloris	Moluccan Bulbul		XXIII.	1887.
3. Eupetes castanonotus	Chestnut-backed Eupetes		XIV.	1883.
4. ,, cærulescens	Blue-bodied Eupetes		III.	1876.
5. Sericornis beccarii	Beccari's Sericornis		XV.	1883.
6. ,, arfakiana	Arfak Sericornis		XV.	1883.
7. ,, minimus	Little Sericornis		I.	1875.
8. Amytis goyderi	Goyder's Striated Wren		II.	1876.
9. Orthonyx novæ-guineæ	Papuan Orthonyx		VII.	1878.
10. Cinclosoma ajax	Orange-sided Ground-Thrush		XII.	1881.
11. Grallina bruijnii	Bruijn's Grallina		XIV.	1883.
12. Megalurus albolimbatus	White-edged Reed-Warbler		XVI.	1884.
13. Drymœdus beccarii	Beccari's Scrub-Robin		XV.	1883.
14. Ephthianura crocea	Yellow-breasted Ephthianura		XXIV.	1888.
15. Pachycare flavo-grisea	Yellow-and-Grey Thickhead		III.	1876.
16. Cracticus rufescens	Rufous Crow-Shrike		XXIII.	1887.
17. Pachycephala collaris	Ramsay's Thickhead		XXI.	1886.
18. ,, fusco-flava	Tenimber Thickhead		XIX.	1885.
19. ,, arctitorquis	Narrow-collared Thickhead		XV.	1883.
20. ,, brunnea	Brown Thickhead		XIV.	1883.
21. ,, hyperythra	Ruddy-breasted Thickhead		XIV.	1883.
22. ,, schlegeli	Schlegel's Thickhead		IX.	1879.
23. ,, christophori	San-Christoval Thickhead		XXIII.	1887.
24. Pachycephalopsis fortis	Grey-throated Thickhead		XVI.	1884.
25. ,, hattamensis	Hattam Thickhead		XIII.	1882.
26. ,, poliosoma	Grey Thickhead		XIII.	1882.
27. Xerophila pectoralis	Chestnut-breasted Xerophila		I.	1875.
28. Sittella albata	White-winged White-headed Sittella		XI.	1880.
29. Climacteris placens	New-Guinea Tree-creeper		I.	1875.
30. Cinnyris maforensis	Mafoor-Island Black Sun-bird		VIII.	1878.
31. ,, mysorensis	Mysore-Island Black Sun-bird		VIII.	1878.
32. ,, sangirensis	Sanghir Black Sun-bird		VIII.	1878.
33. Melirrhophetes leucostephes	White-faced Honey-eater		IV.	1877.
34. ,, ochromelas	Ochraceous Honey-eater		IV.	1877.
35. ,, batesi	Bates's Honey-eater		XXII.	1886.
36. Melidectes torquatus	Pectoral Honey-eater		IV.	1877.
37. ,, emilii	Count Turati's Honey-sucker		XXIV.	1888.
38. Melipotes gymnops	Naked-faced Honey-eater		IV.	1877.
39. Glicychæra fallax	Silky-plumed Honey-eater		XIX.	1885.
40. Melithreptus lætior	Beautiful Honey-eater		II.	1876.
41. Philemon plumigenis	Hoary-throated Honey-eater		XVI.	1884.
42. Stigmatops chloris	Mysol Honey-eater		XX.	1885.
43. ,, kebirensis	Kebir Scaly-throated Honey-eater		XXI.	1886.
44. ,, squamata	Scaly-chested Honey-eater		XXI.	1886.
45. ,, albo-auricularis	Broadbent's Honey-eater		XVII.	1884.
46. Glyciphila subfasciata	Dusky Honey-eater		III.	1876.
47. Ptilotis albonotata	White-marked Honey-eater		XX.	1885.
48. ,, marmorata	Mottled-breasted Honey-sucker		XIV.	1883.
49. ,, frenata	Bridled Honey-eater		II.	1876.
50. ,, flavo striata	Yellow-streaked Honey-eater		II.	1876.

CONTENTS OF VOL. III.

Plate			Part	Date
51.	Xanthotis chrysotis	Golden-eared Honey-eater	XXI.	1886.
52.	Meliarchus sclateri	Sclater's Honey-eater	XXI.	1886.
53.	Euthyrhynchus flavigula	Yellow-tinted Brown Honey-eater	XX.	1885.
54.	,, fulvigula	Buff-throated Honey-eater	XX.	1885.
55.	,, griseigula	Brown Honey-eater	XX.	1885.
56.	Melilestes poliopterus	Grey-winged Honey-eater	XIV.	1883.
57.	,, iliolophus	Long-plumed Honey-eater	XXIII.	1887.
58.	Zosterops uropygialis	Ké-Island White-eye	XIX.	1885.
59.	,, brunneicauda	Brown-tailed White-eye	XIX.	1885.
60.	,, fuscifrons	Dusky-fronted White-eye	XVIII.	1884.
61.	,, longirostris	Heath-Island White-eye	XVII.	1884.
62.	,, delicatula	Grey-sided White-eye	XIV.	1883.
63.	,, rendovæ	Rendova White-eye	XXIII.	1887.
64.	Myzomela wakoloensis	Lake Wakolo Honey-eater	XVIII.	1884.
65.	,, erythrina	New-Ireland Honey-sucker	XVII.	1884.
66.	,, melanocephala	Black-headed Honey-sucker	XVII.	1884.
67.	,, annabellæ	Mrs. Forbes's Honey-sucker	XV.	1883.
68.	,, sclateri	Sclater's Honey-eater	XII.	1881.
69.	,, cineracea	Ash-coloured Honey-eater	XII.	1881.
70.	,, rosenbergi	Von Rosenberg's Honey-eater	X.	1879.
71.	,, cruentata	Red-tinted Honey-eater	V.	1877.
72.	,, nigrita	Black Honey-eater	XXIII.	1887.

GEOCICHLA SCHISTACEA, Meyer.

GEOCICHLA SCHISTACEA, Meyer.

Meyer's Ground-Thrush.

Geocichla schistacea, Meyer, in Madarász, Zeitschr. für Orn. i. p. 211, Taf. viii. (1884).—Forbes, Naturalist's Wanderings, p. 365 (1885).

Two specimens of this strikingly coloured Thrush were procured by Mr. Riedel's hunters in Timor-Laut, and forwarded by him to his friend Dr. Meyer. Mr. Forbes did not apparently meet with the species during his sojourn in Timor-Laut, but he obtained two specimens of a Ground-Thrush which was new to science, and which he called *Geocichla machiki*. We have had to consider whether *G. machiki* is not the female of *G. schistacea*, as it is somewhat singular that both Mr. Forbes's specimens were females and both Mr. Riedel's were males. The nearest ally of *G. schistacea* is undoubtedly *G. wardi* of Southern India; but the latter is very distinct, being black above, with a broad white eyebrow and with no white on the ear-coverts. The difference in the colour of the sexes, however, in *G. wardi* is exactly parallel to the difference between *G. schistacea* and *G. machiki*, and we should not hesitate to unite the two species were it not for the fact that the latter is so very much larger than *G. schistacea*, the measurements being so different that we have not been able to recognize a species wherein the sexes varied so markedly in size, and we have therefore come to consider that they must be really distinct.

The following is a description of the type specimen of *G. schistacea*:—

Adult male. General colour above slaty grey; lesser wing-coverts grey like the back; median and greater coverts black, rather broadly tipped with white, forming a double wing-bar; bastard-wing black, with a small white spot at the end of the outer web; primary-coverts black; quills black, externally slaty grey, lighter on the primaries, the secondaries almost entirely slaty grey; upper tail-coverts like the back; tail-feathers blackish, slaty grey externally, the outermost feathers with a white spot at the end; crown of head slaty grey like the body, the fore part of the crown and forehead black; lores black, surmounted by a broad white eyebrow; feathers round eye and below the latter black, extending on to the fore part of the ear-coverts, which are otherwise white; cheeks, throat, and fore neck black; chest and breast white, spotted with terminal spots of black; abdomen and lower breast white; sides of body pale ashy grey, the lower flanks white; thighs and under tail-coverts white; under wing-coverts white, the lower series blackish; axillaries white; quills below black, white at the base of the inner web. Total length 6·4 inches, culmen 0·85, wing 4·0, tail 2·55, tarsus 1·1.

The Plate represents an adult male of the present species in two positions, the figures being drawn from the typical example lent to us by Dr. Meyer.

[R. B. S.]

CRINIGER CHLORIS, *Finsch*.

CRINIGER CHLORIS, *Finsch*.

Moluccan Bulbul.

Criniger flavicaudus, Gray, Proc. Zool. Soc. 1860, p. 311 (nec Bp.).
Criniger simplex, Wallace, Ibis, 1862, p. 350 (nec Temm.).
Trichophorus simplex, Finsch, Neu-Guinea, p. 168 (1865).
Criniger chloris, Finsch, Journ. für Orn. 1867, p. 36.—Gray, Hand-list Birds, i. p. 274, no. 4021 (1869).—Salvad. Ann. Mus. Civic. Genov. vii. p. 777 (1875), xvi. p. 183 (1880).—Sharpe, Cat. Birds in Brit. Mus. vi. p. 85 (1881).—Salvad. Orn. Papuasia e delle Molucche, ii. p. 376 (1881).—Guillemard, Proc. Zool. Soc. 1885, p. 572.

SPECIMENS of this species were sent by Mr. Wallace from the islands of Batchian and Halmahéra or Gilolo, and were identified by the late Mr. G. R. Gray as *Criniger flavicaudus* of Bonaparte, which is a Ceram species.

Mr. Wallace, in 1862, perceiving Mr. Gray's mistake, described the species under the name of *Criniger simplex*; but this name could not be used, as there was already an African species, described by Temminck as *C. simplex*. Dr. Finsch therefore proposed the name of *C. chloris* for the species in place of that given by Mr. Wallace. He described a specimen in the Leiden Museum from Halmahéra, collected by Dr. Forsten, and Dr. Beccari procured the species both in this island and in Batchian, as Mr. Wallace had already done before him. The latter gentleman likewise records its occurrence in Morotai or Morty Island; but no specimen from this island was in his collection, and the locality requires confirmation. Dr. Guillemard, who obtained a large series of specimens in Batchian, says that the iris was " red-brown, the bill bluish green with lighter edges, and the feet slate-colour."

The following description is copied from the British Museum ' Catalogue of Birds ':—

" *Adult female* (type of *C. simplex*). General colour above dull olive-yellow, the wing-coverts like the back ; quills brown, externally olive-yellow like the back, both the greater coverts and the primaries narrowly edged with brighter yellow ; tail-feathers dark olive-yellow, with a narrow margin of rather brighter yellow ; ear-coverts and sides of face dark olive-yellow, the fore part of the cheeks more dusky, the lores ashy brown ; throat bright yellow, as also the centre of the body and abdomen ; chest and sides of the body olive, the former with indistinct sharp lines of bright yellow ; thighs and under tail-coverts olive with yellow edges ; under wing-coverts and axillaries bright yellow ; quills brown below, yellow along the edge of the inner web : ' bill dusky lead-colour, the margins pale ; feet lead-colour ; iris dark ' (*Wallace*). Total length 8·5 inches, culmen 0·95, wing 3·95, tail 3·6, tarsus 0·75.

" *Adult male*. Similar in colour to the female. Total length 8·3 inches, culmen 0·9, wing 4, tail 3·4, tarsus 0·75."

The Plate represents an adult bird in two positions, the figures being drawn from a specimen in the British Museum.

[R. B. S.]

EUPETES CASTANONOTUS, *Salvad.*

EUPETES CASTANONOTUS, Salvad.

Chestnut-backed Eupetes.

Eupetes castanonotus, Salvad. Ann. Mus. Civic. Genov. vii. p. 966 (1875), xvi. p .187 (1880).—Id. Orn. Papuasia, etc. ii. p. 411 (1881).
Eupetes pulcher, Sharpe, Journ. Linn. Soc., Zool. vol. xvi. pp. 319, 440 (1882).

This beautiful species was discovered by Dr. Beccari in North-western New Guinea, and has recently been sent in some numbers by Mr. A. Goldie from the Astrolabe range of mountains in the south-eastern portion of the island. Owing to a discrepancy in the colouring of the head and in the extent of the black collar with the account of the species as given by Count Salvadori, Mr. Sharpe described the specimens from the last-named locality as new; but having received from the Genoa Museum the types of *E. castanonotus*, he has become convinced that the two species are identical.

The following is a full description of the two sexes:—

"*Adult male.* General colour above rich chestnut, including the hinder head and neck, mantle, and back; lower back, rump, and upper tail-coverts blue; scapulars blue washed with light reddish, with paler shaft-lines, indications of which are also seen on the feathers of the mantle; wing-coverts bright blue; bastard-wing feathers black; primary-coverts black edged with blue; quills black, the primaries slightly, the secondaries more plainly washed with blue externally, the innermost with olive-brown; tail dull blue, brighter blue on the edges of the feathers; base of forehead and lores black, succeeded by a band of dull blue across the forehead and forming a distinct eyebrow, which extends to behind the ear-coverts, which are black; cheeks, lower portion of ear-coverts, and entire throat pure white, surrounded by a very narrow line of black feathers, increasing in extent on the fore neck; remainder of under surface bright blue, the under tail-coverts blue with a broad spot of black at the ends; under wing-coverts and axillaries blackish, the outer ones washed with blue; quills ashy blackish below. Total length 9 inches, culmen 1·05, wing 3·7, tail 4·3, tarsus 1·35.

"*Adult female.* Differs from the male in having the entire upper surface chestnut without any blue; the chestnut colour, however, is much duller than in the male, except on the lower back and rump; wings as in the male; cheeks and throat white; remainder of under surface blue as in the male. Total length 8·5 inches, culmen 1·0, wing 3·65, tail 3·9, tarsus 1·3."

The figures in the Plate represent an adult male and a young bird of this species: they are drawn from the typical specimens now in the collection of the British Museum. [R. B. S.]

EUPETES CÆRULESCENS, *Temm.*

EUPETES CÆRULESCENS, Temm.

Blue-bodied Eupetes.

Eupetes cærulescens, Temm. Pl. Col. ii. pl. 574 (1835).—Müll. Naturl. Geschied. Land- u. Volkenk. p. 22 (1839-44).—Gray, Gen. B. i. p. 208 (1846).—Bp. Consp. i. p. 252 (1850).—Sclater, Proc. Linn. Soc. 1858, p. 158.—Gray, Cat. Mamm. & B. New Guin. p. 25 (1859).—Finsch, Neu-Guin. p. 167 (1865).—Gray, Hand-l. B. i. p. 267 (1869).—Sclater, P. Z. S. 1873, p. 696.

The genus *Eupetes* is no doubt one of the great group of strong-legged Thrushes, or *Timaliidæ*, so plentifully distributed throughout the tropical portions of the Old World; but, like many of the genera contained in the family, its systematic position is extremely difficult to define. Indications, however, of a more intimate acquaintance with these birds are not wanting; for whereas the present species has remained for thirty-five years represented by a single unique specimen in the Leiden Museum, it has been rediscovered by Signor D'Albertis, while the same traveller and his companions, Drs. Bruijn and Beccari, have succeeded in discovering no less than three new species in the northern part of New Guinea. The only remaining member of the genus is the *Eupetes ajax*, of Malacca, a bird plentiful enough in collections.

It cannot be expected that so rare a species should have much of a history attached to it; and, indeed, up to the present time no notes whatever have been published respecting it. D'Albertis met with it in Andai, in North-western New Guinea; and it was originally discovered by the well-known travellers Macklot and Von Mueller in Lobo Bay.

My figure is taken from a specimen kindly lent me by Signor D'Albertis, through Dr. Bennett, of Sydney, to whom I make my best acknowledgments for the loan. It represents an adult bird of the size of life.

As far as I yet know, there is no different colouring in the sexes. The bird which is figured in the Plate is nearly uniform delicate bluish grey; throat white, surrounded by a collar of black, including the ear-coverts and lores; the under part of the tail sooty grey; bill and legs black.

SERICORNIS BECCARII, Salvad.

SERICORNIS BECCARII, Salvad.

Beccari's Sericornis.

Sericornis beccarii, Salvad. Ann. Mus. Civic. Genov. vi. p. 79 (1874), xvi. p. 186 (1880).—Id. Orn. Papuasia, etc. p. 407 (1881).—Sharpe, Cat. Birds Brit. Mus. vii. p. 305 (1883).

THE genus *Sericornis* is mainly Australian, but some representatives of it are met with in New Guinea and the Aru Islands. The present bird is one of these, being found only in the last-named locality.

There are two sections of the genus, one in which the species have a distinct dark subterminal band on the tail-feathers, and a second in which this dark subterminal band is absent. Beccari's Sericornis belongs to the latter division and is closely allied to *S. frontalis*, a widely spread Australian species, but it is distinguished by having the ear-coverts dusky brown and the under tail-coverts dusky with fulvous tips.

As far as we yet know, the present species is peculiar to the Aru Islands, where it was discovered by the celebrated Italian traveller and naturalist whose name it bears.

The following description of the typical example is taken from the 'British Museum Catalogue of Birds':—

"*Adult male* (Aru Islands; Beccari: type of the species). General colour above dark olive-brown, gradually becoming more rufous-brown on the lower back and rump; the upper tail-coverts deep rusty brown; lesser wing-coverts like the back; median and greater coverts and bastard-wing feathers blackish with narrow white tips, the inner greater coverts brown; primary-coverts black; quills dusky brown, with olive edges to the primaries, the secondaries externally rusty brown; tail-feathers brown with dusky bars under certain lights, externally washed with reddish brown; crown of the head a little more dingy than the back, the forehead black, as also the lores, which are surmounted by a white streak; no eyebrow; eyelid above and below the eye white; below the eye a blackish shade; ear-coverts brown; cheeks and throat white, the feathers with narrow blackish margins and spots; remainder of the under surface of the body white slightly tinged with olive-yellow; the fore neck and chest washed with dusky; sides of the body and flanks washed with olive-brown; thighs dusky brown; under tail-coverts yellowish buff, the long ones brown with broad yellowish-buff margins; under wing-coverts dusky brown, the ones near the edge of the wing white, spotted with black; axillaries white; quills below dusky brown, inner edges ashy grey; 'iris cinnabar-red' (*Beccari*). Total length 7 inches, culmen 0·6, wing 2·35, tail 1·65, tarsus 0·85. (*Mus. Civic. Genov.*)"

We are indebted to the kindness of the Marquis of Doria, the director of the Civic Museum at Genoa, for the loan of the typical specimen of the present species, which is contained in the Museum under his care. The Plate represents that bird in two positions, of the natural size.

[R. B. S.]

SERICORNIS ARFAKIANA, *Salvad.*

SERICORNIS ARFAKIANA, Salvad.

Arfak Sericornis.

Sericornis arfakiana, Salvad. Ann. Mus. Civic. Genov. vii. p. 962 (1875), xvi. p. 187 (1880).—Id. Orn. Papuasia, etc. ii. p. 408 (1881).—Sharpe, Cat. B. Brit. Mus. vii. p. 306 (1883).

ALTHOUGH belonging to the same section of the genus *Sericornis* as *S. beccarii*, with no dark subterminal band on the tail-feathers, the present species nevertheless represents a different division of the genus. The species which have the tail marked as above are six in number; but the colour of the throat serves as a good distinguishing character between them. Thus *S. brunnea* has the throat bright rufous, *S. citreogularis* has it yellow, *S. frontalis* and *S. beccarii* have it white, while in *S. magnirostris* and *S. arfakiana* the throat is pale tawny buff like the lores and the base of the forehead.

Sericornis arfakiana differs from its near ally, *S. magnirostris*, in its much darker colour and blacker legs; the colour of the upper surface is dark olive-brown instead of pale ashy rufous; underneath it is deep olivaceous in tint, instead of being pale ashy tinged with olive. It is, as yet, only known to inhabit the Arfak Mountains in North-western New Guinea.

The following description of the typical specimen is taken from the British Museum 'Catalogue of Birds':—

"*Adult male* (Profi, Arfak; Bruijn: type of the species). General colour above dark olive-brown, browner on the lower back and rump, and rusty brown on the upper tail-coverts; lesser wing-coverts and median coverts like the back; greater coverts darker brown, edged with olive-brown, and with tips of dull fulvous forming an indistinct wing-bar; bastard-wing feathers and primary-coverts blackish; quills dark brown, edged with rusty olive, paler along the margin of the primaries; tail-feathers dark brown, washed with reddish brown on their margins; crown of the head more rusty brown than the back; lores and base of the forehead light rusty colour, the latter slightly mottled with dusky tips to the feathers; no eyebrow; feathers round the eye and ear-coverts pale rusty red, the latter with paler shaft-streaks; cheeks and throat pale rusty fulvous; fore neck and remainder of under surface of body pale ashy olive; the chest somewhat washed with rusty; sides of the body and flanks rather deeper olive; thighs and under tail-coverts rusty; under wing-coverts and axillaries dusky olive with somewhat of a reddish tinge; quills below dusky brown, inner edges ashy. Total length 4·5 inches, culmen 0·6, wing 2·35, tail 1·75, tarsus 0·85."

The figures in the Plate are drawn from the original specimen above described, and represent the bird of the size of life and in two positions. The Marquis Doria has been kind enough to lend us the specimen for the purpose of figuring in the present work.

[R. B. S.]

SERICORNIS MINIMUS, Gould.

SERICORNIS MINIMUS, *Gould.*

Little Sericornis.

Sericornis minimus, Gould, MS.

It may, I think, be fairly stated that this particular genus is represented in every portion of Australia: from the outlying Houtmans Abrolhas on the west coast to the brushes of the Clarence on the east, from Tasmania in the south to Cape York in the extreme north, one or another species has already been discovered. That a genus of birds so widely spread over the continent of Australia should not be found in New Guinea is very unlikely; to this time, however, it has not turned up in the scanty collections that have been formed in that country. The nearest allies of this pretty little bird are the *Sericornis frontalis* of the eastern brushes of New South Wales, and the *Sericornis lævigaster* of the northern portion of Australia. From the former it differs in the markings of the face and eye-stripes, from the latter in the total absence of any marks on the tips of the tail-feathers. Nothing has yet been recorded respecting the habits and economy of this species; but, judging from the sleek and silky texture of its feathers, it is an inhabitant of humid sterile places, among grasses and mossy stones, in the wilder portion of the forest.

On reference to the accompanying Plate it will be seen that the male, although generally of the same appearance as *S. frontalis*, has the lores and markings of the face of a different form—a feature I know to be constant; for I have received this bird in considerable numbers from collectors in Northern Queensland and the Cape-York district. I have stated in my history of *S. frontalis* that that bird was the smallest species of the genus; the present is but a trifle larger.

With regard to the reference to the specific name of *minimus*, given by me to this bird some years ago, I regret that I cannot find it at the time of going to press.

The figure may be thus described:—

Male.—Lores black, surmounted in front with a spot of white, a white mark also occupies the hinder portion of both eye-lashes; all the upper surface brown, increasing to rufous on the rump; secondaries and lesser wing-coverts black, with white edges; throat white, passing into yellowish white on the under surface generally; bill light brown, legs fleshy.

The *female* has the lores brown, with the rest of the markings on the face as in the male.

Total length $4\frac{1}{4}$ inches, wing $2\frac{5}{8}$, bill $\frac{5}{8}$, tail $1\frac{3}{4}$, tarsus $\frac{5}{8}$.

The Plate represents both sexes, of the natural size.

AMYTIS GOYDERI, *Gould.*

AMYTIS GOYDERI, *Gould*.

Goyder's Striated Wren.

Amytis Goyderi, Gould, Ann. & Mag. Nat. Hist. 4th series, Oct. 1875, p. 286.

As far as we yet know to the contrary, all the members of this peculiar genus are confined to the continent of Australia, and chiefly, if not wholly, to the interior, their favourite resort being sandy districts covered with rank grass and the usual herbage common to such soils. They are quite terrestrial in their habits, and run over the ground with great celerity. The nearest ally to this species is *Amytis striatus*; the entire throat and chest of the present bird, however, is white, or lightly coloured, which is not the case in *A. striatus* or the two other known species *A. textilis* and *A. macrourus*.

Like the new species of *Melithreptes*, this bird was collected by Mr. Andrews, he being one of the party of the late Lake-Eyre expedition towards the interior of South Australia. The following is a literal copy of my paper in the 'Annals,' which I cannot do better than repeat here, regretting that no information has reached us as to the probable range of the species, nor any account of its habits and economy :—

" General colour fawn-colour, distinctly streaked with white on the upper surface, each white stripe having a dark-brown lateral border on each side, the streaks becoming sandy-coloured on the lower back and disappearing entirely on the rump and upper tail-coverts, which are consequently uniform fawn; tail-feathers brown, with sandy-coloured shafts, the outer feathers rather broadly edged with fulvous; wings brown, the least coverts minutely and the greater series broadly streaked with sandy buff; quills brown, with light sandy-coloured shafts, and externally margined with rufous, the secondaries much more broadly, the innermost with broad fawn-coloured margins all round and streaked down the centre with sandy rufous; lores and sides of face white, the latter narrowly streaked with blackish brown, more distinctly on the cheeks; throat and breast pure white, as well as the centre of the abdomen; the flanks bright fawn-colour, inclining to paler and more sandy rufous on the thighs and upper and under wing-coverts.

" A second specimen sent, probably a female, is tinged with vinous instead of fawn-colour as in the male.

" As Mr. Waterhouse points out in his letter, the new *Amytis* is very closely allied to *A. textilis*; but it differs in being fawn-coloured instead of dull brown, with much broader white streaks to the upper surface; it is also distinguishable at a glance by its white under surface and fawn-coloured flanks."

I cannot close this short account without thanking Mr. Waterhouse, and the directors of the South-Australian Museum, for sending this bird to England that I might include it in the present work.

The figures are the size of life.

ORTHONYX NOVÆ-GUINEÆ, Meyer.

ORTHONYX NOVÆ GUINEÆ, Meyer.

Papuan Orthonyx.

Orthonyx spinicauda, Schlegel, Nederl. Tijdschr. voor de Dierk. iv. p. 47 (1873).
Orthonyx novæ guineæ, Meyer, Sitz. Akad. Wien, lxix. p. 83 (1874).—Beccari, Ann. Mus. Civic. Genov. vii. p. 7 (1875).—Salvad. Ann. Mus. Civic. Genov. vii. p. 935 (1875).—Sclater, Ibis, 1876, p. 247.

The genus *Orthonyx* now contains three species, all of which have been figured by me, viz. :—*O. spinicauda* (or, as Count Salvadori has proved to us that it should be called, *O. temmincki*); *O. spaldingi*, of Ramsay; and, lastly, the New-Guinea representative *O. novæ guineæ*, of Meyer. The first notice of the occurrence of this truly Australasian genus was given by Professor Schlegel, who recorded in 1873 the fact that the Dutch traveller Baron von Rosenberg had met with an adult male specimen of an *Orthonyx* in North-western New Guinea. Prof. Schlegel adds :—" I do not find any sensible differences between it and Australian individuals. It is true that its tail is six lines less than usual in the latter bird; but I refuse to attribute this difference to any other cause than an imperfect moult." In the mean time Dr. Meyer also visited the same part of New Guinea, and procured the female; and on comparing it with *O. spinicauda* he found that it differed sufficiently to show that it belonged to a distinct species. In this conclusion he has been followed by Count Salvadori, who has had six specimens, three males and three females, to examine from the collections formed by Dr. Beccari and Mr. Bruijn in the Arfak Mountains. I must also admit that the species seems to me to be perfectly recognizable. Dr. Meyer writes :—" The reddish brown on the chin and throat is much more restricted than in *O. spinicauda*; and the remarkable black colour on the sides of this reddish brown is altogether wanting; the sides of the body, breast, and belly entirely grey, with only a few white feathers on the latter. Further, the brown of the upper parts is as rich as in *O. spinicauda*, and the black colour more strongly developed. The white spots on the wing are remarkable by their absence."

Count Salvadori also gives a full account of the difference of the sexes, and of the specific difference between the present species and the Australian bird. Any one comparing the Plate now given with the one in my 'Birds of Australia' will see the distinctive characters of the two birds at a glance.

To Dr. Meyer and Count Salvadori I am indebted for the loan of the birds figured in the Plate, which are represented of the natural size.

When this paper had so far gone through the press, I received the following few lines from my friend Dr. Meyer, which I have great pleasure in inserting :—" I only got one female of this new species from the Arfak Mountains, acquired in July 1873, at a height of about 3500 feet. In its general habitus it equals *O. spinicauda* from Australia, but differs in coloration and size.

" Total length 172 millims.; wing 84; tail 67; bill 13."

CINCLOSOMA AJAX.

CINCLOSOMA AJAX.

Orange-sided Ground-Thrush.

Eupetes ajax, Temm. Pl. Col. ii. pl. 573 (1835).—S. Müll. Natuurl. Geschied. Land- en Volkenk. p. 22 (1839-1844).—Gray, Gen. B. ii. p. 208 (1846).—Bp. Consp. i. p. 252 (1850).—Sclater, Proc. Linn. Soc. ii. p. 158 (1858).—Gray, Proc. Zool. Soc. 1858, p. 191.—Id. Cat. B. New Guinea, pp. 25, 56.—Id. Proc. Zool. Soc. 1861, p. 434.—Rosenb. Journ. f. Orn. 1864, p. 119.—Finsch, Neu-Guinea, p. 167 (1865).—Gray, Handl. B. i. p. 267, no. 3913 (1869).—Ramsay, Proc. Linn. Soc. N. S. W. iv. pp. 90, 98 (1879).
Ajax ajax, Less. Compl. Buff., Ois. p. 422 (1838).
Ajax eupetes, Less. Rev. Zool. 1839, p. 226.
Ajax typicus, Less. teste Bp.
Eupetes goldiei, Ramsay, Proc. Linn. Soc. N. S. W. iii. p. 303 (1879).—Salvad. Ibis, 1879, p. 324.
Cinclosoma ajax, D'Alb. & Salvad. Ann. Mus. Civ. Genova, xiv. p. 85 (1879).—Sharpe, Pr. Linn. Soc. xiv. p. 631 (1879).—D'Albert. Proc. Zool. Soc. 1879, p. 218.—Salvad. Ann. Mus. Civ. Genova, xvi. p. 188 (1880).—Id. Orn. Papuasia &c. ii. p. 416 (1881).

This fine species is a native of New Guinea, where it was discovered nearly fifty years ago by Salomon Müller, in Lobo Bay. His specimen remained unique in the Leyden Museum until quite recently, when Signor D'Albertis procured four males on the Fly River in the southern part of that great island; and I have recently seen several individuals obtained in the interior of South-eastern New Guinea by Mr. Goldie and Mr. Charles Hunstein. The original specimen, still existing in the Leyden Museum, was said to be a female; and Count Salvadori, who has examined it, also believes that it has been rightly sexed; but he considers it to be a female in imperfect plumage. I should not hesitate myself to set it down as an immature male, were it not that in the British Museum there appear to be the adults of both sexes and a young male. Unfortunately none of the birds from South-eastern New Guinea in the above institution has been sexed by the collectors; but I fully believe that the birds here figured by me represent the fully adult male and female; and it will be noticed that the latter has the wing-coverts spotted with white (as in the Australian members of the genus), while the male has them totally black. An examination of Temminck's plate, which Count Salvadori states to be a poor representation of the specimen, reveals the fact that in the bird at Leyden the wing-coverts are likewise black. In my collection a bird believed by Mr. Bowdler Sharpe to be the immature male resembles the old female, but has the throat of a dull brown; and this question of the plumage of *Cinclosoma ajax* must be left until some carefully identified specimens reach us from New Guinea.

Nothing is known of the habits of this *Cinclosoma*; but Mr. Kendal Broadbent marked one of his specimens as a new species of "Mountain-Thrush." Mr. Sharpe has very kindly sent me a description of an adult pair of birds in the national collection:—

Adult male.—General colour above dark earthy brown, the head a little lighter than the back; wing-coverts black, excepting the inner greater coverts, which are like the back; primary-coverts black, the inner ones browner; quills dusky, externally like the back, the primaries less strongly marked on the outer web; upper tail-coverts and central tail-feathers rather more olive-brown than the back, with indistinct wavy bars under certain lights; remainder of the tail-feathers black, those near the centre of the tail washed with olive-brown near the base, the three outer feathers on each side tipped with white; lores, eyebrow, feathers below the eye, and ear-coverts black, extending down the sides of the neck, getting narrower as it joins the black throat and encloses a large white patch, which occupies the entire cheeks, widening out behind; throat and chest glossy black; centre of breast and abdomen white bordered with a narrow line of black down each side, the feathers forming this being black on the inner web, white on the outer; sides of breast and flanks orange-rufous, browner on the lower flanks and sides of vent; thighs whitish, ashy-brown behind; under tail-coverts white, the lateral ones varied with black outer webs; under wing-coverts and axillaries white, the edge of the wing black, the inner greater coverts also tipped with black; lower greater coverts ashy like the quills below, which are lighter grey along the edge of the inner web; " bill black; feet whitish; iris yellow" (*D'Albertis*).

Total length 9·3 inches, culmen 1·0, wing 3·9, tail 3·7, tarsus 1·3.

Adult female.—Differs from the male in being duller brown above and in having the wing-coverts brown like the back, the median and greater coverts with triangular spots of black at the tip, the outermost of the former with a white spot at the end; primary-coverts blackish brown; lores and ear-coverts brown, forming a streak along the sides of the head, above which is a tolerably well-defined white eyebrow; cheeks and throat pure white; remainder of undersurface of body from the lower throat downwards orange-chestnut shading into fulvous brown on the flanks; under wing- and tail-coverts as in the male.

Total length 8·75 inches, culmen 0·95, wing 3·75, tail 3·5, tarsus 1·2.

The figures in the Plate represent a male and the supposed adult female, of the natural size; that of the male is drawn from a specimen in my own collection, and the female from a skin now in the British Museum.

GRALLINA BRUIJNI, *Salvad*.

GRALLINA BRUIJNI, Salvad.

Bruijn's Grallina.

Grallina bruijni, Salvad. Ann. Mus. Civ. Genova, vii. p. 929 (1875).—Sharpe, Cat. B. Brit. Mus. iii. p. 273 (1877).—Salvad. Ann. Mus. Civ. Genova, xv. p. 42, n. 4 (1879).—Salvad. Orn. Papuasia, ii. p. 191 (1881).—Sharpe, Journ. Linn. Soc., Zool. vol. xvi. p. 435 (1882).

Pomareopsis semiatra, Oustalet, Bull. Ass. Sci. de France, 1880, p. 173.

The discovery of a species of *Grallina* in New Guinea is of great interest, as the genus was until recently considered to be entirely peculiar to Australia. In general appearance the New-Guinea bird differs considerably from its Australian relative, and is on the whole the handsomer bird of the two; but the same difference in sexes which characterizes the Pied *Grallina* is seen in the species from New Guinea.

The present species was discovered in the Arfak Mountains by the hunters employed by Mr. Bruijn, after whom it is named. More recently it has been sent from the mountains of South-eastern New Guinea by Mr. Goldie, who procured it from the Morocco district at the back of the Astrolabe range. Mr. Goldie states that the native name is *Tada*, and gives the following note:—" These birds are found flying about creeks and hopping about stones; they seem to feed on insects obtained there."

The following descriptions are taken from a pair of Mr. Goldie's birds in the British Museum.

Adult male. General colour above glossy blue-black, including the scapulars; all the wing-coverts pure white, forming a large shoulder-patch; bastard wing, primary-coverts, and quills black with glossy blue-black margins; lower back white; rump, upper tail-coverts, and basal half of the tail creamy buff, the terminal half of the latter black; feathers round the eye and ear-coverts blue-black, with a streak of white above the eye; the lower ear-coverts creamy white, as also the sides of the neck, forming a large patch; throat and entire under surface blue-black; lower flanks and thighs white; under tail-coverts creamy buff; on the sides of the lower back a patch of white feathers; under wing-coverts and axillaries blue-black; quills below black, the inner edges rather browner.

Total length 6·9 inches, culmen 0·9, wing 4·05, tail 3·05, tarsus 1·1.

Adult female. Resembles the male above, but differs as follows. Lores and broad eyebrow white; eyelid, feathers below the eye, cheeks, ear-coverts, and throat blue-black; remainder of under surface creamy white, deeper creamy buff on the flanks; abdomen and under tail-coverts, under wing-coverts, and axillaries white, and base of quills also, broader on the secondaries. Culmen 0·85 inch, wing 4·05, tail 2·9, tarsus 1·05.

The Plate represents an adult male and female of the natural size; and for the loan of the birds I am indebted to Mr. Edward Gerrard, junr. [R.B.S.]

MEGALURUS ALBOLIMBATUS.

MEGALURUS ALBOLIMBATUS.

White-edged Reed-Warbler.

Poodytes albolimbatus, D'Albert. & Salvad. Ann. Mus. Civic. Genov. xiv. p. 87 (1879).—Salvad. op. cit. xvi. p. 189 (1880).—id. Orn. Papuasia e delle Molucche, p. 422 (1881).
Megalurus albolimbatus, Sharpe, Cat. Birds Brit. Mus. vol. vii. p. 129 (1883).

Count Salvadori has placed the present bird in Cabanis's genus *Poodytes*, the type of which was the Australian *Megalurus gramineus*. We cannot, however, separate either the last-named bird or *M. albolimbatus* generically from the other species of *Megalurus*.

In the last-named genus there are two sections, one in which the upper tail-coverts are streaked with dark brown centres, and the other with the tail-coverts uniform. To this latter section belong *M. timoriensis* and the subject of the present article. *M. albolimbatus* is a much smaller bird than *M. timoriensis*, and is distinguished by the conspicuous white edgings to the inner secondaries and the dark brown centres to the tail-feathers.

The following descriptions are from the male and female originally described by Count Salvadori:—

"*Adult male* (Fly River; D'Albertis: type of species). General colour above ferruginous, the head, lower back, rump, and upper tail-coverts uniform, excepting a slight indication of dusky striations on the head; mantle and upper back more fulvous, the feathers broadly centred with black; wing-coverts black, with rufous-buff margins, a little paler on the outer median coverts and bastard-wing feathers; quills black, edged with rufous-buff, the inner secondaries conspicuously bordered with white on both webs, all the other quills narrowly fringed with whitish at the tips; tail-feathers dusky brown, with broad margins of dull rufous; lores and a distinct eyebrow yellowish buff, the latter whiter; feathers round the eye and ear-coverts yellowish buff, rufescent along the upper edge of the ear-coverts; cheeks, throat, and under surface of body white, with a wash of yellowish buff on the fore neck and chest, deepening into tawny fulvous on the flanks and thighs, the under tail-coverts again a little paler; under wing-coverts pale fulvous, the quills ashy brown below, pale rufescent along the edge of the inner web; 'bill brown, the lower mandible whitish; feet fleshy; iris chestnut' (*D'Albertis*). Total length 5·4 inches, culmen 0·65, wing 2·25, tail 3·35, tarsus 0·85.

"*Adult female.* Similar in plumage to the male, but a little duller. Total length 5·3 inches, culmen 0·6, wing 2·15, tail 2·35, tarsus 0·85."

The figures in the Plate represent the male and female birds, of the natural size. They are drawn from the types lent to us by the Marquis Doria, whose kindness in sending us over many valuable specimens for the purposes of the present work we have much gratification in publicly acknowledging.

[R. B. S.]

DRYMŒDUS BECCARII, *Salvad.*

DRYMŒDUS BECCARII, Salvad.

Beccari's Scrub-Robin.

Drymœdus beccarii, Salvad. Ann. Mus. Civic. Genov. vii. p. 965 (1875).—Id. Proc. Zool. Soc. 1878, p. 97.—Sharpe, Proc. Linn. Soc. xiv. p. 633 (1879).—Salvad. Ann. Mus. Civic. Genov. xvi. p. 188 (1880).—Id. Orn. etc. Papuasia, ii. p. 416 (1881).—Id. Report Voy. H.M.S. 'Challenger,' ii. p. 80 (1881).—Sharpe, Cat. Birds Brit. Mus. vii. p. 345 (1883).

The present species affords us another instance of the close relationship between the avifauna of Australia and that of the Papuan subregion; for the genus *Drymœdus* is essentially an Australian form, being found in all parts of that continent, and its presence in New Guinea and the Aru Islands is very interesting.

Discovered by Dr. Beccari in the Arfak Mountains in North-western New Guinea, its existence in the Aru Islands was detected by the naturalists of the 'Challenger' Expedition at Wanumbai. Since that time we have seen numerous examples from South-eastern New Guinea collected by Mr. A. Goldie and Mr. Broadbent, and the species appears to be by no means uncommon in the interior of that part of New Guinea.

The following description is copied from the British Museum 'Catalogue':—

"*Adult.* General colour above chestnut-brown, the head more dusky, and of a deep chocolate-brown; lesser wing-coverts ashy brown, the remainder black, barred with white at the tip, the primary-coverts entirely black; quills blackish, with a white spot at the base of the primaries, which have also a bar of whitish across the middle of the outer web; the secondaries washed with rufous towards the ends of the outer webs; centre tail-feathers chestnut-brown, the remainder blackish, externally chestnut-brown, and tipped with white; forehead blackish; lores and eyelid white, with a black spot above the eye, as well as another broad patch of black below the eye at the base of the ear-coverts; adjoining these black spots the eyelid is also black; a slight ashy shade along the sides of the crown; ear-coverts ashy brown, streaked with white near their bases; cheeks and throat white; remainder of under surface whity brown or whitish washed with brown, the sides of the breast ashy brown, more rufescent on the flanks; under tail-coverts chestnut-brown; axillaries ashy, tipped with white; under wing-coverts blackish, tipped with white, forming broad bars; quills blackish below, with a patch of white near the base of the primaries and outer secondaries; feet pale in skin; bill black. Total length 7 inches, culmen 0·75, wing 3·2, tail 3·5, tarsus 1·7."

The Plate represents an adult bird of the present species, in two positions, of the natural size. The figures have been drawn from a specimen procured by Mr. A. Goldie, and now in the Leiden Museum.

[R. B. S.]

EPHTHIANURA CROCEA, Cast. & Rams.

EPHTHIANURA CROCEA, Cast. & Ramsay.

Yellow-breasted Ephthianura.

Ephthianura crocea, Castelnau & Ramsay, Proc. Linn. Soc. N. S. W. i. p. 380 (1877).—Ramsay, op. cit. ii. p 186 (1878).—Sharpe, Cat Birds in Brit. Mus. vol. vii. p. 669 (1883).

This very pretty little species is one of the recent discoveries in Australian ornithology. It is easily distinguished from the other three species of the genus (*E. albifrons*, *E. tricolor*, and *E. aurifrons*) by its yellow under surface and black pectoral collar.

The species was first found on Norman River in the Gulf of Carpentaria by Mr. Gulliver, and more recently the late Mr. T. H. Bowyer Bower met with it on the Fitzroy River in North-western Australia. In my friend's last journal, which his father has kindly lent to me, I find the following note on this bird :—
" Sept. 15. We had previously observed these beautiful *Ephthianuræ* on the swamp about two miles from our camp, but could not obtain any specimens. I went to try and get a White Heron, leaving Breston to look after the smaller birds. The species is to be seen flitting among some strong reedy grass growing out of the water, and appears to hang therefrom and pick insects off the surface of the water. I only heard a very simple call-note, but no song. Females and young birds are easily obtained, but adult males appear to be rare. When disturbed about the water, they sought safety in some long grass about fifty yards away from the swamp, round which the grass has all been beaten down by sheep."

The following descriptions are taken from a pair of birds presented to the British Museum by Capt. Bowyer Bower, and collected by his son on the Fitzroy River :—

Adult male. General colour above pale olive-brown, with slight indication of dusky centres on the feathers of the back; the lower back washed with yellow; rump and upper tail-coverts bright yellow; wing-coverts brown, edged with pale ashy, whiter on the median and greater series; the lesser coverts washed with yellow; bastard-wing, primary-coverts, and quills brown, margined with pale yellow, the inner secondaries broadly margined and tipped with ashy whitish; tail-feathers dark brown, margined with yellow and with a broad spot of ashy whitish at the ends; crown of head like the back, but washed with olive-yellow, brighter towards the forehead; lores blackish, surmounted by a streak of bright yellow; eyelid yellow; ear-coverts ashy brown washed with yellow; feathers below the eye, cheeks, and throat bright yellow, followed by a crescentic spot of black on the fore neck; remainder of under surface of body yellow, mixed with ashy, the lower abdomen ashy whitish; thighs and under tail-coverts yellow; under wing-coverts and axillaries ashy, with yellow margins; edge of wing yellow; quills below dusky brown, ashy whitish along the inner edge. Total length 4·3 inches, culmen 0·45, wing 2·35, tail 1·35, tarsus 0·7.

Adult female. Differs from the male in being ashy brown above without the wash of olive-yellow, the head like the back; sides of face ashy brown; lores, cheeks, and throat white; fore neck, breast, and sides of body light ashy brown, the latter with a slight yellowish tinge; lower abdomen ashy whitish; thighs and under tail-coverts brighter yellow; rump and upper tail-coverts bright yellow as in the male, but not so brilliant. Total length 4·2 inches, culmen 0·45, wing 2·3, tail 1·35, tarsus 0·7.

The figures in the Plate, which represent an adult pair of birds of the natural size, are drawn from the birds described above.

[R. B. S.]

PACHYCARE FLAVO-GRISEA.

PACHYCARE FLAVO-GRISEA.

Yellow-and-grey Thickhead.

Pachycephala flavo-grisea, Meyer, Sitz. Akad. Wien, lxix. p. 495 (1874).—Salvad. Ann. Mus. Civ. Genova, vii. p. 775 (1875).

In describing this bird as a *Pachycephala* Dr. Meyer seems to have had an idea that the species was by no means a typical one; and I need hardly do more than draw the attention of my readers to the bird figured in the opposite Plate to illustrate the very aberrant style of plumage exhibited by it as compared with the usual black and yellow or green dress of an ordinary *Pachycephala*. In point of fact, not only the plumage, but the form of the bird differs so strikingly that I feel compelled to institute a new generic title for its reception.

In the olden days, when the quinary system had sway, we should doubtless have been told that the present species represented the Sittidæ, or Nuthatches, among the Shrikes, if, indeed, the Pachycephaline birds are to be accounted as belonging to the great group of the Laniidæ. Their position is not well ascertained; and by some ornithologists they are considered to belong to the Ampelidæ, while Mr. D. G. Elliot raised them at one time to the rank of a family (Pachycephalidæ). Although the quinary theory is exploded, and its most potent advocates have passed away, one cannot be surprised at the hold that it once possessed on the minds of many earnest zoologists, as it brought forward characters and affinities which would perhaps have been otherwise neglected; only its advocates were inclined to push their ideas to too great an extreme. The resemblance in the present instance, however, is very striking, not only in the grey Nuthatch-like colour of the back, but even in the two long black stripes (so characteristic of the Sittæ) which run down each side of the neck.

Dr. Meyer obtained his type specimen in the Arfak Mountains, in March 1873; and more recently Count Salvadori has received two specimens from the same locality, sent by Dr. Bruijn. I am pleased to see that this excellent ornithologist is inclined to consider its generic separation expedient.

I am once more indebted to Dr. Meyer's kindness for the opportunity of figuring this interesting addition to the Papuan avifauna; and the following description is drawn up from the typical specimen.

Face, ear-coverts, throat, and undersurface bright yellow; crown of the head and all the upper surface grey; a black mark intervening between the yellow face, ear-coverts, and the grey of the upper surface of the body. The three secondary wing-feathers next the body black, with a distinct square spot of white at their tips; underparts of the wing white; tail grey above, terminating in dusky black.

This is by no means a typical *Pachycephala*, and must receive an appellation as a new genus (PACHYCARE), in which the following characters must be noted:—Plumage lax. Bill stout, with a distinct notch and overhanging tip on the upper mandible; vibrissæ absent. Wing rounded and feeble; first primary short, the fourth the longest. Tail short and rounded. Tarsi moderately long, exceeding the length of the bill, rather slight, as are the toes.

Total length 5 inches, bill $\frac{5}{8}$, wing $3\frac{1}{8}$, tail $2\frac{1}{8}$, tarsi $\frac{3}{4}$.
The figures in the Plate are of the natural size.

CRACTICUS RUFESCENS, Devis.

CRACTICUS RUFESCENS, Devis.

Rufous Crow-Shrike.

Cracticus rufescens, Devis, Proc. Linn. Soc. N. S. Wales, vii. p. 562 (1883).

For the opportunity of describing and figuring this interesting Australian bird we are indebted to the kindness of Mr. T. H. Bowyer Bower, who met with the species in Queensland. At first sight it looks like an immature bird of some of the black or pied species of *Cracticus*, and this was our impression on first reading Mr. Devis's description; but after comparing Mr. Bower's specimens with the immature birds of the other Australian *Cractici*, we find that *C. rufescens* is undoubtedly a good species.

Mr. Devis, who first described it, received his specimens from the Tully and Murray River Scrubs. Mr. Bowyer Bower met with the species in October at Gordon's Camp on the Mulgrave River, twenty-two miles from Cairns, Trinity Bay. It was the only *Cracticus* seen in the neighbourhood and was decidedly scarce.

The following descriptions are taken from an adult pair of birds procured by Mr. Bowyer Bower in Queensland and given by him to the British Museum:—

Adult male. General colour above light rufous, the back streaked with ochreous buff and brown, the feathers being pale in the centre and brown on the margins; upper tail-coverts pale tawny rufous; lesser wing-coverts also pale tawny; median and greater coverts blackish brown, with whitish or pale tawny ends, and a median streak of pale tawny; quills dark brown, externally marked with pale tawny rufous; tail-feathers bronzy brown, the inner webs of all but the centre feathers light rufous; crown of head, hind neck, and upper mantle black, streaked with tawny rufous, more broadly on the latter; lores, a narrow eyebrow, sides of face, and ear-coverts tawny rufous, with dusky brown edges to the feathers; cheeks and under surface of the body pale tawny buff, rather more rufous on the sides of the body, flanks, thighs, and under tail-coverts; axillaries and under wing-coverts clear tawny; quills dusky below, pale tawny along the inner web: "bill bluish lead-colour at base, passing into black at the tips; iris brown" (*T. H. Bowyer Bower*). Total length 14·25 inches, culmen 2·1, wing 6·55, tail 5·2, tarsus 1·6.

The female is a little paler and not so strongly streaked with rufous above. Total length 13 inches, culmen 2·1, wing 6·4, tail 5·1, tarsus 1·55.

The Plate represents a male and female of about the size of life, and the figures are taken from the pair in the British Museum.

[R. B. S.]

PACHYCEPHALA COLLARIS, Ramsay.

PACHYCEPHALA COLLARIS, *Ramsay*.

Ramsay's Thickhead.

Pachycephala collaris, Ramsay, Proc. Linn. Soc. N. S. Wales, iii. p. 74 (1878), p. 281 (1879), iv. p. 99 (1879).—Salvad. Ibis, 1879, p. 324.—Id. Ann. Mus. Civic. Genov. xv. p. 45 (1879).—Id. Orn. Papuasia e delle Molucche, ii. p. 221 (1881).—Gadow, Cat. Birds in Brit. Mus. viii. p. 197 (1883).

The present species is closely allied to *P. melanura*, but differs in its olive-coloured tail, a distinction accurately noted by Count Salvadori, although he had never seen a specimen. Taking advantage of Mr. E. P. Ramsay's visit to England during the year 1883, we borrowed the typical specimens of this bird for illustration in the present work, and have much satisfaction in figuring in the accompanying Plate the actual pair described by Mr. Ramsay, these being, so far, the only ones known in any collection.

The habitat of the species is Courtance Island, off the south-east coast of New Guinea.

The following is a description of the type specimens:—

Adult male. General colour above olive-green, the feathers on the sides of the rump with yellow ends; lesser wing-coverts like the back; median and greater coverts slightly brighter yellow, with dusky bases; bastard-wing and primary-coverts dusky, edged with dull olive-green; quills dusky, edged with yellowish olive, the primaries with ashy grey; tail olive-green, rather yellower on the edges of the feathers; crown of head and nape black, separated from the back by a broad collar of bright yellow; lores, feathers below the eye, and ear-coverts black; cheeks and throat white, separated from the breast by a narrow black collar, which extends up to the sides of the ear-coverts; remainder of under surface of body bright yellow, a little paler towards the vent and under tail-coverts; axillaries and under wing-coverts yellow, white at the bases; quills dusky below, white along the inner edge. Total length 5·7 inches, culmen 0·8, wing 3·65, tail 2·7, tarsus 0·95.

Adult female. Different from the male. General colour olive-brown, a little clearer on the rump and on the hind neck, where there is an indistinct collar; wing-coverts like the back, the greater series with dusky bases; bastard-wing, primary-coverts, and quills dusky, edged with olive-brown, paler on the primaries; tail dull olive-yellowish, the feathers edged with clearer yellow; crown of head dull chocolate-brown, contrasting slightly with the back; sides of face and ear-coverts clearer reddish brown, with the lores and eyelids whitish; cheeks and throat white, the lower throat slightly shaded with brownish; remainder of under surface of body bright yellow; axillaries and under wing-coverts white, edged with yellow. Total length 6·2 inches, culmen 0·85, wing 3·65, tail 2·4, tarsus 0·85.

The Plate represents an adult male and female, of the natural size, drawn from the type specimens lent to us by Mr. Ramsay.

[R. B. S.]

PACHYCEPHALA FUSCOFLAVA, Sclater.

PACHYCEPHALA FUSCOFLAVA, Sclater.

Tenimber Thickhead.

Pachycephala, sp. incog., Scl. Proc. Zool. Soc. 1883, p. 51.
Pachycephala fuscoflava, Sclater, Proc. Zool. Soc. 1883, p. 198, pl. xxviii.—Forbes, tom. cit. p. 589, pl. liii.

THIS fine Thickhead is one of the largest of the genus *Pachycephala*, and was discovered by Mr. H. O. Forbes in the island of Larat in the Tenimber group. In the first collection sent by him to this country were two specimens marked as male and female, from which it was at first surmised that the species belonged to the dull-coloured section of the genus *Pachycephala*, and it was only on the arrival of Mr. Forbes in England that the fully adult male was described, the specimen having been accidentally mislaid by him in packing up his collections.

The male may be described as being very similar in colour to *P. torquata* from the island of Taviuni, but much larger and lighter olive-yellow above, and much paler lemon-yellow below.

The young male being so distinct, we give the following full description of it:—

General colour above olive-greenish, rather more olive-yellow on the rump; head a trifle yellower than the back, from which it is separated by a faintly indicated ring of deep olive-yellow, not pronounced enough to form a distinct collar; lesser and median wing-coverts clearer olive-yellow than the back; bastard-wing, primary-coverts, and primaries blackish brown, externally ashy grey, the secondaries and greater wing-coverts dusky brown, externally olive-yellow; tail-feathers olive-greenish; lores ashy whitish, with a dusky spot before the eye; eyelid rather more yellowish olive; ear-coverts reddish brown, with narrow whitish shaft-lines; cheeks and chin ashy whitish marked with yellow, with a brighter yellow moustachial streak; throat and breast saffron-yellow, washed with brighter yellow in the centre of the breast; the sides of the breast and sides of neck browner; abdomen ashy whitish marked with yellow, the flanks olive-brown; thighs and under tail-coverts bright yellow; axillaries and under wing-coverts pale brownish white, the latter washed with yellow; quills dusky below, ashy along the inner web : " bill black ; legs and feet sooty blue; iris dark brown " (*H. O. Forbes*). Total length 7·3 inches, culmen 0·85, wing 4·15, tail 3·1, tarsus 1·15.

Adult female. Similar to the young male, but a little duller-coloured below, with less yellow on the breast : " bill, legs, and feet black ; iris dark brown " (*H. O. Forbes*). Total length 7·2 inches, culmen 0·85, wing 0·9, tail 3·0, tarsus 1·05.

The figures in the Plate represent the adult male and female of the full size; they are drawn from the typical specimens in the British Museum.

[R. B. S.]

PACHYCEPHALA ARCTITORQUIS, *Sclater*.

PACHYCEPHALA ARCTITORQUIS, Sclater.

Narrow-collared Thickhead.

Pachycephala arctitorquis, Sclater, Proc. Zool. Soc. 1883, p. 55, pl. xiii.

The present species of *Pachycephala* was discovered in the Tenimber Islands by Mr. H. O. Forbes, and belongs to the fourth section of the genus as arranged by Dr. Gadow in the eighth volume of the 'British Museum Catalogue of Birds,' with this reservation, that the character of the back being "dusky black" belongs only to *P. monacha* and not to *P. leucogaster*, which has a grey back.

P. arctitorquis closely resembles *P. leucogaster* in its grey back and black cap, but differs in having a much narrower black collar across the throat and consequently a greater amount of white on the throat, while round the hind neck in *P. arctitorquis* is an indistinct collar of paler ashy, which separates the cap from the mantle.

Mr. Forbes tells us that this species is very common everywhere in the Tenimber Islands, frequenting trees, but by no means unfrequently descending to the ground.

Adult male. General colour above slaty grey, with indistinct dusky shaft-lines on the mantle-feathers; least wing-coverts like the back, median and greater wing-coverts also slaty grey, with blackish shaft-lines and concealed blackish bases; bastard wing, primary-coverts, and quills blackish, externally edged with slaty grey, more broadly on the secondaries; upper tail-coverts rather more dingy than the back, and having narrow blackish shaft-streaks; tail-feathers blackish, broadly edged with slaty grey, and all distinctly waved with transverse cross lines of black; the tips also very narrowly fringed with lighter ashy; crown of head and nape black, forming a cap; lores, feathers round the eye, and ear-coverts also black; cheeks and entire throat white, surrounded by a black pectoral collar, which ascends on the sides of the throat and joins the black ear-coverts; sides of the neck and sides of the fore neck light pearly grey; remainder of under surface ashy white; thigh-feathers dusky, with whitish tips; under tail-coverts, under wing-coverts, and axillaries white, the latter with ashy bases; quills dusky blackish below, ashy white along the edge of the inner web; "bill, legs, and feet black; iris dark brown" (*H. O. Forbes*). Total length 6 inches, culmen 0·6, wing 3·15, tail 2·5, tarsus 0·9.

Adult female. Different from the male. General colour above ashy olive-brown, with somewhat of a rufous tinge; lesser wing-coverts dull rufous brown; median and greater coverts dusky, externally edged with rufous brown, lighter on the latter; bastard wing and primary-coverts blackish, narrowly edged with rufous; quills blackish, edged with rufous, more broadly on the secondaries, the innermost of which are entirely dull rufous; upper tail-coverts pale dull rufous; tail-feathers ashy olive-brown, washed on their edges with rufous; head dull rufous, contrasting slightly with the back; lores ashy whitish, as also the feathers round the eye; ear-coverts and sides of neck dull rufous; cheeks and throat white, narrowly streaked with tiny lines of black; remainder of under surface buffy whitish, longitudinally streaked with black, except the flanks, which are pure white; the sides of the breast slightly tinged with rufous; thighs ashy; under tail-coverts white; under wing-coverts and axillaries dull white; quills dusky below, ashy along the inner web; "upper mandible soot-brown, lower the same at tip, but at base pale flesh-colour; legs and feet lavender-pink; irides dark brown" (*H. O Forbes*). Total length 6 inches, culmen 0·6, wing 3, tail 2·4, tarsus 0·95.

The Plate represents a pair of birds, of the size of life, drawn from the typical specimens lent to us by Dr. Sclater: they are now in the British Museum. [R. B. S.]

PACHYCEPHALA BRUNNEA, *Ramsay.*

PACHYCEPHALA BRUNNEA, *Ramsay*.

Brown Thickhead.

Eopsaltria (?) *brunnea*, Ramsay, Proc. Linn. Soc. N. S. Wales, i. p. 391 (1877).
Pachycephala brunnea, Ramsay, Proc. Linn. Soc. N. S. Wales, iii. p. 382 (1879); iv. p. 99 (1879).—Salvad. Ibis, 1879, p. 324.
Pachycephala dubia, Ramsay, Proc. Linn. Soc. N. S. Wales, iv. p. 99, note (1879).—Salvad. Ann. Mus. Civ. Genov. xv. p. 46 (1879).—Id. Orn. Papuasia &c. ii. p. 228 (1881).

In an early collection of Mr. Goldie's from South-eastern New Guinea occurred a single specimen of this bird, which we identified as *Pachycephala brunnea* of Ramsay, and the Plate was lettered with this name. We regret that at the time we had overlooked the fact that there was already a *P. brunnea* of Wallace, and that therefore Mr. Ramsay's name could not stand. As Count Salvadori has pointed out, the species has also been described a second time by Mr. Ramsay under the name of *P. dubia*, by which title it should be known. The habitat seems to be South-eastern New Guinea, where it has been obtained on the Laloki river, and more recently in the Astrolabe Mountains by Mr. Goldie. It belongs to the plain-coloured section of Thickheads, and appears to be very closely allied to *P. simplex* of Gould.

The following is a copy of the diagnosis of the species given in Count Salvadori's work on the Birds of New Guinea.

Above clear brown tinged with olive, the head darker; inner web of the quills and tail-feathers dusky grey (with the base of the quills whitish), white below; lores dusky; a faintly indicated dusky band above the eye; throat greyish ashy, dusky towards the breast, the latter marked with a transverse band of dusky; sides of the breast and of the abdomen dusky; abdomen, under tail-coverts, and under wing-coverts silky white; tail dusky above, tinged with olive like the wing, below dusky cinereous; shafts of the tail-feathers black above, below white. Total length 5·5 inches, wing 3·3, tail 2·5, bill 0·55.

For the opportunity of figuring this species we are indebted to Mr. Edward Gerrard, jun., who kindly lent us one of Mr. Goldie's specimens from the Astrolabe Mountains. The Plate represents the bird, of the size of life, in two positions.

[R. B. S.]

PACHYCEPHALA HYPERYTHRA, *Salvad.*

PACHYCEPHALA HYPERYTHRA, Salvad.

Ruddy-breasted Thickhead.

Pachycephala hyperythra, Salvad. Ann. Mus. Civ. Genova, vii. sp. 932 (1875); x. p. 142 (1877); xv. p. 47(1879).—
Id. Orn. Papuasia &c. ii. p. 232 (1881).

THE specimens which we identify with the species described by Count Salvadori as *Pachycephala hyperythra* came from the Astrolabe Mountains in South-eastern New Guinea; they agree fairly well with the description given by Count Salvadori; but the colour of the underparts can scarcely be said to be bright rufous, as they are stated to be in *P. hyperythra*.

The habitat of the latter species is also given by Count Salvadori (in his work on the Birds of New Guinea) as the Arfak Mountains in the north-western portion of the island; and it may ultimately turn out that the specimens from South-eastern New Guinea belong to a different species from *P. hyperythra*. But recent experience has so clearly demonstrated the fact that the species inhabiting the Arfak Mountains do in a great many instances range throughout New Guinea into the Astrolabe Mountains, that we do not wish to divide these two birds specifically without an actual comparison of individuals from the two localities, a comparison which as yet we have had no opportunity of making.

The following is a description of one of Mr. Goldie's specimens from South-eastern New Guinea:—

"General colour above rufescent olive-brown or clear earthy brown; wing-coverts like the back, the greater series, bastard wing, primary-coverts, and quills dusky brown, externally like the back, but slightly more tinged with olive; the secondaries externally of the same colour as the back; tail-feathers dull rufous brown; crown of head and nape dark slaty grey, as also the lores; feathers in front of and below the eye blackish; ear-coverts dusky brown; cheeks and throat white; remainder of under surface fawn-brown, a little clearer on the abdomen and under tail-coverts; under wing-coverts brown, washed with olive near the edge of the wing; the axillaries pale isabelline brown; quills dusky below, isabelline along the edge of the inner web; 'bill black; feet pale; iris chestnut.' Total length 5·7 inches, culmen 0·7, wing 3·65, tail 2·6, tarsus 0·85."

The Plate represents an adult male drawn in two positions, and is drawn from a specimen lent to us by Mr. Edward Gerrard, jun.

[R. B. S.]

PACHYCEPHALA SCHLEGELI, Rosenb.

PACHYCEPHALA SCHLEGELI, *Rosenb.*

Schlegel's Thick-head.

Pachycephala schlegeli, von Rosenberg, Nederl. Tijdschr. Dierk. iv. p. 43 (1871).—Sclater, P. Z. S. 1873, p. 697.
—Salvad. Ann. Mus. Civic. Genov. x. p. 141 (1877).

The genus *Pachycephala* is preeminently an Australasian form, being not only widely distributed over the whole of the Australian continent, but ranging over nearly all the Oceanic islands, and reaching its extreme development in the Moluccas. Here almost every group of islands has its peculiar species of *Pachycephala* of the ordinary form, bright yellow underneath, with a black head and collar across the breast. This is the typical characteristic of most of the Thick-heads, though in Australia there are some species of the genus *Pachycephala* which are remarkable for their dull coloration, leading off apparently to the *Eopsaltriæ* (or Large-headed Robins). This is also the case in New Guinea, where both bright and dull-coloured species of Thick-heads are met with.

Professor Schlegel gives the following account of the species:—

"Mr. von Rosenberg has just sent us from the interior of New Guinea a nice series of specimens of a *Pachycephala* evidently new to science. It belongs to the number of species where the male in full plumage has the throat white, the head and chest black, the breast, as well as the belly, and a collar round the neck, bright yellow. Such, for instance, are the *Pachycephala gutturalis* of Australia, recognizable by its very small beak and its tail partly grey and partly black, *P. calliope*, of Timor, with a long bill and green tail, *P. melanura*, of the Moluccas, also with a long bill, but with a black tail.

"*Pachycephala schlegeli* has the bill short, like *P. gutturalis*; the tail, on the other hand, is black, as in *P. melanura*; but it is distinguished from all its allies by its small size, by its pectoral band of black three times as large as usual, its black wings, and finally by the yellow of the breast and abdomen passing into brownish orange.

"Wing 3 inches, tail 2 inches 3 lines; bill from front 5 lines, breadth of bill at forehead 2½ lines; tarsus 10 lines; middle toe 5 lines.

"In the living bird, according to von Rosenberg, the bill is black, the iris dark greyish brown, and the feet bluish grey."

I do not reproduce the entire description of the sexes given by Professor Schlegel, as the characters recorded above sufficiently distingnish the species, which is, indeed, a very well characterized one. The figures in the Plate represent the two sexes, of the natural size, and are drawn from specimens kindly lent to me by Dr. A. B. Meyer, from the Dresden Museum.

PACHYCEPHALA CHRYSTOPHORI, Tristram.

PACHYCEPHALA CHRISTOPHORI, Tristr.

San-Christoval Thickhead.

Pachycephalus christophori, Tristram, Ibis, 1879, p. 441.
Pachycephala christophori, Salvad. Ibis, 1880, p. 131.—Id. Orn. Papuasia e delle Molucche, ii. p. 216 (1881).—Ramsay, Proc. Linn. Soc. N. S. W. vi. p. 178 (1882), vii. p. 25 (1883).—Salvad. Ibis, 1884, p. 323.—Tristr. t. c. p. 398.
Pachycephala astrolabi, juv., Gadow, Cat. Birds in Brit. Mus. ix. p. 200 (1883).

This species was discovered by Lieut. Richards in the island of San Christoval in the Solomon group, and was described by Canon Tristram. It is undoubtedly a distinct species, without the yellow collar round the hind neck or the black head of *P. astrolabi*, which is its nearest ally; it is also a much smaller bird than the last named, which is an inhabitant of Guadalcanar in the Solomon Archipelago. According to Mr. Ramsay, *P. christophori* has also been found in the island of Ugi by Messrs. Morton and Stephens.

Dr. Gadow considered that the present species was merely the young of *P. astrolabi*; but in this he was undoubtedly wrong, as has been shown by Canon Tristram and Count Salvadori.

The following is a description of the typical specimens in Canon Tristram's collection :—

Adult male. General colour dull yellowish green, a little lighter on the lower rump and upper tail-coverts; lesser wing-coverts a little yellower than the back; median and greater coverts dusky blackish, washed externally with olive-yellow; bastard-wing, primary-coverts, and quills also dusky blackish, edged with olive-yellow; the primaries margined with ashy grey; tail-feathers dusky blackish, the centre ones dull olive-yellowish towards the base, the remainder washed externally and tipped with dull olive-yellow; crown of head a little more dusky yellowish green than the back; the lores black, as also a frontal band; fore part of eyelid and feathers below the eye blackish; ear-coverts dull olive-brown with whitish shaft-lines; cheeks and throat bright yellow, followed by a broad black band across the lower throat, fore neck, and chest; breast and abdomen bright yellow, inclining to orange on the upper breast; sides of body and flanks also yellow, very slightly tinged with greenish; thighs and under tail-coverts paler yellow; under wing-coverts and axillaries white, washed with yellow on the edges; quills below dusky, white along the inner edge : "feet ash-colour, iris grey" (*G. E. Richards*). Total length 6·4 inches, culmen 0·8, wing 3·4, tail 2·3, tarsus 1·05.

Adult female. Brighter yellowish green than the male, the head like the back, but with no black forehead; ear-coverts like the crown, tinged with reddish brown, of which colour there is also a shade on the quills; upper tail-coverts and tail, cheeks and entire under surface of body bright yellow, paler on the throat, slightly washed with greenish on the fore neck, chest, sides of body, and flanks, and tinged with orange on the breast; no black pectoral collar; under wing-coverts and axillaries yellow; quills dusky below, reddish along the edge of the inner web. Total length 5·65 inches, culmen 0·8, wing 3·25, tail 2·2, tarsus 0·95.

The young male resembles the old female, but has a few black feathers on the breast, indicating the appearance of the pectoral band.

The Plate illustrates the male and female of the natural size, the figures having been drawn from the type specimens kindly lent to us by Canon Tristram.

[R. B. S.]

PACHYCEPHALOPSIS FORTIS.

PACHYCEPHALOPSIS FORTIS.

Grey-throated Thickhead.

Pachycephala fortis, Gadow, Cat. Birds in Brit. Mus. vol. viii. p. 369 (1883).

We have already had the pleasure of figuring two species of *Pachycephalopsis* in the present work, and we now introduce to our readers a third very interesting species, recently discovered by Mr. A. Goldie during his explorations at the back of the Astrolabe range of mountains in South-eastern New Guinea.

The present species is somewhat allied to *P. hattamensis*, but is strikingly different, being distinguished by the uniform grey throat and breast, the pale yellow under tail-coverts, and the absence of rufous on the wings. It resembles the last-named species, however, in its olive-green back and grey head.

The following is a description of the type specimen in the British Museum :—

Adult. General colour above olive-greenish; wing-coverts like the back, the greater and primary-coverts dusky, externally edged with olive-green, the bastard-wing feathers grey; quills dusky brown, externally olive-green, browner on the primaries, the inner secondaries entirely like the back; tail-feathers dull olive-brown, greener on the edges and barred with dusky under certain lights; head and nape slaty grey, as also the sides of the neck, ear-coverts, and sides of face; lores greyish white; cheeks, throat, and breast pale grey, with narrow dusky shaft-streaks to some of the feathers of the fore neck; abdomen creamy white; sides of body pale brown washed with olive; thighs ashy washed with olive; under tail-coverts yellowish white washed with olive-yellow; under wing-coverts and axillaries pale olive-yellow; quills dusky below, ashy along the edge of the inner web. Total length 6·6 inches, culmen 0·85, wing 3·65, tail 2·75, tarsus 1·05.

[R. B. S.]

PACHYCEPHALA HATTAMENSIS, Meyer.

PACHYCEPHALOPSIS HATTAMENSIS.

Hattam Thickhead.

Pachycephala hattamensis, Meyer, Sitz. k. Ak. Wissensch. zu Wien, lxix. p. 391 (1874).—Salvad. Ann. Mus.
  Civ. Gen. x. p. 142 (1877).—Oust. Bull. Soc. Philom. Paris, 1877.
Pachycephala haltamensis (errore), Sclater, Ibis, 1874, p. 417.
Pachycephalopsis hattamensis, Salvad. Ann. Mus. Civ. Genov. xv. p. 48, no. 47 (1879).—Id. Orn. Papuasia &c.
  ii. p. 236 (1881).

THE shorter tail and longer tarsi in this species have been considered by Count Salvadori sufficient to separate the bird generically from the genus *Pachycephala*; and although I was at first disposed to doubt the admissibility of adopting his genus *Pachycephalopsis* for the species, I am now inclined to think that he was right in effecting this separation. Unfortunately, before I had taken this view, the printing of the Plate had been finished; and hence the discrepancy of the name on the latter with that at the head of the present article.

As far as we know, the present species is only found in North-western New Guinea, having been discovered in the district of Hattam by Dr. Meyer; and it has been met with in the same country and on the Arfak Mountains by the Italian travellers D'Albertis and Beccari. There would appear to be no difference in the sexes; but in the young birds, according to Count Salvadori, the colour is browner and more white on the abdomen, and the external aspect of the quills is rufous.

I add a translation of the description given by the last-named author in his work on the Birds of New Guinea:—Head, neck, and sides of the head ashy; lores white; back and rump greenish olive; chin and throat white; lower throat, breast, and abdomen olive-yellow, duller on the lower throat and the breast; quills dusky, externally margined with brownish rufous; the upper wing-coverts dusky, margined with olivaceous; under wing-coverts rufous; upper tail-coverts and tail brownish rufous; under tail-coverts pale brown; the shafts of the tail-feathers extending a little beyond the web.

Signor D'Albertis states that the iris is chestnut, the bill black, and the feet dusky.

The figures in the Plate represent a pair of birds of the natural size. They are drawn from the typical specimen kindly lent to me by Dr. Meyer.

[R. B. S.]

PACHYCEPHALOPSIS BOLIOSOMA, Sharpe.

PACHYCEPHALOPSIS POLIOSOMA, Sharpe.

Grey Thickhead.

Pachycephalopsis poliosoma, Sharpe, Journ. Linn. Soc. Zool. vol. xvi. p. 318 (1882).

This fine Thickhead was discovered by Mr. A. Goldie in the Taburi district, at the back of the Astrolabe range, in South-eastern New Guinea. Although much resembling the genus *Pachycephala* in the shape of the bill, the present species nevertheless differs in its longer wings and longer tarsi; and I agree with Count Salvadori that, like *P. hattamensis*, it ought to be generically separated. With the last-named bird it agrees entirely in form; but the simple coloration gives it a very different appearance.

Nothing was recorded by Mr. Goldie concerning the habits of this species; but he states that the native name is " Uradaroro."

The following is a translation of the original description published by Mr. Sharpe:—

Above uniform dull ashy grey, the head slightly duller; wing-coverts like the back, quills and tail-feathers rather browner; lores and eyebrows, as well as the ear-coverts, ashy, the feathers before the eye and a streak below the latter black; under surface of body ashy grey, with the lower abdomen and the under tail-coverts slightly whitish; throat whitish brown, the sides washed with ashy; cheeks whitish, lighter than the throat and forming an indistinct moustache; under wing-coverts and axillaries ashy, quills sepia-brown below, edged with pale brown along the inner web. Total length 6·3 inches, culmen 0·8, wing 4·2, tail 2·5, tarsus 1·2.

The figure in the Plate represents this species of the natural size: it has been taken from the typical specimen in the British Museum.

[R. B. S.]

XEROPHILA PECTORALIS, Gould.

XEROPHILA PECTORALIS, *Gould*.

Chestnut-breasted Xerophila.

Xerophila pectoralis, Gould, Ann. N. H. (4) viii. p. 192.

Of the many puzzling forms which Australia produces, the little genus *Xerophila* is one of the most characteristic. The present species has all the appearance of a Finch; and the coloration even approaches that of some Fringilline birds I could mention; to find, therefore, that it had been placed by systematists in the family Fringillidæ is not so very surprising. The late Mr. George Robert Gray, in his 'Hand-list of Birds' (part i. p. 235), has associated *Xerophila* with *Suthora* and *Certhiparus*. My own view of the case, as will be seen on consulting my 'Hand-book of the Birds of Australia,' is very different: notwithstanding its thick bill, I have referred the type to the *Acanthizæ*; and the discovery of the second species does not alter my opinion.

I have to express my regret that absolutely nothing has been recorded respecting this pretty little bird. I am indebted to the kindness of the Director of the South-Australian Institute at Adelaide, and to Mr. Waterhouse, for the opportunity of describing and figuring it. I reproduce the original diagnosis, published by me in 1871.

Face and throat white, passing into greyish white on the ear-coverts; crown and nape hair-brown, mottled with blackish brown, the darker tint occupying the centre of each feather; back chestnut-brown, becoming much darker and richer on the rump; upper tail-coverts hair-brown; two central tail-feathers hair-brown, with lighter edges, the five lateral feathers on each side black, tipped with white; across the chest a well-defined band of cinnamon-brown; under surface white, with a mark of chestnut down the centre of each of the flank-feathers; wings dark brown, the secondaries broadly margined with dull buff; under tail-coverts buffy white; bill and feet black.

Total length $3\frac{7}{8}$ inches; bill $\frac{3}{8}$, wing $2\frac{1}{4}$, tail $1\frac{3}{8}$, tarsi $\frac{5}{8}$.

The type specimen was procured at Port Augusta, South Australia. The figures in the Plate are of the natural size.

SITELLA ALBATA, Ramsay.

SITTELLA ALBATA, *Ramsay*.

White-winged White-headed Sittella.

Sittella leucocephala, Ramsay, Proc. Zool. Soc. 1875, p. 600 (nec Gould).
—————— *albata*, Ramsay, Proc. Zool. Soc. 1877, p. 351.—Id. Proc. Linn. Soc. N. S. Wales, ii. p. 192.

The genus *Sittella* contains the Australian representatives of the Nuthatches of the northern parts of the Old and New Worlds, and is represented on the continent of Australia by seven species, five of which have been described by me; one species is found in New Guinea. It may readily be believed that a considerable number of specimens have passed through my hands; and it is just possible that I may have handled examples of the present new species; but if so, I never noticed the distinction between it and *Sittella leucocephala*. It would, indeed, require a careful examination of a specimen to notice the difference between these two species; for the white spots on the quills (instead of reddish ones, as in *S. leucocephala*) are only seen on spreading the wing. Mr. Waller, to whom I am indebted for lending me a fine specimen of *Sittella albata*, tells me that he was skinning it, when the white spots on the underside of the quills attracted his attention, as he did not remember to have seen them in *S. leucocephala*, with which species he was well acquainted. On pointing out this difference to Mr. Ramsay, that gentleman re-examined his series, and discovered other specimens in his possession from Port Denison, which he described in 1877 as *Sittella albata*.

The geographical ranges of the two species above mentioned do not seem to me to be exactly understood as yet; for Mr. Ramsay gives the range of *S. leucocephala* as the Wide-Bay district, New South Wales, and the interior province, while the new *Sittella albata* is said to come from Port Denison and Rockingham Bay. I understood Mr. Waller to say that he had procured his specimens near Brisbane; and it may be possible that *S. albata* ranges as far south as the neighbourhood of that town.

Mr. Ramsay, in his paper "On the Birds of N.E. Queensland," first identified the present species with *S. leucocephala*, and observes that it is far from being rare there, being usually met with in open forest country over the whole of Northern Queensland as far as Cooktown. Its habits and actions and nidification do not differ materially from those of the other members of the genus. The notes of all closely resemble each other.

The following is Mr. Ramsay's description of the species:—

"Head and neck, a small spot at the base of the primaries on the underside of the wing, a band through the wing as far as the ninth quill, the upper tail-coverts, and the tips of all the tail-feathers except the centre two snow-white; under surface ashy white, with a broad dark brown stripe down the centre of each feather; under tail-coverts of a darker brown, tipped and margined anteriorly with white; back and scapulars brown, darker in the centres of the feathers; wing- and tail-quills blackish brown, the former crossed with a white band as far as the ninth quill; bill at the base, the legs and feet, and skin round the eye yellow; remainder of the bill black. Length 3·7 inches, wing 3, tail 1·5, tarsus 1·7, bill 0·5, bill from gape 0·7."

The figures in the Plate are drawn from a single specimen lent to me by Mr. Waller, whose kindness in showing me many fine species of Australian birds during his recent visit to England I have much pleasure in acknowledging. The birds are represented of the size of life; and I have ventured to introduce into the Plate a representation of the nest of a *Sittella*, which I believe to be that of the present bird, as it was sent to me from the part of the country which this species inhabits. I have never before had the opportunity of figuring one of the nests belonging to any member of the genus; but Mr. Ramsay states that the nidification of all the *Sittellæ* is of a similar character, "the nest being placed in an upright and usually dead fork of some high branch; it is made of fine strips of bark with a large quantity of spiders' webs, with which small scales of bark, resembling that of the branch in which it is placed, are felted on so carefully as hardly to be detected, even at a comparatively short distance; the rim is very thin, the nest open above, and very deep."

CLIMACTERIS PLACENS, *Sclater.*

CLIMACTERIS PLACENS, Sclater.

New-Guinea Tree-Creeper.

Climacteris placens, Sclater, P. Z. S. 1873, p. 693.

"The discovery of a typical species of this Australian genus in New Guinea," writes Dr. Sclater, "is of very great interest. Dr. Schlegel has already recorded the existence of *Sittella* in the same country." In indorsing the opinion of the above-named gentleman, I feel that there is little else for me to say, as each new traveller in the island brings to light some hitherto unsuspected link between the avifaunas of the two countries.

Nothing has been recorded respecting the habits of this Tree-Creeper. The original specimen was obtained by Signor d'Albertis in Atam, near the Arfak mountains.

The following is a transcript of Dr. Sclater's original description :—

"Above mouse-brown; the plumes of the head rufescent, with paler shafts, and narrowly tipped with black; wings black, with a broad basal bar of ochraceous buff traversing the base of the quills; under wing-coverts also ochraceous buff; tips of the quills and the secondaries adjoining the back obscure dusky brown; tail black, tipped with cinereous, the two centre tail-feathers of nearly the same colour as the back; underneath paler, more ashy, the whole belly and vent streaked with black and ochraceous; a patch of feathers under the eye rufous; bill black; feet yellowish; iris black. Total length 5·4 inches, wing 3·2, tail 2·5, tarsus 0·9, hind toe without claw 0·6.

The figures in the Plate are life-size.

CINNYRIS MAFORENSIS.

CINNYRIS MAFORENSIS.

Mafoor-Island Black Sun-bird.

Chalcostetha aspasia, var. *maforensis*, Meyer, Sitz. k. Akad. Wissensch. in Wien, lxx. p. 123 (1874).—Sclater, Ibis, 1874, p. 419.
Hermotimia maforensis, Salvad. Atti R. Accad. Torino, x. pp. 208, 227 (1874), xii. p. 301 (1877).
Cinnyris maforensis, Shelley, Monograph of the Cinnyridæ, part v. (1877).

This species is the representative of the Black Sun-birds in the island of Mafoor, situated in Geelvink Bay. It was discovered by Dr. Meyer during his expedition to New Guinea, and has been allowed to be specifically distinct by Count Salvadori and Captain Shelley. I must say that I feel considerable compunction in admitting these species; they seem to me to be rather races of one form. But a good many specimens of all these different Sun-birds have been procured by the travellers; and the characters, if slight, are constant. As might be expected, there has been nothing written on the habits of this Sunbird; so that I have only to transcribe the single paragraph given by Captain Shelley respecting it:—
"The present species belongs to the division of the '*Hermotimia*' group in which the lower back is green. It only differs from *C. aspasiæ* in the golden colour of the crown, which in this bird is only faintly tinted with green in certain lights. It is also slightly larger. As far as we yet know, it is entirely confined to the small island of Mafoor, in the Bay of Geelvink."

The following description is also taken from Captain Shelley's work.

Adult male. Black with a deep-blue gloss, the entire crown golden; the least median series of wing-coverts, the scapulars, the lower half of the back, and the upper tail-coverts metallic green very slightly shaded with blue; a few of the outer greater wing-coverts and the tail-feathers edged with the same colour; chin and throat violet-shaded steel-blue. Total length 4 inches, culmen 0·75, wing 2·5, tail 1·5, tarsus 0·65.

Two male birds are represented in the Plate, of the size of life. They were lent to me by Dr. Meyer.

CINNYRIS MYSORENSIS.

CINNYRIS MYSORENSIS.

Mysore-Island Black Sunbird.

Chalcostetha aspasia, var. *mysorensis*, Meyer, Sitz. k. Akad. Wissensch. in Wien, lxx. p. 124 (1874).—Sclater, Ibis, 1874, p. 419.
Hermotimia mysorensis, Salvadori, Atti R. Accad. Torino, x. pp. 208, 224 (1874), xii. p. 301 (1877).
Cinnyris mysorensis, Shelley, Monograph of the Cinnyridæ, part v. (1877).

It is now seven years ago since we first began to get some insight into the mysteries of the large islands which are situated in the Bay of Geelvink in North-western New Guinea. During Mr. Wallace's travels in that part of the world he had been unable to penetrate to these islands, the natives of which had a bad reputation for hospitality; and it fell to the lot of Baron von Rosenberg to send home the first birds to the Leiden Museum. Many fine birds from Jobi, Mafoor, and Miosnoum were sent by him; and these researches were still further pursued by Dr. Meyer, who collected largely in these islands and increased our knowledge of their ornithology greatly. Every thing seems to point to the fact that, whereas Jobi is an island which contains a considerable number of peculiar species, it is impregnated with a large number of New-Guinea birds, while Mysore and Mafoor seem to be more distinct as regards their avifauna.

Nothing is known of the economy of the present beautiful little Sunbird; and I transcribe the following remarks of Captain Shelley, who writes in his 'Monograph :'—

"This form is nearly related to *C. aspasiæ*; and I place it in the green-backed division of the '*Hermotimia*' group: yet it much depends upon the light in which we view these feathers, as to whether the green or blue shade predominates; in this respect, therefore, it may be readily distinguished from *C. aspasiæ*, as well as in the greater extent of the metallic throat. It also differs from typical specimens of that bird in having the crown of a darker colour, and the throat almost pure lilac: but these parts in *C. aspasiæ* vary to a certain extent in individual specimens, as may be seen in my article on that species.

"The present bird has only been found on Mysore, the large outer island of the Bay of Geelvink, where it was first collected by Dr. Meyer, who pointed out its specific characters."

The following description of the adult male is given by Captain Shelley:—

Black with a deep-blue gloss; the entire crown metallic bluish emerald-green; the least and median series of wing-coverts, scapulars, lower half of the back, and the upper tail-coverts metallic green very strongly shaded with blue; the tail-feathers and a few of the outer greater wing-coverts edged with the same metallic bluish green; chin and throat rich metallic lilac, and extending further on the chest than in *C. aspasiæ*; bill and legs black; irides dark brown. Total length 4·6 inches, culmen 0·8, wing 2·5, tail 1·7, tarsus 0·65.

The adult female and young male resemble those of *C. aspasiæ*.

In the Plate I have represented two males and a female, from typical specimens lent to me by Dr. Meyer.

CINNYRIS SANGIRENSIS.

CINNYRIS SANGIRENSIS.

Sanghir Black Sun-bird.

Chalcostetha sangirensis, Meyer, Sitz. k. Akad. Wissensch. in Wien, lxx. p. 124 (1874).—Sclater, Ibis, 1874, p. 419.
Hermotimia sanghirensis, Salvad. Atti R. Accad. Torino, x. p. 233, pl. i. fig. 2 (1874).—Id. Ann. Mus. Civic. Genov. ix. p. 56, no. 10 (1876).—Id. Atti R. Accad. Torino, xii. p. 311 (1877).
Cinnyris sangirensis, Shelley, Monograph of the Cinnyridæ, part v. (1877).

THE Sangir or Sanghir Islands are a little group lying to the north-east of Celebes, and between that island and the Philippine archipelago. The ornithology of the group has not yet been worked out completely; but as far as we know the facts at present, it is very closely related to that of Celebes, though at the same time possessing several peculiar forms. The specimens which I figure are lent to me by my excellent friend Dr. A. B. Meyer; and I have followed Captain Shelley, who is our best authority on the family, in the nomenclature of the species. As there is nothing known concerning its habits, I transcribe the account given by the last-named author in his 'Monograph:'—"This well-marked species belongs to that section of the '*Hermotimia*' group which comprises species with the metallic portions of the back blue. It may be distinguished from its allies by the general brownish shade of its plumage, and by the bronzy copper-colour of its throat, in which latter character it appears to be the species which most nearly approaches to *Chalcostetha insignis*. Like the Celebean forms *C. grayi* and *C. porphyrolæmus*, it has the sides of the metallic throat margined with distinct bands of steel-blue, which are not met with in the other members of this group; and they also resemble each other in the absence of the metallic colouring on the scapulars and median series of the wing-coverts; but this latter character is not confined to these three species.

"The present bird was first described by Dr. Meyer from specimens collected by himself at Siao, one of the Sangir or Sanghir islands, a small group situated to the north of Celebes; and it appears to be exclusively confined to that archipelago."

A long account is given by Captain Shelley of a series of birds sent to Count Salvadori from the Sanghir Islands, which it is not now necessary to reproduce, as it concerns the changes of plumage which the bird goes through; and I therefore refer my readers to the Monograph itself.

The following descriptions are from Captain Shelley's book:—

Adult male. Brownish black with a purple gloss; the forehead, crown, and nape metallic golden green; least wing-coverts, a few of the smallest scapulars, the lower back, and upper tail-coverts steel-blue, shaded with violet and green; tail-feathers edged with lilac-bronze; throat coppery bronze, margined with steel-blue on the sides of the chin and upper half of the throat; bill and legs black; irides dark brown.

Total length 4·3 inches, culmen 0·65, wing 2·4, tail 1·75, tarsus 0·6.

Adult female. Upper half of the head and neck, back, and scapulars olive-yellow; upper tail-coverts black; wings dark brown, all the feathers broadly edged with olive-green; tail black, with white tips, broadest on the outer feathers; cheeks yellower than the crown; underparts sulphur-yellow, slightly tinted with olive on the sides of the breast, the under tail-coverts very pale yellow; under surface of the wings brown, with the inner margins of the quills and the coverts white, the latter shaded with yellow; bill, legs, and irides dark brown.

Total length 4 inches, culmen 0·6, wing 2·05, tail 1·4, tarsus 0·6.

The Plate represents two males and a female, of the natural size, for the loan of which I am indebted to Dr. Meyer.

MELIRRHOPHETES LEUCOSTEPHES, Meyer.

MELIRRHOPHETES LEUCOSTEPHES, Meyer.

White-faced Honey-eater.

Melirrhophetes leucostephes, A. B. Meyer, Sitzungsb. der k.-k. Akad. d. Wissenschaften zu Wien, lxx. Juni 1874, p. 111.—T. Salvadori, Ann. del Mus. Civ. di Genova, vol. vii. 1875, p. 776.

This large and attractive Honey-eater must form a conspicuous feature among the flowering trees of New Guinea, much as do the large species of this family so numerously spread over Australia; at least we may surmise this, as not even a few words have been placed on record by its discoverer Dr. Meyer, or by any other person. These meagre descriptions, however, testify to the rarity of the subject, and point out to future travellers over New Guinea how very desirable it would be for them to jot down any such particulars before they leave the forests, or at least as soon after as an opportunity may occur. All I can now say is that my artist, Mr. Hart, has made as faithful a portrait of this new bird as possible, to which I add Dr. Meyer's description very kindly forwarded to me in a letter.

"Feathers of forehead and those which encircle the naked skin surrounding the eye white; throat, crown, and hinder part of head, nape, ears, and a patch which borders the front of the eye black; naked eye-skin and lengthened skin-fold of the mouth yellowish; caruncles of the throat orange; back brownish, feathers of the upper part edged with white and brownish white; under surface black, with some white markings on the breast and abdomen; upper part of wings bright olivaceous; under surface greyish, with light-brown edges on the bases of the inner webs; under wing-coverts blackish, with light brown intermixed; upper part of tail brown, each feather having a lighter edge on the outer web; bill bluish grey, lighter at the tip; feet and tarsi blackish grey, soles of feet lighter."

Total length 11 inches, bill 1⅝, wings 5, tail 5, tarsi 1¼.

According to Salvadori the sexes are alike in colour.

Hab. Arfak Mountains in New Guinea.

The figure in the Plate is of the natural size.

MELIRRHOPHETES OCHROMELAS, *Meyer.*

MELIRRHOPHETES OCHROMELAS, Meyer.

Ochraceous Honey-eater.

Melirrhophetes ochromelas, A. B. Meyer, Sitzungsberichte der k.-k. Akademie der Wissenschaften zu Wien, lxx. Juni 1874, p. 112.

THIS bird is somewhat similar to *M. leucostephes* or the White-faced Honey-eater; however, it has several characters which separate it from that species. First, it is a smaller bird, its bill more slender, whilst the caruncles of the face and naked skin round the eye are less developed. Its general colouring is much darker, and, as its trivial name implies, is of an ochraceous tint. Its legs are more slender and of a pale yellow colour, whilst they are bluish black in *M. leucostephes*.

I wish it were in my power to add some information respecting the habits and economy of this novelty, the discovery of which is due to Dr. Meyer, who has kindly forwarded me his unique example for the purpose of figuring it in the present work, together with the following short note (made, I suppose, at the time the bird was killed):—

"*Melirrhophetes ochromelas* and *M. leucostephes* are closely allied; but the former has no white on the head or under surface, with the exception of some small marks. It has, as before stated, a smaller bill, light-coloured feet, smaller and deeper orange-coloured caruncles at the throat, more lively tints on the outer edges of the wings &c.

"Head black, with a small stripe of the eyebrows and ends of the ear-feathers brownish; naked skin of the eye and lengthened skin-fold at the angle of the mouth yellowish; caruncles of the throat orange-red. Upperside brownish, feathers of the upper surface of the back edged with white; under surface blackish, with some faint white stripes on the abdomen; wings like those of *M. leucostephes*, the only difference being that the edges of the outer webs are of a more ochreous colour. Tail like that of *M. leucostephes*, but with the edges of the outer webs of a lively yellowish tint. Bill bluish grey, with a lighter tip; feet yellowish flesh-colour, tarsi darker."

Total length 9½ inches, bill 1¼, wings 4¾, tail 4½, tarsus 1⅛.

Hab. Arfak Mountains, New Guinea.

The figure is of the natural size.

MELIRRHOPHETES BATESI, Sharpe.

MELIRRHOPHETES BATESI, Sharpe.

Bates's Honey-eater.

Melirrhophetes batesi, Sharpe, Nature, 1886, p. 340.

THE discovery of a species of the genus *Melirrhophetes* in Southern New Guinea is of great interest, as hitherto the genus has been supposed to be confined to the north-western portion of that island. It is being gradually proved, however, that these mountainous faunæ of the north-west and south-east of New Guinea are similar in character, and that either the same species occur throughout the island, or else representative species of the same genus are discovered. In the present instance the *Melirrhophetes* of the Astrolabe Mountains is closely allied to *M. ochromelas* of Meyer, from the Arfak Mountains; but it is evidently distinct, having a tuft of tawny feathers above and below the bare space of the eye, nor does it seem to have the brown shade behind the ear-coverts which is represented in our Plate of *M. ochromelas*, and was drawn from the type specimen. I am not aware of the existence of any example of the last-named species in this country, so we have only had the Plate to compare with, but this leaves little doubt of the distinctness of *M. batesi*.

No particulars accompanied the single specimen sent by Mr. Forbes, who obtained it in the Sogeri district of the Astrolabe range in Southern New Guinea. At his request we have named it after Mr. H. W. Bates, who, a traveller himself, knows how to sympathise with the difficulties which surround the absent explorer.

The following is a description of the typical specimens:—

Adult. General colour above blackish, the mantle and upper back tipped with white or pale tawny buff edges to the feathers; the lower back and rump uniform dark brown; scapulars and lesser wing-coverts blackish with a slight wash of olive, a little more distinct on the median series; the greater and primary-coverts, as well as the quills, blackish brown, edged with olive-yellow, more distinct on the secondaries; bastard-wing dusky blackish; primaries tipped with pale fulvous, extending a little way down the outer web; upper tail-coverts dark brown, with a slight wash of olive; tail-feathers dark brown, edged externally with greenish olive; crown of head black, the occiput and nape browner, with a faint tinge of olive, the hind neck, again, rather blacker; lores black; region of the eye bare, with a band of pale tawny feathers on the sides of the crown along the eye; ear-coverts and feathers below the eye, as well as the cheeks, black, the ear-coverts and hinder cheeks slightly washed with grey, and having a small tuft of tawny feathers behind the former; throat and under surface of body blackish brown, with narrow shaft-lines of ashy white on the breast-feathers, the lower breast and abdomen browner and more sooty; sides of body, flanks, and thighs like the breast; under tail-coverts dusky brown, broadly edged with tawny or yellowish buff; under wing-coverts and axillaries blackish, the latter tipped with yellowish buff; quills below blackish, the primaries with pale tips, the base of the inner webs also pale tawny. Total length 9 inches, culmen 1·3, wing 4·8, tail 4·1, tarsus 1·05.

The figures in the Plate are drawn from the specimen described above, and represent the species of the natural size.

[R. B. S.]

MELIDECTES TORQUATUS, Sclater.

MELIDECTES TORQUATUS, Sclater.

Pectoral Honey-eater.

Melidectes torquatus, Sclater, Proc. Zool. Soc. 1873, Nov. 4, p. 694, fig. 2 & pl. lv.—A. B. Meyer, Sitzungsber. der k.-k. Akademie der Wissenschaften zu Wien, Bd. lxx. p. 128 (1874, Juni 18).

THE late Mr. Strickland used to say that, in his opinion, we had nearly come to the end of species, and that but few more discoveries would be made; yet this is not the case, and, with deference to the opinion expressed by my very worthy and lamented friend, I may state that not only have hundreds of novelties appeared since his premature demise, but we are, even at the present period, yearly in receipt of birds of whose existence we had previously no conception. It is, however, I consider, a piece of good fortune if an ornithologist has an opportunity of describing such a fine bird as the one under consideration. That *Melidectes torquatus* is very different from all others, there can be no doubt; and it gives me great pleasure in being able to assist in giving it publicity.

I must not omit thanking my friend Dr. Meyer for the loan of several examples of this bird in the finest state of plumage. These, when compared with specimens collected by d'Albertis in the Arfak Mountains, present little or no difference in size and appearance. As regards sexes, Mr. Sclater says, "Fem. mari similis;" while all those sent to me from Dr. Meyer were marked female.

Mr. Sclater, in the 'Proceedings of the Zoological Society,' Nov. 4, 1873, says, "This is a most conspicuous new Meliphagine form, not very far from *Ptilotis*, but distinguishable by the bareness of the sides of the face, and bare stripe behind the rictus. These are separated by a scanty line of feathers, extending beneath the eye.

"A pair of these birds were obtained by Signor d'Albertis at Atam in October 1872."

Bill bluish horn-colour, lighter at the tip; throat, crown, ear-coverts, mark down the side of the neck, and pectoral band, black, the space between the pectoral band and the throat white; sides of the neck buff striated with black, upper surface olive black, each feather strongly edged with white; wings brown, edged with olive; tail dark brown, edged with olive on the outer margins; under surface greyish white, spotted with black down the flanks and suffused with reddish buff on the chest; a large bare space of yellow surrounds the eye, above which, separating it from the black, is a line of buff; there is also a small bare space of yellow on the jaw.

Total length $8\frac{1}{4}$ inches, wing 4, tail $3\frac{3}{8}$, tarsi 1, bill $1\frac{1}{4}$.

The figures in the accompanying Plate are of the size of life.

Hab. Atam, apud mont. Papuanos Arfak.—*Sclater.*

MELIDECTES EMILII, Meyer.

MELIDECTES EMILII, Meyer.

Count Turati's Honey-sucker.

Melidectes torquatus (nec Scl.), Sharpe, Journ. Linn. Soc. xvi. p. 438 (1883).
Melidectes emilii, Meyer, in Madarász's Zeitschr. für gesammte Ornithologie, iii. p. 22, Taf. iv. fig. 2 (1886).

THE first occurrence of this species in South-eastern New Guinea was recorded by us in 1883, when Mr. Goldie's collections from the Astrolabe Mountains reached England. We then compared specimens with the plate of *M. torquatus* from the Arfak Mountains, figured in the present work, and were unable to see any differences between them; but Dr. Meyer, who had examples before him from the Horseshoe range of the Astrolabe Mountains, compared them with others from North-western New Guinea, and described the southern bird as *M. emilii*, naming it after Count Emilio Turati of Milan.

We have no doubt that Dr. Meyer, having specimens from both the north-west and south-east of New Guinea, was able to form a more correct judgment than we were, and that he was perfectly right in separating *M. emilii* from *M. torquatus*, although the differences are very slight, consisting in the pale under surface and smaller white throat-spot in the southern bird. We have lately seen several specimens from the Astrolabe range, all of which bore out the characters assigned to *M. emilii* by Dr. Meyer, and we therefore fully believe in the distinctness of the species. Besides the specimens obtained by Mr. Goldie in the Morocco district, where it is called by the natives 'Ugirru,' it has been met with in the Horseshoe range by Mr. Hunstein, and by Mr. H. O. Forbes in the Sogeri district.

It is from specimens obtained in the last-named locality that the figures in the Plate have been drawn.

[R. B. S.]

MELIPOTES GYMNOPS, *Sclater*

MELIPOTES GYMNOPS, Sclater.

Naked-faced Honey-eater.

Melipotes gymnops, Sclater, Proc. Zool. Soc. 1873, p. 695, fig. 3 & pl. 56.—A. B. Meyer, Sitzungsber. der k.-k. Akad. d. Wissenschaften zu Wien, lxx. 1874, p. 128.—T. Salvadori, Ann. del Mus. Civ. di Genova, vol. vii. 1875, p. 776.

The discovery of the present species in the Arfak Mountains of New Guinea makes an important addition to the great group of Honey-eaters—a group almost peculiar to the Papuan Islands and Australia. The size of the bird under consideration is about that of *Melidectes torquatus*; but it differs in its shorter and more robust bill and in its much more naked face. The only remark made by Dr. Sclater is as follows:—

"This form of the Meliphagine family is very distinct on account of the denudation of the whole ocular region, which is fringed below by a narrow caruncle. D'Albertis's notes do not give the colour of these naked parts; but they are probably orange or flesh-colour. The bill is short and rather stout; the nostrils are short and suboval, and situated in a shallow groove near the central feathers."

The following description is taken from a specimen in Dr. Meyer's collection:—Throat and chest blackish, becoming lighter on the flanks and abdomen, the light marks assuming the form of triangular spots; under tail-coverts pale buff; under tail-feathers brownish grey, the shafts, as in the other species of Honey-eaters, being whitish; bill black; feet bluish black; large naked space round the eye wavy, orange-yellow; crown of the head and back blackish brown, lighter on the rump; tail-feathers pale brown, edged with olive; wings, on the upper surface brown, margined as in the tail; under surface of the wings rich buff, passing into brown on the primaries.

Total length $8\frac{1}{4}$ inches, bill 1, wing $4\frac{1}{4}$, tail 5, tarsi $1\frac{1}{4}$.

Hab. Atam, apud montes Papuanos Arfak.

The two birds in the accompanying Plate are of the size of life, from specimens in Dr. Meyer's collection.

GLYCICHŒRA FALLAX, Salvad.

GLYCYCHÆRA FALLAX, Salvad.

Silky-plumed Honey-eater.

Euthyrhynchus, sp., Salvad. Ann. Mus. Civic. Genov. vii. p. 953 (1875), ix. p. 23 (1876).
Glycichæra fallax, Salvad. Ann. Mus. Civic. Genov. xii. p. 335 (1878).—D'Albert. & Salvad. op. cit. xiv. p. 78 (1879).—Salvad. op. cit. xvi. p. 74 (1880).
Glycychæra fallax, Salvad. Orn. Papuasia e delle Molucche, ii. p. 310 (1881), iii. App. p. 542 (1882).—Meyer, Zeitschr. gesammt. Orn. i. p. 288 (1884).
Tephras whitei, Ramsay, Proc. Zool. Soc. 1882, p. 357.—Salvad. Ann. Mus. Civic. Genov. xviii. p. 422, note (1882).
Glycyphila fallax, Gadow, Cat. Birds in Brit. Mus. ix. p. 213 (1884).

The Marquis Doria having very kindly sent over to England some of the rare Papuan species contained in the Civic Museum at Genoa, we have enjoyed the privilege of presenting figures of some of the most interesting forms to our readers. The genus *Glycychæra* at present contains two species, *G. fallax* and *G. poliocephala*, the latter being from Andai, in North-western New Guinea. *G. fallax* also occurs in Andai, but is further distributed over New Guinea, having been sent by D'Albertis from the Fly River and Naiabui in the south-eastern portion of the island. Dr. Beccari also procured a specimen in the Aru Islands, whence Dr. Meyer has likewise recently received a specimen.

The genus *Glycychæra* has certainly no relation whatever with *Glycyphila*, into which it has been merged by Dr. Gadow, without seeing a specimen. It appears to us to be one of the aberrant genera which connect the *Meliphagidæ* and *Dicæidæ*, and would by some ornithologists be placed in the latter family, in the vicinity of *Melanocharis*. The long fluffy plumes on the lower back and on the flanks are most striking, and are not accentuated enough in the Plate which accompanies this description.

Adult male. General colour above dull olive-greenish; the feathers of the lower back and rump long and fluffy and rather lighter olive; lesser and median wing-coverts like the back; greater coverts and quills dull ashy brown, edged with olive like the back; bastard wing and primary-coverts dull ashy brown, narrowly fringed with olive; upper tail-coverts like the back; tail-feathers dull ashy brown, edged with olive; head and neck decidedly more ashy than the back, much greyer on the sides of the face and ear-coverts; round the eye a ring of white feathers; throat whitish, streaked with yellow edges to the feathers; fore neck, breast, and sides of the body ashy, the former streaked, the latter washed with pale yellow; abdomen and under tail-coverts pale yellow; under wing-coverts white, washed with pale yellow, a little clearer on the edge of the wing; quills ashy below, whitish along the edge of the inner web: " bill blackish above, whitish below; feet leaden; iris dull white " (*D'Albertis*). Total length 4·5 inches, culmen 0·5, wing 2·3, tail 1·35, tarsus 0·7.

Adult female. Similar to the male, but with less grey on the head and face, both these parts being duller; the throat, breast, and underparts more distinctly washed with yellow. Total length 4·5 inches, culmen 0·5, wing 2·25, tail 1·55, tarsus 0·7.

The figures in the Plate represent the male and female of this species of the natural size; they are drawn from specimens belonging to the Genoa Museum, and kindly lent to us by the Marquis Doria. The male is from the Fly River, and the female from the Aru Islands. The latter is rather smaller in its dimensions.

[R. B. S.]

MELITHREPTUS LÆTIOR, Gould.

MELITHREPTUS LÆTIOR, Gould.

Beautiful Honey-eater.

Melithreptus lætior, Gould, Ann. & Mag. Nat. Hist. 4th series, Oct. 1875, p. 287.

ALTHOUGH the members of this little but well-defined genus of Honey-eaters are so generally distributed over the great continent of Australia and Tasmania, as yet no single species has been discovered in any other country. What New Guinea will give us, time alone will testify. All the species of the genus *Melithreptus* are of small size, and characterized by being very similarly coloured; yet, with all this, if due attention be given to certain peculiar characters, the specific distinctions are very evident. All have the eyelash thickened and bare of feathers; and in each species this naked skin is differently coloured: for instance, in the larger species inhabiting Tasmania the skin is described by me from the life as being white tinged with bright green; while I have noted (also from the living birds) that the *M. gularis* of New South Wales is of a beautiful bluish green. In the present bird, which is intimately allied to the species just mentioned, the eyelash is bright yellow. After remarking that in the common *M. lunulatus* of New South Wales these same parts are bright scarlet, it will not be necessary to say more on this point with regard to the species found in Western Australia, Port Essington, and Cape York. Every country surrounding Australia has, then, it will be seen, a species of this genus peculiarly its own; and that the more distant interior does not want a representative is evidenced by Mr. F. W. Andrews's discovery of the beautiful bird now under consideration. One thing, I expect, is pretty certain, that wherever there are Eucalypti, such trees will be enlivened by one or another species of the present group. It has always been a supposition on my part that some larger species will yet be discovered, so that *Melithreptus* and the great *Entomyzæ* will become more nearly united than they are now.

In writing to me about this bird, Mr. Waterhouse, to whom I am indebted for a beautiful specimen, says:—"This is the finest species of the genus that I have yet seen. Only four were shot, and I send you one of the best. When alive they had a bright yellow rim round the eyes."

The following is a transcript from my original description as published in the 'Annals':—

Head and nape black, as well as the lores and ear-coverts; the cheeks and a band of feathers round the occiput pure white; back greenish yellow, brighter on the rump and shading off into bright lemon-yellow on the hind neck and sides of the latter; tail brown, with a narrow whitish edging at the tip, all but the outer feathers margined with greenish yellow; wings ashy brown, externally washed with grey, the primaries narrowly margined with whitish; under surface of body white, the breast and flanks shaded with ashy, and the chin black, fading into ashy brown on the throat and producing a distinct chin-stripe; under wing-coverts white, shaded with ashy; naked skin surrounding the eye bright yellow.

Total length 5·5 inches, culmen 0·6, wing 3·4, tail 2·7, tarsus 0·75.

Although very closely allied to *M. gularis*, this species is altogether a much more finely coloured bird. In size it is slightly larger, and is at once to be distinguished by its white under surface and the beautiful lemon-yellow of the neck, as well as by the yellow naked skin surrounding the eye, which part is greenish blue in *M. gularis*. The ashy shade which pervades the entire lower surface of *M. gularis* is not seen in *M. lætior*.

The figures are of the natural size.

PHILEMON PLUMIGENIS.

PHILEMON PLUMIGENIS.

Hoary-throated Honey-eater.

Tropidorhynchus, n. sp., Wall. Ann. & Mag. Nat. Hist. 1857, xx. p. 473.
Tropidorhynchus plumigenis, Gray, Proc. Zool. Soc. 1858, pp. 174, 191.—Id. Cat. Birds New Guinea, pp. 24, 56 (1859).—Id. Proc. Zool. Soc. 1861, p. 434.—Finsch, Neu-Guinea, p. 165 (1865).—Rosenb. Reis. n. Zuidoostereil, p. 79 (1867).—Meyer, Sitz. k. Akad. Wien, lxx. p. 144 (1874).—Rosenb. Malay Arch. p. 365 (1878-79).
Philemon plumigenis, Gray, Hand-list of Birds, i. p. 160, no. 2081 (1869).—Salvad. P. Z. S. 1878, p. 88.—Id. Ann. Mus. Civic. Genov. xiv. p. 655 (1879), xvi. p. 79 (1880).—Id. Report Voy. 'Challenger,' ii. p. 70 (1880).—Id. Orn. Papuasia, etc. ii. p. 353 (1881).—Sclater, Proc. Zool. Soc. 1883, p. 51.

THE Timor Laut examples of this species cannot be said to be strictly identical with the *Philemon plumigenis* of the Ké Islands, for the head is so much paler brown, more like that of *Philemon bouruensis*, which it further resembles in having the light mottling above the eye, on the ear-coverts, and sides of face. There can be no doubt that *P. plumigenis* and *P. bouruensis* are very closely allied—so closely, indeed, that it is not easy to assign distinct specific characters to them.

In our article on *Oriolus decipiens* we have drawn attention to the way in which its plumage mimics that of the present species of Honey-eater, and we would invite our readers to compare the descriptions and figures of the two birds, and to notice how wonderful is the resemblance between them, even to such details as the light hind neck, the appearance of an eyebrow, and the blackish colour of the Oriole's side face, where the bare skin occurs in the Honey-eater. Mr. Forbes tells us that the resemblance is even carried out in their mode of life, and that they are difficult to tell when sitting in the same tree, were it not for the difference of their notes.

We have described here an adult male collected by Mr. Forbes in the Tenimber Islands.

Adult male. General colour above brown, a little paler on the head and mantle; the lower back, rump, and upper tail-coverts darker and more ashy brown; feathers of the crown somewhat lanceolate in shape, with narrow blackish shaft-lines and dusky centres, not sufficiently pronounced to impart a streaked appearance; wing-coverts and quills brown, shaded with ashy on the outer webs of the quills; tail-feathers light brown, rather paler at their ends and crossed with dusky bars under certain lights; sides of the face and region of the eye bare, as well as the ear-spot; ear-coverts and hinder cheeks brown, the fore part of the cheeks washed with hoary whitish; sides of neck hoary grey, extending in a narrow collar round the hind neck; throat and fore neck hoary, the malar line brown; remainder of under surface of body pale ashy brown, the chest with narrow dark-brown shaft-stripes; axillaries and under wing-coverts like the breast, washed with rufous on the edge of the wing; quills dusky brown below; " bill, legs, and feet black; iris dark brown " (*H. O. Forbes*).

The Plate is drawn from one of Mr. Forbes's specimens; the principal figure is of the natural size.

[R. B. S.]

STIGMATOPS CHLORIS, Salvad.

STIGMATOPS CHLORIS, Salvad.

Mysol Honey-eater.

Stigmatops argentauris, pt., Salvad. Ann. Mus. Civic. Genov. xii. p. 336 (1878).
Stigmatops chloris, Salvad. Ann. Mus. Civic. Genov. xii. p. 337 (1878).—Id. op. cit. xvi. p. 76 (1880).—Id. Orn. Papuasia e delle Molucche, ii. p. 325 (1881).—Sharpe, Rep. Voy. H.M.S. 'Alert,' Birds, p. 19 (1884).
Glycyphila ocularis, pt., Gadow, Cat. Birds in Brit. Mus. ix. p. 213 (1884).

Count Salvadori separated the Honey-eater from Mysol from *Stigmatops ocularis* of Australia on account of its greener coloration and whitish auricular spot. Dr. Gadow, in treating of the last-named species, unites not only *Stigmatops chloris* but also *S. subocularis* to *S. ocularis*, stating that intermediate forms frequently occur. We have already had occasion to controvert this reasoning on Dr. Gadow's part, and we do not hesitate to restore to these species of *Stigmatops* the distinct position accorded to them by Count Salvadori, whose work Dr. Gadow has somewhat unreasonably upset.

S. chloris is yellowish both above and below, instead of greyish as *S. ocularis*, and the ear-spot is whitish. Two specimens are in the British Museum from Mysol; and we are indebted to the kindness of our friend Dr. Jentink for the loan of a specimen from the Leyden Museum, collected in Mysol by Hoedt in June 1867, and one of the types described by Salvadori. The following is a description of this specimen:—

Adult male. General colour above dull greyish olive, more distinctly greyish on the neck and scapulars, lighter and tinged with olive-yellow on the rump and upper tail-coverts; wing-coverts like the back, the greater series dusky brown, edged with the same colour as the back, the median and greater coverts indistinctly tipped with ashy olive; bastard-wing, primary-coverts, and quills dusky brown, margined externally with olive-yellow, brighter on the quills; tail-feathers ashy olive, with olive-yellow margins and blackish shafts; crown of head like the back; lores and sides of head above the ear-coverts somewhat more dusky; cheeks pale ashy; below the eye a patch of silvery white dots, followed by a spot of silvery white on the ear-coverts; under surface of body pale ashy washed with light olive-yellow, the centre of the abdomen rather brighter olive-yellow; the feathers of the fore neck and breast with obsolete margins of pale olive-yellow, with a few mesial streaks of the same colour; under tail-coverts pale ashy, margined with light olive-yellow; axillaries and under wing-coverts also pale ashy with light olive-yellow margins; quills dusky below, ashy whitish along the edge of the inner web. Total length 5 inches, culmen 0·7, wing 2·75, tail 2·1, tarsus 0·75.

The Plate represents an adult bird in two positions, the figures being drawn from the typical specimen lent to us by Dr. Jentink.

[R. B. S.]

STIGMATOPS KEBIRENSIS, Meyer.

STIGMATOPS KEBIRENSIS, Meyer.

Kebir Scaly-throated Honey-eater.

Stigmatops kebirensis, Meyer, in Madarasz, Zeitschr. ges. Orn. i. p. 218 (1884).

This species has lately been described by Dr. Meyer from specimens sent to him by Mr. Riedel, and we have thought it advisable to give a figure of it in the present work.

As might be expected, it is very closely allied to *Stigmatops squamata* from the neighbouring group of islands; and it is somewhat remarkable that any difference should be found in specimens from adjacent islands like Kebir and Timor Laut, when examples from the latter island-group are not to be distinguished from others collected in the island of Choor, which is much further distant from Timor Laut than Kebir.

The adult *Stigmatops kebirensis* is a much greyer bird than *S. squamata*, especially on the wings. Underneath, the squamations or mottlings of the plumage are confined to the lower throat and chest, not extending nearly so far down the breast. Beyond this we cannot perceive any differences, and the distinctive characters must be admitted to be very slight.

The pair sent by Dr. Meyer measure as follows:—

	Total length.	Culmen.	Wing.	Tail.	Tarsus.
a. Ad. Kebir (*Riedel*)	5·3	0·75	2·95	2·3	0·80
b. ? Juv. Kebir (*Riedel*)	5·0	0·80	2·90	2·4	0·75

The second specimen, which, in spite of some of its larger dimensions, we take to be a somewhat younger bird, is greener than the type, especially on the wings and sides of the body, but it preserves the more restricted mottling on the breast which distinguishes the species from *Stigmatops squamata*.

The Plate represents an adult and a somewhat younger bird, of the size of life. The figures are drawn from the pair of birds described above and lent to us by Dr. Meyer.

[R. B. S.]

STIGMATOPS SQUAMATA, Salvad.

STIGMATOPS SQUAMATA, Salvad.

Scaly-chested Honey-eater.

Stigmatops squamata, Salvad. Ann. Mus. Civic. Genov. xii. p. 337 (1878).—Id. op. cit. xvi. p. 76 (1880).—Id. Orn. Papuasia e delle Molucche, ii. p. 326 (1881).—Sclater, Proc. Zool. Soc. 1883, p. 198.
Nectarinia, sp. inc. (♀), Sclater, Proc. Zool. Soc. 1883, p. 51.
Glycyphila squamata, Gadow, Cat. Birds in Brit. Mus. ix. p. 217 (1884).
Stigmatops salvadorii, Meyer, in Madarasz, Zeitschr. ges. Orn. i. p. 217 (1884).

Dr. Meyer has very kindly sent us from Dresden all the specimens of *Stigmatops* which have recently been the subject of his studies. We regret that we are unable to follow our learned colleague in all his conclusions, for we cannot find any cause for separating *Stigmatops salvadorii* from *Stigmatops squamata*; and after comparing a series of Timor-Laut specimens with others from Choor collected by Von Rosenberg, we consider them all identical and belonging to one and the same species. Two specimens from the typical series of *S. squamata* were kindly presented to the British Museum by Dr. Jentink, and have been compared by us with several specimens collected by Mr. Forbes in Timor Laut, as well as with those obtained from the same place by Dr. Meyer, and they appear to us to be specifically inseparable, though we must confess to having entertained a different expectation.

The following descriptions are taken from a pair of specimens collected by Von Rosenberg in the island of Choor, and presented to the British Museum by Dr. Jentink, the director of the Leiden Museum. They are from the typical series described by Count Salvadori.

Adult male. General colour above dull olive-greenish, somewhat clearer olive towards the lower back, rump, and upper tail-coverts; wing-coverts like the body, the greater series smoky brown, edged with the same colour as the back, the median coverts margined with pale yellow at the tips; quills dusky brown, externally edged with olive-greenish; tail-feathers pale ashy, washed externally with yellowish olive; crown of head rather more dingy olive than the back; lores and feathers round the eye dull ashy; from the base of the bill below the eye a patch of silvery white dots, with a slight tinge of yellow on the fore part of the patch, followed by a spot of silvery white on the ear-coverts; cheeks dull ashy, as well as the malar line and base of chin; entire under surface of body pale sulphur-yellow, mottled slightly on the throat, but very distinctly on the fore neck and breast, with dusky brown centres to the feathers; under tail-coverts very pale sulphur-yellow with dusky centres; under wing-coverts and axillaries pale yellow with dusky bases; quills dull brown below, ashy yellowish along the inner web. Total length 5·4 inches, culmen 0·75, wing 3·0, tail 2·3, tarsus 0·9.

The female sent by Dr. Jentink seems to be immature, being dull olive-yellowish underneath, with only here and there traces of the squamated feathers on the breast which distinguish the adult male. Total length 5·2 inches, culmen 0·8, wing 2·75, tail 2·1, tarsus 0·85.

The figures in the Plate represent a full-sized adult and a younger bird, the former being drawn from one of Dr. Meyer's specimens of *S. salvadorii*, the immature bird being one of those given by Dr. Jentink.

[R. B. S.]

STIGMATOPS ALBO-AURICULARIS, Ramsay.

STIGMATOPS ALBO-AURICULARIS, *Ramsay*.

Broadbent's Honey-eater.

Stigmatops albo-auricularis, Ramsay, Proc. Linn. Soc. N. S. W. iii. pp. 75, 285 (1879), iv. p. 100 (1879).—Salvad. Ibis, 1879, p. 325.—Id. Ann. Mus. Civic. Genov. xvi. p. 76 (1880).—Id. Orn. Papuasia, etc. ii. p. 324 (1881).
Glycyphila albiauricularis, Gadow, Cat. B. Brit. Mus. ix. p. 217 (1884).

The difficulty of distinguishing the smaller Honey-eaters is so well known that no apology will be needed for offering to our readers a Plate of the present bird, drawn from the only example which has as yet been brought to Europe. The two most recent writers on the Honey-eaters, viz. Count Salvadori and Dr. Gadow, had neither of them ever seen a specimen, and the former author placed it in the vicinity of *Stigmatops ocularis*. Not only, however, is it a much darker bird on the upper surface, but it is easily distinguished from that species by the squamulated appearance of the throat and breast, the feathers of which have dusky centres. This character allies it to *Stigmatops squamata*, in the vicinity of which species it has been more correctly placed by Dr. Gadow; but from the latter bird it is easily recognized by its brown upper surface and by the absence of the yellow on the throat and breast. It was discovered by Mr. Kendal Broadbent in South-eastern New Guinea.

The typical example having been brought to Europe by Mr. Ramsay during his official visit to the International Fisheries Exhibition, we have been enabled to give the following description of it:—

Adult male. General colour above nearly uniform brown, only slightly mottled with obscure dusky centres to the feathers of the head, neck, and mantle; wing-coverts like the back, the greater series, bastard wing, and primary-coverts dusky brown, edged with lighter brown, tinged with olive on the two latter; quills dusky brown, edged with olive-greenish, the inner secondaries with ashy brown; tail-feathers dusky brown, obscurely margined with dull olive; lores dingy brown like the head; an auricular patch of silvery grey, with numerous minute dots of white forming a conspicuous eye-patch; cheeks and chin dull ashy; throat and breast dingy white with an olive tinge, thickly mottled with dusky triangular centres to the feathers; sides of the upper breast nearly uniform brown; sides of body and flanks dull ashy, slightly mottled with dusky white edges; abdomen white; thighs dingy brown; under tail-coverts dull white with brown centres; axillaries and under wing-coverts dull whitish with pale dusky centres; quills dusky below, whitish along the inner web. Total length 5·0 inches, culmen 0·85, wing 2·75, tail 2·3, tarsus 0·7.

The single figure in the Plate is of the size of life, and has been drawn from the type specimen kindly lent to us by Mr. E. P. Ramsay.

[R. B. S.]

GLICIPHILA SUBFASCIATA, Ramsay.

GLYCIPHILA SUBFASCIATA, *Ramsay*.

Dusky Honey-eater.

Glyciphila subfasciata, Ramsay, P. Z. S. 1868, p. 385; 1875, p. 594.

In the present work, which includes the most gorgeous of all birds (I mean the Birds of Paradise and their allies), we shall also figure some very plain-coloured species; but it will in most cases be found that, without brilliancy of plumage to attract the eye, they yet possess some marked feature in their economy which arouses our interest. The present species is a case in point; for it is perhaps the plainest of all the Honey-eaters, many of which are very beautiful birds; but yet we find, from Mr. Ramsay's observations, that it stands unique among that large group for its mode of nest-building. I shall, however, allow him to tell the history of the species in his own words, merely premising that the following notes comprise all that has been as yet discovered about the species. Mr. Ramsay observes:—

"This species, although possessing nothing in its sombre plumage to recommend it, is certainly very interesting on account of its peculiarly shaped nest, being the only one of the Australian Meliphaginæ that I have met with which constructs a dome-shaped nest. It is a neat structure, composed of strips of bark, spiders' webs, and grass, and lined with fine grasses &c. The opening at the side is rather large; but the nest itself is rather deep, being about 4 inches long, and 2½ to 3 inches wide. The eggs I did not obtain; but one taken from the oviduct of a bird is 0·75 inch in length and 0·5 in breadth, pure white, with a few dots of black sprinkled over the larger end.

"The nests were invariably placed among the drooping branches of a species of *Acacia*, always overhanging some creek or running water. All the nests I found were so situated; and my young friend Master I. Sheridan of Cardwell, who has paid considerable attention to objects of natural history, assures me that he has never found them otherwise; and the usual number of eggs for a sitting are two, and frequently without any black dots on the surface. Their note is a sharp, shrill, monotonous cry, oft repeated at intervals; iris reddish brown."

The following is the original description of the species, extracted from Mr. Ramsay's paper:—

Female.—Total length 4·8 inches; bill, from the angle of the mouth 0·6, from forehead 0·5, width at base 0·2, across nostrils 0·1; wing, from flexure, 2·5; tail 2; tarsi 0·65. The whole of the upper surface, sides of the head, and neck glossy brown, a short oblique stripe under the eye white, feathers on the crown of the head centred with dark brown. The whole of the under surface and the extreme tips of the ear-coverts silvery white. The chest faintly barred with lines of brown, which join the sides of the neck above the shoulders; flanks and under coverts of wings tinged with brown; under surface of the wing dark brown, the inner margins of the feathers whitish brown; bill and legs reddish horn-brown."

The sexes differ considerably in size, while in colour there is no difference.

Besides having seen the type specimen in Mr. Ramsay's collection, I have specimens of this bird in my own cabinet, from which the figures in the accompanying Plate are drawn.

PTILOTIS ALBONOTATA, Salvad.

PTILOTIS ALBONOTATA, *Salvad.*

White-marked Honey-eater.

Ptilotis albonotata, Salvad. Ann. Mus. Civ. Genov. ix. p. 33 (1876).—Id. op. cit. xvi. p. 76 (1880).—Id. Orn. Papuasia e delle Molucche, ii. p. 333 (1881).—Gadow, Cat. Birds in Brit. Mus. ix. p. 229 (1884).

THIS species is easily distinguished from its near ally, *Ptilotis analoga*, by its white ear-spot, which is yellow in the last-named bird. It was first discovered by D'Albertis in South-eastern New Guinea, where a considerable series was forwarded from Naiabui. It has also been found in North-western New Guinea, at Ramoi by Dr. Beccari, and at Dorei by Von Rosenberg and Mr. Bruijn's hunters. The Marquis Doria has very kindly sent us a pair of birds for examination, the male being from Naiabui, and the female from Ramoi. We find that these two specimens are certainly of the same species; but the small size noticed in the female bird may be either peculiar to that sex, or may indicate a smaller race existing in North-western New Guinea. At present the series of specimens examined has been too small to decide this question.

Count Salvadori has separated a mountain form from the Arfak range as *Ptilotis montana*; but Dr. Gadow considers it to be the same as *P. albonotata*. As, however, he has never seen a specimen of either species, we think that it would have been wiser to have kept the two distinct, until he had had an opportunity of examining the materials at Count Salvadori's disposal, taking into account the experience and ability of the latter ornithologist.

The following descriptions are taken from the pair of birds lent to us by the Marquis Doria:—

Adult male. General colour above dark olive-green; wing-coverts, quills, and tail-feathers ashy brown, externally light olive-green like the back, especially on the primaries; head like the back; lores and feathers below the eye dusky blackish; ear-coverts dull ashy, with a white spot behind the lower parts; a mark of yellowish white behind the angle of the mouth; cheeks and under surface of body ashy, washed with olive-yellow, the abdomen ashy whitish tinged with yellow; the flanks, sides of body, and thighs browner; under tail-coverts light brown, with olive-yellow margins; under wing-coverts and axillaries pale fulvous, washed with olive-yellow; quills dusky below, yellowish white along the inner web: "bill black; feet ashy; iris ashy" (*D'Albertis*). Total length 6·5 inches, culmen 0·85, wing 3·4, tail 2·75, tarsus 0·9.

Adult female. Similar to the male in colour, but rather smaller. Total length 6 inches, culmen 0·75, wing 3·0, tail 2·45, tarsus 0·8.

The Plate represents the two specimens above described, of the natural size.

[R. B. S.]

PTILOTIS MARMORATA, Sharpe.

PTILOTIS MARMORATA, Sharpe.

Mottled-breasted Honey-sucker.

Ptilotis marmorata, Sharpe, Journ. Linn. Soc., Zool. vol. xvi. pp. 319, 438 (1882).

This large Honey-sucker was discovered by Mr. Goldie in the Astrolabe Mountains in South-eastern New Guinea. It is very closely allied to *P. cinerea* from North-western New Guinea, but is distinguished from it by the whitish edgings to the breast-feathers.

Nothing has been recorded of the habits of these Honey-suckers; but they doubtless do not differ from those of the ordinary species of *Ptilotis*. Mr. Goldie's specimens were obtained in the Morocco district, at the back of the Astrolabe range. He says that the native name is *Eaga*.

The following is a copy of the original description:—

"General colour above dusky brown, the feathers margined with olive, rather lighter on the head, which has a mottled appearance; on the forehead and over the eye a slight shade of ashy; wing-coverts like the back, but the outer median and greater coverts edged with paler olive, inclining to whity brown near the tips; quills and tail dusky, externally edged with yellowish olive, the tail-feathers margined with light rufous on the inner web; sides of face and ear-coverts dusky blackish, with a slight shade of silvery whitish on the ear-coverts, and a streak of dull white from behind the lores under the eye; cheeks dusky blackish, with a slight indication of ashy tips to the feathers; a narrow malar streak of dull yellowish white; throat yellowish white, mottled with dusky bases to the feathers; remainder of under surface of body ashy, the feathers tipped with a white bar and slightly washed with olive; the whole appearance of the under surface mottled, excepting on the lower flanks, which are uniform olive; thighs dusky; under tail-coverts light rufous with dusky bases, the outer ones externally yellowish white, mottled with dusky bases to the feathers; axillaries pale olive-yellowish; under wing-coverts light rufous-buff; quills dusky below, pale rufous along the inner web. Total length 7 inches, culmen 1·05, wing 3·8, tail 3·7, tarsus 1·05."

The figure in the Plate is drawn from one of Mr. Goldie's specimens kindly lent to me by Mr. Edward Gerrard, jun. [R.B.S.]

PTILOTIS FRENATA, *Ramsay.*

PTILOTIS FRENATA, Ramsay.

Bridled Honey-eater.

Ptilotis frenata, Ramsay, P. Z. S. 1874, p. 603.

JUSTIFYING the remarks which I have made on *Ptilotis flavostriata,* the present species is another of the recent additions to the family Meliphagidæ in the Australian continent. I give the following extract from Mr. Ramsay's article, as it comprises all that is at present known respecting the species:—

"Of this new species, for which I beg to propose the name of *P. frenata,* on account of the markings at the base of the bill and round the face, some few individuals were obtained, frequenting the Eucalypti while in blossom, near the margin of a swamp in the Cardwell district.

"The birds were shot by my (then) collector, Mr. Broadbent, who is already well known as an enthusiastic and careful taxidermist. To Mr. Broadbent's researches my collection is also indebted for the first specimen of *Eopsaltria inornata,* nov. sp.

"*Description.*—Whole of the upper surface dull brown; head, lores, and nape of neck blackish brown, the feathers having indistinct lunulate markings and a gloss of olive in certain lights; a semi-bare space below the eye has a few minute buffy white feathers; behind the eye a semilunar patch of white feathers tipped with black, which, extending in a narrow line, almost encircles that organ. Eye-lashes black; ear-coverts black; above them, immediately behind the eye, is a small tuft of bright wax-yellow feathers joining a large triangular patch of light greyish brown feathers on the side of the neck, which has the upper portion of it, nearest the ear-coverts, tinged with olive; a narrow indistinct line of yellow on either side bounding the ear-coverts below, extending obliquely to the lower part of the chin, where, meeting in an angle, they form an indistinct yellow patch on the throat. Chin and remainder of the under surface dull brown, slightly darker in tint on the breast and sides of the neck, lighter on the centre of the abdomen and under tail-coverts; across the chest are indistinct wavy lines of a darker tint, on the flanks indistinct lanceolate markings of the same tint. Under surface of the shoulders, inner margins of tertiaries, secondaries, and the basal portions of inner margins of the primaries light buff. The outer webs of the spurious wing-feathers, the tertiaries, and secondaries, with some of the primaries, are on the upper surface tinged with olive. Bill black, with the basal portion (except the culmen) yellow. The gape, with a narrow fleshy appendage, yellow. Feet and tarsi dark brown.

"Total length 8·5 inches; bill from angle of mouth 1·05 inch, from feathers at the nostrils 0·65, from forehead 1·05, height at nostrils 0·2, breadth 0·2; wing from flexure 4·05; tail 3·6; tarsus 0·35.

"*Hab.* Rockingham Bay.

"Sexes alike in plumage. One specimen, said to be a male, is considerably smaller in all its measurements, which are as follows:—Total length 7·5 inches; wing 3·65; tail 3·25; bill from angle of the mouth 0·95, from feathers at nostrils 0·6 inch, from forehead 0·9, height at nostril 0·2, breadth 0·2."

My figures represent the species of the size of life.

PTILOTIS FLAVOSTRIATA, Gould.

PTILOTIS FLAVOSTRIATA, Gould.

Yellow-streaked Honey-eater.

Ptilopis flavostriata, Gould, P. Z. S. 1875, p. 316.

If Australia is peculiar for its Lyre-birds and for its mound-building Megapodes and Brush-Turkeys, it certainly presents a no less marked feature in the extraordinary development of the family of Meliphagidæ (or Honey-eaters) within its limits. Nor can we suppose that the number of species known to us from that continent is yet exhausted, seeing that not a year passes without our receiving some notable addition to our list of the Australian species of this group of birds. The one which I have now the pleasure to introduce to my readers is from the neighbourhood of Rockingham Bay, in Queensland, and was sent to me for examination by Mr. Waller of Brisbane, to whom it appeared undescribed. Of this there is no doubt; and I have accordingly described the species under the name of *flavo-striata*, on account of the yellow chest-streaks which are such a conspicuous feature in the bird's appearance.

The following is my original description of the species :—

Head and hind neck dusky blackish, with a distinct shade of olivaceous on the crown, leaving a pure black patch on each side of the occiput; hind neck distinctly marked with triangular spots of dull white; mantle straw-yellow, the bases to the feathers dusky brown; scapulars brown, with large triangular whitish spots; lower back and rump brown, with dull olive margins to the feathers; upper tail-coverts and tail brown, paler at tip, the feathers narrowly margined with olive; wing-coverts brown, with large triangular whitish spots as on the scapulars, the greater series dark brown, tipped with whitish and margined with whity brown; quills dark brown, externally edged with olive and slightly tipped with whitish; region of the eye bare and yellow; the few feathers on the ear-coverts hoary; cheeks dull olivaceous buff, running into a distinct tuft of bright yellow; throat greyish white, washed with olive on the lower part; chest olivaceous, distinctly streaked with shaft-lines of bright yellow, the breast paler, the feathers being brown with broad triangular longitudinal spots of white; flanks and under tail-coverts light brown, washed with olive; under wing-coverts yellowish buff; the lower surface of the wings and tail ashy brown, with broad rufous-buff margins to the inner webs. Total length $7\frac{1}{2}$ inches, bill $1\frac{1}{8}$, wing $3\frac{1}{2}$, tail 3, tarsus $1\frac{1}{8}$.

The figures in the Plate are of the size of life.

XANTHOTIS CHRYSOTIS.

XANTHOTIS CHRYSOTIS.

Golden-eared Honey-eater.

Philedon chrysotis, Lesson, Voyage de la Coquille, Zool. i. p. 645, pl. 21 (1826).—Rosenb. Malay Arch. p. 395 (1879).
Myzantha flaviventer, Lesson, Man. d'Orn. ii. p. 67 (1828).
Myzantha chrysotis, Lesson, Traité d'Orn. p. 302 (1831).—Id. Compl. Buff., Ois. p. 594 (1838).
Tropidorhynchus chrysotis, Gray, Gen. B. ii. p. 126 (1846).—Bp. Consp. i. p. 390 (1850).—Sclater, Proc. Linn. Soc. ii. p. 158 (1858).—Gray, Cat. Mamm. etc. New Guinea, pp. 25, 56 (1859).—Finsch, Neu-Guinea, p. 165 (1865).
Xanthotis chrysotis, Bp. Comptes Rendus, xxxviii. p. 262 (1854).—Meyer, Sitz. k. Akad. Wien, lxx. pp. 113, 207 (1874).—Salvad. Ann. Mus. Civic. Genov. viii. p. 401 (1876), x. p. 147 (1877), xvi. p. 78 (1880).—Id. Orn. Papuasia e delle Molucche, ii. p. 347 (1881).
Ptilotis flaviventris, Gray, Proc. Zool. Soc. 1858, p. 190.—Id. Cat. Mamm. etc. New Guinea, p. 55 (1859).—Id. Proc. Zool. Soc. 1861, p. 429.
Ptilotis chrysotis, Gray, Proc. Zool. Soc. 1859, p. 155.—Gadow, Cat. Birds in Brit. Mus. i. p. 238 (1884).—Guillemard, Proc. Zool. Soc. 1885, p. 642.
Xanthotis flaviventris, Reichenb. Handb. Meropinæ, p. 139, fig. 3512 (1862).
Anthochæra chrysotis, Gray, Hand-list Birds, i. p. 159, no. 2070 (1869).—Rosenb. Malay Arch. p. 553 (1879).
Xanthotis rubiensis, Meyer in Madarász, Zeitschr. ges. Orn. i. p. 289 (1884).

The present species is the Papuan representative of the Australian *X. filigera*, and is found in many parts of North-western New Guinea and the neighbouring island of Mysol. It is not found in Waigiou, as stated by Dr. Gadow (*l. c.*), as in this island the allied species *X. fusciventris* of Salvadori takes its place. The specimen in the British Museum from Waigiou belongs to the latter species, which is easily recognizable, being, indeed, allowed by Dr. Gadow, though he refers the only specimen in the Museum to *X. chrysotis*. There is considerable variation in a series of specimens, some showing a more rufous coloration on the under surface than others, and in many specimens the grey ear-spot is nearly obsolete. These differences seem to us to be due sometimes to age, and sometimes to the preparation of the skin, and we cannot perceive any specific characters sufficient to separate *X. rubiensis* of Dr. Meyer (*l. c.*). The latter gentleman has been so kind as to send us over the types of *X. rubiensis*, as well as the birds he considers to be true *X. xanthotis*, and after comparing them with the series in the British Museum, we believe the two species to be inseparable.

Dr. Guillemard has lent us the specimen from which the following description is taken:—

Adult male. General colour above dark brown mottled with yellowish olive; wing-coverts like the back; the outer greater coverts, bastard-wing, primary-coverts, and quills dusky brown edged with brighter olive-yellow, the secondaries margined with dull olive; upper tail-coverts and tail-feathers dusky brown, edged with yellowish olive; crown of head greener than the back; lores dusky; below the eye a bare patch; ear-coverts dark slaty grey, with a tuft of bright yellow below the hinder part; cheeks and sides of face dark olive; throat ashy grey; fore neck and chest olive-greenish, becoming yellower towards the breast, which is dull fawn-brown washed with olive-yellow; abdomen, sides of body, and flanks fawn-brown; thighs and under tail-coverts light brown washed with buff; under wing-coverts and axillaries tawny buff, olive-yellow on the edge of the wing; quills below dusky, tawny along the inner edge: "bill and legs black; iris brown" (*H. Guillemard*). Total length 8 inches, culmen 1·15, wing 4·1, tail 3·5, tarsus 1·0.

The figures in the Plate represent an adult bird of the natural size and in two positions. They are drawn from the specimen above described in Dr. Guillemard's collection.

[R. B. S.]

MELIARCHUS SCLATERI.

MELIARCHUS SCLATERI.

Sclater's Honey-eater.

Philemon vulturinus (nec Reichenb.), Sclater, Proc. Zool. Soc. 1869, pp. 120, 124.
Philemon sclateri, Gray, Ann. and Mag. Nat. Hist. 1870, v. p. 327.—Id. Cruise of the Curaçoa, Birds, p. 362, pl. v. (1873).—Tristr. Ibis, 1879, p. 439.—Gadow, Cat. Birds in Brit. Mus. ix. p. 279 (1884).
Meliarchus sclateri, Salvad. Ann. Mus. Civic. Genov. xvi. p. 75 (1880).—Id. Orn. Papuasia e delle Molucche, ii. p. 322 (1881).

This fine species of Honey-eater appears to be confined to the Solomon Islands, and is probably found only on the island of San Christoval. Here it was procured by the late Mr. Brenchley, and Lieut. Richards also met with it at Makira Harbour, in the same island. Very few specimens exist in European collections, and nothing has been recorded of its habits.

We follow Count Salvadori in referring the present species to a distinct genus, as it appears to us to be by no means a true *Philemon*, excepting under the very elastic definition which Dr. Gadow gives to the latter genus.

The following description is taken from a specimen lent to us by Mr. E. P. Ramsay:—

Adult male. General colour above dull olive on the back and mantle, dull reddish brown on the body, rump, and upper tail-coverts; wing-coverts dark brown, edged with olive; bastard-wing blackish brown; primary-coverts and quills dusky brown, edged with lighter olive-yellow, except on the inner secondaries; tail-feathers reddish brown with olive margins; crown of head pale yellow, streaked with black, the hind neck with dusky; feathers round the eye and eyebrow and the upper part of the ear-coverts yellowish white, with a black spot on the lores; cheeks and lower part of ear-coverts black, streaked with pale yellow edges to the feathers; throat pale ashy grey; breast pale olive-yellow, with dusky centres to the feathers; lower breast and abdomen dull ashy; sides of body and flanks reddish brown washed with olive; thighs and under tail-coverts reddish brown, edged with olive-yellow; axillaries and under wing-coverts dark ashy, washed with olive; quills dusky below, ashy along the inner web: "bill yellowish; feet ash-colour; iris brown" (*G. E. Richards*). Total length 9·5 inches, culmen 1·5, wing 4·5, tail 4·25, tarsus 1·35.

The Plate represents an adult bird of the natural size, and is drawn from the above-mentioned specimen lent by Mr. Ramsay. It was procured at Makira Harbour by Lieut. Richards.

[R. B. S.]

EUTHYRHYNCHUS FLAVIGULA, *Schlegel.*

EUTHYRHYNCHUS FLAVIGULA, *Schl.*

Yellow-tinted Brown Honey-eater.

Euthyrhynchus flavigula, Schl. Nederl. Tijdschr. Dierk. iv. p. 40 (1871).—Salvad. Ann. Mus. Civic. Genov. xii.
    p. 340 (1878).—Rosenb. Malay. Arch. pp. 553, 586 (1879).— Salvad. Ann. Mus. Civic. Genov. xvi.
    p. 78 (1880).—Id. Orn. Papuasia e delle Molucche, ii. p. 341 (1881).
Timeliopsis, nov. sp. ?, Salvad. Ann. Mus. Civic. Genov. vii. p. 964 (1875).
Euthyrhynchus flavigularis, Gadow, Cat. B. in Brit. Mus. ix. p. 287 (1884).
? *Euthyrhynchus griseigularis*, Guillemard, Proc. Zool. Soc. 1885, p. 642.

This species differs from its near ally *E. griseigula* in having a more distinct wash of olive-yellow on the under surface of the body; but under certain lights the two species seem to be almost inseparable. In size it is rather smaller than *E. griseigula*.

    The type of this species is in the Leyden Museum, where it was described by the late Professor Schlegel. It was discovered by Baron Von Rosenberg on the western shores of the bay of Geelvink in North-western New Guinea, and it has been obtained at Sorong by the late Dr. Bernstein, and at Ramoi by Dr. Beccari. Dr. Guillemard's specimen from Andai was also evidently the present bird. He gives *E. flavigula* as a synonym of *E. griseigula* without any comment. These two species may prove to be identical, it is true, but at present the material at our disposal is not sufficient to settle the question.

    General colour above olive, a little duller on the back, tail, and under tail-coverts; the head and neck slightly clearer olive, with the frontal feathers stiffened; wing-coverts dusky, edged with olive, the lesser coverts like the back; the bastard-wing, primary-coverts, and quills pale dusky brown, externally edged with olive like the back; tail-feathers also dusky brown, broadly edged with olive like the back; lores and feathers round the eye a little lighter olive than the head; cheeks and ear-coverts olive like the head, a little paler on the former; under surface of body pale drab-brown, clearer on the flanks and thighs; the throat, breast, abdomen, and under tail-coverts distinctly washed with olive-yellow; under wing-coverts and axillaries pale drab-brown, washed with olive-yellow; quills dusky below, pale ashy rufous along the edge of the inner web. Total length 6·3 inches, culmen 0·8, wing 3·1, tail 2·5, tarsus 0·9.

    The above description has been taken from the specimen lent to us by the Marquis Doria, and the Plate has been drawn from the same bird.

[R. B. S.]

EUTHYRHYNCHUS FULVIGULA, *Schlegel.*

W. Hart del et lith.

Mintern Bros. imp.

EUTHYRHYNCHUS FULVIGULA, *Schl.*

Buff-throated Honey-eater.

Euthyrhynchus fulvigula, Schl. Nederl. Tijdschr. Dierk. iv. p. 40 (1871).—Salvad. Ann. Mus. Civic. Genov. xii.
p. 343 (1875).—Rosenb. Malay. Arch. pp. 553, 586 (1879).—Salvad. op. cit. xvi. p. 78 (1880).—Id. Orn.
Papuasia e delle Molucche, ii. p. 342 (1881).
Timeliopsis acutirostris, Salvad. Ann. Mus. Civic. Genov. vii. p. 964 (1875).

THIS is the most distinct of all the species of the genus *Euthyrhynchus*, and is distinguished not only by the olive-green colour of the upper surface, but also by the amount of olive-green on the lower parts, which causes the tawny buff colour of the throat to appear in strong contrast.

It was first described by Professor Schlegel from a specimen in the Leyden Museum, and has since been discovered in the Arfak Mountains by Mr. Bruijn's hunters, as well as by Dr. Beccari. Count Salvadori remarks on the affinity of this genus to the Australian *Plectorhyncha*, a fact also observed by Mr. Ramsay, who named one of the species *Plectorhyncha fulviventris*.

Adult. General colour above dull olive-green, a little lighter on the lower back and rump; lesser and median wing-coverts, greater coverts, bastard-wing, and primary-coverts as well as the quills dusky brown, edged with olive-green like the back, a little more yellow on the primaries; tail-feathers dusky brown, edged with olive-green; head like the back, but a little more dingy and inclining to dull ashy olive; lores ashy fulvous; eyelid and ear-coverts dull ashy, as well as the feathers below the eye; throat light tawny; breast pale pinkish brown, slightly washed with olive-yellow; sides of body and under tail-coverts pale olive-greenish as well as the thighs; axillaries and under wing-coverts pale fawn-buff; quills dusky below, whitish along the inner web. Total length 5 inches, culmen 0.65, wing 2·65, tail 2·05, tarsus 0·8.

The above description is that of a specimen which the Marquis Doria lent to us; it is figured in two positions, and of the natural size.

[R. B. S.]

EUTHYRHYNCHUS GRISEIGULA, Schlegel.

EUTHYRHYNCHUS GRISEIGULA, Schl.

Brown Honey-eater.

Euthyrhynchus griseigula, Schl. Nederl. Tijdschr. Dierk. iv. p. 39 (1871).—Salvad. Ann. Mus. Civic. Genov. xii. p. 342 (1878).—Rosenb. Malay. Arch. pp. 553, 588 (1879).—Salvad. Ann. Mus. Civic. Genov. xiv. p. 78 (1880).—Id. Orn. Papuasia e delle Molucche, ii. p. 340 (1881).
Timeliopsis trachycoma, Salvad. Ann. Mus. Civic. Genov. vii. p. 963 (1875).
Euthyrhynchus griseigularis, Gadow, Cat. B. in Brit. Mus. ix. p. 287 (1884).

THE genus *Euthyrhynchus* is apparently a very natural one, allied to *Plectorhyncha* of Australia and containing four species; for we cannot admit *Melipotes gymnops* to be a *Euthyrhynchus*, as Dr. Gadow has done. It differs so entirely in its bare ocular region (always a character of importance in the Honey-eaters) that we sympathize with Count Salvadori when he remarks :—" The union of *Melipotes gymnops* in one genus with the species attributed to *Euthyrhynchus* is beyond my comprehension ! "

Having had an opportunity of comparing the four species described, we give a short diagnosis, which will serve to distinguish them :—

E. griseigula.—Olive-brown above ; throat and breast pale pinkish or vinous-brown, with scarcely any shade of olive-yellow. *Hab.* N.W. New Guinea.

E. flavigula.—Olive-brown above ; throat and breast distinctly marked with olive-yellow, this colour pervading the entire under surface. *Hab.* N.W. New Guinea.

E. fulviventris.—Olive-brown above; entire under surface pale pinkish or vinous-brown, with a slight mark of olive-yellow on the throat. *Hab.* S.E. New Guinea.

E. fulvigula.—Olive-green above, paler olive-green below; throat tawny, contrasting with the breast. *Hab.* N.W. New Guinea.

Of these four species, *E. flavigula* and *E. griseigula* are so closely allied that it would never surprise us to hear that their specific identity had been established, and the differences are so slight that the utmost difficulty has been experienced in rendering them in a coloured figure, while at the same time the texture of their plumage is so delicate that to make a drawing of them has been no easy task. *E. fulviventris* is much more recognizable, its vinous-brown under surface being a strongly pronounced character.

The measurements of the present species are as follows :—Total length 6·5 inches, culmen 0·85, wing 3·45, tail 2·7, tarsus 0·9. These dimensions are a little in excess of those of *E. flavigula*.

The Plate has been drawn from a specimen lent to us by the Marquis Doria from the Genoa Museum ; it is represented in two positions, of the size of life.

[R. B. S.]

MELILESTES POLIOPTERUS, Sharpe.

MELILESTES POLIOPTERUS, Sharpe.

Grey-winged Honey-eater.

Melilestes polyopterus, Sharpe, Journ. Linn. Soc., Zool. vol. xvi. pp. 318, 438 (1882).

THE genus *Melilestes* was founded by Count Salvadori for the reception of a little group of Honey-eaters peculiar to the Papuan subregion. The four species comprised under the genus present considerable difference in coloration; but the affinities of the present bird are clearly with *M. novæ guineæ*, a little species which was for a long time thought by naturalists to be an *Arachnothera*. The latter genus is now considered to be exclusively Indian; and although *M. novæ guineæ* and *M. poliopterus* have much the appearance of a Spider-hunter, they would be expected from their habitats to be more nearly allied to the Australian *Meliphagidæ*.

The chief differences which *M. poliopterus* exhibits when compared with *M. novæ guineæ* are the plumbeous wings and head, as well as the yellow spot on the throat.

Mr. Goldie obtained a single specimen of this new species in the Choqueri district, at the back of the Astrolabe Mountains, in South-eastern New Guinea, where it was called by the natives " bererita."

The following description has been taken from Mr. Sharpe's paper on Mr. Goldie's collections :—

"General colour above green, the whole of the crown and nape dark slaty grey; wing-coverts slaty-grey, quills dusky, externally slaty grey, rather lighter along the edge of the primaries, the secondaries with a very aint olive tint on the outer webs; tail-feathers dusky, externally edged with slaty grey and having a small white spot at the tip of the inner web; lores, sides of face and ear-coverts dull slaty grey with a slight wash of green; under surface of body olive-yellow, the chin dusky grey washed with yellow, the lower throat bright yellow; thighs ashy washed with yellow; under tail-coverts yellow, ashy grey along the centre; under wing-coverts and axillaries white, the latter washed with yellow; quills dusky brown, edged with white along the inner web. Total length 4·4 inches, culmen 1·2, wing 2·85, tail 1·55, tarsus 0·7."

The figures in the Plate are drawn from the typical specimen in the British Museum, and represent an adult bird of the natural size in two positions.

[R.B.S.]

MELILESTES ILIOLOPHUS, Salvad.

MELILESTES ILIOLOPHUS, Salvad.

Long-plumed Honey-eater.

Melilestes iliolophus, Salvad. Ann. Mus. Civic. Genov. vii. p. 951 (1875), xvi. p. 75 (1880).—Id. Orn. Papuasia e delle Molucche, ii. p. 316 (1881), iii. p. 543 (1882).—Sharpe, Journ. Linn. Soc., Zool. xvi. p. 437 (1882).
Arachnothera iliolophus, Gadow, Cat. Birds in Brit. Mus. ix. p. 3, pl. i. fig. 2 (1884).

THE genus *Melilestes* has been united by Dr. Gadow to the genus *Arachnothera*; but in our opinion Count Salvadori was right in placing it with the *Meliphagidæ* or family of Honey-suckers, rather than with the Sun-birds or *Nectariniidæ*, where it is located by Dr. Gadow. The long fluffy plumage and the silky tufts on the flanks are characters which ally the genus to the Honey-suckers, though the general appearance of the birds is very much that of the Spider-hunters (*Arachnothera*).

Count Salvadori includes four species in his genus *Melilestes*, of which the present and *M. affinis* (a species summarily suppressed without just cause by Dr. Gadow, who has never seen a specimen) are distinguished by their greyish-olive underparts. A fifth species has been discovered since Salvadori wrote, which is figured in the present work.

The present species was discovered in the islands of Jobi and Miosnoum, in the Bay of Geelvink, by Dr. Beccari, and we cannot find any marked difference between some of the typical examples now in the British Museum and others obtained in South-eastern New Guinea, where it has been obtained by Mr. Goldie and Mr. H. O. Forbes in the Astrolabe Mountains. The following description is from one of Mr. Goldie's specimens:—

Adult. General colour above dull olive-green, the head a little duller than the back; feathers of the lower back and rump very long and silky, and a little lighter than the rest of the back; wing-coverts like the back; the primary-coverts and quills dusky brown, edged with olive-green like the back, the secondaries more broadly; tail dusky black; lores and feathers round the eye ashy olive; ear-coverts lighter olive; under surface of body very pale yellowish, ashy on the cheeks and throat; sides of the body with long silky plumes of paler yellow; under tail-coverts like the abdomen, and washed with pale olive-green; axillaries light yellow like the sides of the body; under wing-coverts light ashy brown, washed with yellowish olive; quills dusky below, whitish along the edge of the inner web. Total length 3·9 inches, culmen 0·85, wing 2·7, tail 1·45, tarsus 0·85.

The figures in the Plate are drawn from two specimens procured by Mr. H. O. Forbes in the Sogeri district of the Astrolabe Mountains.

[R. B. S.]

ZOSTEROPS UROPYGIALIS, Salvad.

ZOSTEROPS UROPYGIALIS, Salvad.

Ké-Island White-eye.

Zosterops uropygialis, Salvad. Ann. Mus. Civic. Genov. vi. p. 78 (1874).—Id. op. cit. xiv. p. 655 (1879) —Id. op. cit. xvi. p. 82 (1880).—Id. Orn. Papuasia e delle Molucche, ii. p. 373 (1881).—Sharpe, Cat. Birds in Brit. Mus. ix. p. 190 (1884).

Dr. Beccari discovered this species of *Zosterops* in Little Ké Island in 1873, and the type is in the Civic Museum at Genoa. We have seen several examples in the Leiden Museum, and one of these was presented to the British Museum by our friend Dr. Jentink, the Director of the Rijks Museum at Leiden. This White-eye belongs to the section of the genus with the under parts entirely yellow, and has been separated by Count Salvadori under a sub-section characterized by the brownish colour of the quills and tail. Since personally examining specimens of the species, we are somewhat inclined to doubt the importance of this colour of the wings and tail-feathers, as to all appearances the fresh-moulted feathers of the White-eyes are always much darker than those which have been well worn.

In the specimen originally obtained by Beccari the white eye-ring, which is the chief character of a *Zosterops*, was wanting; and as it is not present in the specimens at Leiden, its absence is doubtless characteristic of the species. The other characters which distinguish it are the dusky head and yellow rump, the latter contrasting with the yellowish-green back.

The following is a translation of Salvadori's original description :—

"Above yellowish green, the head tinged with dusky; the rump yellowish; entire under surface of body yellow, the under tail-coverts and throat brighter; quills and tail-feathers dusky, margined with the same colour as the back, the former margined internally with yellowish white; under wing-coverts mixed yellow and ashy; bill dusky; feet lead-colour. Total length 4·4 inches, wing 2·5, tail 1·8, bill 0·5, tarsus 0·65."

The Plate represents an adult bird in two positions, and is drawn from a specimen presented to the British Museum by the Leiden Museum.

[R. B. S.]

ZOSTEROPS BRUNNEICAUDA, Salvad.

ZOSTEROPS BRUNNEICAUDA, Salvad.

Brown-tailed White-eye.

Zosterops rufifrons, Salvad. Ann. Mus. Civic. Genov. vi. p. 79 (1874).
Zosterops brunneicauda, Salvad. Ann. Mus. Civic. Genov. xvi. p. 82 (1880).—Id. Orn. Papuasia e delle Molucche, ii. p. 373 (1881).—Sharpe, Cat. Birds in Brit. Mus. ix. p. 190 (1884).

This species was also discovered by Dr. Beccari, at Gesser in Ceram Laut, during his expedition to the East in 1873, and was first named by Count Salvadori *Zosterops rufifrons*. The describer, however, afterwards discovered that the red colour on the forehead was due to a blood-stain, and he therefore very wisely changed the name, as it had become inapplicable to the species. That of *brunneicauda* does not strike us as very appropriate, as many of the allied species appear to have the tail-feathers quite as pale.

Besides being found in Ceram Laut by Dr. Beccari, this White-eye has been met with in the island of Choor by the Dutch traveller Von Rosenberg, and was afterwards procured in the island of Pulo-babi, in the Aru group, by Beccari. We have also seen specimens collected by Dr. Guillemard in the island of Sumbawa, one of the Timor group, which we have been unable to separate from *Z. brunneicauda*.

The following description is translated from Count Salvadori's work:—

"Above yellowish green, conspicuously yellowish; under surface of body deep yellow, the sides of the body greenish; eye-ring snow-white, surrounded below by dusky black; quills and tail-feathers brown, margined with the same colour as the back; under wing-coverts and inner edge of the quills whitish yellow; bill dusky, the lower mandible paler at the base; feet apparently lead-colour. Total length 4·75 inches, wing 2·45, tail 1·65, bill 0·5, tarsus 0·65."

The figures in the Plate represent an adult bird in two positions; they are drawn from a specimen presented to the British Museum by the Leiden Museum.

[R. B. S.]

ZOSTEROPS FUSCIFRONS, Salvad.

ZOSTEROPS FUSCIFRONS, Salvad.

Dusky-fronted White-eye.

Zosterops fuscifrons, Salvad. Ann. Mus. Civic. Genov. xii. p. 339 (1878), xvi. p. 80 (1880).—Id. Orn. Papuasia, etc. ii. p. 365 (1881).—Sharpe, Cat. B. Brit. Mus. ix. p. 201 (1884).

This species comes from Gilolo or Halmahera, in the Moluccas, and is very nearly allied to *Z. atriceps* from the neighbouring island of Batchian; but it has not so much black on the head as that bird, only the sinciput and lores are dusky, the occiput and nape being yellowish olive like the back.

The typical specimens were three in number, and were procured by the late Dr. Bernstein near Galela in the island above mentioned, and were sent by him to the Leyden Museum. During a visit to Leyden in the autumn of 1883, we examined these types and were allowed by Professor Schlegel to bring one of them to England, and have thus been enabled to figure the species in the present work.

The following is a translation of Count Salvadori's original diagnosis, which, by an accidental omission, has not been acknowledged in the 'Catalogue of Birds' (*l. c.*):—

Very similar to *Z. atriceps* of Gray, but having only the sinciput and the lores dusky; the occiput, hind neck, and cheeks yellowish olive, uniform with the back, and not dusky at all; wings and tail dusky, margined with yellowish olive; under surface of body white, with the exception of the under tail-coverts, which are yellow; round the eye a conspicuous ring of white feathers. Total length 4·25 inches, wing 2·1, bill 0·8, tarsus 0·6.

The figures in the Plate represent an adult bird of the natural size in two positions; they are drawn from one of the typical specimens lent to us by the late Professor Schlegel.

[R. B. S.]

ZOSTEROPS LONGIROSTRIS, Ramsay.

ZOSTEROPS LONGIROSTRIS, *Ramsay*.

Heath-Island White-eye.

Zosterops longirostris, Ramsay, Proc. Linn. Soc. N. S. W. iii. p. 288 (1879), iv. p. 100 (1879).—Salvad. Ann. Mus. Civic. Genov. xvi. p. 82 (1880).—Id. Orn. Papuasia, etc. ii. p. 372 (1881).—Sharpe, Cat. B. Brit. Mus. ix. p. 189 (1884).

Mr. Ramsay having brought the type of this interesting White-eye to England, an opportunity was offered us of figuring it, which we gladly accepted, as the difficulty of determining the species of *Zosterops* from descriptions alone is well known to every student of this difficult group. The original specimen of *Zosterops longirostris* was discovered by Mr. Kendal Broadbent on Heath Island, off the south coast of New Guinea, a locality which has not been found by us in any atlas, but which is included by Mr. Ramsay under the heading of the South Cape District of New Guinea and the Louisiades.

This large species of *Zosterops* belongs to the brown-tailed section of the yellow-coloured group of the genus. It is distinguished from *Z. uropygialis* by having the head of the same colour as the back, and from *Z. brunneicauda* by its light brown bill and by the absence of any dusky spot in front of the eye.

The following is a description of the original specimen :—

Adult male (type of species). General colour above dull yellowish olive, a little yellower on the head and rump, the former with indistinct paler shaft-streaks on the forehead and sinciput; wing-coverts a little yellower than the mantle; bastard wing, primary-coverts, and quills dusky brown, externally washed with dull olive-yellow; tail-feathers pale brown, with dull olive-yellow edges; lores yellow; round the eye a ring of whitish feathers; ear-coverts olive-yellow, a little brighter than the crown; cheeks, throat, and centre of breast and abdomen clear yellow, deeper and slightly inclining to orange-yellow on the under tail-coverts; breast somewhat overshaded with greenish; sides of body and flanks decidedly greenish; thighs yellow; under wing-coverts and axillaries pale yellow with whitish bases, brighter on the edge of the wing; quills light brown below, whitish along the edge of the inner web. Total length 4·2 inches, culmen 0·7, wing 2·5, tail 1·8, tarsus 0·75.

The figures in the accompanying Plate represent the species of the natural size.

[R. B. S.]

ZOSTEROPS DELICATULA, Sharpe.

ZOSTEROPS DELICATULA, Sharpe.

Grey-sided White-eye.

Zosterops delicatula, Sharpe, Journ. Linn. Soc., Zool. xvi. pp. 318, 440 (1882).

This pretty little species belongs to the division of the genus *Zosterops* in which the species have a particoloured under surface with the throat and the under tail-coverts yellow. It would appear to be very similar to *Zosterops frontalis* of Salvadori; but in the description of the latter species given by that author in his 'Ornitologia della Papuasia,' vol. ii. p. 368, there is no mention of the grey on the sides of the breast and fore neck, the breast and abdomen being described as white. It will remain therefore for some future comparison of *Z. delicatula* with *Z. frontalis* to determine whether the two birds are absolutely identical as species. There would be nothing surprising if this should turn out to be the case, as there are many instances of the same species occurring in the Aru Islands and in South-eastern New Guinea.

As far as our knowledge goes at present, *Z. delicatula* is only known from the Astrolabe range of mountains, whence several specimens were forwarded by Mr. Goldie.

The following is a description of a specimen kindly lent to me by Mr. Edward Gerrard, jun. :—

"*Adult.* General colour yellowish green, brighter and clearer yellow on the rump and upper tail-coverts; wing-coverts like the back; bastard-wing feathers black; primary-coverts and quills black, externally yellowish green, brighter on the primaries, which are narrowly edged with this colour; the secondaries more broadly margined with the same colour as the back, the innermost secondaries being entirely of the latter colour; tail-feathers blackish, edged with yellowish green near the base; crown of head like the back; forehead black, extending as far as above the middle of the eye; lores and feathers below the eye also black, as well as the fore part of the cheeks and a narrow line skirting the rami of the lower jaws to the base of the chin; round the eye a band of silky white feathers; ear-coverts and hinder cheeks dark green; throat bright yellow; remainder of under surface white with a delicate tinge of ashy grey, extending from the sides of the fore neck down the sides of the breast; thighs white washed with yellow; under tail-coverts bright yellow; under wing-coverts and axillaries white slightly washed with yellow; quills dusky below, ashy along the inner web. Total length 3·8 inches, culmen 0·45, wing 2·3, tail 1·5, tarsus 0·65."

The figures in the Plate represent an adult bird in two positions.

[R. B. S.]

ZOSTEROPS RENDOVÆ, *Tristram.*

ZOSTEROPS RENDOVÆ, Tristr.

Rendova White-eye.

Tephras olivacea, Ramsay, Proc. Linn. Soc. N. S. Wales, vi. p. 180 (1881, nec *Zosterops olivacea* (L.)).
Zosterops ramsayi, Salvad. Ann. Mus. Civic. Genov. xviii. p. 425 (nec Masters).
Zosterops rendovæ, Tristram, Ibis, 1882, p. 135 —Salvad. Orn. Papuasia e delle Molucche, iii. p. 546 (1882).—Ramsay, Proc. Linn. Soc. N. S. Wales, vii. p. 42 (1883).—Sharpe, Cat. Birds in Brit. Mus. ix. p. 188 (1884).

The genus *Zosterops* contains several species which have not the white eyelid upon which the trivial name of "White-eye" is founded, and the present bird belongs to the section in which this peculiar character is not developed. All such species are at least subgenerically distinct from *Zosterops*, and in that case the genus *Tephras* might well be employed, as has been done by Mr. Ramsay in the present instance.

The original specimen was procured by Lieut. Richards in the Island of Rendova in the Solomon Archipelago, and was described by Mr. Ramsay as *Tephras olivacea*. Count Salvadori, considering that *Tephras* was congeneric with *Zosterops*, gave to the Rendova species the name of *Zosterops ramsayi*, as there was already a *Zosterops olivacea* of Linnæus, from the Island of Réunion. Mr. Masters had, however, previously bestowed Mr. Ramsay's name on a species of the genus, and therefore the bird must be called *Zosterops rendovæ*, which title was given to the identical specimen procured by Lieut. Richards when it came into Canon Tristram's hands in England, and the latter author recognized at the same time that the name *Z. olivacea* was pre-occupied in ornithology.

We have copied the description of the type specimen given in our 'Catalogue of Birds':—

"General colour above uniform olive-yellow, a little brighter across the rump; wing-coverts like the back, a little yellower on the median and greater coverts; bastard-wing feathers dusky, washed with olive; primary-coverts and quills blackish, externally olive-yellow, brighter on the edge of the primaries; upper tail-coverts olive-yellow; tail-feathers blackish, washed with olive-green near the base; crown of head and lores like the back; no ring of white feathers round the eye; in front of the eye a dusky spot; ear-coverts, cheeks, and throat olive-yellow, scarcely brighter than the upper surface; fore neck, breast, and abdomen bright yellow, greener on the sides of the body and flanks; thighs and under tail-coverts bright yellow; under wing-coverts and axillaries white, washed with bright yellow; quills below dusky; inner edge of quills ashy white tinged with yellow. Total length 4·6 inches, culmen 1·65, wing 2·55, tail 1·8, tarsus 0·7."

The Plate gives two representations of this species, of the natural size, the figures being drawn from the typical example lent to us by Canon Tristram.

[R. B. S.]

MYZOMELA WAKOLOENSIS, Forbes.

MYZOMELA WAKOLOENSIS, *H. O. Forbes*.

Lake Wakolo Honey-eater.

Myzomela wakoloensis, Forbes, Proc. Zool. Soc. 1883, p. 116.

THIS species of Honey-sucker was discovered by Mr. H. O. Forbes during his expedition to the Malay Archipelago, in the island of Bourou. Even in that island he met with it in one place only, Lake Wakolo, a locality to which few Europeans have ever penetrated before.

In general appearance it resembles *M. cruentata* of Meyer (figured in this work) more than any other species of *Myzomela*, but it is quite distinct specifically. The tint of the red colouring is quite different, and the black wings with their olive margins to the quills, as well as the absence of red on the tail, are good specific characters.

Adult male. General colour above scarlet vermilion; the feathers of the back and scapulars black at base, with broad ends of scarlet vermilion; rump and upper tail-coverts like the back; lesser and median wing-coverts black, the innermost of the latter tipped with scarlet; greater coverts and quills blackish, edged narrowly with olive-yellowish; bastard wing and primary-coverts blackish; tail-feathers blue-black; head like the back; lores black; sides of face, ear-coverts, cheeks, throat, and under surface of body bright scarlet, the feathers of the latter part with dusky bases; vent and under tail-coverts pale olive-yellowish, with dusky centres and somewhat tinged with red; under wing-coverts and axillaries white, with a yellowish tinge along the edge of the wing; quills dusky below, ashy white along the edge of the inner web, with dusky centres; axillaries and under wing-coverts white; the quills dusky below, whitish along the edge of the inner web; " bill black, the lower mandible yellow at base; legs and feet sooty grey; iris dark brown" (*H. O. Forbes*). Total length 3·6 inches, culmen 0·5, wing 2·05, tail 1·3, tarsus 0·65.

Mr. Forbes unfortunately did not procure an adult female, but he managed to get the immature male, which will doubtless give us some idea of the plumage of the old hen bird.

The young male is brown above, with pale streaks on the forehead; wing-coverts dusky brown, edged with paler or whity brown; quills dusky, edged with an olive fringe, the inner secondaries margined with pale brown, like the coverts; tail-feathers dusky brown, with olive-brown edges; lores brown, tinged with red; ear-coverts and cheeks ochreous brown, with a reddish tinge; throat pale yellowish, with a distinct wash of red on the chin; lower throat and fore neck more ashy; abdomen yellowish white, the breast and sides of the body browner; under tail-coverts ruddy brown. Total length 3·7 inches, culmen 0·6, wing 2·1, tail 1·2, tarsus 0·6.

[R. B. S.]

MYZOMELA ERYTHRINA, *Ramsay.*

MYZOMELA ERYTHRINA, Ramsay.

New-Ireland Honey-sucker.

Myzomela erythrina, Ramsay, Proc. Linn. Soc. New South Wales, ii. p. 107 (1877).—Forbes, Proc. Zool. Soc. 1879, p. 270.—Sclater, Proc. Zool. Soc. 1879, pp. 449.—Salvad. Ann. Mus. Civic. Genov. xvi. p. 72 (1880).—Id. Orn. Papuasia, etc. ii. p. 297 (1881).

Myzomela coccinea, Ramsay, Proc. Linn. Soc. New South Wales, ii. p. 106 (1878).—Forbes, Proc. Zool. Soc. 1879, p. 270.—Salvad. Ann. Mus. Civic. Genov. xvi. p. 72 (1880).—Id. Orn. Papuasia, etc. ii. p. 296 (1881).

THE original specimen of this elegant little Honey-sucker came from New Ireland, where it was collected by Mr. Cockerell, and, according to our judgment, is a young bird, probably a male. In the succeeding year Mr. Ramsay described a second species from Duke-of-York Island as *M. coccinea*, and, from an examination of the type, we have little hesitation in saying that it is only the adult male of his *M. erythrina*. Indeed Mr. Ramsay tells us that this was his first impression, but that the collectors insisted that the birds in the two islands were always constant in their colouring. We think, however, that there has been some mistake in the matter.

M. coccinea is closely allied to *M. cruentata* of Meyer, from the Arfak Mountains, but, as Count Salvadori has pointed out, it differs in the red colouring, which is more crimson or rosy, and also in having the head darker red than the back. The bird from New Britain appears to us to belong to an undescribed species which we propose to call *Myzomela kleinschmidti*, after the late well-known collector who discovered it. It closely resembles *M. erythrina*, but differs in its dusky head and throat and in the colour of the under wing-coverts and axillaries; these are smoky brown, washed with crimson, whereas is *M. erythrina* they are pale earthy brown, with scarcely any tinge of red.

The following is a description of the types of *M. coccinea* and *M. erythrina* respectively:—

Adult male. General colour above crimson, rather glossy on the back; head rather more dusky crimson than the back, as also the lores, sides of face, ear-coverts, and throat, becoming brighter crimson again on the lower throat and remainder of under surface of body; under tail-coverts rosy; wing-coverts and quills dusky brown, narrowly edged with crimson, brighter on the margins of the quills; centre tail-feathers dull crimson, the rest brown, dull crimson along the outer web; under wing-coverts and axillaries pale earthy or ashy brown, with a very faint tinge of red; quills dusky below, ashy white along the edge of the inner web. Total length 4 inches, culmen 0·6, wing 2·25, tail 1·6, tarsus 0·55.

Young. Much duller in colour than the male, being obscure rosy on the upper surface, the feathers being brown washed with rosy; head a little brighter; wings and tail as in the adult, but edged with much duller crimson; sides of head like the crown; throat brighter crimson; the remainder of the under surface earthy brown, washed with rosy, deepening into the latter colour on the under tail-coverts. Total length 3·4 inches, culmen 0·55, wing 1·9, tail 1·4, tarsus 0·55.

The figures in the Plate represent the two birds lent to me by Mr. Ramsay, the adult being the type of his *M. coccinea*, and the young being the type of his *M. erythrina*. They are represented of the size of life.

[R. B. S.]

MYZOMELA MELANOCEPHALA, Ramsay.

MYZOMELA MELANOCEPHALA.

Black-headed Honey-sucker.

Cinnyris (?) *dubia*, Ramsay (nec Bechst.), Proc. Linn. Soc. New South Wales, iv. p. 83 (1879).—Reichen. J. f. O. 1879, p. 430.—Salvad. Ibis, 1880, p. 129.
Cinnyris melanocephalus, Ramsay, Nature, 1879, p. 125.—Reichen. *t. c.* p. 101.
Hermotimia melanocephala, Layard, Ibis, 1880, p. 306.
Cyrtostomus melanocephalus, Salvad. Ann. Mus. Civic. Genov. xvi. p. 66 (1880).—Id. Orn. Papuasia, etc. ii. p. 269 (1881).

So peculiar is the coloration of this bird that it is scarcely surprising that some difficulty should have been experienced in finding its natural position. Mr. Ramsay doubted whether it was a true Sun-bird and called it *Cinnyris? dubia*, and he afterwards changed the specific name to *melanocephala*. He also expressed an opinion that the species might belong to the family *Meliphagidæ*; but Count Salvadori considers it to be a true Sun-bird and has placed it in the subgenus *Cyrtostomus*.

On receiving the type specimen from Mr. Ramsay, we showed it to Captain Shelley, who has written such an excellent 'Monograph' of the *Nectariniidæ*, and, after careful consultation, we both agreed that it is in all probability a Meliphagine bird. We say this with all caution, as the question is a difficult one; but we notice that in Sun-birds the nostril is rounded off anteriorly, whereas in the *Meliphagidæ* the anterior edge of the nostril vanishes gradually into the upper mandible, and this character is apparently accompanied by the brush tongue. We shall therefore expect to find that the present species has the last-named peculiarity.

In plumage this bird is not unlike some African Sun-birds, and again resembles some of the Spider-hunters (*Arachnothera*). In the genus *Myzomela* it is singularly out of place as regards it colouring, but appears to belong structurally to this genus.

The following description is taken from the typical specimen, which came from Savo in the Solomon Archipelago, and has been lent to us by Mr. E. P. Ramsay:—

Adult male. General colour above olive-yellow, with a slight appearance of brighter yellow on the mantle; upper tail-coverts dusky brown, edged with the same olive-yellow as the back; wing-coverts and quills dusky brown, narrowly edged with olive-yellow, brighter on the primaries; bastard wing and primary-coverts uniform blackish brown; tail-feathers black, with narrow margins of olive-yellow; crown of head as far as the occiput glossy black; the nape and the sides of the hinder crown olive-yellow, but duller and rather greener than the back; lores, feathers above and behind the eye, cheeks, ear-coverts, sides of face, and throat black like the head, with a slight metallic gloss; remainder of under surface of body dull olive-yellowish, clearer olive-yellow on the fore neck, chest, and sides of neck; breast-feathers mixed with ashy, the bases being of this colour; under tail-coverts ashy brown, edged with olive-yellow; axillary tufts pale olive-yellow, with white bases; under wing-coverts white, those near the edge of the wing blackish, edged with olive-yellow; quills dusky blackish below, white along the edge of the inner web. Total length 4·5 inches, culmen 0·85, wing 2·6, tail 1·9, tarsus 0·7.

Another specimen, apparently younger, has a pale reddish tint on the lower back and rump, as well as on the margin of the wing-coverts, the thighs and under tail-coverts being likewise washed with this colour; the black on the head and throat is also less clearly defined. Total length 4·2 inches, wing 2·5, tail 1·65, tarsus 0·7.

The figures in the Plate are drawn of the size of life, and represent the adult male, lent to me by Mr. Ramsay, in two positions.
[R. B. S.]

MYZOMELA ANNABELLÆ, Sclater.

MYZOMELA ANNABELLÆ, Sclater.
Mrs. Forbes's Honey-sucker.

Myzomela annabellæ, Sclater, Proc. Zool. Soc. 1883, p. 56.

This little *Myzomela* belongs to a small section of the genus which contains *M. boiei* from Banda and *M. adolphinæ* from the Arfak Mountains in North-western New Guinea. The former, however, differs in having the whole of the back scarlet, while according to Count Salvadori's description of the Arfak species, which we have never seen, the back is dusky greyish with a slight olive tint and the rump and upper tail-coverts scarlet, whereas in *M. annabellæ* the mantle is blue-black and the lower back is scarlet as well as the rump and upper tail-coverts.

The single specimen yet known was procured by Mr. Forbes in Loetoe while it was frequenting the cocoa-nut trees from which the natives had been collecting their "Tuak," or palm-wine.

This pretty little species has been named by Mr. Sclater after Mrs. Forbes; and we have great pleasure in figuring a bird bearing the name of this brave lady, who shared the perils of Mr. Forbes's travels in the Moluccas and even accompanied him to Timor Laut, where, in addition to the risks from the climate, their collecting had to be done in a circumscribed area, with the constant dread of an attack from the neighbouring villages.

Adult male. General colour above blue-black, the centre of the back, rump, and upper tail-coverts bright scarlet; wing-coverts blue-black, greater coverts and quills black, with a narrow margin of yellowish olive; tail-feathers blue-black, narrowly fringed with white along the edge of the inner web; crown of head and hind neck scarlet; lores black; sides of face, ear-coverts, cheeks, and throat bright scarlet, succeeded by a black patch across the fore neck; remainder of under surface of the body pale yellowish olive, the flanks mixed with ashy, the bases being of this colour; thighs pale yellowish olive; under tail-coverts white, washed with pale yellowish olive and having faint dusky centres; under wing-coverts and axillaries white, washed with yellow on the edge of the wing; quills blackish, white along the edge of the inner web; " bill black; legs and feet dirty green; iris dark brown " (*H. O. Forbes*). Total length 3·65 inches, culmen 0·55, wing 2·0, tail 1·3, tarsus 0·6.

The figures in the Plate represent the typical specimen, of the natural size, in three positions.

[R. B. S.]

MYZOMELA SCLATERI, Forbes.

MYZOMELA SCLATERI, Forbes.

Sclater's Honey-eater.

Myzomela sclateri, Forbes, Proc. Zool. Soc. 1879, p. 265, pl. 25, fig. 2.—Sclat. ibid. p. 448.—Reichenow et Schal. Journ. f. Orn. 1880, p. 198.—Salvad. Ann. Mus. Civ. Genova, xvi. p. 72, no. 8 (1880).—Id. Orn. della Papuasia &c. ii. p. 298 (1881).

THIS elegant little bird is another of the interesting discoveries made by that well-known collector the Rev. George Brown in New Britain and the neighbouring islands. It was discovered by him on the island of Palakürn, a small island in mid-channel between New Britain and Duke-of-York Island. It was first described by Mr. W. A. Forbes, in his excellent Synopsis of the Meliphagine genus *Myzomela* (published in the 'Proceedings of the Zoological Society' for 1879), where he describes the single male bird at that time known. "At first," writes Mr. Forbes, "I had some doubts as to this individual being adult; but now, from the absence of red feathers on any other part, and from the singularly bright and shining colour of those on the throat, I have little doubt that it has very nearly or quite attained its full plumage. *Myzomela sclateri* hardly admits of being compared with any other species of the group, the entirely dark upperside and the red being confined to the throat rendering it quite unlike any species yet known to us."

In May of the same year Dr. Sclater received a further collection from Mr. Brown, which contained a male and a female of this new *Myzomela*, and amply demonstrated its distinctness from any other known species of the genus. It belongs to the section of the genus where the species have the undersurface either whitish or greyish-olive, and, further, to the division of black-fronted species comprising *M. vulnerata*, *M. jugularis*, and *M. lafargii*; but from all of these it may be told by its blackish head.

The following description is translated from Mr. Forbes's essay above mentioned:—

Adult male. Upper surface of the body, with the wings and tail, dusky blackish, the head darker; the feathers of the lower back yellow at the tip; quills, wing-coverts, and tail-feathers externally edged with olive-yellow; throat bright crimson; undersurface of body greyish yellow, the throat rather duller; under wing-coverts and inner margin of quills white; bill black; feet dusky.

Total length about 4·5 inches, wing 3·65, tail 1·7, bill 0·6, tarsus 0·55.

The *adult female* is described by Mr. Sclater as follows:—Upper surface dark olive; crown of head, wings, and tail blackish, the latter externally edged with olive; underneath greyish yellow; the throat slightly tinged with crimson.

I have been indebted to Dr. Sclater for the loan of male and female birds, from which the life-sized figures in the Plate have been drawn, representing two males and a female.

MYZOMELA CINERACEA, Sclater.

MYZOMELA CINERACEA.

Ash-coloured Honey-eater.

Myzomela cineracea, Sclater, Proc. Zool. Soc. 1879, p. 448, pl. xxxvii. fig. 1 ; 1880, p. 65.—Salvad. Ann. Mus. Civ. Genova, xvi. p. 73, no. 15 (1880).—Id. Orn. della Papuasia &c. ii. p. 304 (1881).
———— *cinerea*, Reichenow et Schalow, Journ. f. Orn. 1880, p. 197.

This plain-coloured Honey-eater was met with by the Rev. G. Brown in New Britain, and belongs to the section of the genus without any bright colouring on the throat, and with the under surface of the body of the same colour as the upper surface. Its nearest ally is my *Myzomela obscura* from Northern Australia, which has also been found to inhabit South-eastern New Guinea. The latter bird, however, in addition to its dusky grey colour, has a slight vinous tinge on the head ; this is absent in *M. cineracea*, which is entirely ash-coloured above and below. Beyond the mere fact of recording the discovery of this species, I regret to say that I have absolutely nothing to add respecting the habits or mode of life. These are doubtless the same as those of the other species of the genus. As Mr. Forbes remarks, " in their habits the *Myzomelæ* seem to resemble the other smaller Honeysuckers, frequenting flowering shrubs and trees, not apparently so much for the sake of the nectar of the flowers as for the insects attracted thereby."

The following short description of the species is a translation of that given by Dr. Sclater in his paper on Mr. Brown's fourth collection :—Uniform dark ashy; under wing-coverts and inner margins of quills white ; bill and feet black. Total length 5·5 inches, wing 2·8, tail 2·4, tarsus 0·8, bill from gape 1·05.

I am again indebted to Dr. Sclater for the opportunity of figuring this species, which has since been added to the collection of the British Museum. The figures in the Plate are drawn from the typical example, and represent the bird in different attitudes, all the figures being of about the size of life.

MYZOMELA ROSENBERGI, *Schleg*.

MYZOMELA ROSENBERGI, Schlegel.

Von Rosenberg's Honey-eater.

Myzomela rosenbergii, Schlegel, Nederl. Tijdschr. Dierk. iv. p. 38 (1871).—Meyer, Sitz. k. Akad. Wien, lxix. pt. i. pp. 211, 212 (1874).—Rosenb. Reistocht. Geelvinksbai, p. 138, pl. xvi. fig. 2 (1875).—Salvad. Ann. Mus. Civic. Genov. vii. p. 776 (1875).—Salvad. & D'Albert. tom. cit. p. 825 (1875).

The genus *Myzomela* contains a number of species of small Honey-eaters, most of them attractive from the mixture of scarlet and black which forms their staple coloration, and spread over the Austro-Malayan, Australian, and Pacific subregions. Some few of them are dull-coloured; but most have a more or less brilliant plumage. Perhaps the most beautiful of all the *Myzomelæ* is the subject of the present article, which was discovered by Baron von Rosenberg in the North-western Province of New Guinea. It was subsequently obtained by Dr. A. B. Meyer during his sojourn in New Guinea in 1873, when it was procured by him near the village of Atam in the Arfak Mountains, at an elevation of about 3500 feet above the level of the sea; and since that time numerous specimens have been procured, by D'Albertis, Beccari, Bruijn, and other collectors, from the same district. It is probably to be found over the mountainous portions of the whole of New Guinea, as Signor D'Albertis procured two mutilated skins from the natives of Mount Epa in the south-eastern part of the island. With this exception, however, all the known specimens of *Myzomela rosenbergi* are from the Arfak district. Nothing has been recorded about the habits or food of this bird. Doubtless, like other *Myzomelæ*, it frequents flowering trees and shrubs for the sake of the nectar of the flowers and the insects attracted thereby. Nor are we yet quite certain as to its changes of plumage and sexual differences; for Dr. Meyer states that, of the specimens killed by himself, the adult males and females are similar in colour, and he treats the bird described and figured here as the female as a young bird which has not yet attained its full plumage. I have, however, followed Count Salvadori in considering that the sexes are dissimilar, as, owing to the large series at his disposal, consisting of forty specimens, his opinion on this point carries great weight.

I am indebted to Mr. W. A. Forbes for the loan of several fine specimens, from which my drawings were made. This gentleman has lately been preparing a monographic revision of the genus *Myzomela*, which he intends to publish. He has also kindly supplied me with the following diagnosis of *M. rosenbergi* :—

Adult male. The back, rump, and upper tail-coverts, together with the neck and breast, shining scarlet, the feathers greyish at their bases; the rest of the body, wings, and tail deep black, with a slight metallic lustre; the quills shining blackish grey beneath; bill black; legs horn-colour. Length about 3·7 inches, culmen 0·65, wing 2·5, tail 1·7, tarsus 0·55.

Young male. General colour reddish brown, mostly so on the back and rump, and paler beneath; the feathers blackish grey at the base, and the shaft lighter. This gives a streaked or flammulated appearance, most noticeable on the back, head, and chest. Wings and tail blackish brown, the greater and lesser wing-coverts tipped with pale chestnut; the quills margined externally with pale olive-yellow, becoming pale chestnut on the innermost secondaries; wings underneath grey, the quills with their inner margins white; bill blackish; feet horn-colour.

Adult female (sex ascertained by Dr. Beccari). Resembles the young male above described; but the forehead, breast, and rump are scarlet, the chin and throat blackish. This sex is also a trifle smaller than the male.

MYZOMELA CRUENTATA, *Meyer.*

MYZOMELA CRUENTATA, Meyer.

Red-tinted Honey-eater.

Myzomela cruentata, Meyer, Sitzungsber. d. k.-k. Acad. d. W. zu Wien, 16. Juli 1874, vol. lxx. p. 202.

Dr. Meyer has been good enough to send me the type of his *Myzomela cruentata*, a charming little bird which he procured in New Guinea. That this country should have contained an undescribed species of this well-defined genus is not surprising, considering that at least ten others frequent the adjacent Papuan Islands, Australia, New Caledonia, Samoa, &c. &c. *M. cruentata* is closely allied to the Meliphagidæ, or, more appropriately speaking, a part of these honey-feeding birds, a little genus which frequents the flowering trees of the forest, particularly the acacias and *Eucalypti*. Sprightly in all their actions, they display their fine colours to the greatest advantage, their prevailing tints often contrasting with the blossoms of the trees upon which they subsist. Their principal food is honey and insects, which their little brush tongues facilitate their gathering. That *Myzomela cruentata* is a very rare species is evidenced by only one specimen being found; and, for my own part, I have never seen a second.

As Dr. Meyer has favoured me with a short note respecting this new species, I have the pleasure of inserting it here. He says:—" *Myzomela cruentata* is distinguished from *M. sanguinolenta*, Gould, 'Birds of Australia,' vol. iv. pl. 63, by the red colour predominating everywhere; besides, the latter has the lores black, and the tail and wings do not possess any red tint, whilst in *cruentata* the parts when closed appear quite red, although somewhat less intense on the back and under surface; furthermore the whole of the underparts are intense red, while in *sanguinolenta* they are brownish yellow. The figure of *Certhia cardinalis*, Aud. and Vieill. Ois. Dor. ii. t. 58, shows the under surface to be quite red, and agrees in this respect with *M. cruentata*; but the wings, tail, and region of the eyes are decidedly deep black.

Male.—" Red, especially on the head, back, and uropygium. Primaries and secondaries dusky black, with red margins; the tertiaries, as well as the upper wing-coverts, deeply tinged with red; underside of wings grey; base of the inner webs whitish; middle tail-feathers rufous; under surface of the tail grey with a reddish tinge; outer webs of the rectrices broadly margined with the same colour. Bill, feet, and claws black. Total length 105 millims., wings 58, tail 42, bill from the front 14."

Hab. Arfak Mountains, New Guinea.

The figures in the accompanying Plate are of the natural size.

MYZOMELA NIGRITA, Gray.

MYZOMELA NIGRITA, Gray.

Black Honey-eater.

Myzomela nigrita, Gray, Proc. Zool. Soc. 1868, pp. 173, 190.—Id. Cat. Mamm. & Birds of New Guinea, pp. 23, 55 (1859).—Id. Proc. Zool. Soc. 1859, p. 155, 1861, p. 434.—Rosenb. Journ. für Orn. 1864, p. 122.—Gray, Hand-list of Birds, i. p. 154, no. 1988 (1869).—Salvad. Ann. Mus. Civic. Genov. viii. p. 400 (1875), xii. p. 334 (1878).—Id. Proc. Zool. Soc. 1878, p. 97.—W. A. Forbes, Proc. Zool. Soc. 1879, p. 265.—D'Albert. & Salvad. Ann. Mus. Civic. Genov. xiv. p. 75 (1879).—Ramsay, Proc. Linn. Soc. New South Wales, iv. p. 469 (1879).—Rosenb. Malay. Arch. p. 553 (1879).—Salvad. Ann. Mus. Civic. Genov. xvi. p. 71 (1880).—Id. Orn. Papuasia e delle Molucche, ii. p. 291 (1881).—Id. Rep. Voy. 'Challenger,' vol. ii. Birds, p. 81 (1881).—Sharpe, Journ. Linn. Soc., Zool. xvi. p. 437 (1883).—Gadow, Cat. Birds in Brit. Mus. ix. p. 139 (1884).

Nectarinia nigrita, Finsch, Neu-Guinea, p. 163 (1865).

Myzomela erythrocephala (nec Gould), Meyer, Sitz. k. Akad. Wien, lxx. p. 204 (1874).

Myzomela meyeri, Salvad. Ann. Mus. Civic. Genov. vii. p. 947 (1875).

Myzomela pluto, Salvad. MSS., Forbes, Proc. Zool. Soc. 1879, p. 266.

This peculiar little Honey-eater was discovered by Mr. A. R. Wallace in the Aru Islands, and was described by the late Mr. G. R. Gray, who omitted to mention in his description anything about the white under wing-coverts, an error corrected by Count Salvadori. It is replaced in the Solomon Islands by a closely allied form, *M. tristrami*, in the Admiralty group by *M. pammelæna*, and in New Ireland by *M. ramsayi*. The latter differs in its smoky black colour, and the first-named by its yellowish bill.

Considerable difference in size is exhibited in a series of this species, and the late Mr. W. A. Forbes, who wrote a most useful memoir on the genus *Myzomela*, has given a table of measurements to show that the birds from New Guinea and the islands in the Bay of Geelvink (Jobi and Miosnoum) are the same as the typical examples from the Aru Islands. Count Salvadori also adopts this view, but Dr. Gadow tries to prove that *M. nigrita* occurs in Western and Southern New Guinea and the Aru Islands, being replaced by *M. pammelæna* on the north coast of New Guinea with its islands, the Admiralty Islands, and the Solomon group also.

Besides the islands of Jobi and Miosnoum, the species has been found at Dorei by Mr. Wallace, and at Rubi in North-western New Guinea by Dr. Meyer. In South-eastern New Guinea Signor D'Albertis met with it on the Fly River, and it has been procured at East Cape by Mr. Huustein. More recently Mr. Forbes obtained specimens on the Astrolabe Mountains.

The following is the description of a pair of birds sent to us by the last-named gentleman:—

Adult male. General colour black, both above and below; quills and tail-feathers black; axillaries and under wing-coverts white, the edge of the wing black: "bill black; legs and feet lavender-blue; iris black" (*H. O. Forbes*). Total length 4 inches, culmen 0·55, wing 2·25, tail 1·55, tarsus 0·55.

Adult female. Different from the male. General colour above dusky brown washed with olive; wing-coverts and quills dusky blackish, edged with olive; upper tail-coverts like the back; tail-feathers blackish; crown of head like the back, the forehead dull crimson; lores and sides of face dull crimson, lighter on the cheeks and throat; ear-coverts dusky like the crown; under surface of body dusky, washed with olive, the centre of the abdomen olive whitish; sides of body and flanks dusky; thighs and under tail-coverts dusky; under wing-coverts and axillaries white; quills below dusky, white along the inner edge: "bill black; legs and feet dark yellow; iris blue-grey" (*H. O. Forbes*). Total length 3·7 inches, culmen 0·55, wing 2·0, tail 1·3, tarsus 0·55.

The Plate represents a male and female of the natural size, drawn from the pair of specimens collected by Mr. H. O. Forbes in the Astrolabe Mountains.

[R. B. S.]

THE
BIRDS OF NEW GUINEA

AND THE

ADJACENT PAPUAN ISLANDS,

INCLUDING MANY

NEW SPECIES RECENTLY DISCOVERED

IN

AUSTRALIA.

BY

JOHN GOULD, F.R.S.

COMPLETED AFTER THE AUTHOR'S DEATH

BY

R. BOWDLER SHARPE, F.L.S. &c.,

ZOOLOGICAL DEPARTMENT, BRITISH MUSEUM.

VOLUME IV.

LONDON:
HENRY SOTHERAN & CO., 36 PICCADILLY.
1875–1888.

[*All rights reserved.*]

PRINTED BY TAYLOR AND FRANCIS,
RED LION COURT, FLEET STREET.

CONTENTS.

VOLUME IV.

Plate			Part	Date
1.	Parus arfaki	New-Guinea Tit	IV.	1877.
2.	Dicæum pectorale	Müller's Flower-pecker	XXI.	1886.
3.	„ tristrami	Tristram's Flower-pecker	XVIII.	1884.
4.	„ æneum	Bronze-shaded Flower-pecker	XVII.	1884.
5.	„ fulgidum	Rosy-flanked Flower-pecker	XV.	1883.
6.	„ geelvinkianum	Geelvink-Bay Dicæum	IX.	1879.
7.	„ eximium	Brown Dicæum	VII.	1878.
8.	Œdistoma pygmæum	Pigmy Honey-eater	XX.	1885.
9.	Rhamphocharis crassirostris	Stout-billed Flower-pecker	XX.	1885.
10.	Pristorhamphus versteri	Verster's Flower-pecker	XIX.	1885.
11.	Urocharis longicauda	Long-tailed Flower-pecker	XIX.	1885.
12.	Macruropsar magnus	Long-tailed Glossy Starling	XIII.	1882.
13.	Calornis gularis	Purple-throated Glossy Starling	XX.	1885.
14.	„ feadensis	Fead-Island Starling	XXI.	1886.
15.	„ crassa	Slaty-grey Glossy Starling	XV.	1883.
16.	Melanopyrrhus anais	Orange-chested Starling	XXII.	1886.
17.	„ orientalis	Robertson's Starling	XXII.	1886.
18.	Lamprocorax minor	Lesser Brown-winged Starling	XXIII.	1887.
19.	Artamus maximus	Meyer's Wood-Swallow	VI.	1878.
20.	„ insignis	New-Ireland Wood-Swallow	VI.	1878.
21.	„ monachus	Hooded Wood-Swallow	VI.	1878.
22.	Munia grandis	Large Rufous-and-Black Finch	XIV.	1883.
23.	„ forbesi	Forbes's Munia	XII.	1881.
24.	Donacicola spectabilis	Orange-rumped Finch	XII.	1881.
25.	„ nigriceps	Black-cheeked Finch	XII.	1881.
26.	„ hunsteini	Hunstein's Weaver-Finch	XXIV.	1888.
27.	Pitta maxima	Great Pitta	II.	1876.
28.	Melampitta lugubris	Black Ground-Thrush	II.	1876.
29.	Pitta cyanonota	Blue-backed Pitta	XI.	1880.
30.	„ forsteni	Forsten's Pitta	X.	1879.
31.	„ concinna	Elegant Pitta	IX.	1879.
32.	„ cæruleitorques	Red-headed Pitta	VII.	1878.
33.	„ maforensis	Mafoor Island Pitta	VII.	1878.
34.	„ celebensis	Celebean Pitta	VII.	1878.
35.	„ rubrinucha	Red-naped Pitta	VII.	1878.
36.	„ novæ-guineæ	New-Guinea Pitta	IV.	1877.
37.	„ rosenbergii	Rosenberg's Pitta	IV.	1877.
38.	Collocalia terræ reginæ	Queensland Edible Swift	I.	1875.
39.	Ægotheles wallacii	Wallace's Goatsucker	XXI.	1886.
40.	Podargus ocellatus	Ocellated Goatsucker	XXIV.	1888.
41.	Microdynamis parva	Dwarf Koel	XXIV.	1888.
42.	Calliechthrus leucolophus	White-crowned Black Cuckoo	XXIII.	1887.
43.	Chalcites meyeri	Meyer's Golden Cuckoo	V.	1877.
44.	Nesocentor milo	Solomon-Island Lark-heeled Cuckoo	XXV.	1888.
45.	Ceyx solitaria	Solitary Kingfisher	III.	1876.
46.	„ gentiana	Gentian Kingfisher	XXII.	1886.
47.	Tanysiptera carolinæ	Blue-breasted Tanysiptera	III.	1876.
48.	„ nympha	Red-breasted Tanysiptera	VI.	1878.
49.	„ danaë	Crimson-and-Brown Kingfisher	XIII.	1882.
50.	„ nigriceps	Black-headed Tanysiptera	VII.	1878.
51.	„ microrhyncha	Port-Moresby Racket-tailed Kingfisher	XXIV.	1888.
52.	Sauromarptis gaudichaudi	Gaudichaud's Kingfisher	XI.	1880.
53.	Melidora macrorhina	Hook-billed Kingfisher	XIII.	1882.
54.	Halcyon tristrami	Tristram's Kingfisher	XIX.	1885.
55.	„ leucopygia	White-backed Kingfisher	XVIII.	1884.
56.	„ quadricolor	Four-coloured Kingfisher	XIII.	1882.
57.	„ nigrocyanea	Black-and-Blue Kingfisher	XI.	1880.
58.	„ stictolæma	Spotted-throated Kingfisher	XI.	1880.
59.	Clytoceyx rex	Spoon-billed Kingfisher	XII.	1881.

PARUS ARFAKI, Meyer.

PARUS ARFAKI, Meyer.

New-Guinea Tit.

Parus? arfaki, Meyer, Sitzungsber. der Isis, April 1875.—Mitth. a. d. k. zool. Mus. Dresden, i. p. 8, 1875.

THE great stronghold of the *Paridæ*, or family of Tits, is the temperate and northern region of the old world; from the boreal countries of Lapland and Siberia to the hot forests of India, its dependencies, and China, the numerous members are found; as we proceed southward from these countries the form gradually disappears, so that in Australia and New Zealand it is totally absent. It was, then, with considerable surprise that I received from Dr. Meyer an undoubted member of this family, collected, I believe, by himself in the northern part of New Guinea; and this unique specimen this gentleman has intrusted to my care for the purpose of figuring in the present work, accompanied by the following notes and description.

This new Tit reminds us in its general appearance of the crested Indian genus *Machlolophus*; but it differs from that form in the length of the first primary, in the rather lengthened pointed ear-feathers, and in the singular gold-brown coloration of the underparts. "Zoogeographically," says Dr. Meyer in a letter to me, "a species of *Parus* from New Guinea is very remarkable. *Parus cinereus*, Vieill., has been seen as far east as Flores (Wallace, Proc. Zool. Soc. 1863, p. 485); but from Australia, New Guinea, or the Moluccas no species of this genus has been known until now."

Dr. Meyer states that he believes that neither Signor d'Albertis nor other New-Guinea explorers met with this bird; neither does Salvadori mention it in his lists.

Forehead, upper and hinder parts of head, chin, throat, and chest black; rump olive green; cheeks and ear-patch bright yellow, the feathers of latter rather elongated and pointed; the whole of the underparts yellow, golden brown in the middle of the belly, merging into yellow on the flanks and under tail-coverts; on the three upper tertiary feathers an oval spot of yellow; bill black; feet light horn-colour, tarsi somewhat darker.

Total length $4\frac{1}{4}$ inches, bill $\frac{1}{2}$, tarsi $\frac{3}{4}$, wing $2\frac{3}{4}$, tail $2\frac{1}{4}$.

Hab. Arfak Mountains, New Guinea.

The figures in the accompanying Plate are of the natural size.

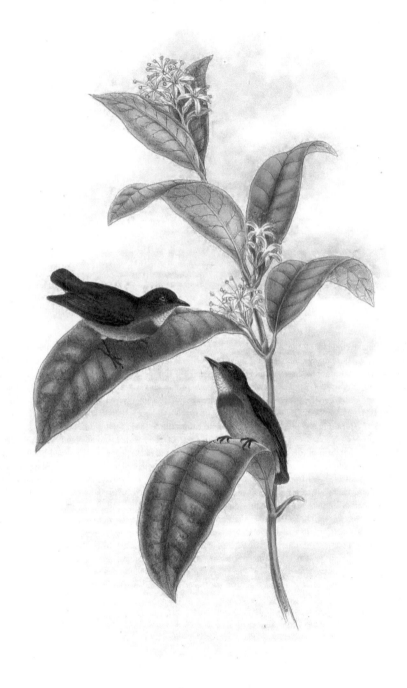

DICÆUM PECTORALE, M.&S.

DICÆUM PECTORALE, Müll. & Schl.
Müller's Flower-pecker.

Dicæum pectorale, Müll. & Schl. Verh. Natuurl. Geschied. Land- en Volkenk. p. 162, note (1839–44).—Hartl. Rev. Zool. 1846, p. 47.—Gray, Gen. B. i. p. 100 (1847).—Bp. Consp. Av. i. p. 403 (1850).—Sclater, Proc. Linn. Soc. ii. p. 157 (1858).—Gray, Proc. Zool. Soc. 1858, p. 190.—Id. Cat. Mamm. etc. New Guinea, pp. 22, 55 (1859).—Id. Proc. Zool. Soc. 1861, p. 434.—Rosenb. Nat. Tijdschr. Nederl. Ind. xxv. p. 237 (1863).—Id. J. f. O. 1864, p. 123.—Finsch, Neu-Guinea, p. 163 (1865).—Meyer, Sitz. k. Akad. Wien, lxx. p. 120 (1874).—Salvad. Ann. Mus. Civic. Genov. xvi. p. 67 (1880).—Id. Orn. Papuasia e delle Molucche, ii. p. 273 (1881).—Sharpe, Cat. Birds in Brit. Mus. x. p. 29 (1885).—Guillemard, Proc. Zool. Soc. 1885, p. 639.

Dicæum erythrothorax, pt., Gray, Hand-list Birds, i. p. 115, no. 1427 (1869).—Rosenb. Malay Arch. p. 553 (1879).

As most of the Flower-peckers in the Malayan Archipelago are confined to a single island, the present species must be considered to have rather a wide range, as it is not only found in North-western New Guinea, but also in the adjoining islands of Salwati, Mysol, Waigiou, and Batanta.

It belongs to the olive-backed section of the genus *Dicæum*, without any red on the rump or on the crown, and with the head of the same colour as the back. Its nearest ally is *D. æneum* of the Solomon Islands, also figured in this work; but *D. pectorale* differs from that species in having no grey on the breast below the red spot, and it also has the breast dull olive like the flanks.

The species was first met with in New Guinea at Lobo Bay by Solomon Müller, and the localities recorded by Count Salvadori in North-western New Guinea are numerous, as Dr. Meyer, Dr. Beccari, and Mr. Bruijn have all procured it during their travels in the Arfak Mountains and their vicinity. Mr. Wallace procured specimens in Salwati, Waigiou, and Mysol, and Mr. Bruijn in Batanta. Nothing has been recorded concerning its habits.

The following description is copied from the British Museum 'Catalogue of Birds,' and is taken from a specimen in the Leiden Museum, procured by the late Dr. Bernstein in the island of Waigiou :—

"*Adult male.* General colour above olive-green, a little more olive-yellow on the head, which is olive like the back; rump and upper tail-coverts also a little more olive-yellow, particularly the latter; wing-coverts and quills dusky, with a steel-green gloss and narrowly edged with olive; tail-feathers blue-black; lores, sides of face, ear-coverts, and cheeks olive like the crown, the hinder cheeks washed with ashy grey like the sides of the neck; throat whitish, with a tinge of olive on the chin; fore neck and chest with a large patch of orange-scarlet; sides of breast ashy, more olive on the flanks; centre of abdomen, thighs, and under tail-coverts pale yellow, the latter with dusky bases; axillaries and under wing-coverts white, the edge of the wing dusky washed with olive; quills dusky below, white along the edge of the inner web: 'bill brownish black; feet dark greyish brown; iris brown' (*Guillemard*). Total length 2·8 inches, culmen 0·45, wing 2·05, tail 1·05, tarsus 0·5."

The Plate gives an illustration of an adult male in two positions, the figures being drawn from an Arfak skin lent to us by Dr. H. Guillemard.

[R. B. S.]

DICÆUM TRISTRAMI, Sharpe.

DICÆUM TRISTRAMI, Sharpe.

Tristram's Flower-pecker.

Dicæum tristrami, Sharpe, Proc. Zool. Soc. 1883, p. 579.

THIS peculiarly coloured Flower-pecker was lent to us by Canon Tristram, and as we had every reason to believe it to be hitherto undescribed, we attached to it our friend's name in recognition of his many services to ornithology. The species was discovered by Lieut. Richards in Makira Harbour, in San Cristoval, one of the Solomon Islands.

It is not necessary to compare Tristram's Flower-pecker with any other members of the genus *Dicæum*, for its style of colouring is quite unique, and does not ally it with any of the species known up to the present time.

The following is a description of the type specimen, which is in Canon Tristram's collection:—

Adult male. General colour above chocolate-brown, the mantle slightly streaked with a few hoary whitish margins to the feathers; wing-coverts darker chocolate-brown than the back; bastard wing, primary-coverts, and quills blackish brown, the inner secondaries chocolate-brown like the back; upper tail-coverts and tail blackish brown, contrasting sharply with the back; head brown, but mottled with blackish-brown centres to the feathers, the plumes of the forehead and vertex margined with hoary whitish; a line of feathers above the eye and ear-coverts hoary white, the latter slightly mottled, with brown bases; lores, eyelid, fore part of cheeks, and base of chin blackish; hinder cheeks, throat, and fore neck hoary white, with brown bases to the feathers; sides of neck like the back; centre of breast, abdomen, and under tail-coverts pure white, the sides of the body ashy; sides of upper breast brown, with hoary whitish edges to the feathers; axillaries and under wing-coverts white; quills dusky below, ashy whitish along the edge of the inner web; " bill black; feet black; iris grey " (*Richards*). Total length 3·5 inches, culmen 0·45, wing 2·3, tail 1·15, tarsus 0·55. (*Mus. H. B. Tristram.*)

The Plate represents the individual above described, of the natural size, in two positions.

[R. B. S.]

DICÆUM AENEUM.

DICÆUM ÆNEUM, Jacq. et Pucher.

Bronze-shaded Flower-pecker.

Dicée bronzée, Hombr. et Jacq. Voy. Pole Sud, Atlas, pl. 22. fig. 4 (1845).
Dicæum, sp., Gray, Gen. B. ii. p. 100 (1847).
Dicæum æneum, Jacq. et Pucher. Voy. Pole Sud, texte, iii. p. 97 (1853).—Hartl. J. f. O. 1854, pp. 165, 168.— Gray, Cat. Birds Trop. Isl. Pacific Ocean, p. 10 (1859).—Id. Hand-l. B. i. p. 115, no. 1434 (1869).— Sclater, Proc. Zool. Soc. 1869, p. 118.—Salvad. Ann. Mus. Civic. Genov. xvi. p. 68 (1880).—Id. Orn. Papuasia, etc. ii. p. 281 (1881), iii. App. p. 540 (1882).
Microchelidon ænea, Reichenb. Handb. Spec. Orn. Scans. p. 244, Taf. 558. fig. 3797 (1853).
Dicæum erythrothorax, Ramsay, Proc. Linn. Soc. N. S. W. iv. p. 77 (1879, nec Less.).—Salvad. Ibis, 1880, p. 129.

The family *Diceidæ*, or Flower-peckers, is distributed over the greater part of the Indian and Australian regions, having a few representatives in the forests of Western Africa. They are particularly abundant in the Austro-Malayan subregion, nearly every island of the Moluccas and Papuasia having its own peculiar representative species of *Dicæum*.

The present bird is a native of Ugi, one of the Solomon Islands; and as most of the species inhabiting this Archipelago are nearly allied to others from New Guinea, it is not surprising to find that *Dicæum æneum* is a representative form of the scarlet-chested Flower-peckers found in the latter country and the neighbouring Moluccan islands. It resembles *Dicæum pectorale* of New Guinea, but is easily distinguished by the bronzy-greenish colour of the upper parts, the lighter blue-grey colour of the sides of the face, extending on to the sides of the fore neck and chest, and forming a large patch on the breast below the scarlet præpectoral spot; the flanks and sides of the body are also olive-yellow.

The following is a description of the adult male and female :—

Adult male. General colour above glossy oil-green, with a bronzy gloss; head like the back; sides of rump with a slight wash of olive-yellow; upper tail-coverts oily green; lesser and median wing-coverts glossy oil-green like the back; the greater coverts, bastard wing, primary-coverts, and quills blackish, externally glossed with oily green; tail-feathers greenish black; lores dull ashy grey; cheeks, ear-coverts, and sides of neck clear ashy grey, descending down the sides of the fore neck and occupying the whole of the breast; throat white, the sides of it ashy grey, blacker at the base of the malar line; a large triangular patch of scarlet occupying the whole of the fore neck; sides of breast and flanks bright olive-yellow; abdomen yellowish white; thighs ashy grey, white on their inner aspects; under tail-coverts white washed with yellow and having dusky bases; axillaries and under wing-coverts white, the edge of the wings blackish; quills blackish below, ashy white along the edge of the inner web. Total length 3·1 inches, culmen 0·45, wing 2·0, tail 1·0, tarsus 0·5.

Adult female. Differs from the male in wanting the scarlet patch on the fore neck, and in not having the patch of ashy grey on the breast; the throat and breast are yellowish white, with a few dusky margins to the lateral feathers of the breast; otherwise the under surface of the body is exactly like that of the male, the sides of the neck being ashy grey, descending on to the sides of the breast, and the rest of the sides of the body being bright olive-yellow; the upper surface resembles that of the male, being entirely oily or bronzy green; but there is a slight loral streak of white, and the base of the lower mandible is pale, characters not seen in the adult male. Total length 3·25 inches, culmen 0·45, wing 1·95, tail 0·95, tarsus 0·5.

For the loan of the two specimens described above, we have been indebted to the kindness of Mr. E. P. Ramsay. The same birds are figured in the Plate, of the size of life.

[R. B. S.]

DICÆUM FULGIDUM, Sclater.

DICÆUM FULGIDUM, *Sclater*.

Rosy-flanked Flower-pecker.

Dicæum fulgidum, Sclater, Proc. Zool. Soc. 1883, p. 56.

Two specimens of this apparently new *Dicæum* were contained in Mr. Forbes's first collection from Timor laut. It differs from *D. ignicolle* from the Aru Islands, with which we have compared it, in having the rosy tint spread over the whole of the under surface, so that there is no olive on the flanks, but the lower abdomen and under tail-coverts are of a pale fawn-brown colour. It must be also very closely allied to *D. keiense* of Salvadori, from the Kei Islands; but this is described as having the abdomen olivaceous, yellow in the middle. Having regard to the way in which the different Molucca Islands possess their own peculiar species of *Dicæum*, it is not unreasonable to suppose that the birds coming from Timor laut will also be found to be specifically distinct from the Flower-peckers of the neighbouring islands.

The following description is from the type specimen, which was kindly lent to me by Dr. Sclater.

Adult male. General colour above dark purplish blue; wing-coverts like the back; quills black, externally, slightly glossed with purplish blue, a little more distinct on the innermost secondaries; tail-feathers purplish blue like the back, but slightly duller; lores black; sides of face, cheeks, and ear-coverts black, with a slight gloss of purplish blue; a very narrow line of black bordering the rami of the lower jaw; under surface of the body pale scarlet, richer on the fore neck and breast, and extending in a rosy tinge over the sides of the body, the lower flanks and vent being of a light fawn-brown; the chin whitish, and the centre of the body also mixed with white, the bases of the feathers being of this colour; in the centre of the breast a faintly indicated blackish spot; thighs black; under tail-coverts pale rose-colour; under wing-coverts and axillaries white; edge of wing and adjacent coverts black; quills blackish below, narrowly edged with white along the inner web; "bill, legs, and feet black; iris black" (*H. O. Forbes*). Total length 3·5 inches, culmen 0·4, wing 2·15, tail 1·25, tarsus 0·55.

The figures in the Plate are drawn from the typical example above mentioned.

[R. B. S.]

DICÆUM GEELVINKIANUM, *Meyer*.

DICÆUM GEELVINKIANUM, Meyer.

Geelvink-Bay Dicæum.

Dicæum geelvinkianum, Meyer, Sitzb. k. Akad. Wiss. Wien, lxx. p. 120.
Dicæum jobiense, Salvad. Ann. Mus. Civic. Genov. vii. p. 945 (1875).

My friend Dr. A. B. Meyer kindly submitted to me the types of the species described by him as *Dicæum geelvinkianum,* and I have figured them on the accompanying Plate. The species was originally described by him from the islands of Jobi, Misori, and Mafoor, in Geelvink Bay; and the following are his remarks on the subject:—

"From the above-mentioned islands come some forms of a *Dicæum* which vary slightly among themselves according to locality, but which I nevertheless unite for the present under the name of *D. geelvinkianum,* as they all three differ from *D. pectorale,* Müller and Schlegel, of the mainland, in the red colour of the forehead, crown, and rump, though they otherwise entirely agree with this species, as well as in size.

"Examples from Mafoor, of which three males are before me, collected by myself in March 1873, and all agreeing perfectly together, have an olive-coloured upper surface, brownish red forehead, crown, and rump, while the breast-spot is large and fiery-red.

"Examples from Misori (of which I have three males, agreeing perfectly together, shot in March 1873) have the upper surface more grey, the head coloured as in specimens from Mafoor, but the rump is somewhat more brilliant red; the breast-spot is small, and of a darker red.

"Lastly, on the island of Jobi, where I obtained a male and female in April 1873, the upper surface is somewhat shining metallic bluish, the crown, forehead, and rump are beautiful deep red, and the breast-spot is of moderate size and of the same colour as the head and rump. The female is uniform greenish grey above, clearer grey below, with a greenish cast, shading into yellowish white on the belly.

"The material at my disposal being all obtained at one season, I am not in a position to decide whether these differences are constant according to locality; nevertheless I consider that they are. Should this turn out so in the course of time, it will prove a not uninteresting example of the different variations of one and the same type in distinct isolated districts such as islands. If one does not admit the isolation as a cause of the difference, one is compelled to allow that the reasons for this variation are utterly unknown."

I have quoted Dr. Meyer's remarks thus fully, as the nomenclature of the species is somewhat involved, by reason of Count Salvadori having received specimens from all three of the islands visited by Dr. Meyer, and having at once separated them as distinct species, giving the names of *Dicæum maforense, D. misoriense,* and *D. jobiense,* and saying, truly enough, that Dr. Meyer has not indicated which of the three species he would retain as *D. geelvinkianum* in the event of the other two proving distinct.

As will be seen from the accompanying Plate, it is the Jobi bird which Dr. Meyer has sent me as his *Dicæum geelvinkianum;* and as such I have figured it, and have added *Dicæum jobiense* of Salvadori as a synonym. I have little doubt that the view of the latter ornithologist is the correct one, and that there are three distinct species confounded under the heading of *D. geelvinkianum* by Dr. Meyer. The figures in the Plate represent a pair of birds, drawn of the natural size, from specimens in the Dresden Museum lent me by Dr. Meyer. I adopt the name of *geelvinkianum* here, as I am forced to do so by the fact of my Plate having been thus lettered, and the copies printed, before I received notice of Count Salvadori's nomenclature for the species.

DICEIUM EXIMIUM, Sclater.

DICÆUM EXIMIUM, Sclater.

Brown Dicæum.

Dicæum eximium, Sclater, P. Z. S. 1877, p. 102, pl. xiv. fig. 2.

THE home of this beautiful little bird, justly named *eximium*, is New Ireland, where it was procured by Mr. George Brown in March 1876. In describing the bird, Dr. Sclater remarked that it was different from any species known to him; and in looking over my series of Dicæidæ, I cannot find any single one which approaches it in style of coloration. Indeed it is seldom that such a very distinct species comes to our notice in these days; and we may fairly suppose that the group of islands where Mr. Brown worked will yield many further discoveries in natural science as they become better explored.

I have copied below the original description given by Dr. Sclater; and I can scarcely agree with him that the *Dicæum æneum* of Messrs. Jacquimot and Pucheran (Voy. Pôle Sud, Zool. iii. p. 97) is likely to turn out to be the female of the present bird. The last-named species is said to have been procured in the Solomon Islands; and perhaps a translation of the description may be useful for comparison. It is as follows:—

Grey, with bronzy reflections on the head, back, and upper wing-coverts; throat white, somewhat tinged with yellow, this yellow colour being a little more pronounced on the thorax; middle of abdomen white, the sides yellowish olive; quills blackish above, with the inner webs edged with clear greyish; under wing-coverts white, and the wing below ashy grey; the grey colour occupies also the sides of the neck, and impinges somewhat on the median line; quills blue-black, duller below; their coverts in the latter sense are yellow, somewhat inclining to whitish; the iris is yellow; bill and feet blue-black, with a little yellow at the base of the lower mandible.

The following is Dr. Sclater's original diagnosis of *D. eximium*:—Above dark bronzy green, the crown and sides of the head browner; rump crimson; under surface of body white, with a narrow pectoral band of crimson; sides of body and middle of belly dusky slate-colour; the flanks bronzy, as on the back; under wing-coverts white. Total length 3·3 inches, wing 2·0, tail 1·0.

The figure in the Plate is taken from the type specimen lent to me by Dr. Sclater, to whom, and to the Marquis of Tweeddale, I am under great obligations for allowing me to figure so many of the fine novelties discovered by Mr. George Brown.

ŒDISTOMA PYGMÆUM, Salvad.

ŒDISTOMA PYGMÆUM, Salvad.

Pigmy Honey-eater.

Œdistoma pygmæum, Salvad. Ann. Mus. Civic. Genov. vii. p. 952 (1875).—Id. op. cit. xvi. p. 74 (1880).—Id. Orn. Papuasia e delle Molucche, ii. p. 312 (1881).—Gadow, Cat. Birds in Brit. Mus. ix. p. 293 (1884).

For an opportunity of figuring this curious little bird we are indebted to the kindness of the Marquis Doria, who was so good as to send us one of the typical specimens from the Genoa Museum. Its small size, slender bill, and sober colouring recall the species of *Melilestes*, and, as in the latter genus, *Œdistoma* has very pronounced silky tufts on the flanks. Count Salvadori considers *Œdistoma* to be very nearly allied to *Melilestes*, but to be distinguished principally by the form of the bill; and this will be easily discernible on a comparison of the Plates given in the present work. We agree with him that the genera *Melilestes* and *Œdistoma*, along with *Glycichæra*, form a natural section of the *Meliphagidæ*, though they show a certain connection with the Sun-birds of the genera *Arachnothera* and *Anthothreptes*.

This species, as far as is known, is confined to the Arfak Mountains in North-western New Guinea.

The following description is taken from the typical specimen belonging to the Genoa Museum:—

Adult female. General colour above dark olive-green, slightly greyer on the head and more distinctly yellowish towards the rump, the feathers of which are long and silky and completely hide the upper tail-coverts; wing-coverts below the back, bastard-wing, primary-coverts, and quills ashy brown, edged with ashy, the secondaries with olive-yellow, lighter than the back; centre tail-feathers olive-green, the remainder dusky edged with olive-greenish; in front of the eyes a dusky spot; eyelid whitish; sides of face and ear-coverts pale ashy, whiter below the eye; throat whitish, with a slight greenish tinge; remainder of under surface pale sulphur-yellow, olive on the fore neck and chest, yellowish white on the centre of the breast and abdomen; axillaries pale sulphur-yellow like the sides of the body; quills dusky below, ashy along the inner web: "bill dusky, the base of the lower mandible dusky below; feet lead-colour" (*Beccari*). Total length 2·8 inches, culmen 0·5, wing 1·9, tail 0·8, tarsus 0·5.

The male, according to Count Salvadori, does not differ from the female in colour.

The figures in the Plate represent an adult bird in two positions, drawn from the typical specimen lent to us by the Marquis Doria.

[R. B. S.]

RHAMPHOCHARIS CRASSIROSTRIS, Salvad.

RHAMPHOCHARIS CRASSIROSTRIS, Salvad.

Stout-billed Flower-pecker.

Rhamphocharis crassirostris, Salvad. Ann. Mus. Civic. Genov. vii. p. 943 (1875).—Id. op. cit. xvi. p. 69 (1880).—Id. Orn. Papuasia e delle Molucche, ii. p. 288 (1881).—Sharpe, Cat. Birds in Brit. Mus. x. p. 84 (1885).

This is at once one of the most curious and at the same time one of the most distinct of the Meliphagine Flower-peckers, by which term we mean those aberrant forms which inhabit New Guinea, and which connect the family *Dicæidæ* with the Honey-eaters or *Meliphagidæ*. It is difficult to say whether such genera as *Rhamphocharis* and its allies should be placed with the last-named family or at the end of the *Dicæidæ*, near the genera *Melanocharis* and *Pristorhamphus*. This is the view adopted by us in the 'Catalogue of Birds;' but as one of the principal characters of the *Dicæidæ* is the absence of a bastard-primary, and *Rhamphocharis* and the allied genera have a very distinct one, it is quite possible that in future re-arrangements of the family they may be left out of the *Dicæidæ* and joined to the *Meliphagidæ*. In any case the genus *Rhamphocharis* must be placed near to *Pristorhamphus*, from which it differs in the form of bill and also in having the wing longer than the tail.

The present species was discovered by Dr. Beccari in the Arfak Mountains in North-western New Guinea; and the following descriptions, copied from the 'Catalogue of Birds,' are taken from the typical specimens, kindly lent to us by the Marquis Doria:—

"*Adult male* (type of species). General colour above olive-green, glossed with oil-green; wing-coverts like the back; greater wing-coverts, bastard-wing, primary-coverts, and quills dusky brown, edged with olive-green, brighter on the quills; upper tail-coverts dusky, washed with olive-green; tail-feathers blackish, edged with olive-green; lores olive dusky; edge of eyelid white; ear-coverts and cheeks light ashy, with the upper parts of the ear-coverts dusky olive; entire under surface of body pearly grey, slightly washed with pale yellow; thighs and under tail-coverts a little darker ashy; axillaries pale yellow; wings white, with a pale yellow wash; quills dusky, ashy grey along the edge of the inner web. Total length 4·6 inches, culmen 0·6, wing 2·6, tail 1·75, tarsus 0·7.

"*Adult female.* General colour above olive-brown, rather greener on the lower back, rump, and upper tail-coverts, the head and back being spotted with minute tips of yellowish white on the feathers; the scapulars olive-brown like the back, similarly tipped with tiny white spots; wing-coverts like the back, with white spots at the end; the greater coverts, bastard-wing, primary-coverts, and quills dusky brown, edged with olive-green, the inner greater coverts with a whitish spot at the ends, the margins of the primaries yellower; tail-feathers blackish, edged with olive-green, with a tiny white spot at the end of the inner web, increasing gradually in extent and forming a well-marked spot on the outer tail-feathers; an indistinct eyebrow of pale olive-brown, lores dusky; eyelid whity brown; ear-coverts dull olive-brown, streaked with dull white; cheeks brown, mottled with white spots; under surface of body yellowish white, mottled all over with dusky-brown centres to the feathers, smaller on the throat and nearly obsolete on the abdomen; thighs ashy tinged with yellow; under tail-coverts yellowish white with dusky centres, the feathers mottled like the breast; axillaries pale yellow; under wing-coverts white slightly washed with yellow, and mottled with dark-brown bases to the feathers; quills dusky below, greyish white along the edge of the inner web. Total length 4·9 inches, culmen 0·6, wing 2·85, tail 1·85, tarsus 0·75."

The Plate represents a male and female of the natural size, and the figures have been drawn from the same pair of birds described above, lent to us by the Marquis Doria.

[R. B. S.]

PRISTORHAMPHUS VERSTERI, *Finsch.*

PRISTORHAMPHUS VERSTERI, *Finsch*.

Verster's Flower-pecker.

Pristorhamphus versteri, Finsch, Proc. Zool. Soc. 1875, p. 642.—Salvad. Ann. Mus. Civic. Genov. vii. p. 940 (1875).—Id. op. cit. xvi. p. 69 (1880).—Id. Orn. Papuasia e delle Molucche, ii. p. 286 (1881).—Sharpe, Cat. Birds in Brit. Mus. x. p. 82 (1885).

This elegant little bird was first described by Dr. Otto Finsch from two female specimens in the Leiden Museum, to which institution it was sent from the Arfak Mountains. Dr. Beccari has likewise procured it in this locality, and M. Laglaise at Karons, but it has not been found up to the present time in any other part of New Guinea.

One of the chief characters of the genus *Pristorhamphus* is its abnormally long tail coupled with a peculiar elongation of the upper and under tail-coverts, while the conspicuous white marking on the tail-feathers renders the male bird easily recognizable; the female is more like the same sex in the genus *Melanocharis*.

The following descriptions are copied from the British Museum 'Catalogue of Birds':—

"*Adult male.* General colour above velvety black with a slight steel-green lustre, which is more distinct on the scapulars; lesser and median wing-coverts glossy steel-green with velvety-black bases; bastard-wing, primary-coverts, and quills ashy blackish, externally glossy steel-green; upper tail-coverts very long, glossy steel-blue; tail-feathers black, waved with dusky under certain lights, glossed externally with steel-blue; the basal half of all the feathers white, only seen when the tail is spread; head glossy steel-green; lores, feathers round the eye, cheeks, ear-coverts, and sides of face velvety black; a little spot of silky white on the upper edge of the eyelid; under surface of body pale slaty blue; thighs and under tail-coverts black, the latter washed with steel-blue; a large tuft of silky white plumes on the side of the body; axillaries and under wing-coverts white; quills blackish, white along the inner web; bill and feet black. Total length 5·8 inches, culmen 0·55, wing 2·45, tail 3, tarsus 0·95.

"*Adult female.* General colour above dull olive-green, slightly yellower on the rump and upper tail-coverts; wing-coverts like the back; bastard-wing uniform dusky brown; primary-coverts and quills dusky brown, narrowly margined with yellowish olive, the secondaries more broadly edged with olive-green like the greater coverts; two centre tail-feathers olive-green, blackish towards the base; remainder of tail-feathers blue-black, browner at the ends, edged with olive-green, the three outer ones with a white spot about the middle of the feather; head like the back; lores, feathers round the eye, cheeks, and ear-coverts dingy olive, with pale shaft-lines on the ear-coverts; under surface of body ashy olive washed with pale yellow, with narrow edges of the latter colour to most of the feathers; thighs dull ashy; under tail-coverts dingy olive; on the flanks a tuft of silky white; axillaries pale yellow; under wing-coverts white, washed with pale yellow; quills dusky below, greyish white along the edge of the inner web. Total length 5·6 inches, culmen 0·55, wing 2·65, tail 2·6, tarsus 0·85."

The figures in the Plate are drawn from a pair of birds lent to us by the Marquis Doria, and represent a male and female of the natural size.

[R. B. S.]

UROCHARIS LONGICAUDA, Salvad.

UROCHARIS LONGICAUDA.

Long-tailed Flower-pecker.

Melanocharis longicauda, Salvad. Ann. Mus. Civic. Genov. vii. p. 942 (1875).—Id. op. cit. x. 144 (1877).—Id. op. cit. xii. p. 333 (1878).
Urocharis longicauda, Salvad. Ann. Mus. Civic. Genov. xvi. p. 69 (1880).—Id. Orn. Papuasia e delle Molucche, ii. p. 286 (1881).—Sharpe, Cat. Birds in Brit. Mus. x. p. 79 (1885).

As Count Salvadori has pointed out, the present species is intermediate in character between *Melanocharis* and *Pristorhamphus*, and he therefore very rightly made it the type of a new genus. From *Pristorhamphus* it is distinguished by the notch on the second primary, and from *Melanocharis* it differs in the form of the bill, which is broader than it is high at the nostrils and has a sharp ridge to the culmen.

At present only three specimens of this bird have been sent from North-western New Guinea, and it would appear to be confined to the Arfak Mountains, where it was met with by D'Albertis and Beccari as well as by Mr. Bruijn's hunters. A full account of these specimens is given by Count Salvadori in his great work on the birds of New Guinea, and we here reproduce the descriptions taken from the typical examples and published in the British Museum 'Catalogue of Birds.'

"*Adult male* (type of species). General colour above glossy steel-black, the feathers with ashy-grey bases, which show rather distinctly on the rump; lesser and median wing-coverts like the back; greater coverts, bastard-wing, primary-coverts, and quills black, edged with steel-black like the back, less distinctly on the primaries; tail-feathers black, slightly glossed on the margins with steel-blue, the outer tail-feather white for three quarters of the outer web; crown of head like the back; eyelid also black; lores and sides of face dusky olive, blackish along the hinder margin of the ear-coverts; throat and chest ashy olive, clearer olive-yellowish on the breast and abdomen, the feathers in most cases margined with pale yellow; thighs dusky olive; under tail-coverts dusky olive, with pale yellowish margins; axillaries pale yellow; under wing-coverts silky white, washed with pale yellow; quills blackish below, ashy white along the edge of the inner web. Total length 4·8 inches, culmen 0·5, wing 2·5, tail 2, tarsus 0·7.

"*Adult female.* Different from the male. General colour above olive-green, with an oily green gloss; wing-coverts like the back; bastard-wing, primary-coverts, and primaries blackish, obsoletely edged with olive-green like the back, the secondaries with broader olive-green margins like the greater wing-coverts; tail as in the male; head olive-green like the back, and with the same oily green gloss; eyelid and sides of face ashy olive, the lores and cheeks ashy grey washed with olive; under surface of body as in the male. Total length 5·1 inches, culmen 0·15, wing 2·45, tail 2·2, tarsus 0·65."

The Plate represents the adult male and female of the size of life, drawn from the type specimens kindly lent to us by the Marquis Doria.

[R. B. S.]

MACRUROPSAR MAGNUS.

MACRUROPSAR MAGNUS.

Long-tailed Glossy Starling.

Lamprotornis magnus, Rosenb. in litt.—Schleg. Ned. Tijdschr. Dierk. iv. p. 18 (1871).—Wald. Trans. Zool. Soc. ix.
    p. 205 (1873).—Rosenb. Reist. naar Geelvinkb. pp. 37, 47, 140 (1875).—Id. Malay Archip. pp. 558,
    591 (1879).
Aplonis magna, Gieb. Thes. Orn. ii. p. 428 (1875).
Lamprotornis major, Salvad. Ann. Mus. Civ. Gen. xii. p. 345 (1878).
Macruropsar magnus, Salvad. Ann. Mus. Civ. Gen. xii. p. 345 (1878); xvi. p. 195, no. 8 (1880).—Id. Orn.
    Papuasia, &c. ii. p. 458 (1881).

THIS peculiar Glossy Starling is to all intents and purposes a *Calornis* with an exceptionally long tail. The last-named genus is very strongly represented in the Papuan region; and one species at least, the *Calornis inornata* of Salvadori, lives in the same island as the *Macruropsar*. The latter bird inhabits the islands of Mafoor and Misori, in the Bay of Geelvink, in both of which it was met with by Baron Rosenberg, the original discoverer of the species. Dr. Beccari also obtained it on both islands; but Dr. Meyer appears to have procured it only in Misori.

Count Salvadori states that the examples from Mafoor, collected by Beccari, had the tail shorter than specimens from Misori; and he considers it not improbable that the birds may be found to constitute two races when a larger series of specimens can be compared.

The adult male is described by Count Salvadori as follows:—

"Entirely dull green with a metallic gloss, the head slightly purplish; wings and tail duller; tail very long and graduated, with obscure barrings under certain lights; bill and feet black."

The female is similar to the male. The young bird is dusky blackish above, white below streaked with black.

The single figure in the Plate represents an adult bird, of about the size of life.

I am indebted to the kindness of Professor Steere, who, on his return journey through London after his travels to the East, lent me the bird from which the life-sized figure in the Plate has been drawn.

[R. B. S.]

CALORNIS GULARIS, Gray

CALORNIS GULARIS, Gray.

Purple-throated Glossy Starling.

Calornis gularis, Gray, Proc. Zool. Soc. 1861, pp. 431, 436.—Id. Hand-list of Birds, ii. p. 27, no. 6385 (1870).—
    Walden, Trans. Zool. Soc. viii. p. 80 (1872).—Sharpe, Ibis, 1876, p. 47.—Rosenb. Malay. Arch. p. 395
    (1879).—Forbes, Proc. Zool. Soc. 1884, p. 433.—Id. Nat. Wand. East. Arch. p. 365 (1885).
Calornis metallica, Sclater, Proc. Zool. Soc. 1883, p. 51 (nec Temm.).
Calornis circumscripta, Meyer, Sitz. u. Abhandl. Gesellsch. Isis, 1884, Abth. i. p. 49.—Salvad. Proc. Zool. Soc.
    1884, p. 579.

In the year 1861 the late Mr. G. R. Gray described a Glossy Starling, under the name of *Calornis gularis*, which had been sent by Mr. A. R. Wallace from the island of Mysol. The type specimen is still in the British Museum and has been variously referred by ornithologists to *Calornis metallica* or to a separate and distinct species. In 1872 the late Marquis of Tweeddale spoke of the species as "apparently nothing but *C. metallica*;" but in a small review of the genus published by us in 1876 we ventured to differ from Lord Tweeddale, and affirmed that *C. gularis* was a distinct species, recognizable by its small bill and purple throat. Count Salvadori, on the other hand, who examined the type specimen, which still remained unique in the British Museum, did not hesitate to unite it to *C. metallica*; and when Mr. Forbes's specimens arrived from Timor Laut, Dr. Sclater identified them as belonging to the last-named species. Dr. Meyer, however, having received some more specimens from Timor Laut, forwarded by Mr. Riedel, considered the *Calornis* from this group of islands to be distinct from *C. metallica*, and described it as *C. circumscripta*; and in this view he has been upheld by Count Salvadori, who does not agree with Mr. Forbes in calling the bird from Timor Laut *Calornis gularis*. Mr. Forbes has published his reasons for considering *C. circumscripta* of Meyer to be synonymous with *C. gularis* of Gray, and he submitted his series to our examination and for exact comparison with the type of *C. gularis*. So convinced were we of the correctness of his identification, that we agreed to figure the latter species from a pair of Mr. Forbes's Timor-Laut skins, and since then Dr. Meyer has lent us some of the typical examples of his *C. circumscripta*. These, however, only confirm the correctness of Mr. Forbes's identification; and we are perfectly certain that if Dr. Meyer and Count Salvadori could re-examine the type of *C. gularis*, they would both be convinced of the absolute identity of *C. circumscripta*. The type of *C. gularis* is labelled by Mr. Wallace, and the locality is in his own handwriting, so that it is unlikely that a mistake in the habitat of the species has been made; but we agree with Count Salvadori that it is curious that the same species should inhabit Mysol and Timor Laut, "so far apart one from the other, while true *C. metallica* lives in so many islands lying between them."

The figures in the Plate represent the male and female of about the natural size; they have been drawn from a pair of birds procured in Timor Laut by Mr. Forbes.

[R. B. S.]

CALORNIS FEADENSIS, Ramsay.

CALORNIS FEADENSIS, Ramsay.

Fead-Island Starling.

Calornis (Aplonis) feadensis, Ramsay, Journ. Linn. Soc., Zool. xvi. p. 129 (1881).—Reichen. & Schalow, Journ. für Orn. 1882, p. 227.
Calornis feadensis, Salvad. Ann. Mus. Civic. Genov. xviii. p. 426 (1882).—Id. Orn. Papuasia e delle Molucche, iii. App. p. 550 (1882).

THE genera *Aplonis* and *Calornis* may ultimately be found to be undistinguishable, and certainly the present bird seems to be a connecting link between the two genera. The species of *Calornis* are of more brilliant coloration, and the bill is not so stout nor so arched as in *Aplonis*; and in the former respect *C. feadensis* agrees with the Polynesian Starlings, while in the shape of the bill it is more like *Calornis*, in which genus we have retained it. Its nearest allies are *C. cantoroides* and that group of *Calornis*, but it possesses a peculiar dull coloration, sufficient to distinguish it at a glance.

As its name implies, it is an inhabitant of Fead Island, one of the Solomon group, and it was here discovered by the Rev. George Brown. Only a single specimen is at present known, and for the loan of this typical example we are indebted to our kind friend Mr. E. P. Ramsay, who described the species.

The following description is taken from the type :—

Adult. Entirely sooty, with here and there a shade of steel-green; wing-coverts, quills, and tail-feathers dusky black, externally washed with steel-green; entire head and under surface of body sooty black, with scarcely any indication of a greenish gloss. Total length 7 inches, culmen 0·85, wing 4·45, tail 2·4, tarsus 1·05.

On the Plate is represented the figure of an adult bird of the natural size, drawn from the specimen above mentioned.

[R. B. S.]

CALORNIS CRASSA, *Sclater.*

CALORNIS CRASSA, Sclater.
Slaty-grey Glossy Starling.

Calornis crassa, Sclater, Proc. Zool. Soc. 1883, p. 56, pl. 14.

The peculiar slaty hue of the plumage of this species of *Calornis* seems to separate it at once from all its allies. We have carefully compared the type specimens with the large series of *Calornis* in the British Museum, and we find that there is no species which approaches *Calornis crassa* in the grey shade of its plumage, which much resembles that of some of the Starlings of the genus *Aplonis*.

Mr. H. O. Forbes discovered the present species during his expedition to the Tenimber Islands, and procured a large series of specimens on the island of Larat in August 1882.

The following is a description of the type specimens lent to us by Dr. Sclater:—

Adult male. General colour above slaty grey, with a greenish gloss, the feathers of the head and neck lanceolate in form; lesser and median wing-coverts like the back; greater coverts, bastard wing, primary-coverts, and quills blackish, externally washed with greenish grey; tail-feathers blackish, with a slight wash of grey on the outer margins; lores black; sides of face and entire under surface of body slaty grey, the plumes of the throat lanceolate; under wing-coverts and axillaries like the breast; quills dusky below, browner along the inner webs; "bill, legs, and feet black; iris dark brown" (*H. O. Forbes*). Total length 7·4 inches, culmen 0·8, wing 4·0, tail 2·9, tarsus 0·95.

Adult female. Different from the male. More ashy grey above, with a less pronounced greenish gloss, the head, neck, and mantle with dusky shaft-lines or blackish centres to the feathers; lesser wing-coverts like the back; remainder of wing-coverts and quills purplish brown, with lighter brown edges; tail-feathers blackish; lores and eyelid blackish; sides of face and ear-coverts slaty grey glossed with green, the cheeks streaked with whitish; under surface of body creamy white, streaked with ashy brown down the centre of the feathers, very narrowly on the throat and fore neck, more broadly on the breast and flanks, and extremely so on the under tail-coverts; the sides of the body washed with slaty grey; thighs slaty grey; abdomen whiter than the breast, and narrowly streaked with blackish; axillaries and under wing-coverts light ashy brown; quills dusky below, browner on the inner webs. Total length 7·3 inches, culmen 0·8, wing 3·8, tail 2·65, tarsus 0·9.

The figures in the Plate represent a male and female of the present species, of the natural size. They are drawn from the typical specimens now deposited in the British Museum.

[R. B. S.]

MELANOPYRRHUS ANAIS.

MELANOPYRRHUS ANAIS.

Orange-chested Starling.

Sericulus anais, Less. Rev. Zool. 1844, p. 44.—Bp. Consp. i. p. 349 (1850).—Id. Compt. Rend. xxxvii. p. 831 (1853).—Sclater, Proc. Zool. Soc. 1857, p. 6.—G. R. Gray, Proc. Zool. Soc. 1858, p. 192, 1861, p. 434.—Rosenb. Nat. Tijdschr. Nederl. Ind. xxv. p. 234 (1863).—Id. J. f. O. 1864, p. 121.
Oriolus? anais, G. R. Gray, Gen. B. App. p. ii (1849).—Id. Cat. B. New Guinea, pp. 26, 57 (1859).
Pastor nigrocinctus, Cass. Proc. Acad. Nat. Sc. Philad. 1850, p. 68.—Sclater, Proc. Zool. Soc. 1857, p. 6.
Melanopyrrhus anais, Bp. Compt. Rend. xxxvii. p. 831 (1853).—Id. Notes Coll. Delattre, p. 9 (1854).—Sclater, Proc. Zool. Soc. 1857, p. 6.—Id. Journ. Linn. Soc. ii. p. 159 (1858).—Salvad. Ann. Mus. Civic. Genov. x. p. 150 (1877), xvi. p. 195 (1880).—Id. Orn. Papuasia e delle Molucche, ii. p. 462 (1881).—Guillemard, Proc. Zool. Soc. 1885, p. 644.
Gracula pectoralis, Wallace, Proc. Zool. Soc. 1862, p. 166, pl. xx.—Sclater, Ibis, 1863, p. 225.—Finsch, Neu-Guinea, p. 174 (1865).
Gracula (Melanopyrrhus) anais, G. R. Gray, Ann. & Mag. Nat. Hist. (3), x. p. 473 (1862).
Gracula anais, Wallace, Ann. & Mag. Nat. Hist. (3), xi. p. 15 (1863).—Schleg. Mus. P.-B. *Coraces*, p. 98 (1867).—Id. Nederl. Tijdschr. Dierk. iv. p. 18 (1871).—Sclater, Proc. Zool. Soc. 1873, p. 697.—Rosenb. Malay. Archip. p. 554 (1879).
Eulabes anais, G. R. Gray, Hand-list Birds, ii. p. 19, no. 6276 (1870).

This beautiful form of Starling is peculiar to New Guinea, and appears to be intermediate between the Mynahs (*Eulabes*) and the Crested Starlings (*Basilornis*). We have therefore followed Count Salvadori in considering it to belong to a separate genus, *Melanopyrrhus*, which he has located in the vicinity of the above-named genera.

The present species is very rare in collections, even at this date, and for many years it was known only by imperfect skins prepared by native hunters, and thus the early descriptions were very inaccurate. In 1862 Mr. Wallace gave a good description for the first time, accompanied by an excellent figure by Mr. Wolf.

The habitat of *M. anais* appears to be North-western New Guinea and Salwati. Mr. Wallace and most recent travellers have met with it near Sorong, and Mr. Bruijn and Dr. Beccari at Ramoi, Mariati, and Dorei-hum.

We transcribe herewith the description given by Count Salvadori in his great work on the 'Birds of Papuasia,' as the series examined by him has far exceeded that at our own disposition:—

Adult. Whole of the head, throat, back, wing-coverts, and abdomen black, the feathers with broad margins of glossy green, some of them changing to violet-blue under certain lights; lower throat and upper breast, rump, upper tail-coverts, and lower abdomen ochreous yellow, with an orange tinge; a cervical collar of pale ochreous yellow, produced laterally on both sides of the head into an occipital band; under tail-coverts yellow, the longer ones white with scarcely any yellow; a white band on the wing in the middle of the quills; under wing-coverts black; a bare space round the eye; feathers of the eyelids black; iris yellow; bill and feet pale yellow.

Young. Similar to the adult, but with the feathers of the breast and abdomen regularly margined with pale yellow; the cervical collar, rump, and upper tail-coverts paler yellow.

The figures in the Plate represent an adult specimen in two positions, and are drawn from an example lent to us by Dr. Guillemard.

[R. B. S.]

MELANOPYRRHUS ORIENTALIS.

MELANOPYRRHUS ORIENTALIS.

Robertson's Starling.

Gracula anais, Schlegel, Nederl. Tijdschr. Dierk. iv. p. 18 (1871, nec Less.).
Gracula anais orientalis, Schlegel, Nederl. Tijdschr. Dierk. iv. p. 52 (1871).—Rosenb. Reis. naar Geelvinkbai, p. 140 (1875).
Gracula orientalis, Beccari, Ann. Mus. Civic. Genov. vii. p. 714 (1875).—Ramsay, Proc. Linn. Soc. New South Wales, iv. p. 99 (1879).
Gracula rosenbergii, Finsch in Brehm's 'Gefangene Vögel,' ii. p. 562 (1876).
Melanophyrrhus orientalis, Salvad. Ann. Mus. Civic. Genov. viii. p. 401 (1876), x. pp. 12, 20, note (1877).—D'Albert. & Salvad. op. cit. xiv. p. 90 (1879).—Salvad. op. cit. xvi. p. 195 (1880).—Id. Orn. Papuasia e delle Molucche, ii. p. 463 (1881).—Guillemard, Proc. Zool. Soc. 1885, p. 644.
Mina robertsonii, D'Albert. Ann. Mus. Civic. Genov. x. pp. 12, 20 (1877).—Id. Ibis, 1877, p. 368.
Gracula affinis, Rosenb. Malay. Arch. pp. 554, 590 (1879).
Eulabes orientalis, Ramsay, Proc. Linn. Soc. N. S. Wales, iii. p. 279 (1879).
Melanopyrrhus robertsonii, Sharpe, Journ. Linn. Soc. xiv. pp. 633, 687 (1879).

THIS species is closely allied to *M. anais*, but differs in having the whole of the head and neck golden yellow. Its habitat is also somewhat different from that of the foregoing species, as it is found chiefly in South-eastern New Guinea, being by no means uncommon in collections from Port Moresby, while Signor D'Albertis procured a large series on the Fly River. It likewise occurs in North-western New Guinea, having been met with at Bondey by Baron von Rosenberg, at Rubi by Dr. Meyer, and at Wandammen by Dr. Beccari.

The specimen figured in the accompanying Plate is, according to the researches of Count Salvadori, not perfectly adult, as it shows a patch of black feathers on the occiput. Sometimes also yellow plumes are seen on the throat, and the breadth of the yellow margins to the abdominal plumes also varies much in proportion to the presence or absence of black feathers on the occiput. Count Salvadori very truly remarks that the great variation in plumage exhibited by the series of thirty specimens examined by him seems to indicate that the species has not yet acquired absolute stability of character.

The following description is translated from Count Salvadori's work on the 'Birds of New Guinea':—

Head and entire hind neck, upper breast, rump, and upper tail-coverts golden yellow; eyebrows, cheeks, and throat black, the latter often more or less varied with yellow; back, wing-coverts, abdomen, and thighs black, with a green gloss on the edges of the feathers; wings and tail black; primary-quills with a white spot in the middle; iris yellow; bill and feet pale yellow.

The Plate represents a mature, but not quite adult, bird in two positions. The figures have been drawn from a specimen lent to us by Dr. Guillemard, and supposed to come from the Arfak Mountains.

[R. B. S.]

LAMPROCORAX GRANDIS.

LAMPROCORAX MINOR, Ramsay.

Lesser Brown-winged Starling.

Sturnoides minor, Ramsay, Proc. Linn. Soc. N. S. Wales, vi. p. 726 (1882).—Id. op. cit. vii. p. 26 (1883).
Lamprocorax? minor, Salvad. Ann. Mus. Civic. Genov. xviii. p. 426 (1882).—Id. Orn. Papuasia e delle Molucche, iii. App. p. 500 (1882).
Calornis fulvipennis, Tristram, Ibis, 1882, p. 137 (nec Jacq. et Puch.).
Sturnoides (Lamprotornis) minor, Ramsay, Proc. Linn. Soc. N. S. Wales, vii. p. 668 (1883).

BOTH Lieut. Richards and Mr. Stephen have met with this species in the island of San Christoval in the Solomon Archipelago, and Canon Tristram has kindly lent us a specimen obtained in that island by the first-named naturalist. It would appear to be a smaller species than *L. grandis*, which it much resembles, and if Mr. Ramsay is correct, both species occur in the island of San Christoval (cf. Proc. Linn. Soc. N. S. Wales, vii. p. 668); but we fancy that he has written "San Christoval" by a slip of the pen instead of Guadalcanar and Lango, in which islands Mr. Cockerell met with it (cf. Ramsay, *op. cit.* iv. p. 76).

The following is a description of the specimen lent to us by Canon Tristram :—

Adult female. General colour above glossy greenish black, each feather with a mesial streak of glossy green; wing-coverts black, edged with glossy green; bastard-wing and primary-coverts black; quills dark brown, internally pale brown, the secondaries much lighter brown, forming a marked contrast to the colour of the back; upper tail-coverts like the back; tail-feathers dark brown; crown of head like the back, as also the sides of the face; ear-coverts, cheeks, throat, and fore neck like the back and similarly streaked; rest of the under surface greenish black, but not so plainly streaked; under wing-coverts and axillaries like the breast; quills below dusky, reddish on the inner web: "bill black; feet black; iris red" (*G. E. Richards*). Total length 8 inches, culmen 0·8, wing 4·35, tail 2·8, tarsus 1·05.

The two figures in the accompanying Plate represent the species of the size of life; they are drawn from the specimen lent to us by Canon Tristram.

P.S.—This Plate was prepared some months before the letterpress was written, and it was lettered *L. grandis* from Canon Tristram's identification of the specimen. Subsequent study has shown us that this identification was wrong; but the mistake was not found out in time to alter the Plate, which was already printed off.

[R. B. S.]

ARTAMUS MAXIMUS, Meyer.

ARTAMUS MAXIMUS, Meyer.

Meyer's Wood-Swallow.

Artamus maximus, Meyer, Sitz. k. Akad. Wien, lxix. part 1, p. 203.—Beccari, Ann. Mus. Civic. Genov. vii. p. 710 (1875).—Sclater, Ibis, 1876, p. 248.

THE genus *Artamus* is one of those peculiar forms of bird life which do not seem to have any immediate allies in the natural system; and for many years it has puzzled ornithologists where these Wood-Swallows should be placed. I can hardly subscribe to the opinion of Mr. Wallace, who places them near the Starlings from their wing-structure; for at the same time their habits are not those of Sturnine birds. But I must reject the arrangement of the late Mr. Gray, who places several Shrikes, such as *Leptopterus* and *Cyanolanius*, in the genus *Artamus*. Whether the curious African genus *Pseudochelidon* is really allied to *Artamus* I cannot determine; but I should think it more probably a Shrike-like form, in which case we should have *Artamus* as a genus inhabiting only the Indian and Australian regions. As we proceed southwards towards New Guinea we find the species increasing in number, until we meet with the metropolis of the genus in Australia, whence it extends to New Caledonia, and even to the Fiji Islands.

When Dr. Meyer described the present species, and bestowed upon it the name of *maximus*, he was scarcely justified in doing so. That it is one of the largest Wood-Swallows discovered, is true; and it will be seen by a comparison of the measurements that it exceeds slightly the newly discovered *Artamus insignis*, which is very nearly its equal in size; but *A. monachus* of Celebes is quite as long as *A. maximus*, and has even a stronger bill. In addition to the large dimensions, the black coloration of Meyer's Wood-Swallow makes it a very characteristic species.

We know very little about the habits of the present bird. Dr. Beccari, in his interesting letter on the ornithology of New Guinea, gives the following note:—"*Artamus maximus* is very common from 3000 to 5000 feet, and has the same habits as *A. papuensis*. It is enough to say that it flies like a Swallow, and sits on the branches of dead trees, especially in the middle of plantations. I have only got one or two specimens, because, through some fatality, I missed all the shots I fired." The above seems to comprise all that has been published respecting this fine Wood-Swallow, which doubtless does not differ in its economy from the other *Artami*.

In a specimen of this bird from Atam, recently lent to me by Mr. A. Boucard, I notice that the bill is almost white, whereas in the typical example it is blue. I fancy that this variation in colour is due to the fading of the bill after death. The following description is taken from the type specimen lent me by Dr. Meyer:—

General colour above black, including the wings and tail, the two latter with a slight slaty gloss; all the feathers slaty grey at base; rump and upper tail-coverts white; sides of face, sides of neck, throat, and fore neck black, like the back; remainder of under surface of body pure white; under wing-coverts and axillaries white, the small coverts along the outer edge of the wing black; quills greyish below. Total length 7·5 inches, culmen 0·8, wing 6·3, tail 2·8, tarsus 0·75.

The principal figure in the Plate represents the species of the size of life, and is drawn from the type specimen kindly lent me by Dr. Meyer.

ARTAMUS INSIGNIS, Sclater.

ARTAMUS INSIGNIS, Sclater.

New-Ireland Wood-Swallow.

Artamus insignis, Sclater, P. Z. S. 1877, p. 101, pl. xv.

I THOROUGHLY indorse the term *insignis* (or "remarkable"), which has been applied to this species by Dr. Sclater; for, in my opinion, it is the handsomest of all the Wood-Swallows. The number of *Artami* has not been greatly increased since I published my 'Birds of Australia;' but at that time, certainly, the two finest species, *A. maximus* and *A. insignis*, were not known. I have therefore great pleasure in presenting my readers with figures of these beautiful birds.

Of the subject of the present article we have very little to record. It is one of the discoveries of Mr. George Brown, who procured it in New Ireland in March 1876; and the two specimens obtained are in the collection of the Marquis of Tweeddale, along with the rest of Mr. Brown's ornithological trophies. Dr. Sclater, who described it, observes that it is closely allied to *A. monachus* of Celebes, but differs in having the wings and tail black, as may be seen by a comparison of the Plates of these two birds given in this work. In his account of the collections sent by Mr. Brown, Dr. Sclater has pointed out in certain of the species a slight indication of Celebesian affinities, and he observes with regard to the *Artamus*:—"Here is a second instance of a repetition on the further side of New Guinea of a Celebesian type, *A. monachus* of Celebes being certainly the nearest known ally of this fine new species. I have examined a specimen of *A. melaleucus* (Forst.), of New Caledonia, in the British Museum, but find it quite distinct, having the upper back black. *A. maximus*, Meyer, of New Guinea, is of the same large size as the present bird, but has the whole back black."

General colour above pure white; the head and neck all round, the wings, and tail black; under surface of the body, from the black throat downwards, pure white; under wing-coverts white, the small ones along the edge of the wing black; bill bluish. Total length 7·3 inches, culmen 1·0, wing 5·65, tail 2·6, tarsus 0·8.

The principal figures in the Plate are of the natural size, and are drawn from the type specimens in the collection of the Marquis of Tweeddale.

ARTAMUS MONACHUS, Temm.

ARTAMUS MONACHUS, Temm.

Hooded Wood-Swallow.

Artamus monachus, Bonap. Consp. Gen. Av. i. p. 343 (1850, ex Temm.).—Wallace, Ibis, 1860, p. 141.—Id. P. Z. S. 1862, p. 340.—Gray, Hand-l. Birds, i. p. 289. no. 4272 (1869).—Walden, Trans. Zool. Soc. viii. p. 67, pl. vi. fig. 1 (1872).

It might be supposed by some of my readers, owing to my having figured the present species perched upon a stranded snag in the middle of a stream, accompanied by a sleepy floating alligator, that this was the usual habitat of one or other of this family of birds. Such, however, is not the case; for I myself have never seen them haunting rivers. But I was reading a short time ago an account of a voyage in the Moluccas; and *Artamus* was described as having been observed in the position drawn by me; so I have endeavoured to reproduce some idea of the scene.

I have always felt an especial interest in the genus *Artamus*, as I have probably seen more species of the genus in their native haunts than any man living; and I have had the good fortune to describe no less than five out of the seventeen or eighteen known. Although the Australian Wood-Swallows are of very varied coloration, they cannot be considered so fine as some of the insular species, such as *A. maximus*, *A. insignis*, or the subject of the present article, *A. monachus*. These are certainly the most remarkable members of the genus *Artamus*, and surpass the other species in size and beauty.

In the 'Birds of Australia' I have given details of the habits of the Wood-Swallows; and doubtless the economy of all is very similar. The present bird is found only in the island of Celebes and the adjoining group of the Sula Islands, which lie to the eastward. Mr. Wallace says that it is found in the mountain districts of North Celebes; and Lord Tweeddale, in his well-known paper on the birds of this island, points out that the Sula specimens do not quite agree with Bonaparte's original diagnosis of the species. I have, however, the good fortune to possess a skin from Celebes itself, sent to me in exchange from the Leiden Museum; and I have compared it with Sula-Island specimens, and cannot find any difference in coloration, though the Celebes bird is rather longer in the wing; there can, I think, be no doubt as to their identity.

The accompanying description is taken from a female bird in my own collection, received from the Leiden Museum, and marked as having been obtained in Celebes by Heer von Duivenbode; but as the sexes are alike, the following colour will suffice for both:—

Above white from the hind neck to the tail, and including the scapulars; head and neck all round, including the throat, light umber brown, darker on the crown; least and median wing-coverts umber brown, the rest of the wing dark ashy brown; tail ashy brown; under surface of body from the fore neck downwards, and including the thighs and under wing- and tail-coverts, pure white; edge of wing ashy brown; quills grey below, whitish along the inner web. Total length 7·5 inches, culmen 1·05, wing 6·3, tail 2·9, tarsus 0·75.

The principal figure in the Plate is drawn from the before-mentioned specimen, and represents the species of the size of life.

MUNIA GRANDIS, Sharpe.

MUNIA GRANDIS, Sharpe.
Large Rufous-and-black Finch.

Munia grandis, Sharpe, Journ. Linn. Soc. (Zool.) vol. xvi. p. 319 (1882).

This large species of *Munia* belong to the widely spread section of Rufous-and-black Finches, of which *Munia rubronigra* is the best-known representative. It would seem to be nearly allied to *M. jagori* of Cabanis, a Philippine species which has been once collected by Dr. Meyer in the island Halmahera or Gilolo. The present bird, however, would seem to differ by its larger size, and by having the whole of the black abdomen so completely joined to the black breast as to leave only a patch of rufous on the sides of the body.

But a single specimen of this species was collected by Mr. Goldie in the Astrolabe Mountains. He procured it in the Taburi district, where it was called by the natives "Quaita."

I give the following description from Mr. Bowdler Sharpe's paper:—

"General colour above light bay, the rump and upper tail-coverts shining straw-yellow; least and median coverts like the back; greater coverts darker and more chestnut; primary-coverts and quills dusky brown, externally chestnut, the innermost secondaries entirely of the latter colour; central tail-feathers straw-yellow, dark brown along the middle; remainder of tail-feathers edged with straw-yellow; entire head and neck all round jet-black, as well as the breast and entire under surface, with the exception of a patch of light chestnut on the sides of the breast and upper flanks; under wing-coverts and axillaries light reddish, the lower series ashy rufous; quills dusky brown below, ashy rufous along the edge of the inner web. Total length 4 inches, culmen 0·5, wing 2·2, tail 1·55, tarsus 0·65."

The two figures in the Plate represent the male bird in different positions and of the natural size. They have been drawn from the typical specimen in the British Museum.

[R. B. S.]

MUNIA FORBESI, *Sclater*.

MUNIA FORBESI, *Sclater*.

Forbes's Munia.

Munia forbesi, Sclater, Proc. Zool. Soc. 1879, p. 449, pl. xxxvii. fig. 3.—Reichenow & Schalow, Journ. f. Orn. 1880, p. 203.—Salvad. Ann. Mus. Civic. Genova, xvi. p. 192 (1880).—Id. Ornitologia della Papuasia &c. p. 438 (1881).

THIS is a very fine and large species of *Munia*, recently discovered in New Ireland by the Rev. George Brown, who procured one specimen only, in the district of Topaio, in September 1878. In the style of coloration it differs greatly from the species of the genus which are found in Australia and the Papuan Islands, and belongs rather to the Indian group, of which *Munia malacca* is the type.

In addition to the large size of the present species, it may be told from any of the other Papuan *Muniæ* by the following characters, which are here quoted from Count Salvadori's new work on the Ornithology of New Guinea. It has the upper tail-coverts, as well as the rump, rufous, the sides of the body rufous, not streaked with black; and the head and under tail-coverts are black, while the breast and entire abdomen are rufous. The nearest ally to the present bird is *Munia jagori* of the Philippines; but that species has the breast and abdomen black like the remainder of the undersurface.

Dr. Sclater gave the following diagnosis of the species, which is named after Mr. W. A. Forbes, the well-known Prosector to the Zoological Society, who has made a special study of the Fringillidæ:—

Rufous, a little paler underneath; head all round, as well as the throat, flanks, and lower part of the belly, including the thighs and under tail-coverts, black, the latter being elongated; bill and feet black; the bill very stout.

Total length 4 inches, wing 2, tail 1·5.

The Plate represents two birds, of the size of life. They are drawn from the unique specimen now in the British Museum, and lent to me by Dr. Sclater before its incorporation in the national collection.

DONACICOLA SPECTABILIS, Sclater.

DONACICOLA SPECTABILIS, *Sclater*.
Orange-rumped Finch.

Donacicola spectabilis, Sclater, P. Z. S. 1879, p. 449, pl. xxxvii. fig. 2.—Salvad. Orn. della Papuasia &c. ii. p. 441 (1881).

THE genus *Donacicola* (as it is now written, instead of my original name *Donacola*, which, I admit, was not classically compounded) contains a very few species of little Finches, all of which are peculiar to the Australian region. The best-known of them is the Chestnut-breasted Finch (*D. castaneothorax*) of Australia; and a very closely allied species is found in South-eastern New Guinea, the *D. nigriceps* of Ramsay. The discovery of a new species in New Britain is of some interest, as showing the Papuan element in Australian ornithology, or, if one prefers it, the Australian element in the Papuan avifauna. It is to be regretted that the Finches are at present in such a neglected state as regards their classification, that the value of the different genera has never been worked out by a competent ornithologist; and therefore it is only fair to state that Dr. Sclater, in his original description of this species, has compared it with my *Donacicola flavoprymna* as its nearest ally, and Mr. G. R. Gray classes the latter species as a *Munia*. I must confess that the present bird is very like a *Munia* in appearance and less like a *Donacicola* than the more typical species of the latter genus.

The single specimen at present known was procured in New Britain by the Rev. G. Brown, to whom we are indebted for the discovery recently of so many fine novelties; and I translate the description given by Dr. Sclater (*l. c.*).

General colour brown; the head, nape, and sides of breast black; upper tail-coverts and margins of the central tail-feathers pale chestnut; under surface of body white; throat, lower part of the belly, and vent, with the thighs, black; under wing-coverts ochraceous white; bill and feet black. Total length 3·4 inches, wing 1·8, tail 1·2.

I am indebted to Dr. Sclater for the loan of the unique specimen of the present bird, which has since passed into the collection of the British Museum.

DONACICOLA NIGRICEPS, *Ramsay*.

DONACICOLA NIGRICEPS, Ramsay.
The Black-cheeked Finch.

Donacola nigriceps, Ramsay, Proc. Linn. Soc. N. S. W. i. p. 393 (1876).—Sharpe, Journ. Pr. Linn. Soc. xiii.
    p. 601 (1877), xiv. p. 688, no. 38 (1878); Ramsay, op. cit. iii. p. 289 (1879), iv. p. 100, no. 149
    (1879).
Donacicola nigriceps, Salvad. Ann. Mus. Civ. Genova, xvi. p. 192, no. 9 (1880).—Id. Orn. della Papuasia &c. ii.
    p. 441 (1881).

This little Finch appears to represent in South-eastern New Guinea the *Donacicola castaneothorax* which I described from Australia, and which, according to Mr. E. P. Ramsay, occurs all over the eastern part of that continent, from New South Wales to Cape York, and is also found in the Gulf of Carpentaria, but whose place is taken in South-eastern New Guinea by the present species. It is very closely allied to *D. castaneothorax*, but may be told at a glance by its black head only slightly spotted with ashy, and still more distinctly by its entirely black cheeks. I have only seen a very few specimens from the immediate vicinity of Port Moresby, where it was procured by Mr. Octavius Stone; and I have not seen any examples from the interior of South-eastern New Guinea. Its habits and mode of life are doubtless similar to those of the other Australian Finches of the genus *Donacicola*.

Adult.—General colour above delicate burnt-sienna, with ashy shading to the feathers of the back; the rump and upper tail-coverts orange, with dusky bases to the feathers; tail-feathers pointed, dark brown with straw-yellow margins, the two centre feathers almost entirely straw-yellow; head and nape chocolate-brown, veined with streaks and spots of ashy whitish, the nape-feathers edged with the latter colour; lores, feathers above the eye, entire sides of face, and throat uniform black; fore neck and chest entirely pinkish fawn-colour, forming a large plastron, succeeded by a band of black across the lower breast; centre of the body and abdomen pure white, the flanks regularly barred with black and white; thighs and under tail-coverts black; under wing-coverts buffy white; the edge of the wing minutely barred with black and white; wing-coverts above sienna-brown; quills light brown, externally washed with sienna-brown, ashy brown below, edged with buff along the inner web. Total length 3·8 inches, culmen 0·4, wing 2, tail 1·55, tarsus 0·6.

The above description has been taken from Mr. Sharpe's account of Mr. Stone's collection; and I am indebted to the latter gentleman for the loan of the specimens from which the figures in the Plate have been drawn. They represent two adult birds, of the natural size.

DONACICOLA HUNSTEINI, Finsch.

DONACICOLA HUNSTEINI, *Finsch.*

Hunstein's Weaver-Finch.

Donacicola hunsteini, Finsch, Ibis, 1886, p. 1, pl. 1.

This very distinct species of Weaver-Finch is one of the most interesting discoveries made by the well-known traveller Dr. Otto Finsch, during his expeditions to New Guinea and the islands of the South Seas. He met with the present bird in New Ireland, where, he says, "it was discovered at the extreme north corner of the island. It lives in the high jungle-grass and is difficult to obtain."

The following is a description of the typical specimens, which are now in the British Museum:—

Adult male. General colour above black, the rump and upper tail-coverts chestnut with a wash of golden yellow; wing-coverts black like the back; bastard-wing, primary-coverts, and quills rather paler blackish brown edged with golden; tail-feathers blackish brown, the centre ones golden towards the ends, like the upper tail-coverts; crown of head, nape, and hind neck hoary grey, mottled with blackish bases to the feathers; lores and feathers below the eye, eyelid, and fore part of cheeks black; ear-coverts hoary grey like the head, a shade of the same colour overspreading the hinder cheeks; throat and entire under surface of body velvety black; under wing-coverts pale tawny buff, the edge of the wing black; quills below blackish, pale tawny buff along the inner edge: "bill and feet black; iris dark" (*O. Finsch*). Total length 3·5 inches, culmen 0·4, wing 2·0, tail 1·15, tarsus 0·6.

Adult female. Similar to the male in colour. Total length 3·5 inches, culmen 0·4, wing 1·9, tail 1·2, tarsus 0·55.

Young. Brown, without any of the chestnut on the rump and tail, and only a slight indication of grey here and there on the head; side of face, cheeks, throat, and chest dark chocolate-brown, the breast and abdomen isabelline buff.

The figures in the Plate represent an adult male and female, as well as a young bird. They are drawn from the typical examples described above.

[R. B. S.]

PITTA MAXIMA, *Müll and Schleg.*

PITTA MAXIMA, Müll. & Schl.

Great Pitta.

Pitta maxima, Müll. & Schl. Verh. Nat. Gesch. Ned. Ind. Zool. p. 14.—Westerm. Bijdr. Dierk. p. 45, *Pitta*, pl. 1.—Gray, Gen. B. i. p. 213.—Wallace, Ibis, 1859, p. 112, 1860, p. 197.—Schl. Vog. Nederl. Ind. *Pitta*, p. 30.—Gray, Hand-l. B. i. p. 296.
Brachyurus maximus, Bp. Consp. i. p. 253.—Elliot, Monogr. Pittidæ, pl. 12.
Gigantipitta maxima, Bp. Consp. Vol. Anisod. p. 7.
Pitta gigas, Wallace, Malay Arch. ii. p. 3.

THERE are several species of Ant-Thrushes which are nearly, if not quite, equal to the present bird in size; so that the specific name of *maxima* would be by no means justified, if naturalists were content to class all these birds under the heading of genus *Pitta*. By many writers, however, the large Ant-Thrushes of Malaisia and the eastern Himalayas are generically separated as *Hydrornis*; and Mr. George Robert Gray was inclined to range the present bird under the same heading. In this I cannot agree; I think that it should be kept along with the true *Pittæ*, of course in the short-tailed group. Whether the latter section should be regarded as constituting a separate genus is quite another matter; I consider this much more feasible. Throughout the present work, however, I have retained these particoloured Ant-Thrushes under the genus *Pitta* in preference to *Brachyurus*, and therefore adhere to it in the present instance.

The habitat of this beautiful bird is the Moluccan island of Gilolo. Very little has been recorded of its habits. Mr. Wallace, in his 'Malay Archipelago,' writes that during his stay in the above-mentioned island his boy Ali shot "a pair of one of the most beautiful birds of the east—*Pitta gigas*, a large Ground-Thrush, whose plumage of velvety black above is relieved by a breast of pure white, shoulders of azure blue, and belly of vivid crimson. It has very long and strong legs, and hops about with such activity, in the dense tangled forest bristling with rocks, as to make it very difficult to shoot." From the above short note of Mr. Wallace's we can imagine what a beautiful sight it must be to see this finely plumaged bird in its native forests; and even in a tropical island like Gilolo, where brilliantly coloured birds abound, there can be few to compete with the subject of our present article.

No description of the bird is necessary, as it stands alone among the Pittidæ, and has no near allies. The Plate gives a correct idea of the plumage; and the principal figure is full-sized.

MELAMPITTA LUGUBRIS, Schl.

MELAMPITTA LUGUBRIS, Schlegel.

Black Ground-Thrush.

Melampitta lugubris, Schl. N. T. D. iv. p. 47.—Sclater, P. Z. S. 1873, p. 696.

What are the natural affinities of this most curious bird? is a question which will exercise the ingenuity of ornithologists for some time to come. The generic appellation *Melampitta*, or "Black Ground-Thrush," bestowed upon it by Professor Schlegel, shows that by that eminent ornithologist the bird was evidently considered a near ally of the genus *Pitta*; and this is the position which I myself would assign to it. But the interesting aspect of the question still remains with regard to the affinities of the Mascarene genus *Philepitta*, another systematic puzzle to ornithologists. As the name of the latter genus implies, it was considered to be a relation of the Pittidæ, in which family it has generally been included; but Mr. Sharpe has referred it to the Paradiseidæ—an indication of the difficulty presented by the structural peculiarities of the bird.

An important link between *Pitta* and *Philepitta* seems to be offered in the present species, which unites the general appearance of a true Ground-Thrush with something of that velvety plumage for which *Philepitta* is famous; and therefore I cannot but regard this discovery of Baron von Rosenberg's as of the highest interest to the ornithologist, not only as uniting genera whose affinities were doubtful, but also as exhibiting another of the mysterious links which unite the fauna of Madagascar and certain portions of the Malayan archipelago.

Nothing is known of the present species beyond the fact that it was discovered in the northern peninsula of New Guinea, and was afterwards met with by d'Albertis in the Atam district.

The entire bird is black with a slight bluish tinge, the feathers of the forehead, region of the eye, lores, base of mandible, and chin having a velvety appearance. The length is about six inches.

My Plate is taken from Signor d'Albertis's Atam specimen, and represents the species of the size of life.

PITTA CYANONOTA, Gray.

PITTA CYANONOTA, Gray.

Blue-backed Pitta.

Pitta cyanonota, Gray, Proc. Zool. Soc. 1860, p. 351.—Schl. Mus. Pays-Bas, *Pitta*, p. 8.—Id. Vog. Nederl. Indië, *Pitta*, pp. 18, 35.—Finsch, Neu-Guinea, p. 168.—Gray, Hand-l. B. i. p. 296, no. 4380.—Schl. Mus. Pays-Bas, Revue *Pitta*, p. 13.
Brachyurus cyanonotus, Elliot, Monogr. Pittidæ, pl. xx.—Id. Ibis, 1870, p. 418.

This species belongs to the section of the genus *Pitta* which is called by some ornithologists *Erythropitta*, containing certain red-breasted species from the Indo-Malayan islands, the Moluccas, and New Guinea, which form a very natural group. The present bird is one of the best-defined species of the section, being distinguished at a glance by its blue back. It was discovered by Mr. Wallace in the small island of Ternate; and later on Dr. Bernstein also met with it in the same locality, to which for some time it was supposed to be confined. The latter ornithologist, however, afterwards procured the species in the island of Guebeh; so that, as Professor Schlegel remarks, it appears to represent in these two islands the *Pitta ruficentris* of Halmahera (or, as we English naturalists miscall the island, Gilolo). It is distinguished from the last-named bird by its blue back. The Dutch travellers Bernstein and Von Rosenberg procured a good series of specimens in Ternate, meeting with the species apparently all the year round. Sixteen specimens from this island alone are preserved in the Leiden Museum, having been killed in the months of May, June, August, and November. On the 6th of May, 1871, Von Rosenberg took two nestlings; so that this month may be taken as indicating the breeding-season, though it is evident that the eggs must be deposited in the month of April.

The following is a description of the species:—General colour of the upper surface blue, the crown dull reddish, brighter red on the nape, hind neck, and sides of crown; lores, sides of face, and throat dusky brown washed with reddish; fore neck and chest bright blue, forming a band; remainder of the under surface scarlet; tail a little duller blue than the back; wing-coverts blue, like the back, with a small spot of white on the shoulder, formed by white marks near the base of the outer web of some of the smaller coverts; quills blackish, washed with the same blue as the back on the outer web, broader on the secondaries, the innermost of which are like the back; the third primary marked with a spot of white near the base of the inner web only, the fourth primary having a white spot on both outer and inner web.

The figures in the Plate are drawn from specimens in my own collection; they represent the birds of the natural size. It should be noticed that the white shoulder-spot which is conspicuous in one individual, is absent in another.

PITTA FORSTENI, *Mull.*

PITTA FORSTENI.

Forsten's Pitta.

Pitta melanocephala, Müll. & Schl. Verh. Natuurl. Geschied., Zool., *Pitta*, p. 19 (1844, ex Forsten, MSS., nec Wagler).—Westerm. Bijdr. Dierk. Amsterd. (folio), i. part vi. (1854), p. 46, pl. 2.—Schlegel, Mus. Pays-Bas, *Pitta*, p. 4 (1863).—Id. Vog. Nederl. Indië, *Pitta*, pp. 5, 30, pl. ii. fig. 1 (1863).—Id. Mus. Pays-Bas, *Pitta*, Revue, p. 9 (1874).

Brachyurus forsteni, Bonap. Consp. Gen. Av. i. p. 256 (1850).—Elliot, Ibis, 1870, p. 419.

Melanopitta forsteni, Bp. Consp. Volucr. Anis. 1854, p. 7, no. 195.—Walden, Trans. Zool. Soc. viii. p. 62 (1872).

Brachyurus (Melanopitta) forsteni, Elliot, Monogr. Pittidæ, pl. xxiv. (1863).

Pitta forsteni, Gray, Hand-list of Birds, i. p. 295, no. 4363 (1869).

THE genus *Pitta* has been divided by some ornithologists into various subgenera, founded for the most part on the prevailing style of coloration; and so strongly characterized are these differences of coloration, that I am inclined to admit them as generic characters of no small value. To take, for instance, the section to which Forsten's Pitta belongs, and to which the subgeneric title *Melanopitta* has been given, how unmistakable a character is the black head! while at the same time it is accompanied by a green plumage strongly varied by a red vent and under tail-coverts and a lustrous green shoulder-patch. As a rule, too, the black-headed Pittas are remarkable for their white quills, which must form a very conspicuous feature when the birds are alive; and many of the species depend upon the amount of white on the wing-feathers for their separation one from the other. It may be taken, therefore, as a character of the greatest importance that Forsten's Pitta has the quills entirely black; and it is on this account nearest allied to *P. novæ-guineæ*, which has only a concealed white spot on the fourth, fifth, and sixth quills. The last-named bird, moreover, differs in having a blue shade bordering the black ventral patch, the absence of which in *P. forsteni* is compensated for by a broad band of metallic greenish blue across the upper tail-coverts. This band is found in most of the black-headed Pittas; but in *P. novæ-guineæ* it is scarcely distinguishable, being represented only by a slight metallic green tip to a few of the upper tail-coverts.

As far as we know at present, Forsten's Pitta is found only in the island of Celebes, and seems, indeed, to be confined to the northern parts of that island. It was found by Dr. Forsten at Kema and at Tondano. I have specimens in my collection from Menado, collected by Dr. Meyer; and the following description is taken from one of these.

Adult. General colour above bright grass-green with somewhat of a metallic lustre; lesser and median wing-coverts bright metallic greenish cobalt, forming a shoulder-patch; greater series green; primary-coverts and quills black, the secondaries black, externally green, like the back, the innermost entirely green; most of the upper tail-coverts metallic greenish cobalt, forming a transverse band; longer upper tail-coverts and tail-feathers dull green; head and hind neck, sides of face and ear-coverts, cheeks and throat black; rest of under surface, from the lower throat downwards, bright green, with the lower abdomen, vent, and under tail-coverts bright scarlet, bordered above with an abdominal patch of black, which descends slightly on each side of the scarlet patch, the lower feathers being black broadly tipped with scarlet; thighs brown; under wing-coverts and quill-lining black.

Total length $7\frac{3}{8}$ inches, culmen $\frac{7}{8}$, wing $4\frac{5}{8}$, tail $1\frac{7}{8}$, tarsus $1\frac{5}{8}$.

Since the above meagre description of a fine bird was in press, I have received a few additional remarks from Dr. Meyer, to whom I wrote a few days since, and who says of *Pitta forsteni* :—

"This species only occurs on the island of Celebes, and is represented in Borneo by *P. mülleri*, on Sangi Island by *P. sanghirana*, on Mindanao by *P. steerii*, and on other islands of the Philippine group (as well as on Mindanao) by *P. sordida*.

"All Pittas may be said to be rare birds everywhere, and are only met with singly or a pair at a time. I met with but *one* Pitta that was plentiful, viz. *P. rosenbergi*, on the island of Mysore, in the north of Geelvink Bay. Besides, the black-headed Pittas are still rarer than the red-bellied Pittas; and so it is also on the island of Celebes with *P. forsteni* in relation to *P. celebensis*. In the southern parts of Celebes I did not procure a single specimen of *P. forsteni*, as far as I remember, and also am not aware that specimens collected by any one else have reached Europe from there; whereas *P. celebensis* appeared to be less rare there than in the northern parts. Here I got specimens of *P. forsteni* in the Minahassa and in the district of Gorontalo.

"Pittas are shy birds, as I have said before; but their flute-like cry once heard, the specimen can nearly always be got with patience and quietness; imitating its voice, the bird can be called up till it is close to the hunter's gun. The rareness of the black-headed Pitta on Celebes is proved by the fact that its colours shelter it even less than the colours of *P. celebensis* shelter that species, or the bright blue shoulder-patches of *P. forsteni*, which always glitter on the ground, which it never quits. The colour of the iris is dark, the feet dusky, the bill black. It feeds on insects of all kinds. Its name in the Minhassa is 'Mopo idiu,' that is to say, 'Green Mopo,' 'Mopo' being the name for *P. celebensis*, the meaning of which word in the Alfuro language I have explained in my 'Field-notes on the Birds of Celebes' ('Ibis,' 1879, p. 126), where I also narrated the story which the natives attached to this bird."

PITTA CONCINNA, Gould.

PITTA CONCINNA, *Gould*.

Elegant Pitta.

Brachyurus vigorsi, Bonap. Consp. Gen. Av. i. p. 255 (1850, nec Gould).
Pitta concinna, Gould, P. Z. S. 1857, p. 65.—Wallace, P. Z. S. 1863, p. 485.—Schlegel, Vog. van Nederl. Indië, pp. 12, 32, pl. iii. fig. 1 (1863).—Id. Mus. Pays-Bas, *Pitta*, p. 10 (1865).—Id. op. cit., Revue *Pitta*, p. 14 (1874).
Pitta mathildæ, J. & E. Verreaux, Rev. et Mag. de Zool. 1857, p. 303, pl. xi.
Brachyurus concinnus, Elliot, Monogr. Pittidæ, pl. x. (1863).—Id. Ibis, 1870, p. 416.

THIS species of *Pitta* was first published by me in 1857, when I described it from specimens obtained by Mr. A. R. Wallace in the island of Lombock. It appears that it also inhabits the island of Sumbawa, as specimens from the latter locality were contained in the Leiden Museum many years before I described the bird as new, and one of these specimens was wrongly identified by Bonaparte as *Pitta vigorsi*, figured by me in the 'Birds of Australia.' Professor Schlegel, however, in his list of the Pittas in the Leiden Museum, has corrected the error of Bonaparte, which an examination of the specimen described by the latter enabled him to do, and has placed the species in its correct position. In the supplementary list of the Leiden Pittas, Professor Schlegel records two specimens from the island of Flores; so that its range is now known to include the three islands of Flores, Lombock, and Sumbawa, to which it will probably be found to be confined.

The characters by which *P. concinna* may be distinguished are its small size and the tint of the brown on the head, which is much clearer than in *P. strepitans* and only extends as far as the occiput, where it is prolonged into a streak of bluish white.

I regret to say that nothing has been written respecting the habits of this bird; and I can only add that the name *concinna*, published by me, has a slight priority (only of a few days, according to Mr. Elliot) over the name *mathildæ*, given by MM. Verreaux in the same year.

The following is a copy of the original description:—

"Head, back of the neck, cheeks, chin, and stripe down the centre of the throat velvety black; from the nostrils over each eye a broad mark of deep buff, posterior to which is a narrower one of pale glaucous blue; back, tail, and wings dark grass-green; lesser wing-coverts and a band across the rump glossy verditer blue; primaries and secondaries black, the fourth, fifth, and sixth of the former crossed by a band of white near their base, and all the primaries tipped on the external web with olive grey; upper tail-coverts black; under surface delicate fawn-colour, becoming much paler where it meets the black of the cheeks and throat; centre of the abdomen black; vent and under tail-coverts fine scarlet; bill black; feet fleshy.

"Total length 6 inches, bill 1, wing 4, tail 1½, tarsus 1⅜."

The figures in the Plate, representing the two sexes about the natural size, are drawn from the typical specimens, still in my possession.

PITTA CÆRULEITORQUES, *Salvad.*

PITTA CÆRULEITORQUES.

Red-headed Pitta.

Pitta cæruleitorques, Salvad. Ann. Mus. Civic. Genov. ix. p. 53 (1876-77).—Rowley, Ornithological Miscellany, part viii. (1877).

In an interesting communication made by Dr. Meyer to Mr. Dawson Rowley's 'Ornithological Miscellany,' that gentleman points out the distribution of the red-breasted *Pittæ* in the Malayan archipelago, and shows how each of the species, which I consider should be kept under the heading of *Erythropitta* in the present work, has its own separate area of distribution, however closely they may be allied as species. Thus *Pitta celebensis* is the species of Celebes, *P. palliceps* of Siao, *P. cæruleitorques* of Sangi (Sanghir), *P. erythrogastra* of the Philippines, *P. cyanonota* of Ternate, *P. rufiventris* of Batchian and Gilolo, and *P. mackloti* of Papua and its islands, as well as the northern part of Australia. Many other instances of a similar distribution could be brought forward.

Count Salvadori, in his original description of the present species, writes as follows:—" This species, and the *P. erythrogastra* of the Philippines, are the only two species of the subgenus *Erythropitta* which have a blue band on the neck; and *P. cæruleitorques* differs from the above-named bird principally in the more uniform red colour of the head, which becomes much brighter on the neck, by the absence of the two dull bands on the side of the crown, by the reddish-brown colour of the sides of the head and throat, by the blue colour of the breast being more extended crosswise, and separated from the red of the abdomen by a well-marked black band, and by the somewhat larger dimensions."

Dr. Meyer obtained several examples of the blue-ringed Pitta from Sangi, at Tabukan, on the north-east coast of the island, no difference being observable in the colour of the sexes; and I give the following extract from his remarks communicated to the 'Ornithological Miscellany:'—

"This species inhabits the largest island of the Sangi group; and it is an interesting one, because it is more closely allied to *Pitta erythrogastra* from the Philippines, in the north, than to the two species from islands immediately to the south (viz. *Pitta palliceps* from Siao, and *Pitta celebensis* from Celebes), and therefore presents a good example of variation of species in consequence of separated insular habitat. Good examples for the same point of view are, amongst others, *Pitta cyanonota* from Ternate, and *Pitta rufiventris* from Halmahera, in their relation to the species from the neighbouring islands (New Guinea, Celebes, the Sangi, and Philippine Islands). *Pitta palliceps* on Siao is as slightly different from *Pitta celebensis* on Celebes as *Pitta cæruleitorques* on Sangi is from *Pitta erythrogastra* on the Philippines. That insular separation is a reason for such variations is not to be doubted, in my opinion; nevertheless we cannot examine this subject more closely at present. *Pitta celebensis*, for instance, does not show the least difference over the whole extent of the island of Celebes. My specimens from the neighbourhood of Makassar resemble exactly those from Manado (nearly the north and south points of this long island); whereas when we cross over to the closely neighbouring island of Siao, immediately a variation appears in *Pitta palliceps*. Whether this variation has specific value or not is of no importance at all upon this part of the question. Authors do not agree, and never will agree, at least for some time to come: one says it has, the other says it has not; but all see that a difference exists; and this is of value, notwithstanding its smallness, because it is a constant one. That insular separation does not always produce constant differences is known; and I only mention it here for this reason—that it refers to a closely allied species, *Pitta macklotii*. I got a large series of specimens on New Guinea in different places, viz. at Dore, Andei, Passim, Inwiorage, Rubi, and on the Elephant Mountains, and some on the island of Jobi in the north of Geelvink Bay. I first thought that the Jobi specimens differed by brighter colours in general, and noted this difference in my diary; but now, in the cabinet, I do not see the slightest difference in several of the New-Guinea specimens."

The following description is a translation of the original one given by Count Salvadori:—

Head above red, the latter colour perceptibly brighter towards the hind neck; sides of head and throat brownish red; a very broad patch of black on the lower throat; a band round the hind neck, another very broad one on the breast, the latter succeeded by a band of black; the wings, upper tail-coverts, and tail bluish lead-grey; back, scapulars, and sides of breast olivaceous; abdomen and under tail-coverts very bright red; tips of the longer under tail-coverts blue; primaries marked in the middle with a white spot; a white spot near the bend of the wing; bill and feet dusky.

Total length 6 inches, culmen 1·0, wing 4⅝, tail 1½, tarsus ⅞.

I owe the opportunity of figuring this species to the kindness of two friends, Count Salvadori and Mr. George Dawson Rowley, both of whom lent me their specimens.

PITTA MAFORENSIS, Schleg.

PITTA MAFORENSIS, *Schlegel.*

Mafoor-Island Pitta.

Pitta novæ guineæ mefoorana, Schlegel, Revue Coll. *Pitta* Mus. Pays-Bas, p. 8 (1874).—Meyer, in Dawson Rowley's 'Ornithological Miscellany,' pt. vii. p. 268 (1877).

This species was separated by Professor Schlegel in 1874 in his Review of the Pittas in the Leiden Museum under the trinomial title above quoted; and he apparently regards it as nothing but a race of *Pitta novæ guineæ.* In this conclusion I am unable to agree, as it seems to me to be a thoroughly well-marked species. It is nearly allied to the last-mentioned bird and to *Pitta rosenbergi,* but is distinct from both. It is of about the same size, and has the colour of the chest, breast, and nape of a fine glistening greenish white, as it exists in *P. novæ guineæ,* but more extended, the green of the chest blending into green and blue on the flanks.

Dr. Meyer did not get a *Pitta* on the island of Mafoor; but the Leiden Museum possesses four specimens, killed there in January and February 1869 by Von Rosenberg. Dr. Beccari also managed to procure some examples.

In describing the species, Professor Schlegel says that the Mafoor bird is similar to *P. novæ guineæ,* but has the tail-feathers more or less tipped with dirty green, the large upper tail-coverts black, with a fine blue edging, the smaller upper tail-coverts of a fine metallic whitish green, and the blue of the abdomen darker, the quills being without white spots. This constitutes, as far as I know, all that has been published respecting the present bird. The Plate represents a pair of these birds of the natural size. They form part of the rich collections made by Dr. Beccari in New Guinea and the islands of Geelvink Bay. They were kindly lent to me by Count Salvadori during his visit to this country; and to him I have once more to express my great appreciation of his kindness.

Total length $6\frac{1}{4}$ inches, wing $5\frac{1}{4}$, tail $1\frac{3}{8}$, tarsus $1\frac{3}{4}$.

I must apologize for the oversight by which the name *maforensis* instead of *mafoorana* was printed on the Plate, as before I had discovered the mistake the whole impression had been printed off; and I thought it best in this instance to keep the name at the head of this article to harmonize it with that of the Plate, though I regret the *lapsus calami* which caused the error.

I regret that no further information should have reached me respecting this beautiful species, which finds a near ally in *P. novæ guineæ.* In size it is much the same; but the green of the under surface is suffused with luminous glistening green.

PITTA CELEBENSIS, *Forst.*

PITTA CELEBENSIS, *Forst.*

Celebean Pitta.

Pitta celebensis, Müller & Schlegel, Verh. Nat. Geschied. Zool. Aves, p. 18. no. 16 (1839-44, ex Forster MS.).—Gray, Genera of Birds, i. p. 213 (1846).—Westerman, Bijdr. tot de Dierkunde, folio, i. p. 46, pl. iii. (1848-54).—Wallace, Ibis, 1860, p. 142.—Schlegel, Mus. Pays-Bas, *Pitta* p. 6 (1863).—Id. Vog. van Nederl. Indië, *Pitta*, pp. 17, 34, pl. iv. figs. 4, 5 (1863).—Wallace, Ibis, 1864, p. 105.—Gray, Hand-list of Birds, i. p. 296. no. 4377 (1869).—Schlegel, Revue *Pitta* Mus. Pays-Bas, p. 10 (1874).—Salvad. Ann. Mus. Civic. Genov. vii. p. 663 (1875).—Meyer in Rowley's Ornith. Misc. part viii. (1877).
Brachyurus celebensis, Bonap. Consp. Gen. Av. i. p. 253 (1850).—Elliot, Monogr. Pittidæ, pl. xvii. (1863).—Id. Ibis, 1870, p. 418.
Erythropitta celebensis, Bonap. Consp. Volucr. Anisod. p. 7 (1854); Walden, Trans. Zool. Soc. viii. p. 62 (1872).

I HAVE already, in one of my other articles, spoken of the distribution of the red-breasted Pittas in the Malay archipelago, and I have quoted the remarks of Dr. Meyer on this subject; I therefore need only say that the present species is the representative of this section of the genus on the island of Celebes, to which it appears entirely restricted. It is true that, in his Review of the Pittas contained in the Leiden Museum, Professor Schlegel enumerates several examples from the island of Siao in the Sanghir archipelago; but these no doubt belong to the species since named *Pitta palliceps* by the late Dr. Brüggemann. Certain differences, indeed, seem to have struck Professor Schlegel at the time; for he says that in the birds from Siao the rufous colour of the head is paler than in examples from Celebes, and often replaces the black bordering the blue stripe on the head.

Mr. Wallace found the species scarce in Northern Celebes, which appears to be the only part of the island where it has yet been found. I may be mistaken in this, as the localities Modélido, Négri-lama, and Boné, mentioned in the list of specimens at Leiden, do not occur in any of the maps I have examined. The other places, however, Menado, Gorontalo, and Toudano are situated in the northern part of Celebes; and Dr. Beccari, although he collected at Buton, in the south-west corner of the island, only met with the *Pitta* at Kema, in the north. The presumption at least is, that, even if it is found all over the island, it is more abundant in the northern portion.

Count Salvadori mentions that the specimen shot by Dr. Beccari at Kema had the outermost of the smaller wing-coverts close to the bend of the wing marked with white—a feature not previously noted or figured in the plates of the species which have at present appeared.

Mr. Elliot, in his latest revision of the genus *Pitta* (Ibis, 1870, p. 418), gives the following diagnosis of the species, which I translate:—

Adult. Green: head rufous, with a vertical band of bluish; wings and tail blue; pectoral band cobalt-blue, the throat rufous. In addition to these characters the white spots on the quill-feathers and the scarlet breast are common to the other allied species.

The soft parts are noted by Mr. Wallace to be as follows in freshly-killed specimens:—" Bill blackish-horny; feet dusky lead-colour; iris pale olive."

The figures in the Plate are of the size of life, and are drawn from examples in my own collection.

PITTA RUBRINUCHA, Wall.

PITTA RUBRINUCHA, *Wall.*

Red-naped Pitta.

Pitta rubrinucha, Wallace, P. Z. S. 1862, p. 187.—Id. P. Z. S. 1863, p. 25.—Gray, Hand-list of Birds, i. p. 297. no. 4382 (1869).—Schlegel, Revue Coll. *Pitta* Mus. Pays-Bas, p. 12 (1874).—Salvad. Annali Mus. Civic. Genov. viii. p. 375 (1876).
Brachyurus rubrinucha, Elliot, Monograph of the Pittidæ, pl. xviii. (1863).—Id. Ibis, 1870, p. 418.

ALTHOUGH bearing a general resemblance to the other red-breasted species of the genus *Pitta*, the present bird possesses such a well-marked character in its red nape-spot, that it can be easily distinguished at a glance from all its allies. Another conspicuous and peculiar mark exists in the blue colour on the ear-coverts. It was discovered by Mr. Wallace in the Moluccan island of Bouru, where it has since been met with by the hunters of Mr. Bruijn, who sent four specimens to the museum at Genoa.

Professor Schlegel, in his Review of the Ant-thrushes contained in the Leiden Museum, records a single specimen from the island of Ceram, collected there by Von Rosenberg—a new locality for the species, and one which we should almost suppose would require confirmation, when one thinks of the absolute manner in which these red-breasted Pittas are confined each to his own locality. This seems to be also Count Salvadori's opinion.

Nothing whatever has been written of the habits or mode of life of the red-naped Pitta, which still remains one of the rarest of the genus. I am indebted to Count Salvadori for the opportunity of figuring this scarce species, and I herewith return him my hearty thanks for the loan of the pair which are represented in the accompanying Plate, and which, I believe, are from the collection of the Genoa Museum. I add a short description.

General colour above olive-green, with a broad scarlet patch on the nape; head chestnut-brown, with a blue patch on the crown; sides of face and throat light reddish grey, the ear-coverts bluish; lower throat blackish brown, succeeded by a broad chest-band of pale blue, the rest of the under surface being scarlet; primaries black, with a white spot on two or three of them; bill brown; legs light grey.

The sexes are alike in plumage.

Total length 7 inches, bill $4\frac{3}{4}$, wing $1\frac{1}{8}$, tarsus $2\frac{1}{8}$.

PITTA NOVÆ-GUINEÆ, Müll. & Schleg.

PITTA NOVÆ GUINEÆ, Müll. & Schleg.

New-Guinea Pitta.

Pitta novæ guineæ, Müll. & Schleg. Verh. Nat. Gesch. Ned. Ind., no. 21.
Brève à tête noire, Quoy et Gaim. (nec Cuv.) Voy. Astrol. pl. 8. fig. 3.
Pitta novæ guineæ, G. R. Gray, Proc. Zool. Soc. 1858, p. 175. sp. 39.
Brachyurus novæ guineæ, Bon. Cons. Av. vol. i. p. 256. sp. 24 (1850).
Brachyurus (Melanopitta, Bon.) *novæ guineæ*, Elliot, Mon. of the Pittidæ (1863).
Melanopitta novæ guineæ, Bon. Cons. Voluc. Anisodact. p. 7, no. 197 (1854).

[The above synonyms are taken from Mr. Elliot's 'Monograph of the Pittidæ.']

However rare this *Pitta* might have been when Mr. Elliot wrote his 'Monograph,' it has since become very common, and there are but few collections of birds in which it is wanting. When the 'Astrolabe' visited New Guinea it was only known to come from one locality; since then it has been discovered in the Aru Islands, whence Mr. Wallace brought many examples, while other collectors have greatly added to our stores. One of the most striking features by which the present species may be distinguished from its congeners is the beautiful silvery white line which separates the black of the throat from the peculiar oil-green colour of the chest and flanks. In size it is about equal to *Pitta mackloti*, and rather like, in its general appearance, the splendid bird which bears the name of *rosenbergi*. As regards colour, but little difference occurs in the plumage of the sexes, while in size the male is a trifle larger than the female in all its admeasurements.

Independently of the mainland of New Guinea, where it was discovered by the naturalists of the 'Astrolabe,' and the Aru Islands, as before stated, Mr. Gray adds the island of Salawatti.

"The New-Guinea Pitta," says Mr. Elliot, "was discovered by Messrs Quoy and Gaimard during the first voyage of the 'Astrolabe;' and their type (from which I took my description) was labelled 'Triton Bay.' They state that it has also been killed on the Bay of Dorcy, where, however, it is very rare. It was not considered by them as a distinct species, but merely supposed to be the *P. atricapilla* of Cuvier, from which it can readily be distinguished by the entire absence of blue on the rump, and by having a small white spot only on its primaries.

"It is a rare bird; and my plate of it was executed in Paris, under the direction of my friend Mons. J. P. Verreaux, from the type now contained in the Museum of the Jardin des Plantes."

Head and neck black; back and wings dark green; lesser wing-coverts light blue; primaries dark brown, a white spot in the centre of the fourth, fifth, and sixth; tail brownish green; breast light green, with metallic reflections; abdomen black; crissum and under tail-coverts deep red; bill dark brown; feet and tarsi very light brown.

Total length of male $6\frac{1}{2}$ inches, wing $3\frac{1}{2}$, tail $1\frac{3}{4}$, bill 1, tarsi $1\frac{1}{4}$.

Hab. New Guinea.

The figures in the accompanying Plate are of the size of life.

PITTA ROSENBERGII, *Schleg.*

PITTA ROSENBERGII, Schlegel.
Rosenberg's Pitta.

Pitta rosenbergii, Schlegel, Obs. Zool. v., Ned. Tijdschr. voor de Dierk. iv. p. 16, 1873.

In the richness of its colouring and broad sweeping tints on the under surface *Pitta rosenbergii* will ever rank among the finest of this gorgeous group of birds. The nearest ally to it is the *Pitta novæ-guineæ*; but the differences which occur between them may be easily recognized in the figures of the accompanying Plates. In Rosenberg's Pitta the lively collar of silvery white on the lower part of the throat, so conspicuous in *P. novæ-guineæ*, is wanting; on the other hand, the blue colouring of the flanks is much richer. In size the two birds are about the same, as is also the black colouring of the head and green of the upper surface. Having made these remarks, I will now state all that is known of its history. But on this head I must necessarily be brief; for it is only of late that we have become acquainted with it—it having been first described by my friend Professor Schlegel in 1873, *loco suprà citato*.

Dr. Meyer informs me that it is very restricted in its geographical distribution, and represents *P. novæ-guineæ* in the island of Mysore; where, as he remarks, it is a very interesting insular deviation from the mainland form.

The sexes are similar in their colouring, which may be described as follows:—

Crown of the head, nape, and throat black; back and upper surface generally brownish olive-green; the same colour also pervades the chest, where it borders on the black of the throat; this brownish green colouring of the chest gradually passes into deep blue on the flank; centre of the abdomen and the under tail-coverts rich scarlet; shoulders and a broad mark on the rump beautiful silvery green. Primaries and tail-feathers black; some of the former have a small white spot near their bases. Bill black; tarsi and toes fleshy brown.

Total length 7 inches; wing $4\frac{3}{4}$, tail $1\frac{1}{2}$, tarsus 2, bill $1\frac{1}{2}$.

Hab. Island of Mysore, in the north of Geelvink Bay.

The figures are of the size of life.

COLLOCALIA TERRÆ REGINÆ.

COLLOCALIA TERRÆ REGINÆ.

Queensland Edible Swift.

Cypselus terræ reginæ, Ramsay, P. Z. S. 1874, p. 601.

AUSTRALIA has long been known to possess true Swifts, Swallows, and Martins; but hitherto the genus *Collocalia*, of which numerous species inhabit the surrounding islands, has been conspicuous by its absence. Specimens of an Edible Swift were, indeed, procured on Dunk Island by the late Mr. John Macgillivray during the voyage of the 'Rattlesnake;' but it was not till two years ago that they became known as inhabitants of the continent of Australia. As might have been expected, the Dunk-Island birds are identical with the Australian; and, after careful comparison, I have decided on keeping this new Swiftlet distinct from *Collocalia spodiopygia* of Peale, to which it bears undoubted affinity. The addition of Australia as a habitat for the genus *Collocalia* is of great interest, as its range is decidedly peculiar, extending as it does over the different islands of Oceania, Malasia, India, and it even occurs in Mauritius.

For the acquisition of a specimen of this new Australian bird I am indebted to Mr. Waller, of Brisbane. In a short note this gentleman states:—"This Swallow was collected by Mr. Broadbent on the coast-range of Rockingham Bay. First seen at Dalrymple's Gap, in the morning they appeared to come from the north and returned again in the evening. Before rain, this bird assembles in large flocks, and skims over the ground with great rapidity. They were all leaving about the latter part of June."

The following is Mr. Ramsay's description, copied from the 'Proceedings of the Zoological Society,' where the name of *terræ reginæ* was first bestowed:—

"Whole of the upper surface, except the rump, very dark sooty brown tinged with metallic lustre, being of a darker brown on the outer webs and paler on the inner webs of the wing-feathers; across the rump a greyish-white band having a narrow line of dark brown down the shaft of each feather; whole of the under surface dull greyish brown, of a silky texture and somewhat glossy; under surface of wings and tail and the under tail-coverts of a darker tint, the basal half of all the feathers on the body nearly black; bill black; feet blackish brown; iris dark brown.

"Total length from 4 to 4·2 inches; bill from the nostril 0·1, from forehead 0·2, from angle of the mouth 0·45; wing from flexure 4·4; tail 2·1 to 2·4; tarsi 0·35.

"The sexes are alike in plumage and size. The texture of the plumage is remarkably soft, and to the touch resembles the fur of a Bat.

"This species frequents the north-east coast-ranges near Cardwell, Rockingham Bay, where it is tolerably plentiful, but very difficult to procure, from its small size and swift flight. Small flocks may be seen flying to and fro over the clearer parts of the lower spurs of the coast-ranges; and frequently the same troop returns to the same open ground day after day; towards evening others may be found sweeping over the tops of the scrubs and about precipitous sides of the rocky ridges, where they doubtless breed. I found several young or immature-plumaged birds; and none amongst those I obtained had the tail fully grown. I have never seen this species in any other part of Australia than near Rockingham Bay. It was observed in the neighbourhood of Cardwell during October 1873, and when I left in April 1874 was still numerous there. For the first knowledge of this and several other new and rare species I am indebted to Inspector Robert Johnstone, of the police force on the Herbert river near Cardwell, as well as for much valuable information on the natural history of that interesting region."

The figures in the accompanying Plate are rather under the size of life.

ÆGOTHELES WALLACII, Gray.

ÆGOTHELES WALLACII, Gray.
Wallace's Goatsucker.

Ægotheles wallacii, Gray, Proc. Zool. Soc. 1859, p. 154.—Id. Proc. Zool. Soc. 1861, p. 433.—Finsch, Neu-Guinea, p. 162 (1865).—Schlegel, Nederl. Tijdschr. Dierk. iii. p. 340 (1866).—Gray, Hand-list Birds, i. p. 55, no. 603 (1869).—Sclater, Proc. Zool. Soc. 1873, p. 696.—Meyer, Sitz. k. Akad. Wien, lxix. p. 75 (1874).—Salvad. Ann. Mus. Civic. Genov. x. p. 310 (1877).—Id. Proc. Zool. Soc. 1878, p. 94.—Id. Orn. Papuasia e delle Molucche, ii. p. 526 (1880).—Id. Report Voy. H.M.S. 'Challenger,' p. 77 (1882).—Meyer in Madarász, Zeitschr. ges. Orn. i. p. 278, pl. xvii. fig. 4 (1884).—Guillemard, Proc. Zool. Soc. 1885, p. 630.

? *Caprimulgus brachyurus*, Schl. Nederl. Tijdschr. Dierk. iii. p. 340 (1866, ex Rosenb. MSS.).—Rosenb. Nat. Tijdschr. Nederl. Ind. xxix. p. 143 (1867).—Id. Reis. naar Zuidoostereil. p. 37 (1867).

This interesting Goatsucker was discovered by Mr. A. R. Wallace during his travels in the east, at Dorei in New Guinea, and the type specimen is in the British Museum. It has since been met with in the same locality by Mr. Bruijn's hunters, as well as on the Arfak Mountains by Dr. A. B. Meyer, and in Atam by Signor D'Albertis. It appears to represent in North-western New Guinea the Australian *Ægotheles novæ-hollandiæ*, and it is replaced in South-eastern New Guinea by *Æ. bennetti*. It differs from the latter, and consequently also from *Æ. novæ-hollandiæ*, in its dark coloration, and in having the fore part of the crown varied with rufous, and in exhibiting some whitish-red spots on the scapular feathers.

A Goatsucker from the Aru Islands has been described by Baron von Rosenberg as a distinct species under the name of *Caprimulgus brachyurus*. Count Salvadori, who has examined the type in the Leyden Museum, is of opinion that it is a young bird of the genus *Ægotheles*, with an imperfectly developed tail, probably referable to *Æ. wallacii*; but he has also seen a second specimen from the Aru Islands collected during the 'Challenger' expedition, and he believes that the Aru bird is probably distinct, by reason of its smaller dimensions and more minute vermiculations. We have examined the last-named specimen and find that it fully bears out Count Salvadori's opinion; but it will be better to wait for a larger series of specimens before venturing to separate the Aru bird specifically, as Goatsuckers vary so much in the intensity of their coloration.

Dr. Meyer has recently received a specimen from the same group of islands, where the bird is called by the natives "Tatar faffu." The iris was greyish brown and the feet dark flesh-colour. The egg, which is figured by Dr. Meyer, is cream-coloured, with scribblings of dusky greyish.

We do not give a detailed description of this species, the characters having been well pointed out above.

The Plate represents two adult birds of the natural size, the figures having been drawn from a specimen lent to us by Dr. Guillemard.

[R. B. S.]

PODARGUS OCELLATUS, Quoy et Gaim.

PODARGUS OCELLATUS, Quoy & Gaim.
Ocellated Goatsucker.

Podargus ocellatus, Quoy et Gaim. Voy. Astrol. i. p. 208, pl. 14 (1830).—Less. Compl. de Buffon, Ois. p. 435 (1838).—Gray, Gen. B. i. p. 45 (1846).—Bp. Consp. Av. i. p. 58 (1850).—Id. Parall. Cant. Fissir. Vol. Hianti e Nott. ovvero Insidenti, p. 8 (1857).—Scl. Journ. Linn. Soc. ii. p. 155 (1858).—Gray, Proc. Zool. Soc. 1858, pp. 170, 189.—Id. Cat. B. New Guin. pp. 17, 54 (1859).—Id. Proc. Zool. Soc. 1861, p. 433.—Rosenb. J. f. O. 1864, p. 117.—Finsch, Neu-Guinea, p. 162 (1865).—Schl. Nederl. Tijdschr. Dierk. iii. pp. 340, 341 (1866).—Rosenb. Reis naar Zuidoostereil. p. 36 (1867).—Gray, Hand-list of Birds, i. p. 54, no. 588 (1869).—Meyer, Sitz. k. Akad. der Wissensch. Wien, lxix. p. 209 (1874).—Salvad. Ann. Mus. Civ. Gen. ix. p. 23 (1876), x. p. 309 (1877).—D'Alb. et Salvad. op. cit. xiv. p. 54 (1879).—Salvad. Uccelli di Papuasia e delle Molucche, i. p. 518 (1880).

Podargus superciliaris, Gray, Proc. Zool. Soc. 1861, pp. 428, 433, pl. 42.—Finsch, Neu-Guinea, p. 162 (1865).—Schl. Nederl. Tijdschr. Dierk. iii. p. 341 (1866).—Gray, Hand-list of Birds, i. p. 51, no. 590 (1869).—Meyer, Sitz. k. Akad. Wissensch. Wien, lxix. p. 209 (1874).

Podargus marmoratus, Gould, B. Austr. Suppl. pl. 4.—Gray, Proc. Zool. Soc. 1859, p. 154.—Id. op. cit. 1861, p. 433.—Finsch, Neu-Guinea, p. 162 (1865).—Schl. Nederl. Tijdschr. Dierk. iii. p. 341 (1866).—Gray, Hand-list of Birds, i. p. 54, no. 589 (1869).—Meyer, Sitz. k. Akad. Wissensch. Wien, lxix. p. 209 (1874).—Ramsay, Proc. Linn. Soc. N. S. Wales, iii. p. 264 (1878), iv. p. 97 (1879).—Salvad. Ibis, 1879, p. 322.

It is so difficult to describe in words the exact differences between the plumage of the various species of Goatsuckers, that we do not attempt to give a detailed description of the present species. Its nearest ally is probably *Podargus papuensis*, from which it is easily recognized by its diminutive size, as it is not half the bulk of the former bird.

It has been found in all the parts of New Guinea yet visited by naturalists, and it has also occurred in the islands of the Bay of Geelvink, having been procured in Jobi by Dr. Meyer, and by Dr. Beccari in Miosnom. Mr. Wallace met with the species in Waigiou and also in the Aru Islands, where Baron von Rosenberg and Dr. Beccari likewise obtained specimens. During his explorations in South-eastern New Guinea, Signor D'Albertis found the species near Naiabui and also on the Fly River. Mr. H. O. Forbes has recently obtained several specimens in the Sogeri district of the Astrolabe Range of mountains in the interior of South-eastern New Guinea.

A full account of the variation in plumage in the present species will be found in Count Salvadori's 'Uccelli di Papuasia;' and from the series sent by Mr. Forbes it is evident that there are two distinct phases of plumage, one thickly mottled with white, and the other more uniform rufous-brown. Apparently the latter are the female birds; and this assumption is confirmed by the specimens in the British Museum, where several individuals are emerging from the uniform rufous stage into that of the white-spotted dress of the adult male. On comparing examples from Northern Australia with others from New Guinea, we fail to find any differences to warrant their specific separation, and we have therefore unhesitatingly added *Podargus marmoratus* of Gould as a synonym of *P. ocellatus*.

Baron von Rosenberg states that the "Gongaboel," as it is called by the natives, is the least rare of the Goatsuckers found on the Aru Islands, where they were seen in some numbers at a little distance from Dobbo, in a small wood composed of low shrubs, above which towered some giant Casuarinas, and where some open bare spaces alternated with marshy spots covered with loose grass. According to the same author, the "Gongaboel" is a strictly nocturnal bird, which sleeps during the daytime in the hole of a tree or perched upon some large branch of the *Casuarina*, in which case they are always seated lengthwise and not across the bough. They feed on *Phalenæ, Phasmæ*, &c., which they capture on the wing.

Our illustration represents the adult male of this species of the natural size, and the figure is drawn from a specimen procured by Mr. H. O. Forbes in the Astrolabe Mountains. If Baron von Rosenberg's note as to the way in which the bird sits is correct, then the perching attitude in which our artist, Mr. Hart, has represented the species must be wrong; but he has only followed the traditions of the 'Birds of Australia,' and we suppose that the late Mr. Gould must have seen many *Podargi* at rest on a tree.

[R. B. S.]

MICRODYNAMIS PARVA.

MICRODYNAMIS PARVA.

Dwarf Koel.

Eudynamis parva, Salvad. Ann. Mus. Civic. Genov. vii. p. 986 (1875).
Microdynamis parva, Salvad. Ann. Mus. Civic. Genov. xiii. p. 461 (1878).—Id. Orn. Papuasia e delle Molucche, i. p. 371 (1880).
Rhamphomantis rollesi, Ramsay, Proc. Linn. Soc. N. S. Wales, viii. p. 24 (1884).—Salvad. Ibis, 1884, p. 354.

WE regret greatly that we have not been able to obtain a specimen of the adult male of this very interesting species of Cuckoo, but recognizing the importance of figuring in the present work as many peculiar Papuan genera as possible, we have deemed it better to figure the immature birds than to omit the species altogether.

The male has the head glossy black, and the general aspect of the bird, as well as the colour of its plumage, proclaims its relationship with the Koels or Black Cuckoos of the genus *Eudynamis*. Its short curved bill, however, distinguishes it from the typical Koels, while its small size is a striking peculiarity.

The history of *Microdynamis* is somewhat involved, for the original specimen was sent by Dr. Beccari, along with a number of others, from the island of Tidore in the Moluccas; but with the consignment were one or two New Guinea species, which led Count Salvadori to suspect that the type of *Microdynamis parva* might also have come from New Guinea rather than from Tidore. In this surmise we expect him to be correct, as there can be little doubt that the specimens figured in our Plate are of the same species as the bird described by Count Salvadori. There is no question also that they are the same as the *Rhamphomantis rollesi* described by Mr. Ramsay from Mount Astrolabe. Although closely allied to, and in appearance much resembling, *Rhamphomantis megarhynchus*, the latter has a differently formed bill, and belongs to another section of the family *Cuculidæ*.

Mr. H. O. Forbes has procured two specimens in the Sogeri district of the Astrolabe Mountains, viz. a female (which we presume to be adult) and a young bird, of which the following are descriptions:—

Adult female. General colour above brown, glossed with greenish bronze, with indistinct traces of rufous margins to the feathers; wing-coverts like the back, but a little more rufous, with the rufous margins more pronounced; bastard-wing dusky brown, edged with rufous; primary-coverts and quills brown, edged with rufous and glossed with greenish bronze, especially on the secondaries; upper tail-coverts and tail-feathers brown, glossed with greenish bronze and edged with rufous, the outer feathers slightly freckled with rufous on the inner web; crown of head like the back, and spotted with rufous, with a band of glossy black across the nape; lores dusky; below the eye a streak of white from the base of the bill across the ear-coverts, which are otherwise like the crown; cheeks black, forming a broad band bordering the throat, which is tawny rufous; remainder of under surface of body ashy brown, washed with rufous, with faint indications of dusky cross bars; thighs dusky brown; under tail-coverts pale ashy, washed with rufous and faintly barred with dusky; under wing-coverts and axillaries pale rufous, the latter with indistinct dusky bars; quills below dusky, rufous on the inner edge: " bill blue-black; legs and feet lavender-blue; iris with a red ring" (*H. O. F.*). Total length 7·2 inches, culmen 0·65, wing 3·85, tail 3·4, tarsus 0·7.

The young bird differs in having scarcely any black on the nape or cheeks; the upper surface is more distinctly washed with rufous, especially on the head, which is also plainly barred with dusky; the wings are more rufous, and the mottlings on the inner webs of the tail-feathers are more marked; the throat is ashy, with faint dusky cross bars.

The type specimen, described by Count Salvadori, is probably an adult male, and has the whole of the head and hind neck black. Mr. Ramsay's description also agrees with this; but in his account of the female, which appears to be immature, there seem to be some misprints, as we cannot understand the description as it stands.

The figures in the Plate represent the female and young bird of the natural size; they are drawn from the specimens obtained by Mr. Forbes, and described above.

[R. B. S.]

CALLIECHTHRUS LEUCOLOPHUS.

CALLIECHTHRUS LEUCOLOPHUS.

White-crowned Black Cuckoo.

Cuculus leucolophus, S. Müll. Verh. Land- en Volkenk. p. 22, note, p. 233 (1839-44).—Schl. Handl. Dierk. i. p. 204, pl. iii. fig. 33 (1857).—Gray, Proc. Zool. Soc. 1858, p. 195.—Id. Cat. Mamm. etc. New Guinea, pp. 44, 60 (1859).—Id. Proc. Zool. Soc. 1861, p. 437.—Schl. Mus. Pays-Bas, *Cuculi*, p. 16 (1864).—Gray, Hand-list Birds, ii. p. 216, no. 9012 (1880).—Beccari, Ann. Mus. Civic. Genov. vii. p. 715 (1875).—Id. Ibis, 1876, p. 253.

Simotes albivertex, Blyth, Journ. As. Soc. xv. pp. 15, 283 (1846).—Id. Cat. B. As. Soc. Mus. p. 75 (1849).

Cuculus albivertex, Gray, Gen. B. iii. App. p. 23 (1849).

Symotes leucolophus, Blyth, Cat. B. As. Soc. Mus. p. xix (1852).

Hierococcyx leucolophus, Bonap. Consp. Av. i. p. 104 (1850).—Id. Consp. Volucr. Zygod. p. 7 (1854).—Sclater, Proc. Linn. Soc. ii. p. 166 (1858).—Rosenb. J. f. O. 1864, p. 117.

Calliechthrus leucolophus, Cab. & Heine, Mus. Hein. iv. p. 31 (1862).—Salvad. Atti R. Accad. Torin. xiii. p. 313 (1878).—Id. Ann. Mus. Civic. Genov. xiii. p. 461 (1878).—D'Albert. & Salvad. op. cit. xiv. p. 43 (1879).—Iid. in D'Albert. New Guinea, ii. p. 405 (1880).—Salvad. Orn. Papuasia e delle Molucche, i. p. 358 (1880).

Eudynamis leucolophus, Finsch, Neu-Guinea, p. 159 (1865).

This is a peculiar species of Cuckoo, having the black coloration of a Koel (*Eudynamis*), but with the nostrils of a true Cuckoo (*Cuculus*). Its bill, however, is abnormally broad, and it forms an interesting link between the two genera above mentioned.

It was originally discovered at Lobo in New Guinea by the well-known traveller Solomon Müller, and in the north-western portion of the same island it has been met with at Mum by Dr. Meyer, at Andei by Baron von Rosenberg, and at Warbusi by Dr. Beccari. The latter naturalist says that it is one of the rarest of birds in the north-western portion of New Guinea. Mr. Bruijn has received it from Salwati, and in the south-eastern part of New Guinea Signor D'Albertis met with it on the Fly River, and Mr. Forbes has procured specimens in the Astrolabe Mountains, at Moroka (alt. 5000 feet), and in the Sogeri district at a height of 2000 feet.

So far as is known, the present species is only found in New Guinea and Salwati. Dr. Finsch gives Mysol as a habitat, but apparently in error, as no specimens from this locality are in the Leiden Museum. The late Mr. Blyth described the species as from Borneo; but he afterwards corrected this, and stated that it was from "an islet off the coast of Waigiou," where, however, no recent traveller has obtained it. There is at the same time no improbability in the occurrence of the species in either of the above-mentioned islands.

The following is a description of the pair of birds procured by Mr. H. O. Forbes:—

Adult. General colour above glossy blue-black; quills and tail black, with a gloss of blue-black externally; a broad line of white feathers along the centre of the crown to the nape; sides of face and under surface of body black, the breast and abdomen more ashy, the long under tail-coverts barred near the end and tipped with white; under wing-coverts black, with a few white bars: "bill black; feet blackish lead-colour; iris chestnut-brown" (*D'Albertis*). Total length 12·5 inches, culmen 1·25, breadth at gape 0·55, wing 6·6, tail 6·0, tarsus 0·85.

Young. Differs from the adult in being more dingy black, and in having white bars on the breast, under tail-coverts, and under wing-coverts, and a white tip to the tail-feathers. Total length 12 inches, culmen 1·1, wing 6·3, tail 5·8, tarsus 0·85.

The figures in the Plate represent an adult and young of this curious Cuckoo, of about the natural size; they are drawn from the above-mentioned specimens collected by Mr. Forbes.

[R. B. S.]

CHALCITES MEYERI.

CHALCITES MEYERI.

Meyer's Golden Cuckoo.

Chrysococcyx splendidus, Meyer, Sitz. k. Akad. der Wissenschaften zu Wien, lxix. p. 81 (1874).
Chrysococcyx meyeri, Salvad. Ann. Mus. Civic. Genova, vii. pp. 82, 762 (1875).
Lamprococcyx meyerii, Salvad. Ann. Mus. Civic. Genova, vii. p. 912 (1875).

The little Golden Cuckoos form a natural section of the *Cuculidæ*, and are found all over Africa, India, the Malayan archipelago, Australia, and New Zealand. The African species are certainly the most brilliant; and none of the eastern ones can approach the Emerald Cuckoo (*Chalcites smaragdineus*) for beauty of plumage. In the plate I have endeavoured to illustrate a very common scene in the life of the Australian Golden Cuckoos; and I have no doubt that this New-Guinea species is parasitic on some of the small Warblers which are found in the same country, such as the *Malurus alboscapulatus*, which is the species I have ventured to introduce into my Plate. Dr. Buller, in the 'Birds of New Zealand,' gives a very interesting account of the little Cuckoo of that country (*Chalcites lucidus*) and its breeding-habits; and other notes on these birds will be found in my own and other authors' works.

The peculiar fiery bronze colour of the present species is one of its special characteristics. It was discovered by Dr. Meyer in the Arfak Mountains; and, as the latter gentleman observes, it " has the same brilliant gloss as the African *Chrysococcyx klaasii*, while the other known Golden Cuckoos of the east are not so entirely metallic. It is, moreover, distinguished from the other known species by the absence of any grey or white over the eye and on the cheeks, but is especially remarkable for the fine rust-brown colour of the wings, which in some degree call to mind the ' rufous tint of the upper surface ' of *Chrysococcyx russata* of Gould."

Dr. Meyer named this bird *C. splendidus*; but as that title had already been applied by the late Mr. Gray to a South-African species, Count Salvadori very properly changed it to *C. meyeri*, one of the best and most appropriate names which could have been selected, in my opinion, in acknowledgment of Dr. Meyer's services to science in his celebrated voyages to the East. I should have followed Salvadori in calling this species a *Chrysococcyx*; but having placed all the Golden Cuckoos in my previous works under the genus *Chalcites*, I am obliged, for the sake of uniformity, to relegate this species to the same genus.

The following is a translation of Meyer's original description:—

" Head, cheeks, neck, back, wing-coverts, uropygium, and upper tail-coverts splendidly metallic green and copper-red; only behind the eyes, on the sides of the neck, a large white patch; chin, throat, breast, belly, and under tail-coverts with bands just as brilliant as the upper parts; under wing-coverts also striped, but the stripes brownish grey; wings on the upperside, at the base and at the ends blackish, in the middle reddish brown, and more vividly coloured on the outer webs than on the inner; underside of the wing at the ends grey, elsewhere light reddish brown; upperside of the tail, the two middle rectrices metallic green and copper-coloured, but not as brilliant as the upperside of the body; the other rectrices metallic only on the outer webs, the outermost very feeble, inner webs blackish with a white patch at the end; the outermost rectrice bears on the inner web, on a black ground, five white spots, on the outer web six, the last very narrow; underside of the tail greyish black, with whitish tips, the outermost rectrices on the inner web black, with five white spots, on the outer web lighter, with six white spots.

Total length 160 millims.; bill from the front 12, wings 91, tail 70.

The Plate represents, of the natural size, the type specimen of this species, kindly lent to me by my friend Dr. Meyer.

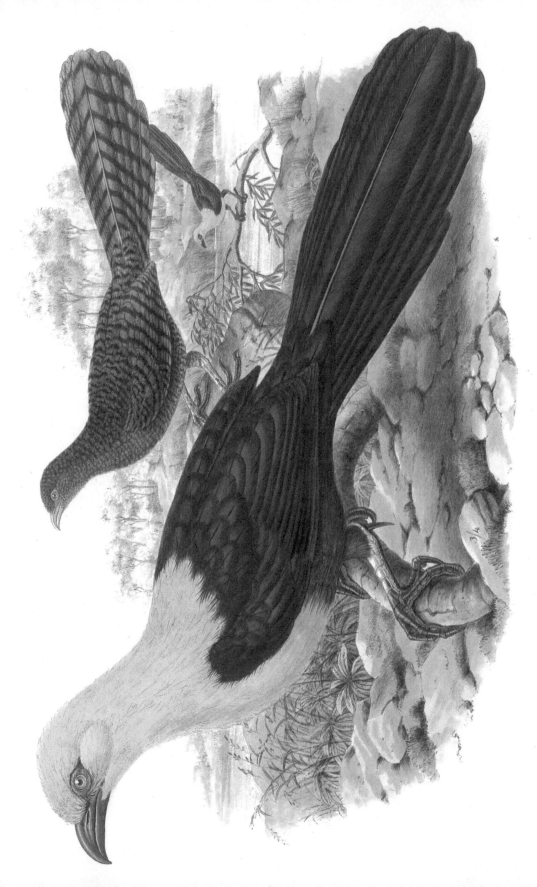

NESONYMPHA

NESOCENTOR MILO.

Solomon-Islands Lark-heeled Cuckoo.

Centropus milo, Gould, Proc. Zool. Soc. 1856, p. 136.—Gray, Cat. B. Tropical Isl. Pacific Ocean, p. 34 (1859).
—Sclater, Proc. Zool. Soc. 1869, p. 124.—Gray, Hand-l. B. ii. p. 213, no. 8974 (1870).—Ramsay, Proc. Linn. Soc. N. S. Wales, iv. p. 69 (1879).
Nesocentor milo, Cab. & Heine, Mus. Hein. iv. p. 120 (1862).—Salvad. Ann. Mus. Civic. Genov. xiii. p. 463 (1878).—Id. Orn. Papuasia e delle Molucche, i. p. 385 (1880).—Grant, Proc. Zool. Soc. 1888, p. 191.

The genus *Nesocentor* was founded in 1862 by Drs. Cabanis and Heine for the reception of several Lark-heeled Cuckoos from the Austro-Malayan Subregion; but apart from their sombre style of coloration, we can see no reason for separating these birds from the genus *Centropus*, though in the present instance we have adopted the nomenclature of Count Salvadori, the leading authority on Papuan ornithology.

The type specimen of the present species was discovered by the late John Macgillivray on the island of Guadalcanar, where it has since been met with by Mr. Woodford and other travellers. The typical example is not quite adult, and was described by Count Salvadori in his work on the birds of New Guinea. He appears afterwards to have entertained some doubt as to the specimen described by him in England having been really the type; and on requesting Dr. Sclater to re-examine the specimen, he was assured by the latter gentleman that it did not exist in the Museum. How this mistake arose we cannot say, and we have no immediate recollection of a visit from Dr. Sclater to examine the specimen in question; it may have been temporarily mislaid during the removal of the Natural History collections down to South Kensington, but we are happy to say that it is quite safe in the national collection.

As far as is known, the present species is only found in the island of Guadalcanar, in the Solomon group, where it replaces the smaller *N. ateralbus* of New Ireland, which is a violet-black bird with a white head. The latter is also said to inhabit the Solomon Islands on the faith of a collection sent by Mr. Krefft to Dr. Sclater in 1871. So many birds in this collection really came from New Ireland, and not from the Solomons, that we may fairly suppose that the locality for *N. ateralbus* is wrongly recorded.

In *N. milo* the adult male is black, sides of the body greenish black with a steel-green gloss; the head, neck, mantle, throat, and breast creamy white; the abdomen black. Total length 28 inches, culmen 2·3, wing 10·1, tail 13·5, tarsus 2·65. The young is rufous streaked with black, and somewhat resembles the adults of other Lark-heels.

The figures in the Plate are taken from an adult male and a young female shot on Guadalcanar by Lieut. Reginald Tupper, R.N., and presented by him to the British Museum. He says that the iris was yellow or orange. Mr. Woodford gives the iris as red, and the bill and feet black, in an adult male from Aola. Another adult male had a brown iris and grey feet, while in an adult female and an immature bird the iris was dark grey and brown respectively.

[R. B. S.]

CEYX SOLITARIA.

CEYX SOLITARIA.

Solitary Kingfisher.

Ceyx meninting, Less. Voy. Coquille, Zool. i. p. 691 (1826).—Cuv. Règne An. i. p. 444 (1829).—Less. Traité, p. 241 (1831).—Pucher. Rev. et Mag. 1861, p. 345.
Ceyx solitaria, Temm. Pl. Col. 595. fig. 2 (1836).—Gray, P. Z. S. 1858, p. 172; 1859, p. 155; 1861, p. 433.—Sharpe, P. Z. S. 1868, p. 271.—Id. Monogr. Alced. pl. 38 (1868).—Salvad. Atti R. Accad. Torino, iv. p. 474 (1869).—Gray, Handl. B. i. p. 95 (1869).—Salvad. Ann. Mus. Civ. Genoa, vii. p. 765 (1875).
Alcyone solitaria, Gray, Gen. B. i. p. 82 (1847).—Bp. Consp. i. p. 158 (1850).—Reich. Handb. *Alced.* p. 7, taf. cccxcviii. fig. 3067 (1851).—Bp. Consp. Vol. Anis. p. 10, sp. 358 (1854).—Rosenb. J. f. O. 1864, p. 118.
Alcedo solitaria, Schl. M. P. B. *Alcedines*, p. 17 (1863).—Id. Voy. Nederl. Ind. *Alced.* pp. 12, 48, pl. 3. fig. 5 (1864).—Id. Revue Alced. p. 9 (1874).

This little three-toed Kingfisher is an inhabitant of New Guinea and the neighbouring islands. The original type came from the Bay of Lobo; and the Leiden Museum has specimens from Sorong, Dorey, and Andai, collected by the late Dr. Bernstein and by Von Rosenberg. In the island of Salwatty, Bernstein collected examples at Kalwal and Sailolo; and it has been sent from the island of Batanta by Von Rosenberg: in the last-named locality Dr. Beccari also met with it. Mr. Hoedt is our authority for the occurrence of the bird in Mysol, five individuals being in the Leiden Museum from Waaigama and Kasim. In his original 'Catalogue of the Kingfishers of the Leiden Museum,' Professor Schlegel recorded a specimen from Ceram; but Mr. Wallace at the time doubted the occurrence of the species in that island, and it now seems that the specimen in question was a Mysol one. Lastly, it occurs in the Aru Islands, whence we saw a large series in the collection recently formed there by Mr. Cockerell. Von Rosenberg's localities are Wokam and Wonoumbai. Mr. Cockerell is likewise our authority for the occurrence of the present species in Australia, as specimens were in his last collection from Cape York, and he assures me that Mr. Jardine had obtained them in the neighbourhood of Somerset. Beyond this, I have never in all my experience heard of the species nor seen a single example from the Australian continent.

All that is at present written concerning its economy is contained in the following brief note, communicated by Mr. Wallace to Mr. Sharpe's 'Monograph.' He observes that " the stomachs of those he killed contained the remains of water-beetles and other insects."

The present species is distinguished first of all by having only three toes, a peculiarity of the genera *Ceyx* and *Alcyone*; and from all the members of the latter it differs by its black bill and yellow belly.

In the Plate the bird is represented of the natural size, drawn from an Aru-Island specimen in my collection.

CEYX GENTIANA, *Tristram*.

CEYX GENTIANA, *Tristram.*

Gentian Kingfisher.

Ceyx gentiana, Tristram, Ibis, 1879, p. 438, pl. xi.—Salvad. Orn. Papuasia e delle Molucche, i. p. 423 (1880).

CANON TRISTRAM has kindly lent us the type specimen of this lovely Kingfisher, which was discovered by Lieut. Richards in Maikara Harbour, in the island of San Christoval, one of the Solomon group. The original specimen is in the collection of Canon Tristram, who pointed out the characters of the species in 'The Ibis' (*l. c.*). The plate by Mr. Smit, accompanying his paper, gives a poor representation of the species, the colouring of the bird being quite wrong in tint, and the long scapulars being left out altogether. The species is, however, a very distinct one, the only other *Ceyx* with a black bill being *C. solitaria*, which is yellow below.

The following is a description of the typical specimen:—

Adult female (type of species). Mantle and scapulars black, washed with purplish blue, the feathers being mesially streaked with dark cobalt; entire back, rump, and upper tail-coverts bright silvery greenish cobalt, a little more purplish on the latter; lesser, median, and inner greater coverts black, tipped with purplish blue, with a narrow mesial line of cobalt; remainder of the greater coverts, bastard-wing, primary-coverts, and quills black, slightly shaded with purplish blue externally; tail-feathers black, edged with purplish blue; crown of head black, all the feathers tipped with purplish blue, each feather with a central line of brighter cobalt; the spots thicker on the nape and hind neck, which appear almost entirely purplish blue; lores and base of forehead black, with a white loral spot; feathers round eye, ear-coverts, and cheeks black, the feathers mesially streaked with purplish blue; throat and under surface of body white, as also a longitudinal patch on the sides of the neck; the upper breast with a patch of black feathers washed with blue; flanks with black streaks; thighs blackish; under tail-coverts white, tipped with purplish blue; under wing-coverts and axillaries white, with a black bar across the former; quills below dusky, white along the edge of the inner web: "bill black; feet flesh-colour; iris bluish black" (*Richards*). Total length 5·5 inches, culmen 1·55, wing 2·55, tail 1·1, tarsus 0·45.

The Plate illustrates the species, of the natural size, in two positions. The figures are drawn from the specimen lent to us by Canon Tristram.

[R. B. S.]

TANYSIPTERA CAROLINÆ, *Schl.*

TANYSIPTERA CAROLINÆ, Schlegel.
Blue-breasted Tanysiptera.

Tanysiptera carolinæ, Schl. N. T. D. iv. p. 13 (1873, ex Von Rosenb. MS.).—Id. Mus. P.-B. Revue *Alcedines*, p. 42 (1874).—Von Rosenb. Reist. Geelvinkbaai, p. 137, pl. 14. fig. 1 (1875).

No one can mistake this splendid Kingfisher for any species previously described; for it differs conspicuously from all the other long-tailed *Tanysipteræ*. Of the genus *Tanysiptera* Mr. Sharpe admits in his 'Monograph' twelve species; and this is the only new one which has been discovered since he completed his labours on the group. Of the twelve species enumerated by him, ten have white breasts, while *T. nympha* (from New Guinea) and *T. sylvia* (from North-east Australia) have the breasts vermilion and cinnamon-colour respectively. It needs, therefore, no more than a glance at the blue under surface of the present bird to see that in this respect it is not approached by any other member of the genus; and one cannot but congratulate Baron von Rosenberg on his discovery of such a fine novelty.

The following is the account given by Professor Schlegel in his description of the bird:—

"This new species was discovered by M. von Rosenberg in the island of Méfoor, situated in the great Bay of Geelvink, and is distinguished at a glance from all the other species known up to the present day by the under surface being of the same deep blue as the upper, instead of being white or reddish. It will be sufficient to point out the following facts to give an idea of this beautiful bird.

"Size, form, beak, and feet as in *Tanysiptera dea*.

"*Male and female adult.*—General colour a bluish violet, passing to purple cobalt on the upper part of the head, to blackish on the quills and under surface of the wings, and to pure white on the rump and anal region; tail-coverts and tail white, but the contracted part of the tail-feathers blue; this part is very narrow and constantly with the webs almost totally worn; beak of a lively red.

"*Young of both sexes.*—Bill blackish; upper parts of the bird with the sides of the head and the lower surface of the wings coloured as in the adult; lower parts clear rufous, varied with blackish excepting on the chin and the middle of the throat; rump and upper tail-coverts pure white more or less washed with rufous, with blackish streaks; under tail-coverts white; tail-feathers white for the length of their shafts, the rest pale black passing on the upper surface to more or less dark blue; the two centre feathers are but slightly elongated, a little contracted, but with very distinctly indicated spatules. Wing from 3″ 10‴ to 4″ 2‴, centre tail-feathers 6″ to 7″ 3‴, bill from front 7‴. It is a noteworthy fact that M. von Rosenberg does not appear to have met with *Tanysipteræ*, during his voyage to the Geelvink Islands, in Méosnoum, or in Jobie."

Bill, in adult, fine red; feet purplish brown.

My Plate is drawn from a fine specimen sent to me in exchange from the Leiden Museum; and the bird is represented of about the natural size.

TANYSIPTERA NYMPHA, G.R. Gray.

TANYSIPTERA NYMPHA, *G. R. Gray*.

Red-breasted Tanysiptera.

Tanysiptera nympha, Gray, Ann. & Mag. Nat. Hist. 1841, p. 237.—Id. Gen. B. i. p. 78 (1846).—Id. Cat. Fissirostres Brit. Mus. p. 59 (1848).—Bp. Consp. Gen. Av. i. p. 157 (1850).—Gray, P. Z. S. 1860, p. 347.—Wallace, P. Z. S. 1862, p. 165, 1863, p. 24.—Schl. Nederl. Tijdschr. Dierk. ii. p. 250 (1866).—Sharpe, P. Z. S. 1869, p. 631.—Id. Monogr. Alced. p. 269, pl. civ. (1869).—Sclater, P. Z. S. 1873, p. 697.—Schl. Mus. Pays-Bas, Alced. Revue, p. 41 (1874).—Salvad. Ann. Mus. Civic. Genov. x. p. 127 (1877).

Dacelo dea, pt., Schl. Mus. Pays-Bas, Alced. p. 43 (1863).—Id. Vog. Nederl. Ind. Alced. pp. 33, 62 (1864).

Alcedo dea, Martens, J. f. O. 1864, p. 18.

The first specimen of this beautiful Kingfisher, described nearly thirty years ago from a native skin, still exists in the British Museum with the wings of a *Halcyon* attached to it. The majority of skins which reach us even to this day are prepared by the native collectors, and arrive in a flat state, without wings. The first person who appears to have obtained a perfect skin was Mr. Wallace, who, in 1869, contributed the following note to Mr. Bowdler Sharpe's 'Monograph.' He writes:—" This rare species was obtained by my assistant, Mr. Allen, in the interior of the north-west peninsula of New Guinea; and the single specimen he obtained is, I believe, the only complete and authentic one in Europe. New Guinea is the only island which possesses more than two species of this genus, and may therefore be looked upon as its metropolis, and as more likely than any other part of the world to produce new forms of Kingfishers." Since Mr. Wallace's 'Voyage' one or two mutilated skins have been seen in this country. It appears to be entirely confined to North-western New Guinea; and the first idea, that it was a Philippine bird, is now admitted on all sides to have been a palpable mistake. The Leiden Museum has received a single specimen from Sorong; and D'Albertis procured a couple of individuals during his stay in North-west New Guinea. Beccari also met with it, and observes:—"Of *Tanysiptera nympha* I could only get one specimen. It is not very rare among the mangroves near Ramoi, and in the low places surrounding it. On several occasions it was met with by my hunters without their being able to kill it. It is wanting near Dorey, but reappears at Rubi, in the south of the Bay of Geelvink, a locality which seems interesting, and which I was sorry not to have visited, because several species which are only known from Salwatti and Sorong are found there, among others *Seleucides alba*."

The present species is distinguished at once from all the other *Tanysipteræ* by its red rump and under surface. I give the following description from Mr. Sharpe's 'Monograph.'

Head rich ultramarine, inclining to cobalt on the nape; ear-coverts and upper part of the back and scapulars jet-black; lower part of the back, rump, and upper tail-coverts rich vermilion; wing-coverts black, edged with ultramarine, the innermost ones rich cobalt; quills black, the inner web light rufous from the base, the outer web very narrowly washed with blue; tail-feathers deep blue, the interior ones tipped with white, the two middle tail-feathers rich blue, the spatula long and white, with a faint blush of rose; entire under surface rich vermilion, a little paler on the abdomen; bill and feet sealing-wax-red; eyes black. Total length 11 inches, of bill from front 1·2, from gape 1·7, wing 3·6, tail 2·8, middle rectrices 0·6, tarsus 0·3, middle toe 0·6, hind toe 0·2.

The figure in the Plate is life-sized, and is drawn from one of the specimens obtained by Signor D'Albertis, and I have been indebted for the loan of it to Dr. Sclater, through whose hands D'Albertis's collection passed.

TANYSIPTERA DANAE, *Sharpe.*

TANYSIPTERA DANAE, *Sharpe.*

Crimson-and-brown Kingfisher.

Tanysiptera nympha juv., Ramsay, Proc. Linn. Soc. N. S. Wales, iv. p. 4 (1879).
Tanysiptera danae, Sharpe, Ann. & Mag. Nat. Hist. fifth series, vol. vi. p. 231 (1880).

I am not surprised that my friend Mr. E. P. Ramsay, when he first received a specimen of this bird, took it to be the young of the beautiful *Tanysiptera nympha*; and had I seen but one example, I believe I should have done the same. In the collection, however, which was sent by Mr. Charles Hunstein to this country there were at least twenty specimens, and both old and young birds were represented in plenty, and Mr. Sharpe was no doubt right in considering it to be a new species.

The home of this beautiful bird is the interior of South-eastern New Guinea; and here it represents the *Tanysiptera nympha*, which inhabits the north-western portion of the island. It is somewhat allied to the last-named bird, having the under surface of the body crimson, as also the lower back and rump, characters which are also possessed by *T. nympha*. It differs, however, in having the head and back brown instead of black, and by the rufous-brown colour of the side face.

We owe the discovery of this species to the energy of a young German naturalist, Mr. Charles Hunstein, who has been collecting in South-eastern New Guinea, round East Cape, Milne Bay, and the adjacent parts, where he discovered likewise the wonderful new Kingfisher, named by Mr. Sharpe *Clytoceyx rex*, and figured in the present work.

We know nothing of the habits of this new *Tanysiptera*; but there is no reason to suppose that they would differ from those of its congeners. The measurements of the adult are as follows, according to Mr. Sharpe:—
Total length 10·5 inches, culmen 1·25, wing 3·45, tail 6·4, tarsus 0·6.

On the arrival of the specimens at the British Museum, Mr. Sharpe kindly showed them to me; and I was thus able to draw the accompanying Plate from the typical specimens. It represents two old birds and a young one, all being of about the natural size. [R. B. S.]

TANYSIPTERA NIGRICEPS, Sclater.

TANYSIPTERA NIGRICEPS, Sclater.
Black-headed Tanysiptera.

Tanysiptera nigriceps, Sclater, P. Z. S. 1877, p. 105.—Salvadori, Ann. Mus. Civic. Genov. x. p. 303.

WRITING in 1871, the late Mr. Blyth insisted strongly on the fact that the Cape-York Peninsula of Australia ought to be united in a zoo-geographical sense to the Papuan subregion as we now understand it. The main reason was, of course, the presence of a Cassowary in the above-named peninsula; but, in addition to this, there are also some very remarkable forms of birds which are of a Papuan type and were known to occur only in New Guinea and in the north-eastern peninsula of Australia, such, for instance, as *Pitta mackloti* and *Tanysiptera sylvia*. It is more particularly of the latter that I would now speak.

The last-named beautiful Kingfisher, which I had the pleasure of describing as long ago as 1850, differed at once from all the other *Tanysipteræ* in its fawn-coloured under surface; and for many years it was a very rare bird in European museums. More recently, however, numerous specimens have been collected in the Cape-York Peninsula, and forwarded to this country; but its occurrence in New Guinea was until lately somewhat conjectural. All doubt on this point has now been set at rest by the researches of Signor D'Albertis, who procured two specimens in Hall Bay, in South-eastern New Guinea; for, although Count Salvadori mentioned to me certain differences of plumage observable in the New-Guinea birds, I believe that these are not any of specific importance, and that they are indications of immaturity in the examples examined by him.

Now, as most of my readers are aware, there lie to the eastward of New Guinea certain large and important islands (New Britain, New Ireland, and the Solomon Islands); and the more we know of the zoology of these islands the more do we see that their fauna is Papuan in its character, having little in common with Australia or with Oceania. Of New Britain and New Ireland we know very little ornithologically; but our knowledge has been increased during the past twelve months by the very interesting investigations of Mr. George Brown, C.M.Z.S., who sent a collection of birds from the above-named localities and from Duke-of-York Island to Dr. Sclater, by whom an account of the collection has been published in the 'Proceedings of the Zoological Society' for 1877. The specimens themselves have now become the property of the Marquis of Tweeddale, to whom I am indebted for the opportunity of figuring some of the most interesting novelties. Most of the species sent by Mr. Brown are either Papuan or at least are closely allied to New-Guinea species; and it was highly interesting to find in the Kingfisher here figured a representative, not of the brilliant blue-and-white forms, such as *Tanysiptera galatea*, *T. nais*, &c., but of *T. sylvia*, the Cape-York species. Like the latter, it wants the great racket or spatule at the end of the long tail-feathers, such a pronounced feature in most of the *Tanysipteræ*, and it has the same tendency to cross these two long rectrices; in fact it is very closely allied to *T. sylvia*, but is at once recognizable by its black head.

The following is a description of the type specimen:—

Head, neck, mantle, and scapulars jet-black, as also the lores, sides of face, and ear-coverts; in the centre of the back a large dorsal patch of white; rump also white; upper tail-coverts particoloured, white on the inner web, black on the outer one; lateral tail-feathers black, washed with blue on the outer web; the two long centre feathers white, with slightly dilated ends, the shafts white, the outer edges blue; the next feather on each side black, the shafts black, white for the greater part of the inner web, washed with blue on the outer; wings blue, more brilliant on the least coverts adjoining the back, the quills black on the inner webs; cheeks and under surface of body light cinnamon; the under tail-coverts whitish; flank-feathers somewhat varied with black edgings, as also the thigh-feathers; sides of the lower back black, bordering the white rump; under wing-coverts cinnamon-colour, like the breast, some of the outermost and greater series edged with blackish; quills dusky below, whitish along the base of the inner web; bill coral-red.

Total length 11·7 inches, culmen 1·5, wing 3·65, tail 7·0, tarsus 0·6.

The figure in the Plate is drawn from the typical example, and represents the species of about the natural size.

TANYSIPTERA MICRORHYNCHA, Sharpe.

TANYSIPTERA MICRORHYNCHA, *Sharpe*.

Port-Moresby Racket-tailed Kingfisher.

Tanysiptera galatea, var. *minor*, Salvad. et d'Albert. Ann. Mus. Civic. Genov. vii. p. 815 (1875).—Salvad. op. cit. ix. p. 21 (1876).—Id. Proc. Zool. Soc. 1876, p. 752.—D'Albert. & Salvad. Ann. Mus. Civic. Genov. xiv. p. 49 (1879).—Ramsay, Proc. Linn. Soc. N. S. Wales, iv. p. 90 (1879).
Tanysiptera microrhyncha, Sharpe, Proc. Linn. Soc. xiii. pp. 311, 493 (1878), xiv. pp. 629, 686 (1879).
Tanysiptera galatea, pt., Salvad. Orn. Papuasia e delle Molucche, i. p. 439 (1880).—Finsch, Vögel der Südsee, p. 24 (1884).

The Racket-tailed Kingfisher from South-eastern New Guinea is closely allied to the ordinary *T. galatea* of the north-western portion of the island, but it is decidedly smaller, and has the back black with less blue. Whether the latter is a good character remains to be proved, as we find that it varies somewhat, a specimen collected by Mr. Broadbent having the back strongly washed with purplish blue, contrasting with the head and wing-spot, which were rich silvery cobalt. It is probable that this is the very old male bird, but the full changes of plumage in the species require to be worked out.

In South-eastern and Southern New Guinea the present species seems to have a wide range. Mr. Hunstein met with it at East Cape, and the Rev. W. G. Lawes at Walter Bay. It also occurs in the interior portions of the Port Moresby district, as Dr. Finsch records it from the Goldie and Laloki Rivers, Mr. Goldie in the Morocco district, and Mr. H. O. Forbes in the Sogeri district of the Astrolabe Mountains. Signor D'Albertis procured specimens at Naiabui, Yule Island, Mount Epa, and Hall Bay; he also met with the species on the Fly River.

The following description is a copy of the one originally given by us (*l. c.*):—

Adult female. Crown of head and nape cobalt-blue, a little brighter round the edge of the crown and on the nape, and slightly dashed with silvery cobalt over the eye, the forehead rather duller blue; lores, feathers below the eye, and the ear-coverts black, slightly washed with deep blue, as also are the sides of neck, mantle, back, and scapulars, on which, however, the blue shade is not so apparent; wings black, the wing-coverts and secondaries externally deep purplish blue, the innermost of the greater and median coverts bright cobalt, forming a shoulder-patch; lower back, rump, and upper tail-coverts pure white, the latter with an indistinct narrow fringe of dark blue; tail white, the outer feathers more or less edged with blue towards the tip, the two long centre feathers blue, with a long white spatula, the base of these two plumes irregularly white in the centre and along the inner edge, the next two feathers with remains of blue near the base of the shaft; moustache and entire under surface of body pure white, the sides of the upper breast black; under wing-coverts white, the lesser plumes on the upper band of the wing blue-black, shading into brighter blue on the edge of the wing itself; bill red; feet dusky. Total length 15·5 inches, culmen 1·3, wing 4, tail 3·9, centre feathers 9·7, tarsus 0·6.

The specimen figured in the Plate represents an adult bird procured by Mr. H. O. Forbes in the Sogeri district in South-eastern New Guinea.

[R. B. S.]

SAUROMARPTIS GAUDICHAUDI.

SAUROMARPTIS GAUDICHAUDI.

Gaudichaud's Kingfisher.

Dacelo gaudichaudi, Quoy & Gaim. Voy. de l'Uranie, p. 112, pl. xv.—Gray, Cat. Fissirostres Brit. Mus. p. 52.—
Id. Gen. B. i. p. 78.—Bonap. Consp. Av. i. p. 154.—Cass. Cat. Halcyonidæ Philad. Mus. p. 14.—Sclater,
Proc. Linn. Soc. ii. p. 155.—Schlegel, Mus. Pays-Bas, Alced. p. 20.—Id. Vog. Nederl. Indië, Alced.
pp. 13, 49, pl. iv.—Gray, Hand-list of Birds, i. p. 89, no. 1063.—Sharpe, Monogr. Alced p. 295, pl. cxvi.
—Id. Journ. Linn. Soc. xiii. pp. 313, 493.—Ramsay, Proc. Linn. Soc. N. S. Wales, i. p. 389.—D'Albert.
Ann. Mus. Civic. Genov. x. p. 19.—Sharpe, Linn. Soc. Journal, xiv. p. 686.
Choucalcyon gaudichaudi, Lesson, Traité d'Orn. p. 248.—Bonap. Consp. Volucr. Anis. p. 9.
Monachalcyon gaudichaudi, Reichenbach, Handb. Alced. p. 37, Taf. ccccxxv. fig. 3156.
Sauromarptis gaudichaudi, Cab. & Heine, Mus. Hein. Th. ii. p. 164.—Salvad. Ann. Mus. Civic. Genov. vii. p. 765.
—Id. & D'Albert. tom. cit. p. 816.—Salvad. op. cit. viii. p. 398.—Id. op. cit. ix. p. 21.—Id. op. cit. x.
pp. 128, 306.—D'Albert. & Salvad. op. cit. xiv. p. 53.

No greater proof could be given, of the great progress which ornithology has made during the last ten years, than a comparison of the localities which this species was known to inhabit in the year 1869 with the list of habitats which are enumerated by Count Salvadori in his 'Prodromus' of the Kingfishers of New Guinea and the Papuan Islands. Originally discovered in New Guinea and Guébeh by Messrs. Quoy and Gaimard during the voyage of the 'Uranie,' the localities of Waigiou, Mysol, and the Aru Islands were added by Mr. Wallace. Mr. Sharpe also includes Ceram, on the authority of specimens in the Leiden Museum; but Count Salvadori omits this island from his list, and it seems doubtful if the species has ever really occurred there. Count Salvadori has examined one hundred and twenty specimens collected in the Papuan Islands by the recent Italian explorers Beccari and D'Albertis, as well as the Dutch voyagers Bruijn, Bernstein, and von Rosenberg; and the following places are given by him as habitats for this fine species :—N.W. New Guinea, Dorey, Mansinam, Andai, Warbusi, Wairoro, Dorei-Hum, Sorong, Kukuladi, Lobo, and the following islands—Salawati, Batanta, Waigiou, Guébeh, Jobi, Miosnom, Mysol, and the Aru Islands; while it has also been found in South-eastern New Guinea. In this part of the great Papuan island D'Albertis has found the species on the Fly River and in Hall Bay, the Rev. Mr. Lawes at Hood Bay (60 miles east of Port Moresby), Dr. James at Nicura, and Mr. Stone on the Laloke river.

Messrs. Quoy and Gaimard give the following note on this Kingfisher in their original account of the bird :—"This species, to which we have given the name of our friend and colleague the botanist attached to the expedition, inhabits the woods of the Papuan Islands. The aborigines call it *Mangrogone* and *Mankinetrous*; the inhabitants of Guébé call it *Salba*,—these being the names employed by the islanders for all Kingfishers. It is not shy, and is easily approached. The individuals that we killed had their beaks still covered with the earth in which they had been digging to procure their food." Mr. Wallace states that he found the species not uncommon in swampy jungle, where its curious loud barking was often heard and was sometimes mistaken for that of a dog. It feeds on Crustacea, butterflies, Mollusca, and Myriopoda.

The following descriptions are given by Mr. Sharpe in his 'Monograph :'—

"*Adult male.*—Crown of the head, cheeks, ear-coverts, upper part of the back, and scapularies deep black; in very old birds a few of the feathers edged with bright blue; a patch of feathers along the base of the upper mandible, a stripe behind the eye, and a collar round the neck ochre; a spot on the occiput white; wing-coverts black washed with bright cobalt; quills blackish, the inner web light ochre from the base, the outer web edged with deep indigo, more especially on the secondaries; lower back and upper tail-coverts bright silvery blue; tail deep indigo above, black underneath; throat pure white; sides of neck and under wing-coverts white tinged with light ochre; rest of under surface of body deep chestnut; bill light yellow, the upper mandible tinged with black; feet black. Total length 11·8 inches, culmen 1·8, wing 5·3, tail 4·0, tarsus 0·85.

"*Female.*—Similar to the male, but having the colours not quite so bright, and the tail reddish."

MELIDORA MACRORHINA, Less.

MELIDORA MACRORHINA.
Hook-billed Kingfisher.

Dacelo macrorhinus, Less. Voy. Coquille, Zoologie, Atlas, pl. 31 bis, fig. 2 (1826).—Id. Féruss. Bull. Sci. Nat. xii. p. 131 (1827).—Id. Voy. Coquille, Zool. i. p. 692 (1828).—Id. Man. d'Orn. ii. p. 94 (1828).— Gray, Cat. B. New Guinea, p. 19 (1859).—Id. Proc. Zool. Soc. 1859, p. 154.

Melidora euphrosiæ, Less. Traité d'Orn. p. 259 (1831).—Id. Compl. Œuvres de Buffon, Oiseaux, p. 653 (1838).— Bp. Consp. Gen. Av. i. p. 150 (1850).—Finsch, Neu-Guinea, p. 160 (1865).

Dacelo macrorhynchus, Less. Traité d'Orn. p. 249 (1831).

Melidora macrorhyncha, Gray, List Gen. Birds, i. p. 10 (1840).—Id. List Gen. Birds, p. 14 (1841).—Id. Handl. B. i. p. 89, no. 1067 (1869).—Ramsay, Proc. Linn. Soc. New S. Wales, iii. p. 252 (1878), iv. p. 97 (1879).

Dacelo macrorhynchus, Gray, Genera of Birds, i. p. 78 (1846).

Melidora macrorhina, Reichenb. Handb. spec. Orn. *Alced.* p. 41, sp. 99, Taf. 428. figs. 3166-7 (1851).—Sclater, Journ. Linn. Soc. ii. p. 156 (1858).—Sharpe, Mon. Alcedinidæ, p. 120 (1871).—Beccari, Ann. Mus. Civ. Genova, vii. p. 708 (1875).—Salvad. tom. cit. p. 766, viii. p. 398 (1876).—Cab. & Reichenow, Journ. f. Orn. 1876, p. 323.—Salvad. Ann. Mus. Civ. Genova, xi. pp. 128, 303; xiii. p. 319 (1878).—Id. Orn. Papuasia, &c. i. p. 500 (1881).

Melidora euphrosinæ, Reichenb. loc. cit.; Rosenb. Nat. Tijdschr. Nederl. Ind. xxv. p. 230 (1863).—Id. Journ. f. Orn. 1861, p. 117.

Melidora euphrasiæ, Bp. Consp. Vol. Anisod. p. 9 (1854).

Dacelo macrorhynchus, Gray, Proc. Zool. Soc. 1858, p. 189.—Id. Cat. B. New Guinea, p. 54 (1859).—Id. Proc. Zool. Soc. 1861, p. 433.

Dacelo macrorhina, Schlegel, Mus. Pays-Bas, *Alcedinidæ*, p. 22 (1863).—Id. Vog. Ned. Ind. *Alced.* pp. 17, 51, pl. 4. fig. 1 (1864).—Id. Mus. Pays-Bas, *Alced.* (Revue), p. 18 (1874).—Giebel, Thes. Orn. ii. p. 7 (1875).

Melidora goldiei, Ramsay, Proc. Linn. Soc. New S. Wales, i. p. 369 (1876).

Melidora collaris, Sharpe, Journ. Linn. Soc. xiii. p. 313 (1877).

It is to Count Salvadori that we owe our knowledge of the plumages through which this extraordinary form of Kingfisher passes. Mr. Bowdler Sharpe, in his Monograph of the family, seems to have been acquainted with the female bird only; and at the time that he wrote, very few specimens existed in European museums. More recently, however, Mr. Bruijn and the well-known Italian traveller Signor D'Albertis have forwarded to the Genoa museum a large series of specimens, while it has also been obtained by other naturalists in South-eastern New Guinea. The late Dr. James met with it near Hall Bay; and Mr. Ramsay has recorded it from the interior of the country near Port Moresby. We have also seen specimens collected near East Cape by Mr. Hunstein, and by Mr. Goldie on the Astrolabe range.

The south-eastern specimens were described independently by Mr. Ramsay and Mr. Sharpe as a distinct species; but Count Salvadori compared the type of *Melidora collaris* with others from North-western New Guinea, and feels certain that it is nothing but the ordinary adult male of *M. macrorhina*.

The present species was discovered by Lesson during the voyage of the 'Coquille,' near Dorey, where also Mr. Wallace and Von Rosenberg met with it. It has likewise been found by the latter collector, and by Signor D'Albertis at Andai; while Dr. Beccari met with it at Warbusi, and D'Albertis at Ramoi. It inhabits likewise the islands of Salawati, Batanta, Waigiou, and Mysol. Professor Schlegel states that the specimens in the Leyden Museum from Waigiou differed in having the spots on the back and wings clearer, and of a bright greenish yellow; but Count Salvadori could not find any points of difference in individuals collected in that island by Beccari and Bruijn. We are still without any exact information as to the habits of this peculiar Kingfisher, whose hooked bill would seem to be adapted for some special purpose in the capture of its prey.

I here translate the description of the sexes given in Count Salvadori's work, from which also the synonymy of the species has been derived.

Adult male. Head black, feathers margined with blue, a spot on each side of the forehead, extending above the eyes, rufous; cheeks and ear-coverts black, the former separated from the latter by a whitish band starting from the angle of the mouth; round the neck a black collar, succeeded by a white one; remainder of the upper surface dusky brown, with broad ochraceous yellow margins to the feathers, those of the rump and upper tail-coverts more yellow; quills dusky, the secondaries externally margined with rufous, internally with broad fulvous edges; beneath dull white, with very narrow dusky edges to the feathers of the throat and upper breast; under tail-coverts rufescent; tail-feathers brownish tipped with ochraceous; bill black above, white underneath; feet greenish olive; iris black.

Female. Similar to the male, but with the head black with obscure yellowish ochre margins to the feathers, surrounded by a band of blue; collar round the hind neck whitish and rufous; vent white, very slightly rufescent; throat rufescent with dusky black margins to the feathers. Young birds resemble the old female, and have the underparts of a rufous colour. It must be noted that the figures in Mr. Sharpe's work have the soft parts very erroneously coloured, as might have been expected at a time when all information on this subject was wanting.

The Plate represents a male and a female, of the natural size, the latter being the upper figure. They have been drawn from a pair of birds in my own collection.

[R. B. S.]

HALCYON TRISTRAMI, Layard.

HALCYON TRISTRAMI, Layard.

Tristram's Kingfisher.

Halycon tristrami, Layard, Ibis, 1880, pp. 299, 460, pl. xvi.—Ramsay, Proc. Linn. Soc. N. S. W. vi. p. 834 (1882).—Tristr. Ibis, 1882, p. 609.
Sauropatis tristrami, Salvad. Ann. Mus. Civic. Genov. xviii. p. 420 (1882).—Id. Orn. Papuasia e delle Molucche, iii. App. p. 524 (1882).

Considering that scarcely five years have elapsed since this species was discovered, its history has already been sufficiently complicated. The first mention of the bird is in 'The Ibis' for 1880; and Mr. E. L. Layard there states that he has a specimen of a Kingfisher from the Solomon Islands, which he proposes to name *H. tristrami*, but he does not describe it. Later on, however, in the same year he refers to this bird as being probably from Makira Harbour, and compares it with *H. vagans* of New Zealand. In his list of the birds of the Solomon Islands, published in 'The Ibis' for 1882 (pp. 141-146), Canon Tristram omits all mention of the species as an inhabitant of the group; but in the meantime Mr. Ramsay makes some remarks upon the species, and calls attention to the absence of the nape-patch in the plate and description given by Mr. Layard (*l. c.*). Canon Tristram, noticing this paper in 'The Ibis' for 1882 (p. 609), gives the dimensions of *H. tristrami* for the first time, and states that he considers it to be "further removed from *H. vagans* than from any other of the group." He likewise states that it has no occipital patch whatever, and that he "possesses the type specimens." No wonder, then, that Count Salvadori finds that he has been a little puzzled with regard to the history of the species.

First of all, there cannot be more than one type specimen of *H. tristrami*, which must be the supposed Solomon-Island specimen first mentioned by Mr. Layard; and with regard to the want of the occipital patch, it is clearly visible on a close examination in the specimen described by us below, and therefore its absence in Canon Tristram's specimen must be purely accidental. The British Museum possesses a nestling which has it plainly developed.

Lastly, we cannot quite understand why Canon Tristram should object to the close resemblance of *H. tristrami* to *H. vagans*, as in our opinion it is only distinguished from the latter species by its larger size, more vivid coloration, and by the deep cinnamon-buff colour of the underparts.

It is possible that there is some error respecting the occurrence of this bird in the Solomon Islands, for Mr. Layard does not appear to have been very certain about the origin of his type specimen. Three examples which have fallen under our notice have been from New Britain; but in all probability it is likewise found in Duke of York Island, as Mr. L. C. Layard does not mention particularly that it is confined to the former of the two (see 'Ibis,' 1880, p. 294). He states that it was "only got in the thick parts about the mountain-slopes; we never observed it mixing among its smaller brethren (*H. sanctus*) on the open shore. Their habitats being so different, their food was different also. The large one ate beetles, locusts, and small lizards, and the lesser one contented himself with fish and sea-worms. Native name for both 'Akiki.'"

Adult male. General colour above dark green, with more or less of a bluish tinge; the whole of the back, rump, and upper tail-coverts brighter cobalt, leaving the mantle and scapulars dull green; wing-coverts greenish cobalt, brighter blue on the greater series, and deepening into ultramarine on the outer aspect of the quills, which are otherwise black; the innermost secondaries above greenish; tail-feathers deep blue, the shafts below, as well as the edge of the inner web, blackish; crown of head green, slightly washed with blue; a loral spot of ochreous buff; feathers in front of and round the eye black, with a half-concealed white spot below the latter; on the sides of the crown above the eye a few small whitish streaks; sides of the crown brighter blue, meeting on the nape; a concealed occipital patch of ochreous buff; a streak from below the gape joining the feathers below the eye and the ear-coverts bluish green, the hindermost of the latter black, joining a collar which surrounds the nape and separates the head from a very broad band of ochreous buff, which runs round the hind neck and joins the breast; this ochreous-buff band is separated from the mantle by a narrow shade of blackish; cheeks ochreous buff; throat whiter; fore neck and breast ochreous buff, deepening into cinnamon-buff on the lower breast, sides of body, and under tail-coverts; under wing-coverts and axillaries also deep cinnamon-buff; quills blackish below, yellowish buff along the edge of the inner web: "bill black; legs ash-coloured; iris brown" (*L. C. Layard*). Total length 9·3 inches, culmen 2·05, wing 4·3, tail 2·8, tarsus 0·55.

The specimen described and figured is a male, obtained in Blanche Bay, New Britain, by Captain Richards on the 2nd of July, 1879.

HALCYON LEUCOPYGIA.

HALCYON LEUCOPYGIA.

White-backed Kingfisher.

Cyanalcyon leucopygius, Verr. Rev. et Mag. de Zool. 1858, p. 385.—Salvad. Ann. Mus. Civic. Genov. x. p. 305 (1877).
Halcyon leucopygia, Gray, Cat. B. Trop. Isl. Pacific Ocean, p. 7 (1859).—Sclater, Proc. Zool. Soc. 1869, p. 119.—Gray, Hand-l. B. i. p. 92, no. 1109 (1869).—Sharpe, Monogr. Alced. pl. 74 (1871).—Ramsay, Proc. Linn. Soc. N. S. W. iv. p. 67 (1879).—Salvad. Ibis, 1880, p. 127.
Todirhamphus leucopygius, Sclater, Proc. Zool. Soc. 1869, p. 124.
Cyanalcyon leucopygia, Salvad. Ann. Mus. Civic. Genov. x. p. 305 (1877).—Id. Orn. Papuasia, etc. i. p. 456 (1880).

This beautiful Kingfisher was until recently one of the rarest of the family in European collections; for when the 'Monograph of the Kingfishers' was written but one specimen, the type in the British Museum, was known to naturalists, and this was figured in the work above mentioned. Since then the only naturalist who has met with the species in its native haunts has been Mr. Cockerell, who procured a large series in Guadalcanar in the Solomon Archipelago. His collection was described by Mr. Ramsay, who was the first to point out the difference in the colouring of the sexes, the white back, from which the species derives its name, being apparently the sign of the male, as the female has the lower back beautiful blue. The lilac colour which is so distinct on the sides of the lower back in both sexes will always be considered one of the peculiar characters of this fine species of *Halcyon*.

Nothing has been recorded concerning the habits of this species, which, so far as we know, is only found in the Solomon group of islands.

The following is the description of the type specimen, transcribed from the 'Monograph':—

Adult male. Head, scapulars and wing-coverts, and upper part of the back rich ultramarine; a collar round the neck, the entire back except the interscapulary portion, and the under surface of the body pure white; cheeks black; upper tail-coverts ultramarine; lower part of the flanks bordering the rump and vent lilac shaded with purple; quills and tail black, washed with blue above, greyish black underneath; bill entirely black; feet olive-brown. Total length 8·2 inches, of bill from front 1·6, from gape 2·1, wing 3·3, tail 2·5, tarsus 0·5.

Adult female. Similar to the male, but having the lower back blue instead of white. Total length 7·5 inches, wing 3·3, tail 2·3, tarsus 0·5.

The figures in the Plate are drawn from a pair lent to us by Mr. E. P. Ramsay, and belong to the Australian Museum; they represent the male and female of the natural size.

[R. B. S.]

HALCYON QUADRICOLOR, *Oustalet*.

HALCYON QUADRICOLOR.

Four-coloured Kingfisher.

Cyanalcyon quadricolor, Oustalet, Le Naturaliste, 1880, p. 323.

This beautiful species was discovered by Mr. Bruijn's hunters on the west coast of New Guinea, and appears to be a very distinct bird. Count Salvadori has, indeed, suggested to me that it may be the young of *Halcyon nigrocyanea*; but this I do not think likely to be the case. Mr. Sharpe examined the type specimen in the Paris Museum, and assures me that the bird is adult, and quite distinct from the last-named species. Both Professor Schlegel, in his Revue of the Kingfishers in the Leyden Museum, and Count Salvadori, in his 'Ornitologia della Papuasia,' record specimens of *H. nigrocyanea*, which they consider to be immature, as being rufous or rufous-brown underneath; but I can scarcely think that there can be much similarity between these young birds and the specimen figured in my Plate.

Dr. Oustalet describes the present bird as follows:—

"Some months ago I had occasion to draw attention to the presence of a new species of *Talegallus* (*Talegallus* or rather *Æpypodius bruijnii*) among a collection of birds killed by Mr. Bruijn's hunters on the west coast of New Guinea and in the neighbouring islands. In this same collection, which was acquired by the Paris Museum, was a beautiful Kingfisher of moderate size belonging to the small group which is known by the name of *Cyanalcyon*, and much recalling by its proportions, and by the coloration of the upper part of the body, of its beak, its feet, throat, and breast, the species coming from the northwest coast of New Guinea, which was described by Mr. Wallace in 1862 under the name of *Halcyon nigrocyanea* (Proc. Zool. Soc. 1862, p. 165, pl. xix.). The individual before me has, like the bird figured by Mr. Wallace, the bill black, with a little white in the middle of the lower mandible near its base; the feet black; the upper part of the head dark blue, passing into ultramarine on the nape and towards the eyebrows, and contrasting strongly with the black colour which occupies the middle of the back and the two large black spots which cover the cheeks, the feathers below the eye, and the sides of the throat. The wings are of a dark blue, with the scapular feathers of a bright blue; the under wing-coverts black, crossed by a white band; the tail is blue above, black below; the upper tail-coverts of a clear and intense cobalt-blue; the throat pure white, bordered below by a broad blue band. But the abdominal region presents an entirely different coloration: in fact, in the female which was figured in the 'Proceedings,' the belly is of a pure white, with black flanks; here, on the contrary, the belly is of a very pronounced cinnamon-rufous, with some black and blue feathers on the sides, and this rufous colour is separated from the blue band by a somewhat narrow but well-defined line of white. We may not attribute this difference in colour to a difference in sex, because, according to Mr. Bruijn's indications, in which one can place confidence, the individual acquired by the Museum is a female like Mr. Wallace's type. We know, moreover, as a natural fact that the male of *Halcyon nigrocyanea* has the belly azure-blue. In short, although I find in Schlegel's 'Catalogue des Martins Pêcheurs du Musée des Pays-Bas' (Revision, 1874, p. 31) this note—'A young male killed on the 7th May 1870, at Andai, by Von Rosenberg, is remarkable for having the blue of the under surface replaced by rufous brown,' I certainly cannot consider the individual before me as being young. It has, in fact, the dress of a perfectly adult bird, with the colours pure and brilliant, and does not show any sign of spots, or of the grey or rufous bars which are the sign of immaturity. I propose, therefore, to make it the type of a new species, to be called *Cyanalcyon quadricolor*, to draw attention to the four colours (blue, white, rufous, and black) which are spread over its plumage."

During a recent visit to Paris, Mr. J. G. Keulemans painted me a picture of the type specimen in the Paris Museum, which has been reproduced by Mr. Hart in the accompanying Plate. The principal figure is life-size.

[R. B. S.]

HALCYON NIGROCYANEA, *Wallace.*

HALCYON NIGROCYANEA, *Wallace*.

Black and Blue Kingfisher.

Halcyon nigrocyanea, Wallace, Proc. Zool. Soc. 1862, p. 165, pl. xix.—Gray, Hand-list of Birds, i. p. 93 (1869).—
    Sharpe, Monogr. Alced. p. 201, pl. 75 (1870).—Beccari, Ann. Mus. Civic. Genov. vii. p. 708 (1875).
Dacelo nigrocyanea, Schlegel, Nederl. Tydschr. Dierk. iii. p. 250 (1865).
Cyanalcyon nigrocyanea, Salvad. Ann. Mus. Civic. Genov. x. p. 127 (1877).—Id. op. cit. p. 305 (1877).

THE present species is the finest of the little group of Kingfishers to which the generic title of *Cyanalcyon* has been applied, and which includes *Halcyon macleayi* of Australia, *Halcyon diops* of the Moluccas, *H. lazuli* of Ceram and Amboina, *H. leucopygia* of the Solomon Islands, and lastly the beautiful species from the Fly River, *H. stictolæma*, which I figure in the present part. I do not wish to deny for a moment that the little group above enumerated does not constitute a distinct genus or, rather, subgenus; but for the sake of uniformity I keep the species in the genus *Halcyon*, in which the allied species have been included in all my former works.

Mr. Wallace originally discovered the subject of these remarks in North-western New Guinea, but only managed to procure a single hen bird, which was figured in the 'Proceedings' of the Zoological Society, and, again, in Mr. Sharpe's 'Monograph of the *Alcedinidæ*.' The latter work contained a representation also of the male sex, which had been collected by Von Rosenberg at Andei, and was in the Museum at Leiden. Since the year 1870 more examples have been procured by the travellers to New Guinea; and Count Salvadori enumerates fourteen specimens as belonging to the Civic Museum at Genoa or examined by him during his study of the Papuan Kingfishers. The localities given by him are Dorei (*Bruijn*), Andei (*Von Rosenberg*, *D'Albertis*, *Bruijn*), Warbusi (*Beccari*), Sorong (*Bernstein*, *Bruijn*), Batanta (*Beccari*, *Bruijn*). These localities are all situated on the mainland of North-western New Guinea, or are islands closely adjacent to the N.W. peninsula. Beccari found it nowhere common; but beyond this we know nothing of its habits; these, however, without doubt are similar to those of other species of the genus *Halcyon*.

The following descriptions are taken from Mr. Bowdler Sharpe's Monograph of the Kingfishers:—

"*Adult male*. Head intense ultramarine, brighter on the sides, a line of brilliant ultramarine commencing at the back of each eye and encircling the nape; middle of the back and scapulars deep velvety black, a blue lustre being apparent here and there on the latter; wing-coverts deep ultramarine, the innermost greater coverts more brilliant, inclining to cobalt; quills black, the outer web washed with deep ultramarine; lower portion of the back and rump brilliant cobalt; upper tail-coverts deep ultramarine; tail deep ultramarine above, black beneath, cheeks and ear-coverts jet black; chin dusky black; throat and a narrow band across the centre of the breast white; rest of the under surface of the body deep ultramarine, becoming black on the sides of the body and lower abdomen; under wing-coverts black; bill black, yellow at the extreme base; feet black. Total length 8·5 inches, culmen 1·9, wing 3·5, tail 2·8, tarsus 0·5.

"*Adult female*. Upper surface as in the male. Entire under surface white, with the exception of a broad pectoral band of deep ultramarine; sides of the body black; under wing-coverts black; some white; bill black, with more yellow on the under mandible than the male. Total length 9 inches, culmen 1·9, wing 3·6, tail 2·8, tarsus 0·5."

The specimens figured in the accompanying Plate are in my own collection. They represent the male and female of the natural size.

HALCYON STICTOLÆMA.

HALCYON STICTOLÆMA.

Spotted-throated Kingfisher.

Cyanalcyon stictolæma, Salvad. Ann. Mus. Civic. Genov. ix. p. 20 (1876).—Id. op. cit. x. p. 304 (1877).—D'Albert. & Salvad. op. cit. xiv. p. 51 (1879).
Halcyon nigrocyanea, D'Albert. (nec Wallace), Ibis, 1876, p. 360.
Cyanalcyon nigrocyanea, D'Albert. (nec Wallace), Ann. Mus. Civic. Genov. x. pp. 10, 19 (1877).

VERY similar to the *Halcyon nigrocyanea* of North-western New Guinea, in my opinion the present species is nevertheless quite distinct. Nobody examining the two species could, I believe, hesitate to separate them, notwithstanding the fact that Signor D'Albertis, who discovered this new Kingfisher in South-eastern New Guinea, considered it to be precisely the same as *Halcyon nigrocyanea*, a bird he had killed himself in North-western New Guinea.

The present species is like *Halcyon nigrocyanea*, but differs at a glance by the want of the white band across the breast, and by having the throat almost entirely blue, mottled with white bases to the feathers. The female bird is nearly the same as the female of the allied species; but *H. stictolæma* has much less white on the throat and abdomen, while the pectoral band is much broader. The habitat of the species is, as far as is known at present, only the vicinity of the river Fly, where Signor D'Albertis obtained one specimen during his first expedition in 1875. On his second excursion, in 1877, he managed to procure six examples, which are fully described by himself and Count Salvadori in the fourteenth volume of the 'Annali' of the Civic Museum of Genoa. Three of these were most kindly lent to me by Signor D'Albertis for the purposes of the present work; and I have to acknowledge my obligations to this gentleman for his assistance on this and other occasions.

Mr. Sharpe, who has seen the specimens, has supplied me with the following descriptions:—

"*Adult male*. General colour above black, the head deep ultramarine, the sides of the crown more brilliant ultramarine inclining to cobalt, forming an eyebrow which borders the blue crown and encircles the nape; wing-coverts ultramarine, the lesser ones slightly more brilliant; quills blackish, externally washed with dull blue; scapulars black, washed with blue at the ends; lower back cobalt, deepening into ultramarine on the longer feathers of the rump; upper tail-coverts deep ultramarine; tail-feathers dark blue; lores, feathers round the eye, cheeks, ear-coverts, sides of neck, sides of breast, flanks, thighs, vent, and under tail-coverts black, the latter tipped with ultramarine; chin black; throat ultramarine mottled with white bases to the feathers; breast and abdomen rich ultramarine; under wing-coverts and axillaries black; edge of wing greenish blue; quills dusky blackish below, ashy along the inner web. Total length 9 inches, culmen 2·25, wing 3·25, tail 3, tarsus 5.

"*Adult female*. On the upper surface entirely like the male, but differs below in having the throat and abdomen white, separated by a broad pectoral band which is ultramarine in the middle, black at the sides; sides of body and flanks black; under wing-coverts black, with a band of white running down the middle, many of the median and greater under wing-coverts being tipped with white; under tail-coverts as in the male.

"*Young*. Resembles the old female, but is much more dusky black, the crown being also black, with an ultramarine eyebrow; wing-coverts black, tipped with blue; lower back bright ultramarine, but not so brilliant as in the adults; sides of face and sides of body, vent, and lower abdomen dusky black; throat dull white, the feathers obscured with dusky blackish tips; centre of abdomen pale ochraceous brown, separated from the throat by a broad blackish band, the central feathers washed with blue; under tail-coverts blackish, tipped with blue."

Count Salvadori describes some of the males as having a residue of rusty feathers on the abdomen, the remains of the young plumage. The adults, according to Signor D'Albertis, have the bill black the feet clear plumbeous, and the iris chestnut-brown; and in the younger birds the iris is black, and the feet are very dusky plumbeous. The food of the species consists of Crustacea.

The figures in the Plate represent an adult male, adult female, and a young female, drawn to about the natural size from three of the typical specimens, lent to me by Signor D'Albertis.

CLYTOCEYX REX, Sharpe.

CLYTOCEYX REX, Sharpe.

Spoon-billed Kingfisher.

Clytoceyx rex, Sharpe, Ann. Nat. Hist. (5th ser.) vi. p. 231 (1880).

The remarkable bird which is figured in the accompanying plate is a native of South-eastern New Guinea, where it was obtained by Mr. Charles Hunstein, who has been collecting in the interior about Milne Bay, East Cape, and the neighbouring localities. Having already in the course of the present work had occasion to express some disappointment that the southern portion of the great Papuan island had not produced the number of new species which one might reasonably have expected, I feel bound to qualify this opinion when I see before me such an extraordinary form of bird as the present. It is evident that the avifauna of the lowlands has too much resemblance to that of the adjoining continent of Australia and to that of the Aru Islands for us to expect, until the mountains are reached, any thing strikingly different from the birds of these two localities.

In the MS. list of birds sent by Mr. Hunstein he speaks of this species as the "Spoon-billed" Kingfisher; and I have adopted this English name, not so much on account of its absolute correctness from an ornithological point of view, but because it represents the first impression of the original collector. To ornithologists the epithet of "spoon-bill" recalls the flattened and spatulated bill of the orthodox Spoonbill (*Platalea*), or that of the spoon-billed Sandpiper (*Eurhinorhynchus*); but the beak of this large Kingfisher more resembles the bowls of two spoons placed in opposition to each other.

We have as yet no details as to the habits of the present species; but they doubtless resemble those of the large "Jackasses" of Australia. The bills of the specimens sent were covered with dried earth, as if the birds had been grubbing for food on the ground. Although the species is, no doubt, generically distinct from the Laughing Jackasses of Australia, there can be no doubt that in the genus *Dacelo* it will find its nearest allies. Different as the bill is, there is one character which betrays this affinity; and that is seen in the difference of the sexes, the male having a blue tail, and the female a rufous one; this, as is well known, is one of the leading features in a true *Dacelo*.

I translate the original description given by Mr. Bowdler Sharpe.

Male. Head brown; feathers surrounding the eye and sides of the face brown; ear-coverts black, extending backwards onto the sides of the neck and forming a broad band; a stripe above the eye, the lower cheeks and a broad band on the neck ochraceous buff; interscapular region black; scapulars and wing-coverts brown, the latter margined with ochraceous, the outermost of the least series washed with greenish blue; primary-coverts and quills dark brown, externally washed with dull green; lower back and rump silvery cobalt; upper tail-coverts and tail dark brown washed with green; throat white; rest of the body underneath, with the under wing-coverts, ochraceous buff; quills dusky below, with the inner web margined with pale ochraceous. Total length 12 inches, culmen 1·95, wing 6·35, tail 4·7, tarsus 0·9.

Female (immature). Differs from the male in its reddish tail. The hind neck and the undersurface having dusky margins to the feathers, show that the bird is not quite adult.

The figures in the Plate are drawn from the pair in the British Museum.

THE
BIRDS OF NEW GUINEA

AND THE

ADJACENT PAPUAN ISLANDS,

INCLUDING MANY

NEW SPECIES RECENTLY DISCOVERED

IN

AUSTRALIA.

BY

JOHN GOULD, F.R.S.

COMPLETED AFTER THE AUTHOR'S DEATH

BY

R. BOWDLER SHARPE, F.L.S. &c.,
ZOOLOGICAL DEPARTMENT, BRITISH MUSEUM.

VOLUME V.

LONDON:
HENRY SOTHERAN & CO., 36 PICCADILLY.
1875–1888.

[*All rights reserved.*]

PRINTED BY TAYLOR AND FRANCIS,
RED LION COURT, FLEET STREET.

CONTENTS.

VOLUME V.

Plate			Part	Date
1.	Cyclopsitta occidentalis	Western Perroquet	XIX.	1885.
2.	„ cervicalis	Southern Ringed Perroquet	X.	1879.
3.	„ diophthalma	Double-eyed Perroquet	IX.	1879.
4.	„ aruensis	Aru Perroquet	IX.	1879.
5.	„ suavissima	D'Albertis's Perroquet	VII.	1878.
6.	„ melanogenys	Black-cheeked Perroquet	VII.	1878.
7.	„ maccoyi	M'Coy's Perroquet	I.	1875.
8.	„ coccineifrons	Astrolabe-Mountain Perroquet	XXIII.	1887.
9.	Aprosmictus callopterus	Yellow-winged King-Parrot	X.	1879.
10.	„ insignissimus	Beautiful King-Parrot	I.	1875.
11.	Charmosyna margaritæ	Duchess of Connaught's Parrakeet	XVIII.	1884.
12.	„ josephinæ	Josephina Parrakeet	III.	1876.
13.	„ pulchella	Pectoral Lorikeet	III.	1876.
14.	„ papuensis	Papuan Lorikeet	II.	1876.
15.	„ stellæ	Stella Parrakeet	XXIV.	1888.
16.	Psitteuteles rubronotatus	Red-backed Lorikeet	V.	1877.
17.	„ subplacens	Green-backed Lorikeet	V.	1877.
18.	„ arfaki	Arfak Lorikeet	III.	1876.
19.	„ wilhelminæ	Wilhelmina Lorikeet	III.	1876.
20.	„ placens	Beautiful Lorikeet	III.	1876.
21.	Nasiterna pygmæa	Pygmy Parrot	VI.	1878.
22.	„ maforensis	Mafoor Pygmy Parrot	VI.	1878.
23.	„ misoriensis	Misori Pygmy Parrot	VI.	1878.
24.	„ bruijnii	Bruijn's Pygmy Parrot	VI.	1878.
25.	„ beccarii	Beccari's Pygmy Parrot	VI.	1878.
26.	„ pusio	Solomon-Islands Pygmy Parrot	VI.	1878.
27.	„ keiensis	Ké-Island Pygmy Parrot	VI.	1878.
28.	Geoffroyius heteroclitus	Yellow-headed Parrot	VIII.	1878.
29.	„ simplex	Blue-collared Parrot	V.	1877.
30.	„ timorlaoensis	Tenimber Parrot	XXIII.	1887.
31.	Eclectus polychlorus	Green Lory	VIII.	1878.
32.	„ riedeli	Riedel's Parrot	XVI.	1884.
33.	Dasyptilus pesqueti	Pesquet's Parrot	XIV.	1883.
34.	Eos fuscata	Banded Lory	XXI.	1886.
35.	„ reticulata	Blue-streaked Lory	XV.	1883.
36.	Lorius flavo-palliatus	Yellow-mantled Lory	XXIV.	1888.
37.	„ tibialis	Blue-thighed Lory	XXV.	1888.
38.	Chalcopsittacus scintillatus	Red-fronted Lory	XIV.	1883.
39.	Psittacella brehmi	Brehm's Parrot	IV.	1877.
40.	„ madaraszi	Madarász's Parrakeet	XXII.	1886.
41.	Trichoglossus goldiei	Goldie's Perroquet	XIV.	1883.
42.	„ musschenbroekii	Van Musschenbroek's Lorikeet	V.	1877.
43.	Loriculus aurantiifrons	Orange-crowned Loriculus	V.	1877.
44.	Cacatua triton	Triton Cockatoo	XX.	1885.
45.	„ ophthalmica	Blue-eyed Cockatoo	XX.	1885.
46.	„ gymnopis	Naked-eyed Cockatoo	XIX.	1885.
47.	„ ducorpsi	Ducorps's Cockatoo	XIX.	1885.
48.	Phlogœnas jobiensis	White-chested Pigeon	VII.	1878.
49.	„ johannæ	Mrs. Sclater's Ground-Dove	VII.	1878.
50.	Œdirhinus insolitus	Knob-billed Fruit-Pigeon	VIII.	1878.

CONTENTS OF VOL. V.

Plate			Part	Date
51.	Ptilopus nanus	Tiny Fruit-Pigeon	II.	1876.
52.	,, solomonis	Solomon-Island Fruit-Pigeon	XIX.	1885.
53.	,, richardsi	Richards's Fruit-Pigeon	XVIII.	1884.
54.	,, lewisi	Lewis's Fruit-Pigeon	XVII.	1884.
55.	,, wallacei	Wallace's Fruit-Pigeon	XV.	1883.
56.	,, fischeri	Fischer's Fruit-Pigeon	XI.	1880.
57.	,, speciosus	Lilac-bellied Fruit-Pigeon	IX.	1879.
58.	,, bellus	Purple-bellied Fruit-Pigeon	IX.	1879.
59.	,, rivolii	Massena Fruit-Pigeon	IX.	1879.
60.	Gymnophaps pœcilorrhoa	Rusty-banded Fruit-Pigeon	XI.	1880.
61.	Otidiphaps cervicalis	Grey-naped Otidiphaps	XIII.	1882.
62.	Eutrygon terrestris	Papuan Ground-Pigeon	XIII.	1882.
63.	Carpophaga rubricera	New-Ireland Fruit-Pigeon	XIX.	1885.
64.	,, van-wyckii	Van Wyck's Fruit-Pigeon	XXV.	1888.
65.	,, finschi	Finsch's Fruit-Pigeon	XVII.	1884.
66.	,, subflavescens	Yellow-tinted White Fruit-Pigeon	XXV.	1888.
67.	Ianthœnas albigularis	White-throated Pigeon	XXV.	1888.
68.	Megapodius brenchleyi	Brenchley's Megapode	XXII.	1886.
69.	Megacrex inepta	New-Guinea Flightless Rail	XI.	1880.
70.	Rallicula forbesi	Forbes's Rail	XXIII.	1887.
71.	Gymnocrex plumbeiventris	Grey-bellied Rail	XXIV.	1888.
72.	Sternula placens	Torres Straits Tern	III.	1876.
73.	Casuarius bicarunculatus	Two-wattled Cassowary	XII.	1881.
74.	,, picticollis	Painted-throated Cassowary	V.	1877.
75.	,, westermanni	Westermann's Cassowary	V.	1877.

CYCLOPSITTA OCCIDENTALIS.

CYCLOPSITTA OCCIDENTALIS, Salvad.

Western Perroquet.

Opopsitta desmarestii (part.), Sclater, Proc. Linn. Soc. ii. p. 166 (1858).—Gray, Cat. B. New Guinea, p. 42 (partim, 1859).—Rosenb. (nec Garn.), Journ. für Orn. 1862, p. 63.—Id. Nat. Tijdschr. Ned. Ind. xxv. pp. 143, 226 (partim, 1863).—Sclater, P. Z. S. 1873, p. 697.

Cyclopsitta blythi (part.), Wallace, Proc. Zool. Soc. 1864, p. 285.

Psittacula desmarestii (part.), Schlegel, Mus. Pays-Bas, *Psittaci*, p. 75 (1864).—Finsch, Die Papageien, ii. pp. 620, 957 (partim, 1868).—Schleg. Mus. Pays-Bas, *Psittaci*, Revue, p. 32 (partim, 1874).

Cyclopsittacus desmarestii (part.), Salvad. Ann. Mus. Civic. Genov. vii. p. 754 (1875).

Cyclopsittacus occidentalis, Salvad. Ann. Mus. Civic. Genov. vii. p. 910 (1875).—Id. op. cit. x. pp. 27, 119 (1876).—Id. Orn. Papuasia e delle Molucche, i. p. 152 (1880).

ALTHOUGH this species has been in European collections for some time, the differences between it and *C. desmaresti* were overlooked by naturalists until Count Salvadori separated the two birds specifically. It has been referred by Mr. Wallace to *Cyclopsitta blythi*, but lacks the blue spot under the eye which distinguishes the latter species.

Count Salvadori gives its habitat as Western New Guinea, near Sorong, Dorei-Hum, and also the islands of Salawati and Batanta. Two specimens, said to have been collected by the Dutch traveller Hoedt in Mysol, are in the Leyden Museum, but Count Salvadori thinks that these may have come from Salawati.

This species may be briefly described as similar to *C. desmaresti*, but distinguished by its golden-yellow cheeks and ear-coverts, and by the paler blue of the spot under the eye, which has more or less of a greenish shade, by the absence of the blue occipital spot, and by having the head more tinged with red.

The Plate represents an adult and an immature bird in two positions, drawn from specimens in the Gould collection.

[R. B. S.]

CYCLOPSITTA CERVICALIS, *Salvad. & D'Albert.*

CYCLOPSITTACUS CERVICALIS, Salv. & D'Alb.

Southern Ringed Perroquet.

Cyclopsittacus cervicalis, Salvad. & D'Albert. Ann. Mus. Civic. Genov. vii. p. 811 (1875).—Salvad. tom. cit. p. 911.—Id. op. cit. ix. p. 12 (1876); x. p. 28 (1877).—Sharpe, Proc. Linn. Soc. xiii. p. 310 (1878).—D'Albert. & Salvad. op. cit. xiv. p. 30 (1879).

This beautiful species was first discovered by Signor D'Albertis, in South-eastern New Guinea, on Mount Epa. It is, as Count Salvadori has rightly pointed out, a representative in the southern part of New Guinea of *C. desmaresti*, which comes from Doray. It is, perhaps, more strictly allied to *C. blythii* of Mysol, like which species it has no blue spot under the eye. The differences are pointed out by Count Salvadori; and the diagnosis which he assigns to the present species reads as follows:—"Cheeks reddish orange; on the breast a single band of blue only; hinder neck entirely blue." It would seem, however, that even at the time when he wrote this description he was not in possession of a perfectly adult bird, as the blue on the hind neck is now known to disappear with age. For a knowledge of the different stages of plumage through which the present species passes we are indebted to the recent labours of Signor D'Albertis, who procured a series of some thirty examples from the Fly River; and an exhaustive account has been written on the progress of this bird to maturity by Count Salvadori in his description of D'Albertis's Fly-River collections. The Count remarks that this species, which is more beautiful than any of its allies, is remarkable for the extreme variability of its plumage—a fact which is not observable in any of the others. From individuals having the hind neck blue with a pectoral band of dull blue and the hinder ear-coverts red above and blue below, a perfect passage is found to specimens which have the hind neck orange-red like the head, with a cervical collar of fine yellow, and with a pectoral band of clear sky-blue and the ear-coverts entirely yellow. The latter is the adult dress; and between the young ones and this stage D'Albertis's collection contained a number of examples in different stages of transition.

Beyond the specimens of this Perroquet which Signor D'Albertis possesses, I have seen but one other, which was obtained by the late Dr. James in the Eucalyptus-range on the mainland of South-eastern New Guinea to the east of Yule Island. This specimen was fully described by Mr. Bowdler Sharpe, and is now in my collection, but it is evidently not quite adult.

I translate herewith the descriptions which Count Salvadori has recently published of the adult and young plumages of this beautiful species:—

"*Adult.* Green; the head and neck reddish orange, a collar round the hind neck beautiful yellow, sides of the head yellow, more or less tinged with reddish orange; a band across the fore part of the breast, and the sides of the breast, pale blue; upper breast tinged with beautiful orange, a concealed red spot on the inner quill-feathers.

"*Young.* Green; the head reddish orange, clearer yellow behind; the hind neck beautiful blue; the fore part of the sides of the head reddish orange, the hinder part yellow; the posterior ear-coverts above reddish orange, below blue; a band across the fore part of the breast bright blue; the sides of the breast pale blue; a concealed yellow spot on the inner quills.

"Total length 7·1 inches, culmen 1, wing 4·6, tail 2·1, tarsus 0·5."

According to Signor D'Albertis the bill is black, the feet greenish, the iris has an inner circle of chestnut-brown and an outer one of red.

The food consists of fruits.

I have to offer my best thanks to Signor D'Albertis for having placed at my disposal for perfecting the present Plate six magnificent examples of this fine Parrot, for which liberality I feel, as will also, I am sure, the public who are interesting themselves in this New-Guinea work, greatly obliged—all such discoveries going to confirm the opinion frequently expressed by myself, and others that "New Guinea is the country left for the researches of ornithologists," the natural productions being so marvellous.

CYCLOPSITTA DIOPHTHALMA.

CYCLOPSITTA DIOPHTHALMA.

Double-eyed Perroquet.

Psittacula diophthalma, Hombr. et Jacq. Annales des Sciences Naturelles, 2nd series, xvi. p. 318 (1841).—Gray, Gen. B. ii. p. 423 (1846).—Schlegel, Mus. Pays-Bas, Psittaci, p. 75 (1864).—Id. Nederl. Tijdsch. Dierk. iii. p. 331 (1866, pt.).—Finsch, Die Papageien, ii. p. 628 (1868, pt.).

Cyclopsitta double œil, Hombr. et Jacquinot, Voyage Pôle Sud, Atlas, pl. 25 bis, figs. 4, 5 (1842-1853).

Cyclopsitta diophthalma, Jacq. et Pucheran, Voy. Pôle Sud, texte, iii. p. 107 (1853).—Bonap. Rev. et Mag. de Zool. 1854, p. 154.—Sclater, Proc. Linn. Soc. 1858, p. 106.—Wallace, P. Z. S. 1864, p. 284 (pt.).

Opopsitta diophthalma, Sclater, P. Z. S. 1860, p. 227.—Rosenb. J. f. O. 1864, p. 115.—Sclater, P. Z. S. 1873, p. 697.

Cyclopsittacus diophthalmus, Salvad. Ann. Mus. Civic. Genov. x. p. 28 (1877).—Id. tom. cit. p. 120 (1877).

Our knowledge of the present beautiful little species is unfortunately very limited; and at present we are unaware of any thing connected with its habits and economy, although these are doubtless precisely similar to those of the other *Cyclopsittæ*. Its chief habitat appears to be the north-western portion of New Guinea, where it has been collected by Signor D'Albertis and by Dr. Meyer near Andai, and in the Arfak Mountains by Mr. Bruijn and Dr. Beccari, while a good many specimens are contained in the Leiden collection from Kalwal in Salwati, where they were obtained by the late Dr. Bernstein. It appears likewise to be plentiful in the island of Mysol, where Mr. Hoedt obtained a considerable series from the neighbourhood of Kasim and Waaigama. Count Salvadori also mentions its occurrence in the island of Koffias, on the authority of specimens in Count Turati's museum. The original examples, procured by MM. Hombron and Jacquinot, are said to have come from Southern New Guinea.

The following is a description of an adult male bird from Mysol:—

General colour above bright grass-green; wing-coverts a little darker green than the back, the outermost of the lesser coverts blue, as well as the outer greater coverts, primary-coverts, and primaries, the latter black on the inner webs, the secondaries green, with a bright orange patch on the inner web of the dorsal secondaries; tail green, forehead and sinciput red, fading into yellow on the hinder crown, lores and sides of face also red, the ear-coverts with a blue streak across the middle, separating the red face from the hinder ear-coverts and sides of neck, which are green; in front of the eye a spot of cobalt-blue; under ur face of body bright grass-green, lighter on the centre of the body, the flanks with dashes of bright yellow; under wing-coverts bluish green; quills blackish below, yellow on the inner webs, as also across the lower greater coverts. Total length 5·8 inches, wing 3·6, tail 2·0.

According to Professor Schlegel, there is no difference in the coloration of the sexes; but the young birds always have the forehead as far as the top of the head tinged with red, duller than in the adults, while the red colour of the cheeks is replaced by clear yellowish brown, changing to red more quickly in the males than in the females. A young male is also described by Count Salvadori as being similar to the old female, but much smaller, with the red colour on the upper part of the head tinged posteriorly with yellow, while the cheeks are greyish buff margined below and behind with azure blue, and from the forehead a thin red line extends underneath the eyes to the region of the ear-coverts. Signor D'Albertis gives the colour of the eyes as black.

My own specimens of this bird are in very fine condition, having been obtained from Mr. Wallace's expedition; so that I have been able to institute exact comparisons between the present species and *C. aruensis*, the result being that I am perfectly convinced of the specific distinctness of these two little Perroquets, and have given figures of the two in the present work.

The Plate represents an adult and a young male of the natural size, drawn from skins in my own collection.

CYCLOPSITTA ARUENSIS, *Schleg.*

CYCLOPSITTA ARUENSIS, Schleg.

Aru Perroquet.

Psittacula diophthalmus (partim), G. R. Gray, P. Z. S. 1858, p. 195.
Psittacula diophthalma, G. R. Gray, List Psittacidæ Brit. Mus. p. 90 (1859, nec Hombr. et Jacq.).—Id. Cat. Mam. & B. New Guinea, pp. 42, 60 (1859, pt.).—Id. P. Z. S. 1861, p. 437.—Schlegel, Mus. Pays-Bas, *Psittaci*, p. 75 (1864, pt.).—Id. Nederl. Tijdschr. Dierk. iii. p. 831 (1866, pt.).—Finsch, Papag. ii. p. 628 (1868, pt.).
Opopsitta diophthalma, Rosenb. Journ. für Ornith. 1862, p. 65.—Id. Natuurl. Tijdschr. voor Nederl Indië, 1863, p. 226 (partim).
Cyclopsitta diophthalma, Wall. P. Z. S. 1864, p. 284.
Psittacula diophthalma aruensis, Schlegel, Mus. Pays-Bas, *Psittaci*, Revue, p. 33 (1874).
Cyclopsitta aruensis, Salvad. Ann. Mus. Civic. Genov. vi. p. 73 (1874).
Cyclopsittacus aruensis, Salvad. Ann. Mus. Civic. Genov. x. p. 28 (1877).

This Perroquet was for a long time confounded with *Cyclopsitta diophthalma*, but is now admitted by ornithologists to constitute a separate species. Almost at the same time Professor Schlegel and Count Salvadori separated it under the name of *aruensis*; and the former gentleman remarks as follows :—" This bird, which represents the *Psittacula diophthalma* in the Aru group, presents us with a curious fact, viz. that the adult male is distinguished only by very subtle characters from the adults of both sexes of *P. diophthalma*, while the females and young are distinguished by very sensible differences. On comparing the adult male of *P. aruensis* with the adult of both sexes of *P. diophthalma*, one sees at once that the red on the head is a little clearer and does not pass into yellow on the top, while the blue spot above the eye is much more restricted and is of a green colour, differing little from the prevailing tint of the bird, and, lastly, the blue colour behind the region of the ears is prolonged underneath the chin. The young male has absolutely the same colours as the adult, with the exception that the red of the head is paler. In the females, both young and old, the parts of the head which are red in both sexes of *P. diophthalma* are, on the contrary, in *P. aruensis* rather clear blue without the least trace of red. The species would appear to occur in most of the islands of the Aru group, having been observed by Mr. Wallace at Dobbo in Wammer, and at Wonoumbai by Von Rosenberg and Hoedt, as well as by Dr. Beccari in Lutor. Mr. Wallace refers to his specimens as having been 'shot while feeding on the fruit of a *Ficus*, close to the trading-town of Dobbo.'"

Beyond this trifling note, I believe that nothing whatever has been written or said concerning this most elegant little bird.

The figures on the Plate are of the size of life, and represent adult birds.

CYCLOPSITTA SUAVISSIMA, Sclater.

CYCLOPSITTA SUAVISSIMA, Sclater.

D'Albertis's Perroquet.

Cyclopsitta suavissima, Sclater, P. Z. S. 1876, p. 520.—Sharpe, Journ. Linn. Soc., Zool. xiii. p. 491 (187 8).
Cyclopsittacus suavissimus, Salvadori, Annali Mus. Civic. Genov. ix. p. 12 (1876-7).—Id. op. cit. x. p. 28 (1877).

OF the four Papuan species of *Cyclopsitta* two have blue foreheads, and two have the forehead brown; the blue-fronted birds are *C. gulielmi III.* and *C. suavissima*; the brown-fronted ones *C. melanogenys* and *C. fuscifrons*. The present bird was discovered by Signor D'Albertis in the neighbourhood of Naiabui, in South-eastern New Guinea; Mr. Octavius Stone also met with it in the neighbourhood of Port Moresby; and the former naturalist tells us that it feeds on fruits and on seeds. Although closely allied to *C. gulielmi III.*, it is, according to Count Salvadori, a much smaller bird, and is distinguished by the less-brilliant orange of the breast, by its white lores, by the black patch on the face much bigger and more constant; and he gives a very full account of the species in his paper on D'Albertis's collection.

I copy from the last-named essay the full diagnosis given by the Count.

Adult male. Green, darker on the upper surface; forehead and a spot behind the eye blue; lores white, the sides of the head and the throat whitish yellow; cheeks broadly black; lower throat and breast orange; abdomen and under tail-coverts pale green, slightly inclining to yellow; quills dusky, the primaries blue on the outer web; secondaries and upper wing-coverts green, uniform with the back; carpal edge of the wing bluish; lesser and median under wing-coverts greenish yellow, the greater ones dusky; quills yellowish towards the base of the inner web; tail green; bill, feet, and iris black.

Female. Similar to the male, but having the cheeks blue, the ear-coverts orange, the breast scarcely orange, but rather greenish yellow, the sides of the breast on each side with a longitudinal yellow mark.

Young male. Similar to the old female, but having the forehead only slightly tinged with blue, the lores whitish yellow, the cheeks greenish, tinged on the upper part with blue, the ear-coverts yellow, the lower throat and upper breast yellowish.

The figures in the Plate represent the typical pair of birds described by Dr. Sclater, and now in Italy. I am indebted to Count Salvadori for permission to figure them here.

Whatever time may bring in regard to the discovery of new Parrots, we can scarcely expect, even in the splendid genus *Cyclopsitta*, a more beautiful bird than the one here figured.

CYCLOPSITTA MELANOGENYS.

CYCLOPSITTA MELANOGENYS.

Black-cheeked Perroquet.

Psittacula melanogenia, Rosenberg, Tijdschr. voor Nederl. Indië, xxix. p. 142 (1866).—Schlegel, Nederl. Tijdschr. voor de Dierk. iii. p. 330 (1866).—Rosenberg, Reis naar de Zuidoostereilanden, p. 49 (1867).—Gray, Hand-l. of Birds, ii. p. 168 (1870).—Schlegel, Mus. Pays-Bas, *Psittaci*, p. 35 (1874).
Psittacula melanogenys, Finsch, Papag. ii. p. 627 (1868).
Cyclopsittacus melanogenys, Salvadori, Annali Mus. Civic. Genov. ix. p. 14 (1876-7).—Id. *op. cit.* x. p. 29 (1877).

I HAVE figured in my different works several species of these little Perroquets, which appear to form a small group inhabiting only New Guinea and the adjacent Papuan islands, and extending into North-eastern Australia.

During the course of the present work several new species have been discovered; and the most of these belong to the orange-breasted section of the genus *Cyclopsitta*.

As Count Salvadori remarks, the distribution of these birds is truly remarkable, especially as regards the present bird and its two nearest allies: thus *C. melanogenys* is found in the Aru Islands, and is replaced in South-eastern New Guinea, where so many of the birds are identical with Aru species, by *C. suavissima*; while on the Fly river, which is an intermediate locality, occurs *C. fuscifrons*. Von Rosenberg procured examples of the present bird in the three islands of Wokam, Wonoumbai, and Mikor (all in the Aru group), where they are known to the natives by the name of Joa. Professor Schlegel gives a full description of the bird, and makes the following remarks, which I extract from his paper:—

"The Aru group produces a little *Psittacula* which has escaped the researches of Mr. Wallace, but of which M. von Rosenberg has furnished us with four individuals, viz. an adult male and female, and two specimens marked as males but wearing the livery of the females. This species is allied by its general form and its system of coloration to our *Psittacula gulielmi III.*, which inhabits Salawati and the neighbouring coast of New Guinea; and it appears to replace it in the Aru archipelago. It is, however, much smaller in size, all its colours are less vivid, the black bar on the ear-coverts is proper to both sexes, the yellow of the loral region is replaced by white, and the blue of the forehead and the superciliary streak are also blackish; the greater under wing-coverts, instead of being uniform blackish, are yellowish and only tipped with blackish; the quills have all of them, excepting the first two, a very large yellowish band on their inner web, whereas in *P. gulielmi III.*, this band is either in no way pronounced, or slightly indicated, passing insensibly to blackish, and confined to the secondary quills. The chest is always tinged or washed with yellow or orange-red."

The following is a translation of the description given by the learned Professor :—

Adult male. General colour grass-green, passing to yellowish green on the lower parts and to blackish on the inner webs of the quills. Forehead, region of the eye, and the whole of the posterior portion of the ear-coverts and the moustachial plumes of a slightly pronounced black; behind the region of the ear a very large patch extending onto the chin, of a white colour, washed with orange-yellow; the chest of a dark orange-rufous, not very bright; lower edge of wing, as also the outer edge of the primaries, blue; lesser and median under wing-coverts yellowish green, passing into blue towards the edge of the wing; greater under wing-coverts yellowish, but blackish at the tip; inner web of the quills, with the exception of the first two, having a very large yellow band.

The female and the male in imperfect plumage are distinguished from the adult male in the colour of the large patch behind the region of the ear, which is not white, but of a lively orange-yellow, the part of which occupying the chin passes into a greyish blue; lastly the chest is simply washed with orange-yellow. Von Rosenberg gives the colours of the soft parts as follows :—"Bill, feet, and iris dark greyish brown."

The figures in the accompanying Plate are of the size of life, and are a male and a young bird from the Aru Islands.

CYCLOPSITTA MACCOYI, Gould.

CYCLOPSITTA MACCOYI, Gould.

M'Coy's Perroquet.

Cyclopsitta Maccoyi, Gould, P. Z. S. 1875, p. 314 (April).
—————— *Leadbeateri*, M'Coy, Ann. N. H. (4) xvi. p. 54 (July 1875).

AUSTRALIA now possesses two species of *Cyclopsitta*, a genus unknown to inhabit the continent until a few years ago, when the *C. Coxeni* was discovered. New Guinea has three species, the Aru Islands and the Philippines two each, while *C. Blythi* is confined to the island of Mysol. The distribution of this little genus is therefore very remarkable, and its absence from Celebes and the Halmahéra group of islands is not what we should have expected.

I am indebted to Mr. Waller, of Brisbane, for the loan of the specimens from which my original description was taken; and I felt great pleasure in adopting the suggestion of that gentleman that I should confer upon the species the name of Professor M'Coy, to whom so much of the progress of science in the Australian colonies is due. Unfortunately this little bird is already burdened with a synonym; for nearly at the same time that I described it, Professor M'Coy himself sent a description of the species, proposing for it the name of *C. Leadbeateri*. He gives the characters as follows:—" The general size, shape, and colouring is nearly like that of *C. Coxeni*; but it is somewhat smaller, and has in both sexes an oblong patch of red on the forehead, just over the cere. It differs also in habitat, frequenting the scrubs more than *C. Coxeni* does. It seems to be rather rare at Cardwell, where the specimens described were collected by Mr. Broadbent."

Male.—General colour green, the face having all the fantastic colours of the Harlequin; on the forehead a band of bright scarlet, surrounded by cobalt, a shade of the same colour encircling the latter, narrow above, broader below; on the cheeks, from the base of the bill to the tips of the ear-coverts, a band of scarlet like that on the forehead; and below this is an obscure band of purplish blue, gradually fading off into the green of the neck; flank-feathers tipped with bright lemon-yellow; quills externally blue, the inner secondaries green, with the usual tiny patch of scarlet adjoining the back; tail green; under wing-coverts green, the outer ones washed with blue; quills blackish below, diagonally crossed near the base with a yellow band. Total length $5\frac{1}{2}$ inches; wing $3\frac{3}{8}$, tail $1\frac{3}{4}$, tarsus $\frac{3}{8}$.

Female.—Differs from the male in wanting the scarlet cheeks; in other respects similar.

All the members of this little genus roam about in flocks, but, perched among the leaves and flowers of the Eucalypti, are excessively silent and seldom betray their presence until the whole company simultaneously burst forth into the open air and wing their way to other trees.

The figures in the Plate are of the size of life.

CYCLOPSITTA COCCINEIFRONS.

CYCLOPSITTA COCCINEIFRONS.

Astrolabe-Mountain Perroquet.

Cyclopsittacus coccineifrons, Sharpe, Journ. Linn. Soc., Zool. xvi. pp. 318, 426 (1882).—Salvad. Ann. Mus. Civic. Genov. xviii. p. 418 (1882).—Id. Orn. Papuasia e delle Molucche, iii. p. 516 (1882).

This species was discovered by Mr. Goldie in the Morocco district of the Astrolabe Mountains in Southeastern New Guinea. Mr. H. O. Forbes has recently met with it in the Sogeri district on the same range of mountains.

We separated the present species from *C. diophthalmus* on account of the darker colour of the red on the crown, but now that we have had an opportunity of examining a second specimen, and that a male bird, we do not find that this distinction is upheld, but the species is nevertheless different from *C. diophthalmus*. It may in fact be recognized by the greater breadth of the yellow band across the crown, which is little more than a narrow line in the last-named species.

Neither Mr. Goldie nor Mr. Forbes have given any information respecting the habits of this pretty little Perroquet, but the former gentleman states that it is called 'Ciguri' by the natives of the Astrolabe Mountains.

The following is a description of the sexes of this species:—

Adult female. General colour above grass-green; lesser wing-coverts like the back, becoming a little brighter on the median and greater coverts, the outermost of which are slightly washed with blue; bastard-wing and primary-coverts deep blue; quills black, the primaries externally deep blue; the secondaries green like the back; the innermost marked with scarlet on the inner web, as also the innermost greater coverts; tail-feathers darker green than the back, washed with blue near the base and on the inner web; a broad band across the forehead and sinciput, deep scarlet or dull crimson, succeeded by a well-defined band of yellow, the occiput and nape and hind neck a little more yellowish green than the back; lores crimson, as also a streak below the eye; above the fore part of the eye a broad streak or longitudinal spot of greenish turquoise; ear-coverts and cheeks pale drab-brown, washed with greenish blue below, and with a narrow line of purplish blue behind the ear-coverts; under surface of body light emerald-green, with a broad streak of golden yellow down the sides of the body; thighs and under tail-coverts light emerald-green, the long ones edged with yellow; under wing-coverts emerald-green, those near the edge of the wing deep bluish green; axillaries emerald-green with a darker green centre; lower coverts and quills below dusky blackish, crossed with a double bar of pale yellow. Total length 5·6 inches, culmen 0·55, wing 3·65, tail 1·55, tarsus 0·3.

The adult male differs from the type, which is evidently a female, in having the cheeks and sides of face scarlet: "bill bluish carneous, black on the tips of the mandibles; legs and feet pale blue; iris orange-red" (*H. O. Forbes*). Total length 5·5 inches, wing 3·25, tail 1·45, tarsus 0·35.

The Plate represents an adult male in two positions, the figures being drawn from a specimen procured by Mr. Forbes in the Astrolabe Mountains.

[R. B. S.]

APROSMICTUS CALLOPTERUS, D'Albert & Salvad.

APROSMICTUS CALLOPTERUS, D'Albert. & Salvad.

Yellow-winged King Parrot.

Aprosmictus callopterus, D'Albert. & Salvad. Ann. Mus. Civic. Genov. xiv. p. 29 (1879).

In a former part of the present work I gave a figure of a very fine King Parrot from Queensland, which I called *Aprosmictus insignificus*. At the time of my describing this some of my friends believed the bird to be a hybrid, and not a true species. That such a supposition is somewhat untenable is proved by the bird which I now figure in the accompanying Plate having similar characteristics. Although scarcely as large as the *A. insignificus*, it is evidently of the same group, and is distinguished by the lengthened brilliant shoulder-patch which formed such a conspicuous feature in that species.

The *Aprosmictus callopterus* is one of the recent discoveries of Signor D'Albertis, during his exploration of the Fly River, in New Guinea; and more recently Mr. Kendal Broadbent has procured a similar Parrot in the mountains to the interior of Port Moresby. The latter bird has been separated by Mr. Sharpe, and called by him *Aprosmictus broadbenti*, for the following reasons:—It is, says that gentleman, a smaller bird, and has the blue patch of the mantle continued upwards onto the crown, whereas in *A. callopterus* the entire head and neck are red. The tail-feathers in *callopterus* are conspicuously tipped with rosy red, whereas, of the two specimens I have seen (one in the collection of the British Museum, the other in my own), the faintest spot only of this colour is observable in one feather of the bird I possess. Without wishing to disparage so great an authority as Mr. Sharpe as to the two birds being distinct, it may be possible that age may have something to do with the matter, and the *A. callopterus* is the younger bird; for, although it is not only in Parrots that I could point out where such is the case, I can hardly believe in the existence of two species of so limited a genus as *Aprosmictus* existing in such close proximity. At the same time it must be remembered that such is apparently the case in the Crowned Pigeons, where the *Goura albertisi* of Port Moresby is distinct from the *Goura sclateri* of the Fly River. This is a question which can only be solved by further explorations; and I am compelled to leave the matter as it stands for the present, and to allow *A. broadbenti* to stand or fall when more positive proofs have been obtained.

I have received the following note on the subject from Signor D'Albertis:—

"Dear Mr. Gould,

"I am very sorry I cannot give you the amount of information you want about the beautiful Parrot I discovered during my last expedition up the Fly. I found the bird on the 28th of June, 1877. I saw a pair. The male being the brighter-coloured, was first killed; and the female was never seen again. They were feeding on a small soft berry that grows on very high and thick trees. I was delighted with my capture, believing at the time I had got a new species, and, showing it to my men, offered a large reward for any other specimen they might bring me; but many days passed, and the bird could not be obtained by either of us. The country where I found it was hilly, and the forest magnificent, abounding in the most beautiful tropical plants. Perhaps our having anchored further up the river, where the country was flat, was the reason we did not find any more for a time; but when we again reached the hills another specimen, a female, was procured. The first two were adult birds. Late in October three more were secured; but they were all young; so that I should imagine the nesting-season to be from June to October or, perhaps, November. It is a shy bird, slow, and not noisy, as most Parrots are; and this may account for the difficulty of finding it in the thick forest. I think it also prefers the hilly districts, which would explain its scarcity in the prevailing flat country on the banks of the Fly. I am sorry not to be able to tell you any more about this bird; but its rarity did not allow me to study its habits."

Signor D'Albertis mentions that in the living bird the bill is black, excepting for the reddish spot on the upper margin of the base of the upper mandible; the feet are black; and the iris is yellow or orange-yellow.

For the opportunity of figuring this new and handsome species I have to thank Signor D'Albertis, who most kindly lent me a pair of specimens for the purpose. He informs me that the red tips to the tail-feathers are by no means a constant character; but he admits that he never saw a specimen of *A. callopterus* similarly coloured to the two males of *A. broadbenti* above referred to.

APROSMICTUS INSIGNISSIMUS, Gould.

APROSMICTUS INSIGNISSMUS, *Gould*.

Beautiful King-Parrot.

Aprosmictus insignissimus, Gould, P. Z. S. 1875, p. 314.

I wish it were in my power to write a complete history of the splendid Parrot figured in the accompanying Plate. All that is at present known is that it was shot, in 1874, a few miles north of the village of Dalby, on the Darling Downs, in Queensland. My first knowledge of its existence was through a life-sized sketch from the hands of the son of Mr. Waller. Since then the actual specimen has been forwarded to me by Mr. Coxen, who has purchased it for the infant museum of Brisbane.

Those ornithologists who have paid attention to the Parrots of Australia, either in a state of nature or in the cabinet, will at once perceive the affinity of this bird with the ordinary King-Parrot of the brushes of New South Wales (*Aprosmictus scapulatus*). In structure it is very similar to that bird, while in colour it greatly partakes of *Ptistes*. In their habits and local habitations, however, *Ptistes* and *Aprosmictus* widely differ. *Aprosmictus* is almost solely an inhabitant of the thick brush, while *Ptistes* is as exclusively a frequenter of the thinly timbered open plains; the former is dull, quiet, and slow in its movements, while the latter passes over the tops of the highest trees of the plains with a strong vigorous flight. Each of them has a well-developed os furcatorium, a bone not found in *Platycercus* and allied genera.

The King-Parrot has a stout bill, the lower half of which is black, the upper red; while in *Ptistes* both mandibles are highly coloured; the former has also a shorter and less ample wing than the latter. I mention this particularly to show the improbability of this bird being a lusus, or hybrid, between the birds above mentioned; and it may be that it is an accidental visitor from New Guinea.

The following is a repetition of what is published in the 'Proceedings of the Zoological Society' for 1875:—

"Head emerald green, excepting the centre of the crown and a patch on the nape, these parts being scarlet, the green forming a narrow frontal line between the nostrils and the crown; round the hind neck a narrow collar of emerald green; back, including the mantle and scapulars, deep grass-green, each feather obscurely edged with darker colour; lower back and rump shining blue; upper tail-coverts bright grass-green; all the tail-feathers above deep grass-green, with indistinct narrow bars of darker colour on every one of them; the underside of the tail uniform purplish black; wings green, with a broad longitudinal patch of yellow, many of these feathers edged with scarlet, this mark being very indistinct and similar to that seen in the male of *Ptistes*; under wing-coverts greenish blue; inner lining of quills purplish black like the lower surface of the tail; under surface of body scarlet, with dashes of bright green on the flanks; under tail-coverts green, fringed with scarlet; both mandibles of the bill bright orange-red; legs blackish. Total length $15\frac{1}{2}$ inches, wing 9, tail $7\frac{3}{4}$, tarsus $\frac{7}{8}$."

It is pretty evident that this specimen had not quite completed its fully adult livery. If it had done so, the green feathers on the back of the head would have been red.

In conclusion I must thank the authorities of the Brisbane Museum for the extreme courtesy which induced them to send so valuable a bird to England for the purpose of this work. I have thus been enabled to give a life-size illustration of this remarkable and interesting species.

CHARMOSYNA MARGARITÆ, Tristr.

CHARMOSYNA MARGARITÆ, Tristr.

Duchess of Connaught's Parrakeet.

Charmosyna margarethæ, Tristram, Ibis, 1879, p. 442, pl. xii.
Charmosynopsis margaritæ, Salvad. Orn. Papuasia, i. p. 319 (1880).—Id. Ann. Mus. Civic. Genov. xviii. p. 419 (1882).—Id. Orn. Papuasia, etc. iii. App. p. 520 (1882).
Trichoglossus (Charmosyna) margarithæ, Ramsay, Proc. Linn. Soc. N. S. W. vi. p. 720 (1881).

This beautiful little Parrakeet belongs to the subgenus *Charmosynopsis* of Salvadori, along with *C. pulchella*, to which it is allied. It differs from the latter bird, however, in many particulars, in the yellow collar round the hind neck, and in the broad yellow band on the chest, which takes the place of the yellow streaks in *C. pulchella*. In the latter species also there is no purplish-black band across the breast as there is in *C. margaritæ*.

The present bird, named by Canon Tristram in honour of H.R.H. the Duchess of Connaught, was discovered in the Solomon Islands by Lieut. Richards, R.N., and Mr. Ramsay has likewise received specimens from the same Archipelago. He has very kindly lent us a pair of birds from the island of Ugi, the one with the red on the sides of the rump being marked "male" and the yellow-sided bird "female."

The following is a description of the specimens lent to us by Mr. Ramsay :—

Adult male. General colour above bright grass-green, the rump and upper tail-coverts washed with golden ; sides of the lower back crimson; wing-coverts green, like the back ; primary-coverts and quills black, externally green like the back ; tail-feathers crimson slightly tipped with yellow, which increases into a large terminal mark on all the other feathers, which have black shafts, and an inner broad margin of black and a narrower external one of grass-green ; forehead and sinciput crimson, separated by a broad bar of deep purplish black on the vertex from the occiput and hinder neck, which are also crimson, and are succeeded by a narrow band of purplish black, followed by a somewhat broader one of orange-yellow, which divide the green back from the crimson neck; lores, sides of face, cheeks, ear-coverts, sides of neck, and entire throat crimson, followed by a narrow band of purplish black, which is succeeded by a broad band of orange-yellow, both these bands being continuous with the similarly coloured bands round the hind neck ; remainder of under surface of body crimson, with a band of purplish black across the breast ; thighs and basal under tail-coverts crimson, the longer under tail-coverts grass-green ; axillaries and under wing-coverts crimson ; edge of wing grass-green. Total length 7·5 inches, culmen 0·7, wing 4·25, tail 3·25, tarsus 0·4.

The female exactly resembles the male, but has less tinge of golden on the rump, and is further distinguished by the sides of the lower back being orange-yellow, with crimson tips to a few of the feathers. Total length 7·6 inches, culmen 0·65, wing 4·1, tail 3·5, tarsus 0·4.

The figures in the Plate are taken from the pair of birds mentioned above as lent to us by Mr. Ramsay : they represent the male and female of about the size of life.

[R. B. S.]

CHARMOSYNA JOSEPHINÆ.

CHARMOSYNA JOSEPHINÆ.

Josephina Parrakeet.

Trichoglossus josephinæ, Finsch, Atti R. Accad. Torino, 1872, p. 42, tav. 7.—Meyer, J. f. O. 1874, p. 56.

The Josephina Parrakeet was first described by Dr. Otto Finsch, from a specimen preserved in the collection of Count Turati at Milan. By the latter gentleman it was obtained from M. Laurent de Gréaux, a natural-history dealer in Marseilles, having been originally purchased by him in London, along with two specimens of the rare *Psittacula gulielmi III*. Dr. Finsch rightly concluded that the habitat of this beautiful species would prove to be the island of New Guinea; for he found in the Bremen Museum some native head-dresses of plumes composed of the tail-feathers of this Parrakeet and of *C. papuensis*, which Von Rosenberg assured him were worn by the inhabitants in the interior of the island. None of the Dutch collectors managed to obtain specimens for the Leiden Museum; and it is to Dr. Meyer that we are indebted for our authentic information respecting it. He obtained specimens during his last voyage, and he was thereby enabled to determine that the species had been founded by Dr. Finsch on a female bird; Dr. Meyer was therefore the first to discover the male. I am indebted to him not only for the loan of the specimens which I now figure, but also for the following note on the species:—" I found this bird on my voyage to New Guinea in June 1873, on the west coast of Geelvink Bay, where it was seen near the sea-shore in large flocks; but in no other spot, during my residence in the island, did I meet with it. The nearly allied *C. papuensis* has, up to the present, only been found far in the mountainous districts of New Guinea; and perhaps this smaller species represents it in the lowlands. Young birds have the breast undulated with blackish and green; and in some females the black of the belly is strongly mixed with olive green. In life the bill, feet, and irides are deep orange-red, the claws and the naked part round the eye greyish black. In the stomach I only found juice of plants."

Dr. Finsch thus describes the female :—

"Head, neck, and under surface of body fine crimson-red; an occipital spot obscure bluish lilac, posteriorly surrounded with a black band. Back and wings green. Belly and thighs black, with a violet lustre. Rump and flanks yellow. Two centre tail-feathers red; the rest green, with the inner webs red, the tips yellow, preceded by an obscure subterminal mark of bluish."

The male differs from the female in being red on the rump, where the latter is yellow.

I have figured a male and female on the accompanying Plate, of about the natural size. In general size this species is rather smaller than *C. papuensis*, and has a coarser and stronger bill, while the long tail-feathers are stiffer and less flexible.

CHARMOSYNA PULCHELLA, *Gray.*

CHARMOSYNA PULCHELLA, Gray.

Pectoral Lorikeet.

Charmosyna pulchella, Gray, List Psittac. B. M. p. 877 (1859).—Wall. P. Z. S. 1864, p. 292.—Salvad. Ann. Mus. Civ. Genova, vii. p. 813 (1875).
———— *pectoralis*, Von Rosenb. J. f. O. 1862, p. 64.—Id. Nat. Tijdschr. Nederl. Ind. 1863, p. 144.—Id. J. f. O. 1864, p. 112.
Eos pulchella, Schl. Dirent. p. 69 (1864).
Trichoglossus pulchellus, Finsch, Papag. ii. p. 877 (1868).—Gray, Hand-l. B. ii. p. 157 (1870).—Meyer, Sitz. Akad. Wien, lxix. pp. 74, 76.

DR. OTTO FINSCH, our greatest authority on the family Psittacidæ, places the present beautiful little species near the end of his comprehensive genus *Trichoglossus*, in the vicinity of *Charmosyna papuensis*, which he does not consider generically separable. Its near allies he admits also to be the elegant little birds which I have figured in the present work as *Psitteuteles placens*, &c.; but I contend that its general form and coloration make it rather a *Charmosyna*, in which genus it was placed by the late Mr. Gray, and also by its original discoverer Mr. A. R. Wallace. As Dr. Finsch remarks very truly, it does not show the extremely long tail-feathers which adorn *C. papuensis*; but I think that any one comparing the plates in this volume will see that in general system of coloration it agrees best with the last-named species.

I regret to say that nothing whatever has been recorded respecting its habits and economy; and it does not seem to be a common bird even in the country which it inhabits. This is the northern part of New Guinea, Mr. Wallace having discovered it at Dorey, and Signor d'Albertis having also met with it in Atam.

The following is Mr. Gray's original description:—

"Total length 7½ inches, wings 3½. Head and underpart of the body carmine; the breast-feathers green at base, with some small streaks of bright yellow; crown and spot on the lower part of back purplish black; nape, back, wings, and base of tail-feathers green; sides of rump bright yellow; thighs green, streaked with bright yellow; knees purplish black; middle tail-feathers mostly carmine, with the tips yellow; lateral feathers green, with the inner web carmine and the tips bright yellow."

Not only is great elegance of form conspicuous among the species of this genus, but how strikingly does the tint of red prevail not only in the pretty *Charmosyna*, but in most of the Lories and Lorikeets and other birds of the rich region of Papuana!—a feature which must have forcibly struck any person who has made a study of comparing the birds of Asia and India, on the one hand, with those of Australia and New Zealand on the other. With the exception of Mr. Wallace's specimen now in the British Museum, no examples, as far as I am aware, have been brought to this country.

The figures in the accompanying Plate represent a supposed male and female, of the size of life.

I am indebted to the courtesy of Dr. Meyer in sending these birds to England.

CHARMOSYNA PAPUENSIS.

CHARMOSYNA PAPUENSIS.

Papuan Lorikeet.

Papuan Lory, Lath. Gen. Syn. i. p. 215 (1781).
Psittacus papuensis, Gmel. S. N. i. p. 327 (1788).—Lath. Ind. Orn. i. p. 88 (1790).—Kuhl, Consp. Psitt. p. 33.
Psittacus papou, Scop. Del. Flor. et Faun. Insubr. ii. p. 86 (1786).
La Perruche Lori papou, Levaill. Perroq. ii. p. 9, pl. 77 (1805).
Psittacus lichtensteini, Bechst. Kurze Uebers. p. 82 (1811).
Palæornis papuensis, Vieill. N. Dict. d'Hist. Nat. xxv. p. 336 (1819).—Vigors, Zool. Journ. ii. p. 56 (1826).
Lorius papuensis, Less. Tr. d'Orn. p. 195 (1831).—Schl. Mus. P.-B., *Psittaci*, p. 130 (1864).
Charmosyna papuensis, Less., Sclater, Proc. Zool. Soc. Nov. 1873.—Bp. Rev. et Mag. Z. 1854.—Gray, List Psitt. 1859.
Trichoglossus papuensis, Finsch, Papag. ii. p. 878 (1868).

When the collection of birds formed by Signor d'Albertis was submitted to me, I was highly pleased to find an exquisite specimen of the present bird from Atam. Skins denuded of the head, legs, and wings, as skins of the Birds of Paradise formerly were, had, it is true, been before my eyes for the last forty years; but until now I had not seen it in a perfect state. It is, however, if we may depend upon the older authors, one hundred years since a knowledge of its existence was acquired by ornithologists. As a matter of course its synonymy is not only extensive, but in such a state of confusion that it would be out of place, in a folio work such as I am now penning the letterpress for, to attempt to unravel it. Those who desire further information on the subject would do well to consult the very excellent and learned work on the Parrots, 'Die Papageien,' by Dr. Otto Finsch, who says:—" This extraordinarily rare Parrot, distinguished for its splendid colours, was well known to the old author Seba, and was first figured by him. His single figure represents the bird as a skin in the way in which they are prepared by the Papuan natives. Seba, who calls every bird with decorative plumage a Bird of Paradise, takes this bird also for one, and gives its habitat quite correctly as New Guinea. This species is very perplexing as regards its synonymy; but it is in any case erroneous for Wagler and Gray to place as synonyms of *Trichoglossus papuensis* the very doubtful species *Psittacus japonicus* of Linnæus, and *parvus* of Bontius, of my Appendix of doubtful species.

"Concerning the native country of this bird there has been hitherto some indecision; but we learn through Wallace and Von Rosenberg that the species really exists in the northerly and north-westerly parts of New Guinea. Near Doreh it is very rare; the long tail-feathers, however, are seen in the feather ornaments of the natives of this place. Schlegel gives the eastern part of New Guinea as its habitat."

That *Charmosyna papuensis* will be regarded as one of the finest of the New-Guinea birds is certain; the researches of Dr. Meyer, however, have brought under my notice two or three other exquisitely coloured species, which will hereafter be figured in sequence to this, the head of the genus.

The figure in the accompanying Plate is of the size of life.

CHARMOSYNA STELLÆ, Meyer.

CHARMOSYNA STELLÆ, Meyer.

Stella Parrakeet.

Charmosyna josephinæ (nec Finsch), Sharpe, Journ. Linn. Soc., Zool. xvi. p. 428 (1882).
Trichoglossus papuana, Ramsay, Proc. Linn. Soc. N. S. Wales, x. p. 244 (1885).
Charmosyna stellæ, Meyer, in Madarász, Zeitschr. ges. Orn. iii. p. 9, pl. ii. (1886).

This splendid Parrakeet is the southern representative of *C. papuensis*, which inhabits North-western New Guinea. It was described by Dr. Meyer from specimens procured by Mr. Hunstein in the Owen Stanley Mountains; and Mr. Forbes has sent a fine series from the Astrolabe range, where it would seem to be abundant, if only sufficient altitude is reached. Mr. Goldie sent home some years ago a pair of mutilated skins, which we recognized as different from *C. papuana*, but erroneously referred them to *C. josephinæ*. Mr. Ramsay says that he has also received specimens of both species from the Astrolabe Mountains, the *C. papuana* mentioned by him being of course *C. stellæ*.

The present species is easily distinguished from both *C. papuana* and *C. josephinæ* by the absence of the yellow patches on the sides of the body, and also in the different arrangement of the nuchal patch, the greater part of the crown in *C. stellæ* being crimson, followed by a large lilac-blue patch on the occiput and nape, whereas in *C. papuana* the occiput is black, separated from the crimson forehead by a line of lilac-blue feathers.

Mr. Goldie, when he procured the first mutilated skins from the natives, wrote as follows :—" The feathers from the tail have been frequently obtained along the coast. The natives said that the bird was only to be obtained at a considerable distance from Morocco inland on the mountains." The native name is given by him as " Divu."

Adult male. General colour above dark grass-green, the back and upper tail-coverts rich crimson, with a patch of lilac-blue on the rump; a broad collar of deep crimson separating the head from the green mantle, the adjacent feathers being crimson with green tips; wing-coverts grass-green like the mantle; bastard-wing, primary-coverts, and quills dusky blackish, externally grass-green; tail-feathers green, the long centre feathers passing into red and thence into yellow at the ends; the remainder of the feathers yellow towards the ends, the base of the inner web rich crimson; crown of head crimson, the nape black, the occiput with a patch of long lilac-blue feathers; entire sides of face, throat, and breast deep crimson, with a small patch of green feathers at the sides of the upper breast; lower breast and abdomen blackish with a purplish-blue gloss, more distinct over the thighs; sides of body and flanks deep crimson; lower abdomen, vent, and under tail-coverts deep crimson; under wing-coverts and axillaries crimson; edge of wing green; lower coverts and quills below dull blackish. Total length 15 inches, culmen 0·8, wing 5·6, tail 4·6 (longest feathers 10·4), tarsus 0·55.

Adult female. Differs from the male in having the centre of the back golden yellow instead of crimson, and in having the blackish colour of the abdomen extending on to the flanks. Total length 14·5 inches, culmen 0·85, wing 5·2, tail 5·5 (longest feathers 10·2), tarsus 0·55.

The young male resembles the old male, and, like it, has a crimson back; but the crimson feathers of the hind neck and breast are margined with purplish blue, and the blackish abdomen is washed with green and has blue endings to the feathers.

The Plate represents an adult male and female of this Parrakeet, the figures being drawn from specimens collected by Mr. H. O. Forbes.

[R. B. S.]

PSITTEUTELES RUBRONOTATUS.

PSITTEUTELES RUBRONOTATUS.

Red-backed Lorikeet.

Coriphilus rubronotatus, Wallace, P. Z. S. 1862, p. 165.
Charmosyna rubronotata, Wallace, P. Z. S. 1864, p. 293.
Trichoglossus rubronotatus, Finsch, Pap. ii. p. 876 (1868).
Nanodes rubronotatus, Schlegel, Mus. P.-B. Revue Psitt. p. 51 (1874).
Trichoglossus rubronotatus, Meyer, Sitzb. d. k.-k. Akad. d. W. zu Wien, lxix. p. 400 (1874).
Trichoglossus kordoanus, Meyer, Verh. d. k.-k. zool.-bot. Ges. zu Wien, 1874.
Charmosyna kordoana, Salvadori, Ann. Mus. Civ. di Genova, vii. 1875, p. 212.

Dr. Meyer, ever anxious to advance Papuan ornithology, has obligingly forwarded to me seven specimens of a little Lorikeet, which he considers to be two different birds; and he believes that his *Trichoglossus kordoanus* should stand as a species, and not as a synonym to Mr. Wallace's *Coriphilus rubronotatus*. But, after having given the subject my closest attention and, moreover, after consulting the opinions of others, I can come to no other conclusion than that the specimens sent to me are identical or have insufficient characters to separate them. Trifling differences, it is true, occur among the seven specimens; but they are not more, nay, not even so much as those that have occurred to me many times before—differences attributable, perhaps, to locality or to some cause equally unimportant. As Dr. Meyer, however, still clings to his opinion and has, moreover, taken the trouble to forward me a translation of his paper on the subject with additional remarks, it is right that I should give his views on the subject.

He remarks:—"*Trichoglossus rubronotatus* was discovered by Mr. Wallace on the island of Salwati at the north-west extremity of New Guinea, and has since been procured by Dr. Bernstein in the same district, but was never figured before, specimens even now being very rare. It was only in one place in New Guinea, in May 1873, that I obtained this lovely bird, viz. near Rubi, the southern point of the great Geelvinks Bay; and in 1874 I published some notes respecting it, which I beg to be allowed to reproduce here.

"Dr. Finsch says of *rubronotatus*:—'A beautiful bird, representing the *Trichoglossus placens* on Salwati and the north-west coast of New Guinea;' but, inasmuch as the two species occur together (mine were both obtained near Rubi), they must rather be considered parallel forms and not representations of each other, the adult male only being known till lately, and described by Dr. Finsch in detail, although he does not mention the sex of the bird.

"A short time before I published these notes I had described a similar form from the island of Mysore, in the north of the Geelvinks Bay, under the name of *Trichoglossus kordoanus*, a female, and the only one that had come under my notice. I then stated the differences between the females of the two forms to be as follows:—*T. kordoanus*. Cheeks and sides of the neck bright bluish green, whereas in *T. rubronotatus* the cheeks are of a darker tint, with yellow stripes, and the sides of the neck light grass-green, like the under surface of the whole body; besides which the red of the uropygium in *T. kordoanus* is much more vivid than the same colour in the males of *T. rubronotatus*. The irides are bright yellow; bill and cere bright red; feet of a somewhat lighter tint; claws grey. The tongue is a regular brush; the stomach contained only flower-sap.

"I imagined that the males would prove to be still more distinct, as the difference between the females was so marked; this, however, did not prove to be the case. The Dresden Museum possesses a male specimen of the Mysore bird; the chief points of difference between this and *T. rubronotatus* are as follows:—Under surface, especially on throat, bluish green instead of the yellowish tint in *T. rubronotatus*; the red of the uropygium brighter and more extensive than in the New-Guinea bird, where it only consists of a few feathers even in the male when in full plumage; besides which the red on the sides of the breast appears to extend more to the centre in *T. kordoanus* than in *T. rubronotatus*; and, finally, the green of the upper surface of the former is of a more brilliant hue than in the latter.

"These differences are not great, yet they are conspicuous and cannot be overlooked or underrated; their real significance, however, lies in the fact that they coincide with the geographical separation, and in their constancy. In our Darwinian days but few naturalists will dispute that insular forms are to be regarded as directly derived from continent ones, nor will the value of the small differences be disputed. When sufficient materials of similar deviations from a parent stock are collected, the reasons of these variations may, perhaps, be rendered more clear, and the laws by which they sometimes take place be more fully recognized."

The principal figures are of the size of life.

PSITTEUTELES SUBPLACENS.

PSITTEUTELES SUBPLACENS.

Green-backed Lorikeet.

Trichoglossus subplacens, Sclater, Proc. Zool. Soc. 1876, p. 519.

THE present lovely Lorikeet, lately discovered by Signor D'Albertis on the south coast of New Guinea, may dispute the palm of beauty with the well-known *Psitteuteles placens*, so generally considered one of the finest of the family. As regards colour, *P. subplacens* has even a more beautiful tail than its near ally; and the markings of this organ are still more vivid. The greatest difference, however, is in the colouring of the back, which is uniform green, while on the back of *P. placens* a conspicuous mark of blue breaks the uniformity of this part; lastly, although the crown in both is of a different colour from the hinder neck, it is much brighter and better-defined in the new bird. We have now five very distinct Lorikeets which are not second in beauty when compared with any other section of the family. These five species are *Psitteuteles placens*, *P. subplacens*, *P. arfaki*, *P. rubronotatus*, and the elegant *P. wilhelminæ*. A question now arises in my mind, Are there others of the genus yet to be discovered? or is five the normal number, as is the case with many other little groups to which generic appellations have been given? With regard to the sexual differences in colour, we find the usual distinctive marks in the male and female; that is, the ear-coverts are blue in the former, whilst they are yellow in the latter, which may be readily seen by reference to the Plates.

The following is taken from Mr. Sclater's paper in the 'Proceedings' of the Zoological Society as above quoted:—" Extracts were read from letters received from Signor L. M. D'Albertis, C.M.Z.S., dated Sydney, March 27th," 1876.

He " also exhibited a small collection of bird-skins collected at Yule Island and on the adjoining coast of New Guinea, which that gentleman had transmitted to him for examination.

" Mr. Sclater stated that, the collection having been only just received, he had not had time to examine it carefully, but took the opportunity of pointing out the characters of two apparently new species of Parrots," of one of which the description is as follows:—

" TRICHOGLOSSUS SUBPLACENS, sp. nov.

" *Psittaceo-viridis, pileo summo flavicante; macula magna auriculari utrinque cærulea; subtus dilutior, lateribus et alarum tectricibus inferioribus coccineis; remigum pagina inferiore nigra fascia flava intersecta; rectribibus ad basin coccineis, inde nigris flavo terminatis, harum duabus mediis supra viridibus subtus nigris, linea media coccinea versus apicem occupatis; rostro rubro; pedibus rubro-flavidis: long. tota* 6·5, *alæ* 3·5, *caudæ* 2·8.

" *Hab.* Mountains of Naiabui, south of New Guinea (*D'Albertis* et *Tomasinelli*)."

The Plate represents a male and a female, of the natural size; and I am indebted to the kindness of Dr. Sclater for shewing me these birds and allowing me to figure them.

PSITTEUTELES ARFAKI.

PSITTEUTELES ARFAKI.

Arfak Lorikeet.

Trichoglossus (Charmosyna) arfaki, Meyer, Verh. z.-b. Gesellsch. Wien, 1874, p. 37.
──────── *arfaki*, Meyer, Sitz. k. Akad. Wien, lxix. p. 74 (1874).—Rowley, Orn. Misc. part 3 (1876).

THE minute size and singular coloration of the outer tail-feathers, which make this tiny Parrot resemble at first sight one of the Indian Minivets (*Pericrocotus*), caused me no little astonishment when I first uncovered the specimen which Dr. Meyer was so kind as to send over to me from Dresden for the purpose of figuring; but I further found that, in addition to these characteristics, it differed from all its allies in having *fourteen* tail-feathers instead of twelve. This peculiarity would almost be sufficient to place it in a separate genus; but this I cannot bring myself to do, in the face of its evident affinity to *P. placens* and *P. wilhelminæ* : and when we consider that certain Snipes are still retained in the genus *Gallinago* which present similar variations in the number of the tail-feathers to that exhibited by these little Lorikeets, it is not unreasonable to keep the latter in one and the same genus.

Only three specimens were procured by Dr. Meyer during his trip to the Arfak Mountains, one old male in full livery and two young ones; one of the latter passed into the collection of Count Turati, of Milan, and the other into that of Mr. Dawson Rowley, of Brighton, by whom it was figured in his interesting work the 'Ornithological Miscellany.' I am much indebted to this gentleman for the loan of his specimen, which is depicted in the adjoining Plate along with the adult male retained by Dr. Meyer for the Dresden collection and sent over by him to England. The plate in the 'Ornithological Miscellany' represents the bird to be a female; but I am assured by Dr. Meyer that some error must have occurred by the displacement of the original label; for he himself dissected the three specimens shot, and proved them to be males, one being adult, as before noted, and the other two immature : these had the colouring incomplete except as regards their tails; but in size, as is often the case with young birds, they somewhat exceeded the adult. It is one of these young birds that has been figured by Mr. Dawson Rowley.

That this is a honey-eating Parrot I have no doubt, its general characteristics uniting it with this group of the Parrots; but its colours are strikingly anomalous. Fancy a little bird, scarcely bigger than a Bearded Reedling (*Calamophilus biarmicus*) with a tail like that of a Minivet (*Pericrocotus*), and exhibiting a silvery tear-mark running down a cheek of smutty blue, and we have some of the peculiarities of this curious form. The Plate, however, gives a better idea of the bird than can be gained from any remarks of mine. The following is a more detailed description of the old male :—

The half of the crown nearest the bill, the under part of the shoulders, flanks, centre of the abdomen, and apical half of the outer tail-feathers bright red, while the base of each of the feathers last mentioned is jet-black; the rest of the plumage both above and below fine grass-green; the margins of the primaries washed with blue; sides of face dull blue, with a silvery streak running down the centre; the under surface of all the primaries and secondaries jet-black, crossed by a V-shaped band of bright yellow; bill and feet black.

Total length 6 inches, wing $2\frac{3}{4}$, tail $3\frac{5}{8}$, bill $\frac{1}{2}$, tarsi $\frac{1}{4}$.

Hab. Arfak Mountains, N.W. Guinea.

The figures represent the old and young birds, of the size of life.

PSITTEUTELES WILHELMINÆ.

PSITTEUTELES WILHELMINÆ.
Wilhelmina Lorikeet.

Trichoglossus wilhelminæ, Meyer, J. f. O. 1874, pp. 56, 57, 111.—Id. Sitz. k. Akad. Wissensch. lxix. p. 74 (1874).

IF it were possible to use such a term in speaking of a bird, I should describe this exquisite little Lorikeet as being an example of the *multum in parvo* principle as regards colouring. Its diminutive size, together with the great variety of its tints, combine to render it one of the most striking of the birds obtained by Dr. Meyer during his recent travels to New Guinea and the adjacent islands. Although for the present I associate *P. arfaki*, *P. placens*, and this species under one genus, it must not be overlooked that they differ in one important character, namely in the possession of fourteen tail-feathers by *P. arfaki*, whereas *P. placens* and *P. wilhelminæ* have only twelve, like other Parrots. Its beautifully coloured graduated tail is, in fact, the only character which it possesses in common with *P. arfaki*, the fashion of the coloration being very similar.

In naming this species *P. wilhelminæ* Dr. Meyer has paid a graceful compliment to his wife, who has been a constant companion of his travels, not only in Europe, but also to the far distant Molucca Islands; and I must confess that it gives me great pleasure in figuring the species in an early number of my work, and thereby assisting to perpetuate the name of a lady so heroic.

That other species of this little genus remain to be discovered one can hardly doubt, seeing the vast accessions to our knowledge of ornithology which each successive collection from the northern part of New Guinea brings to us.

On the accompanying Plate two males and a female are depicted, as nearly of the natural size as possible; but it will perhaps be desirable to add a detailed description, since the female differs in some points from the male as regards the colour and disposition of the markings; but the most prominent is the entire absence of red on the under surface of the wing. Her other tints, though similar to the male, are not so fine.

The following is a description of a male sent:—" Bill yellow; face, throat, and under surface generally, lively pale green, with a large patch of lengthened narrow stripes of yellow on the breast, each stripe being margined with a dark colour; crown of the head and nape reddish brown, with narrow blue feathers interspersed down the latter; mantle and upper surface green; lower part of the back fiery red, succeeded by a patch of purple, blending into green on the upper tail-coverts; two central tail-feathers black at their bases, passing into greenish at their tips; the bases of all the external feathers fiery red, succeeded by black, and green tippings; on the flank a small patch of yellow joining the red of the back; the feet appear to have been grey.

"Total length of male $4\frac{3}{4}$ inches, wing $2\frac{3}{8}$, tail $2\frac{3}{8}$, tarsi $\frac{3}{8}$.

"*Hab.* New Guinea, passim."

PSITTEUTELES PLACENS.

PSITTEUTELES PLACENS.

Beautiful Lorikeet.

Psittacus placentis, Temm. Pl. Col. iv. pl. 553 (1835).—Müll. & Schl. Naturl. Gesch. Land- en Volkenk. p. 23 (1839-44).
Conurus placens, Bourjot St.-Hilaire, Perroq. pl. 46 (1839).
Coryphilus placentis, Gray, Gen. B. ii. p. 417 (1845).—Id. List Psitt. B. M. p. 59 (1859).—Id. Cat. Mamm. &c. N. Guin. p. 41 (1859).—Von Rosenb. Reis. Zudoostereilanden, p. 87 (1867).
Psitteuteles placens, Bp. Rev. et Mag. de Zool. 1854, p. 157, et Tabl. Syst. in Naumannia, 1856.
Trichoglossus placens, Scl. Pr. Linn. Soc. 1858, p. 164.—Finsch, Papag. ii. p. 872 (1868).
Coryphilus placens, Schl. Dirent. p. 78 (1864).—Finsch, Neu Guin. p. 158 (1865).
Nanodes placens, Schl. Mus. P.-B. *Psittaci*, p. 113 (1864).—Id. Rev. Psitt. p. 50 (1874).
Charmosyna placentis, Wall. P. Z. S. 1864, p. 292.
Trichoglossus placentis, Gray, Handl. B. ii. p. 157 (1870).

In contrast to the very restricted range of *P. arfaki* and *P. wilhelminæ*, the bird now before us has rather a wide distribution, being found in nearly every one of the Papuan Islands, and Professor Schlegel gives the following localities in which the species has been known to occur:—" Halmahera, Ternate, Ambaou (an island to the south of Bouru), Ceram, Amboina, Poulo-Padjang (of the group of Ceram Laut), Great Key, Aru Group, Mysol, Salwatti, and Guebéh, as well as in the western part of New Guinea." The Leiden Museum possesses a series of no less than sixty-four examples, obtained by the well-known travellers Bernstein, Hoedt, and Von Rosenberg.

Our knowledge of this beautiful little bird is extremely limited; and I believe that there is nothing known on the subject of its manners and general economy. A single note is given by Mr. Wallace in his interesting work on the Malay Archipelago (i. p. 314), where he says:—

"In September 1858, after my return from New Guinea, I went to stay some time at the village of Djilolo, situated in a bay on the northern peninsula. Here I obtained a house through the kindness of the Resident of Ternate, who sent orders to prepare one for me. The first walk into the unexplored forests of a new locality is a moment of intense interest to the naturalist, as it is almost sure to furnish him with something curious or hitherto unknown. The first thing I saw here was a flock of small Parroquets, of which I shot a pair, and was pleased to find a most beautiful little long-tailed bird ornamented with green, red, and blue colours, and quite new to me. It was a variety of the *Charmosyna placentis*, one of the smallest and most elegant of the brush-tongued Lories. My hunters soon shot me several other fine birds; and I myself found a specimen of the rare and beautiful day-flying moth *Cocytia d'Urvillei*."

It would be difficult to imagine a bird more variously coloured than the present; and it is in consequence by no means easy to describe. The bill is red, the feet yellow, the entire face bright scarlet, the ears blue, surrounded by lively green. The upper surface of the body is also of the latter colour; a bright blue spot, however, vies with the mark of this tint on the ear-coverts, while the under surface of the body from the chest to the under tail-coverts is light yellowish green, relieved on the flanks by a brilliant patch of scarlet; on raising the wing a brilliant scarlet mass also occupies a part of the shoulders, while a triangular-shaped mark of yellow crosses the primaries and some of the secondaries. Its graduated and somewhat cuneate tail is much diversified, particularly on the under surface, the bases of the feathers being red, the middle black, and the tips yellow.

The female differs from the male in having no red on the cheeks or blue on the ear-coverts, the latter being striped with yellow and dark brown; neither has she the bright scarlet on the flanks and under the shoulders.

Total length of male 7½ inches, wing 3¾, tail 4¼.

On the Plate are figured a male and female, of the natural size.

NASITERNA PYGMÆA.

NASITERNA PYGMÆA.

Pygmy Parrot.

Psittacus (Psittacula) pygmæus, Quoy & Gaim. Voy. de l'Astrol. Zool. i. (1830) p. 232, pl. 21. fig. 1 (♂ juv.), fig. 2 (♀).
Nasiterna pygmæa, Wagler, Monogr. Psitt. p. 631 (1832).—Gray, Gen. B. ii. p. 423 (1846).—Bp. Consp. i. p. 6 (1850).—Id. Rev. et Mag. de Zool. 1854, p. 156.—Id. in Naumannia, 1856, Extrah. iv.—Sclater, Proc. Linn. Soc. ii. p. 166 (1858).—Gray, P. Z. S. 1858, p. 195.—Id. List Psitt. Brit. Mus. p. 52 (1859).—Id. Cat. B. New Guinea, pp. 43, 60 (1859).—Id. P. Z. S. 1861, p. 437.—Schl. J. f. O. 1861, p. 377.—Rosenb. *op. cit.* 1862, pp. 63, 64, 68.—Wall. P. Z. S. 1862, p. 165, 1864, pp. 281, 293.—Rosenb. Tijdschr. Ned. Ind. xxv. p. 226 (1863).—Id. J. f. O. 1863, p. 226.—Bernst. Tijdschr. Ned. Ind. xxvii. p. 297 (1864).—Id. N. T. D. ii. p. 327 (1865).—Finsch, Neu-Guinea, p. 158 (1865).—Schl. N. T. D. iii. p. 331 (1866, pt.).—Finsch, Papag. i. p. 325 (1867).—Gray, Hand-l. B. ii. p. 168 (1870).—Schl. N. T. D. iv. pp. 5, 7 (1871, pt.).—Id. Mus. P.-B. Psittaci, Revue, p. 71 (1874).—Meyer, Sitz. Isis Dresd. 1875, p. 76.—Rosenb. Reist. naar Geelvinkb. p. 56 (1875).—Salvad. Ann. Mus. Civic. Genov. vii. p. 985 (1875).—Rowley, P. Z. S. 1875, p. 470.—Id. Orn. Miscell. 1876, p. 154, pl. xix. (♂), pl. xx. (♀).—Finsch, *t. c.* p. 161 (1876).—Salvad. Ann. Mus. Civic. Genov. x. p. 25 (1877).
Micropsitta pygmæa, Less. Compl. Buff. Ois. p. 607, pl., fig. 2 (1838).
Micropsites pygmæus, Bourj. Perroq. pl. c. (1837-38).
Psittacus pygmæus, S. Müll. Verh. Land- en Volkenk. pp. 22, 107 (1839-44).—Schl. Handl. Dierk. i. p. 480, pl. iii. no. 37 (1858).
Psittacus (Nasiterna) pygmæus, Schl. Handl. Dierk. i. p. 185 (1858).
Psittacula pygmæa, Schl. Dirent. p. 67, cum fig. (1864).—Id. Mus. P.-B. Psittaci, p. 74 (1864).

Of the very singular group of diminutive Parrots the *N. pygmæa* is the oldest known; and a considerable interval of time elapsed before an addition was made by the discovery of the *N. pusio*. The later explorations, however, that have been made in New Guinea and its surrounding satellites have rewarded travellers with at least four or five others, to which Professor Schlegel, Dr. Meyer, Count Salvadori, and others have applied specific appellations. That more still remain hidden in the forests of Papuana is probable. Up to this time (1878) I consider there are seven species only for me to deal with in the present work. These have many characteristics in common:—first, their very diminutive size and disproportionately large bills when contrasted with the body; the structure of their feet and greatly prolonged outer hind toe are peculiarities, as is also the spiny terminations of the four or six central tail-feathers. The group to which the *Nasiternæ* are most nearly allied appears to me to be *Cyclopsitta*, a section of little Parrots inhabiting the same country. Although all the species of *Nasiterna* are remarkable for their tiny size, and a somewhat general resemblance reigns throughout the whole group, yet differences are to be detected. These distinctions may be readily seen by turning to the various Plates representing the species, and be rendered more intelligible than by written explanations.

Count Salvadori, who has in the most liberal manner forwarded me all the information and synonymy of the present group of birds, extracted from his forthcoming work on the ornithology of New Guinea, has also kindly sent me notes on the several species. "This Parrot (*Nasiterna pygmæa*) inhabits New Guinea, Salwatty, Waigiou, Guebeh, Mysol, and Koffiao. In New Guinea it has been found near Dorey, where it was first discovered, as well as in the neighbourhood of Andai, Dorey-Hum, and Mtanata. I have seen specimens from all these localities. The late Mr. G. R. Gray mentions the Louisiade archipelago as one of the habitats; but if a *Nasiterna* really occurs there, it would doubtless be specifically distinct. The same may be said of the island of Jobi, where Von Rosenberg asserts the existence of *N. pygmæa*, though no specimen of a *Nasiterna* from that island is contained in the Leiden Museum.

"I lately examined the two specimens in the latter collection from the Aru Islands, mentioned by Schlegel as differing from those of New Guinea in the ochraceous yellow colour of the pileum. These specimens, which were in rather bad condition, seemed to me to be not different from my *N. keiensis*; and if this proves to be the case, the latter name is not a very appropriate one for the species.

"*Nasiterna pygmæa* seems to be not a very uncommon bird in the countries it inhabits; but, from its small size and green coloration, it is difficult to discover and to obtain. Beccari says that, if one can discover their bowers, it is not hard to catch them. According to him they have the habit of climbing up the trunks of the trees; he says also that the natives often catch them alive inside the hollow trees, where it seems that they nest. I have heard from D'Albertis that he once obtained from a native a living bird, which after some time succeeded in escaping.

"It was likewise reported to Mr. Wallace by his assistant, Mr. Allen, that this Pygmy Parrot nests in hollow trees, and lays eggs like those of the South-American *Psittaculæ*."

Dr. Meyer has written me a note as follows:—"I got this bird only near the foot of the Arfak Mountains in New Guinea, where, at Andai, I procured specimens in the middle of the day. There this lovely little Parrot was sleeping on the lower branches of the trees, and could be whipped off with a stick. This is also the case with other Parrots which are allied to the *Cacatua* group. I may mention *Cyclopsitta lunulata*, from the Philippine Islands, the individuals of which species sleep in the middle of the hot tropical day in rows, under the shade of the foliage, when one after another can be shot down without the survivors attempting to fly away. It may be imagined how soundly they sleep when the noise of the shot does not disturb them; and it is the same with *Nasiterna*. At other times of the day it is difficult to procure, as it lives in the high trees, where its small size and green plumage form a sufficient protection."

The following description of the species has also been given to me by Count Salvadori:—

"*Adult male.* Green, the underparts lighter; pileum yellowish, faintly tinged with reddish towards the forehead; cheeks brown; middle of the abdomen red; under tail-coverts yellow, the lateral ones tinged with green; two middle tail-feathers blue, the lateral ones black, edged externally with green, the three outer tail-feathers with a yellow spot at the tip of the inner web.

"*Female and young male.* Differs from the adult male in wanting the red colour on the middle of the underparts, which are yellowish green. Total length 3·1 inches, wing 2·3, tail 1·0, culmen 0·4."

The figures are of the size of life.

NASITERNA MAFORENSIS.

NASITERNA MAFORENSIS.

Mafor Pygmy Parrot.

Nasiterna pygmæa geelvinkiana (pt.), Schl. N. T. D. iv. p. 7 (1871).—Rosenb. Reist. naar Geelvinkb. p. 137 (pt., 1875).

———— *geelvinkiana* (pt.), Schleg. Mus. P.-B. Psittaci, Revue, p. 71 (1874).—Meyer, Sitz. Isis zu Dresden, 1875, p. 76 (pt.).—Rowley, P. Z. S. 1875, p. 470.—Id. Orn. Misc. p. 153, pl. xviii. (1876).—Beccari, Ann. Mus. Civic. Genov. vii. p. 714 (1875, pt.).

———— *pygmæa*, Rosenb. Reist. naar Geelvinkb. p. 36 (1875, nec Q. & G.).

———— *maforensis*, Salvad. Ann. Mus. Civic. Genov. vii. p. 908 (1875).—Sclater, Ibis, 1876, p. 358.—Salvad. Ann. Mus. Civic. Genov. x. p. 26 (1877).

As Dr. Meyer during his visit to the Island of Mafor does not appear to have procured any *Nasiternæ*, I have been indebted to Count Salvadori and Mr. Dawson Rowley for the four examples which have been examined by me; and to the former gentleman I am under special obligations for an elaborate note on the species, as well as its complete synonymy.

The present species is very nearly allied to *N. misoriensis*; but under existing circumstances they must, in my opinion, be considered distinct species. In both of them the bills are disproportionately large, giving their faces an *outré* and unpleasant aspect; and the bill and feathers of the face are frequently covered with a dirty glutinous substance, which does not improve their appearance.

Count Salvadori writes to me as follows:—"This species is peculiar to the island of Mafor. Both male and female have the crown blue; and in this respect *N. maforensis* differs from *N. misoriensis*. When I discriminated the two species I would willingly have left Schlegel's name of *geelvinkiana* to the bird of Mafor; but as the latter name, which belongs equally to both species, would have been a constant source of confusion, I thought it better to give a new title to the Mafor bird also.

"This is the species figured by Mr. Dawson Rowley in his 'Ornithological Miscellany;' but the description given by Dr. Finsch in the same work, under the heading of *N. geelvinkiana*, probably applies to the female of the Misori bird.

"Von Rosenberg has the following remarks on the present species:—'The small *Nasiterna* is common; and the natives, who are very clever, brought me several specimens, alive and dead, of the adult and young. This pretty little bird is especially abundant in the neighbourhood of Roemsaro. It nests in the holes of trees; and the female lays two eggs, not larger than those of our Long-tailed Titmouse (*Parus caudatus*). They breed in January and February, which is the breeding-season also for other Parrots.'"

The following diagnosis of the species has been sent to me by Count Salvadori:—

"*Adult male*. Green; pileum brown, the feathers edged with bright blue; cheeks brown, with the edges of some of the feathers bluish; nape green like the back, with a yellowish-green spot in the middle, not very conspicuous; centre of the breast and of the abdomen ochraceous yellow; under tail-coverts and a spot at the tip of the three outer tail-feathers yellow, the lateral ones black, the outer ones edged with green.

"*Female*. Like that of *N. misoriensis*, but the feathers of the crown edged with brighter blue. Dimensions nearly the same as those of *N. misoriensis*."

The figures in the Plate are those of a male and a female, of the natural size, drawn from birds lent to me by Mr. Dawson Rowley and Count Salvadori.

NASITERNA MISORENSIS.

NASITERNA MISORIENSIS.

Misori Pygmy Parrot.

Nasiterna pygmæa geelvinkiana (pt.), Schl. N. T. D. iv. p. 7 (1871).—Rosenb. Reist. naar Geelvinkb. p. 137 (1875, pt.).

────── *geelvinkiana* (pt.), Schl. Mus. P.-B. Psittaci, Revue, p. 71 (1874).—Meyer, Sitz. Isis zu Dresden, 1875, p. 76 (pt.).—Beccari, Ann. Mus. Civic. Genov. vii. p. 714 (1875, pt.).—Finsch, Orn. Misc. p. 160 (1876).

────── *misoriensis*, Salvad. Ann. Mus. Civic. Genov. vii. p. 909 (1875), x. p. 26 (1877).

────── *misorensis*, Sclater, Ibis, 1876, p. 358.

For a comparison of the present species with *N. maforensis*, I may refer my readers to the Plates of the two, where it will be seen that these Pygmy Parrots, although closely allied, appear to possess good specific characters. I owe to the courtesy of Dr. Meyer two mounted specimens of the Misori bird, the male having a widely spread mark of orange-yellow down the abdomen, a feature apparently wanting in the opposite sex; and Count Salvadori has also favoured me with a sight of the type specimen obtained by Beccari. All the males had brown heads, with a distinct lunate mark of yellow on the nape, while the females had the brown suffused with bluish.

Count Salvadori writes to me:—"This species is peculiar to the island of Misori, where Von Rosenberg was the first to collect specimens. He obtained a male and two females, which Prof. Schlegel referred to his *Nasiterna pygmæa geelvinkiana* along with the Mafor specimens. When Beccari sent home examples from both localities, I determined those from Misori as belonging to a different species from those of Mafor.

"The adult male is easy to distinguish by reason of his brown head with a very conspicuous yellow spot on the nape.

"Dr. Finsch has described (*l. c.*) two specimens of this bird collected by Dr. Meyer in Misori, which Dr. Finsch seems to think is the same as Mafor. I should say that his description of the supposed male has been taken from a female, and that the description of the supposed female was that of a young bird; one thing is quite certain, that neither the one nor the other of the specimens described by him were adult."

I have received the following note from Dr. Meyer:—"I noticed in my diary, under the heading of *N. pygmæa geelvinkiana* from Mysore (Kordo):—'Colour of the eyes orange-yellow; feet and claws bluish grey, the underparts of the feet somewhat yellowish. Bill bluish grey, the base of the lower mandible white. Cere black, the nostrils surrounded by an elevated thick fleshy ring. Head covered with many little white parasites (lice). In the stomach the remains of fruit.'"

I subjoin a description of the species, sent to me along with the synonymy by Count Salvadori.

"*Male.* Green; head brown, with a conspicuous yellow spot on the nape; round the neck there is a slight indication of a blue collar; middle of the breast and abdomen bright ochraceous yellow; under tail-coverts and a spot at the tip of the inner web of the three outer tail-feathers pure yellow; the two middle tail-feathers blue, with a small black spot, scarcely visible, near the tip; the outer tail-feathers edged externally with green, the spiny tip of the rectrices rather long; bill strong, of a greyish colour; iris orange-red. (*Beccari.*)

"*Female.* Green; pileum brown, with the feathers of the vertex edged with dull blue; cheeks brownish green; underparts yellowish green; under tail-coverts yellow; tail as in the male.

"Total length 3·6 inches, culmen 0·45, wing 2·1, tail 1·2."

I have figured on the Plate two males and a female, of the size of life; and for the opportunity of figuring these I have to thank Dr. Meyer, who, with his usual liberality, sent me over his fine specimens for the purpose.

NASITERNA BRUIJNII, *Salvad*

NASITERNA BRUIJNII, Salvad.

Bruijn's Pygmy Parrot.

Nasiterna bruijnii, Salvad. Ann. Mus. Civic. Genov. vii. pp. 715 (note), 753. sp. 13, pl. xxi. p. 907 (1875).—Sclater, Ibis, 1876, p. 255.—Salvad. Ann. Mus. Civic. Genov. x. p. 25. sp. 8 (1877).

This is the most beautiful of the Pygmy Parrots, and is altogether a lovely little bird. It seems to have escaped the observations both of Dr. Meyer and Signor D'Albertis, but was discovered by the collectors of Heer Bruijn, whose name is now so famous for the wonderful novelties collected by his means in North-western New Guinea. But I must let Count Salvadori tell the story of his little favourite, though I here take the opportunity of returning him my most cordial thanks for an act of liberality not easily to be forgotten. During a recent visit to London he showed me a series of these Pygmy Parrots; and on my asking permission to figure them in the present work he not only acceded to my request, but furnished me with the full synonymy of the species, as it is about to be published by him in his work on the ornithology of the Moluccan and Papuan Islands.

Concerning *Nasiterna bruijnii* he writes to me:—"This species was discovered by the men employed by Mr. Bruijn in the Arfak Mountains, where they first obtained a male bird: afterwards Beccari and Bruijn's hunters got some more male specimens, and also succeeded in procuring the female. Both sexes have been described by me; and I have seen altogether nine specimens.

"The males, according as they are more or less adult, vary a little as regards the red colour of the pileum and of the cheeks, being more or less brilliant on these parts; some (among them the type) have these parts dull fulvous tinged with rosy red: a young bird, not sexed, is like the females, only smaller. The bill of this species is very small.

"Nothing is known about the habits of this the most beautiful species of the genus *Nasiterna*; but in all probability it is found in the mountainous districts only."

The following description is from Count Salvadori's MSS.:—

"*Male*. Green, the feathers narrowly edged with black; primaries and anterior secondaries blackish, edged with green; wing-coverts black, with rather wide green edges; pileum, cheeks, and middle of the under parts, with the under tail-coverts bright red, the red pileum changing into brown towards the occiput, and surrounded by a bright blue band, which, from the nape, encircling the red cheeks, extends down to the throat, whence it descends on the sides of the breast; the two middle tail-feathers blue, with a round black spot near the tip; the other rectrices are black, with a red-orange spot on the tip of the inner web; the outer tail-feathers have the outer web partly bluish; bill and feet horny grey.

"*Female*. The upper parts green; the underparts also green, but inclining to yellowish along the middle and on the under tail-coverts; pileum bright blue; forehead whitish; cheeks pale reddish, the throat slightly washed with blue; the wings and the tail as in the male; but the lateral tail-feathers have the spots at the tip orange-yellow. Total length 3·7 inches, wing 2·9-2·7, tail 1·2-1·1, bill from the forehead 0·4."

With regard to the Plate of this bird I have indulged my memory a little, having seen thousands of pairs of not distantly allied species in the space of as many yards in the interior of Australia, breeding in the spouts of the decayed branches of the gum-tree. The upper pair of birds in this case are supposed to be already mated, while the lower pair are represented in the act of courtship.

NASITERNA BECCARII, Salvad.

NASITERNA BECCARII, Salvad.

Beccari's Pygmy Parrot.

Nasiterna beccarii, Salvad. Ann. Mus. Civic. Genov. viii. p. 396 (1876), x. p. 26. sp. 13 (1877).

Of this bird, which is closely allied to *N. pusio*, Count Salvadori has lent me the single specimen at present known; and I have had the satisfaction of figuring it here for the first time. It differs from all the other *Nasiternæ* which I have seen. On the head is a well-defined cap of dull indigo-blue surrounded by brown, the remainder of the plumage being green, as in the other members of the genus, with black spots on the shoulder moderately strongly indicated. It is certainly one of the smallest species; and the spines on the tail are but little developed; the bill is small for a *Nasiterna*, and of a light horn-colour. There appears no indication of an orange-red patch in the centre of the body, as is the case with *N. pygmæa* and *N. bruijni*; but, as the only specimen at present in our hands is a female, it is impossible to speak very positively on this point.

I owe to the kindness of Count Salvadori the following note on the present species:—"I only know of a female of this Parrot, which was collected by Beccari near Wairoro, on the coast of Geelvink Bay, more than one degree to the south of Dorey. I have compared this specimen with *N. pusio*, to which it is nearly allied; and it differs from the latter species in the blue colour of the crown being more extended and brighter, in the brown colour of the cheeks being darker, and in the smaller dimensions. The bill is rather small, as in *N. pusio* and *N. pygmæa*."

For the accompanying description I am also indebted to the Count:—

"Green, the underparts being much lighter; middle of the crown dull blue, the edges of the feathers black; forehead and sides of the head brown, the sides of the nape greenish brown; the two centre tail-feathers blue with a black spot near the tip; lateral rectrices black, edged externally with green and with a yellowish orange spot at the tip of the inner web. Size of *N. pygmæa*."

The Plate contains a life-sized representation of the present species in two positions, taken from the typical specimen. I have also introduced into the picture a figure of *Eupholus bennetti*, a magnificent beetle recently described from South-eastern New Guinea; and I have figured this fine insect with the greatest pleasure, in compliment to the gentleman whose name it bears, Dr. Bennett of Sydney, who has for many years been known to naturalists as an active patron and promoter of science in his adopted country.

NASITERNA PUSIO, *Sclater.*

NASITERNA PUSIO, Sclater.

Solomon-Islands Pygmy Parrot.

Nasiterna pusio, Sclater, P. Z. S. 1865, p. 620, pl. 35.—Finsch, Die Papag. i. p. 327 (1867).—Sclater, P. Z. S. 1869, pp. 124, 126.—Gray, Hand-l. B. ii. p. 168. no. 8382 (1870).—Schlegel, N. T. D. iv. p. 5 (1871).—Ramsay, Tr. Linn. Soc. N. S. W. i. p. 67 (1876).—Rowley, Orn. Misc. p. 155, pl. xxi. (1876).—Finsch, *tom. cit.* p. 163 (1876).—Sclater, P. Z. S. 1877, p. 108. no. 38.—Salvad. Ann. Mus. Civic. Genov. x. p. 26. no. 14 (1877).
Nasiterna pygmæa solomonensis, Schl. N. T. D. iv. p. 7 (1877).

For a great many years *Nasiterna pygmæa* remained the only representative of the genus known; and the present kind was the second discovered—the forerunner, as it has proved, of no less than five others with which we are now acquainted. The original specimens were sent over in spirits by Mr. Gerard Krefft to Dr. Sclater; and the typical example is now preserved in the British Museum, where I have myself examined it. It has lately been noticed by Dr. Sclater in his paper on the birds collected by Mr. George Brown in Duke-of-York Island and New Ireland; but no locality was given for the specimen, and so we do not know the exact origin of this individual, though there can be no doubt that it came from one or the other of the above-mentioned islands. The typical birds were said to be from the Solomon Islands; but we shall want confirmatory evidence on this point, as the recent collections from this locality have not contained any examples of a *Nasiterna*, and it is just possible that they may have come from New Ireland; on the other hand, the species may possibly be found in both groups of islands.

Count Salvadori writes:—"This species comes very near to *N. beccarii*, from which it differs in the blue colour of the crown being less extended and of a duller shade, in the somewhat ochraceous brown colour of the forehead and sides of the head, in the more yellowish colour of the middle of the belly, and in the longer wings. I have seen four specimens of this Pygmy Parrot, viz.:—the type in the British Museum; one in the Zoological Museum of Turin, received from Mr. Krefft during the voyage of the 'Magenta;' another in the Museum of Berlin, also received from Mr. Krefft; and, lastly, a specimen received from Mr. Brown without indication of the locality."

The description given below was also forwarded to me by Count Salvadori:—"Green, the underparts lighter and a little tinged with yellowish down the centre of the belly; crown dull blue, the forehead and sides of the head brown, with a slight ochraceous tinge; tail and wings as in *N. beccarii*. Size a little larger than the latter bird." Total length 3·4 inches, culmen 0·4, wing 2·5, tail 2·2.

To the above description of Count Salvadori I have only to add that the black spot on the centre tail-feathers is larger than in most of this species.

The lower figure in the Plate was drawn from the British-Museum specimen; and the upper figure is taken from Mr. Brown's bird above mentioned. They are of the natural size.

NASITERNA KEIENSIS, Salvad.

NASITERNA KEIENSIS, Salvad.

Ké-Island Pygmy Parrot.

Nasiterna keiensis, Salvad. Ann. Mus. Civic. Genov. vii. p. 984 (1875), x. p. 26. sp. 10 (1877).
?*Nasiterna pygmæa* (pt.), Schl. N. T. D. iii. p. 331 (1866, spec. from Aru Islands).—Rosenb. (nec Q. & G.), Reis naar zuidostereil. pp. 48, 49 (1867, Aru).—Schl. N. T. D. iv. pp. 5, 7 (pt., 1871).—Id. Mus. P.-B. Psittaci, Revue, p. 71 (pt., 1874).
?*Nasiterna aruensis*, Salvad. Ann. Mus. Civic. Genov. vii. p. 985 (1875, ex Schlegel), x. p. 25, note 2 (1877).

I HAVE reproduced exactly the synonymy which my friend Count Salvadori has sent to me, as it explains so thoroughly the history of the species, and makes clear the following note with which he has favoured me:—" The Ké-Island Pygmy Parrot is very like the female of *N. pygmæa*; but it is larger, has the pileum more conspicuously ochraceous yellow, and the underparts more greenish, and without the yellow tint. Male and female scarcely differ. Total length 3·9-3·6 inches, culmen 4·5, wing 2·2-2·6, tail 1·2-1·1.

"Besides the three typical specimens, two males and a female, collected by Beccari on the Ké Islands and described by me, I have seen in the Leiden Museum two examples from the Aru Islands, collected by Von Rosenberg, which have been spoken of by Professor Schlegel. They are in rather bad condition; and although they seem to resemble the birds from the Ké Islands, with which I have compared them, I am not quite sure that they really belong to the same species."

Count Salvadori left with me for the purpose of this work a fine male and female; and as these appear to be fully adult, we may conclude that the species has none of the richer colour on the breast as in *N. pygmæa*, from which it also differs in the more conspicuous spotting of the shoulder. I have no doubt as to the specific value of *N. keiensis*.

The following description is drawn up from the typical specimens:—

Crown dirty yellow, forming a well-defined cap; face suffused with brown, which gradually blends into green on the cheeks; all the upper and under surface green, the back being of a deeper hue than the lower parts; on the chest of the male in certain lights is a very delicate wash of blue. Primaries blackish brown, each feather slightly margined with green; secondaries and, especially, the feathers of the shoulder conspicuously spotted with black; two centre tail-feathers blue on the upper surface, with the shaft black, the four or five outer feathers on each side with the usual spot of yellow on the tips of their inner webs; the spines of the centre feathers very fine, but little prolonged, and without the spot of black found in some of the species; under tail-coverts yellow, stained with green.

The figures in the Plate are taken from the type specimens, and are of the natural size. It will be seen that the sexes are alike in colour.

GEOFFROYIUS HETEROCLITUS, (Hombr. & Jacq.)

GEOFFROYIUS HETEROCLITUS, *Hombr. & Jacq.*

Yellow-headed Parrot.

Psittacus geoffroyi heteroclitus, Hombr. et Jacq. Ann. Sci. Nat. xvi. p. 319 (1841).
Psittacus heteroclitus, Gray, Gen. B. ii. p. 421, no. 8 (1846).—Id. List Psittacidæ, Brit. Mus. p. 73 (1859).— Id. List Birds Tropical Islands of the Pacific Ocean, p. 34 (1859).—Id. Hand-l. B. ii. p. 160, no. 8277 (1870).
Pione heteroclite, Hombr. et Jacq. Voy. Pôle Sud, pl. 25 bis, figs. 1, 2 (1842-53).
Pionus heteroclitus, Jacq. et Pucher. Voy. Pôle Sud, iii. p. 103 (1853).
Pionus cyaniceps, Jacq. et Pucher. Voy. Pôle Sud, texte, iii. p. 105 (1853).
Geoffroyius heteroclitus, Bonap. Rev. et Mag. de Zool. 1854, p. 155.—Souancé, op. cit. 1856, p. 218.—Sclater, P. Z. S. 1869, p. 122.—Salvad. Ann. Mus. Civ. Genov. x. p. 30 (1877).
Geoffroyius cyaniceps, Bonap. Rev. et Mag. de Zool. 1854, p. 155.—Sclater, P. Z. S. 1877, p. 107.
Pionias heteroclitus, Finsch, Papageien, ii. p. 390 (1868).

This beautiful Parrot is so rare in collections, that when Dr. Finsch wrote his celebrated work on the Psittacidæ he had not seen an example; and it is only within the last year that I have had the pleasure of examining it myself. The original specimens were three in number, and were brought from the islands of St. George and Ysabel in the Solomon group by the French Expedition to the South Pole. More recently it has been rediscovered by Mr. George Brown in New Britain, as recorded by Dr. Sclater *l. c.*; and I am indebted to this gentleman for the loan of the specimen from which my Plate has been drawn.

When discovered by MM. Hombron and Jacquinot, a second form with a blue head was also procured, which these naturalists considered to be the female of the yellow-headed bird. Dr. Pucheran, in his account of the birds procured by the expedition, thought otherwise—and believing that the blue-headed specimen was a distinct species, named it *Pionus cyaniceps*; but Dr. Finsch agreed with the first opinion, and made it the female of *G. heteroclitus*. I am unable to say for certain whether this is right or wrong, as I have not yet seen more than one example of the blue-headed form. Recently Dr. Sclater has considered it probable that there are two species inhabiting the New-Ireland group; as, however, Mr. Brown has lately sent over the yellow-headed and blue-headed specimens figured by me, I incline to the opinion that they are the same species.

The colouring of this species is so very distinct, that the figures in the Plate will serve to distinguish it from all its allies. Dr. Finsch places it in the genus *Pionias*, along with a great many American and African species, which, according to my views, belong to distinct genera (*Prioniturus, Pœocephalus,* &c.). But taking his comprehensive view of *Pionias*, the present species belongs to the first section, with green under tail-coverts, sky-blue under wing-coverts, and blackish wing-lining. Its yellow head and cheeks, coupled with the blue band round the neck, are sufficient to distinguish it at a glance.

As before mentioned, the specimens here figured were from Mr. George Brown's third collection. The principal figure is of the size of life; and a slightly reduced figure is placed in the back-ground, and is supposed to be that of a young female.

GEOFFROYIUS SIMPLEX, Meyer.

GEOFFROYIUS SIMPLEX.

Blue-collared Parrot.

Pionias simplex, Meyer, Mitth. zool.-botanischen Gesellsch. Wien, xxiv. p. 39 (1874).
Geoffroyius simplex, Salvad. Ann. Mus. Civic. Genov. vii. p. 759 (1875).

In a paper contributed by him to the Zoological and Botanical Society of Vienna, Dr. Meyer describes this new species of Parrot discovered by him in New Guinea; and I am now able to give correct figures, thanks to his kindness in sending me the type specimens for that purpose.

After giving a full description of the bird, the learned Doctor proceeds as follows:—"I obtained in the same locality a female of a *Pionias* (the one described being a male bird), which I think may be the female of this species, as it resembles the male in general aspect and in some particular characters. It differs from the male in wanting the blue collar; but at the same time it exhibits a slight but well-pronounced bluish green shade on the head and cheeks. The whole upper surface is uniform green; but, on the other hand, the yellowish brown spot on the wing-coverts is more strongly indicated than in the male. The underparts are uniform light green. The tail is not so pale-coloured below as in the male."

"As the female bird exhibits a bluish shade on its head, and the male has nothing of this on the same part, it might be imagined that the male now before me is not quite in full plumage; but this is not likely, as it has the blue collar so well developed; therefore, from the above-mentioned differences, it is by no means improbable that we have here two separate species, the male belonging to one and the female to another. As I only obtained two specimens, and as the occurrence of several very closely allied but specifically well distinguished species in the same locality is nothing uncommon among the birds, and especially the Parrots, of New Guinea, I am not able to affirm or deny the fact with certainty; but in my own mind I have good reason to believe that they are male and female of one species."

The following is a translation of Dr. Meyer's description of the male:—"Green; the back brownish; rump washed with blackish; under surface of body lighter green; round the neck a collar of light blue, somewhat shaded with lilac under certain lights, and being broader and less defined on the nape; wings green, the inner webs of the quills black; under wing-coverts sky-blue, this colour descending somewhat on the sides of the chest; under surface of the quills blackish, the secondaries with a pale yellow spot on the inner web; on the edge of the wings a yellowish white spot; wing-coverts above and below with a slight patch of yellowish brown; cheeks and chin paler green like the under surface; abdomen shaded with brown; under tail-coverts pale green shaded with yellowish; upper surface of the tail green, below greenish yellow. Bill and cere, feet and nails black."

"I obtained this new species in July 1873 on the Arfak Mountains in the north-west of New Guinea, about 3000 feet above the level of the sea, and I name it *simplex* on account of its plain coloration."

The figures in the accompanying Plate are of the size of life.

GEOFFROYIUS TIMORLAOENSIS, Meyer.

GEOFFROYIUS TIMORLAOENSIS, Meyer.

Tenimber Parrot.

Geoffroyius keyensis, Salvad.; Sclater, Proc. Zool. Soc. 1883, pp. 51, 200.—Forbes, Naturalist's Wanderings, p. 356 (1885).
Geoffroyius timorlaoensis, Meyer, Vögel, Nester und Eier aus dem Ostind. Archipel, p. 15 (1884).

The first specimens sent from Timor Laut by Mr. H. O. Forbes were referred by Dr. Sclater to *Geoffroyius keyensis*, the species from the Ké Islands. Dr. A. B. Meyer, the well-known Director of the Dresden Museum, received no less than eleven specimens from Timor Laut, collected by Mr. Riedel's hunters, and he came to the conclusion that the species from the two groups of islands above named were not identical, and he named the bird from the Tenimber Islands *Geoffroyius timorlaoensis*.

The differences referred to by Dr. Meyer consist of the smaller size of *G. timorlaoensis* and the green instead of blue colour on the external aspect of the first primary. We have compared four specimens of the Timor-Laut birds with two specimens of *G. keyensis* from the Ké Islands, and we must confess that the characters for their separation are of the very slightest. We cannot see the smallest difference between the two species as regards the blue on the first primary, and the only character is the slightly smaller size of *G. timorlaoensis*. Mr. Forbes has also written a critique on the species in his entertaining narrative of his expedition to Timor Laut, and his conclusions are the same as our own.

In deference to Dr. Meyer's kindness in lending us the specimens, we have given figures of the species and add a description.

Adult male. General colour above grass-green, a little lighter towards the under tail-coverts; wing-coverts like the back, the innermost reddish, forming a shoulder-patch; bastard-wing, primary-coverts, and quills dusky blackish, externally dark grass-green, the secondaries entirely of the latter colour; the first primary edged with richer and deeper green than the others; upper tail-coverts lighter green than the back; tail-feathers glistening yellow, edged with emerald-green; crown of head plum-coloured or purplish lilac, with a frontal band of scarlet, which colour extends over the lores and entire sides of the face, being tinged with lilac on the ear-coverts and sides of the head; throat also scarlet, the lower throat, fore neck, and remainder of under surface bright grass-green, paler towards the vent and under tail-coverts; under wing-coverts and axillaries cobalt-blue, some of the long axillary plumes green with blue tips; the edge of the wing green; quills below blackish. Total length 9·5 inches, culmen 1·05, wing 7·0, tail 3·5, tarsus 0·6.

Adult female. Similar to the male, but lacks all the brilliant colouring of the head and face, the head being of an olive-yellowish tint all round, including the throat, the crown greener and more like the back. Total length 9·5 inches, culmen 1·05, wing 6·85, tail 3·55, tarsus 0·6.

The Plate represents an old male and female of about the natural size, the figures being drawn from the two birds lent to us by Dr. Meyer.

[R. B. S.]

ECLECTUS POLICHLORUS.

ECLECTUS POLYCHLORUS.

Green Lory.

Psittacus polychlorus, Scopoli, Del. Faun. et Flor. Insubr. ii. p. 87 (1786).
Psittacus sinensis, Gmelin, S. N. i. p. 337 (1788).
Psittacus magnus, Gmelin, S. N. i. p. 344 (1788).
Psittacus viridis, Latham, Index Orn. i. p. 125 (1790).
Psittacus lateralis, Shaw, Gen. Zool. viii. p. 490 (1811).
Mascarinus prasinus, Lesson, Traité, p. 188 (1831).
Eclectus linnæi, Wagler, Monogr. Psittac. p. 571, pl. xxii. (1832).
Eclectus polychlorus, Gray, Genera of Birds, ii. p. 418 (1845).—Id. List Psittacidæ Brit. Mus. p. 66 (1859).
Eclectus puniceus, Bonap. P. Z. S. 1849, p. 142.—Rosenb. J. f. O. 1864, p. 114.
Eclectus grandis, Bonap. Consp. Gen. Av. i. p. 4 (1850).—Id. Rev. et Mag. de Zool. 1854, p. 155.
Eclectus westermanni, Bp. Consp. i. p. 4 (1850).
Eos puniceus, Lichtenstein, Nomencl. Av. p. 71 (1854).
Polychlorus magnus, Sclater, P. Z. S. 1857, p. 226.
Mascarinus polychlorus, Finsch, Nederl. Tijdsch. Berigten, p. xvi. (1863).
Psittacodis magna, Rosenb. Tijdschr. Nederl. Indië, 1863, p. 226.—Id. J. f. O. 1864, p. 114.
Psittacus linnæi, Finsch, Neu-Guinea, p. 157 (1865).

WERE any thing required to assure the student of ornithology that there is still plenty of work to do in the science, the history of the present bird would afford a text for a discourse on that subject. A cage-bird in every menagerie of any repute, described and figured over and over again during the last hundred years, and for the last twenty years by no means rare in collections, the present species might have been supposed to have been well understood. No one, therefore, was prepared for the astounding assertion made by Dr. Meyer in 1874, that the Lories of the Moluccas, considered by everybody to represent many distinct species, were nothing but the males and females of perhaps three. I candidly confess that I was for a long time extremely sceptical on the subject; but after examining specimens sent me by Dr. Meyer, I must admit that he is perfectly right, and that this curious fact must be accepted by ornithologists. The story comes better from Dr. Meyer himself than from me; and I therefore give the note which he has just forwarded:—

"'When crossing the sea from the Island of Mafoor to the Island of Mysore, in Geelvink Bay, in the year 1873, having spread out before me, on board of my small vessel, the ornithological harvest which I had reaped on Mafoor, it struck me that all the specimens of *Eclectus polychlorus*, Scop., were labelled as males, and all those of *E. linnæi*, Wagler, as females; and I had six green males (*polychlorus*) and nine red females (*linnæi*). The suspicion then arose in my mind that it could not have been by chance that I had only shot the males of *E. polychlorus* and the females of *E. linnæi*.'

"With these words of introduction I commenced the first paper which I published on the sexual differences in the genus *Eclectus*, in the year 1874 (Verh. d. zool.-bot. Ges. Wien, 1874, p. 179, and Zool. Garten, May 1874, p. 161). Since then I have been obliged to write three more notes on the same subject, because at first the opinion that the green parrots are indeed the males of the red ones was almost universally contested. Nevertheless I already said in my first paper (*l. c.*), 'The fact, discovered by myself, is thoroughly ascertained, and cannot be doubted.'

"In my second note (Mitth. d. k. zool. Mus. Dresden, i. p. 11, 1875) I chiefly disputed Prof. Schlegel's view, who had promulgated the following opinion, and supposed it to be well founded:—'Adopting this hypothesis, we should be obliged in the meantime to accuse of negligence four of our most experienced travellers; and to establish among the parrots the quite exceptional case of a singular sexual difference would be the more remarkable, as it would besides offer in the females variations constant according to the localities' (Mus. d'Hist. Nat. des Pays-Bas, Psitt. 1874, p. 17). By the four most experienced travellers of the Leiden Museum, Prof. Schlegel meant, as far as I am aware, Salomon Müller, Dr. Bernstein, Hoedt, and Von Rosenberg. But I proved that even the facts published in the Catalogues of the Leiden Museum show that I am right, as, for instance, among seven specimens of *E. intermedius* (green) six are marked as males and only one as a female; and, on the other hand, among fourteen specimens of *E. grandis* (red) twelve are designated as females and only two as males. I further adduced in favour of my discovery that, taking my six green males and nine red females (from Mafoor) by themselves, it could be mathematically demonstrated that the probability of a really existing sexual difference is as 32700 : 1.

"In my third note (Proc. Zool. Soc. of London, 1877, p. 800, pl. 79) I figured the tail of a specimen in the Dresden Museum, which is half red, half green, and drew attention to some young individuals in several Museums, which are partly green, partly red, proving that the young male is coloured like the female, whereas the well-known savant, Dr. Beccari, wrote that the young ones offer the same differences as the adult birds (Ann. Mus. Civ. Gen. vii. p. 715).

"In my fourth note on the same subject (Orn. Centralblatt, 1878, p. 119) I chiefly showed that Dr. Brehm, who is of opinion (Illustrirtes Thierleben, 2nd ed. vol. iv. p. 68) that the existence of green females and of red males has been proved, was misled by inaccurate statements, and I further drew attention to other facts, which confirm the statement that the young ones of both sexes are red.

"Now, after the lapse of more than four years, I do not hesitate to state that almost every ornithologist admits the fact of the sexual differences in *Eclectus*, and that all objections and doubts which came from the most different quarters are silenced. The only thing at which I am surprised is, that this sexual difference could have been so long overlooked, red and green *Eclecti* having even been several times placed in two different genera. On my last visit to London (August of this year), I saw in the bird-galleries of the British Museum a specimen (which has already been a very long time there) labelled as *E. westermanni*, but which is nothing else than a young male of *E. polychlorus* changing from its red dress into its green one, and which has not yet acquired the red spots on the breast; but it is covered all over its back with red spots, the residue of the first dress; the bill, too, proves it to be a young bird. I only ask, How can the existence of such a specimen (green, with red spots) be understood, if not as the result of a sexual difference?

"Apropos of *E. westermanni*, I am still of the opinion which I expressed in 1874, that this is not a good species, but that the specimens known are only individuals which have not acquired their full plumage, in consequence of the unnatural conditions incident to a state of captivity. All the specimens hitherto known, of which I have seen those at Copenhagen, at Bremen, in the British Museum, and at Leiden, are those of birds that lived in captivity. The same remark *perhaps* applies to *E. corneliæ*, which is rarer than *E. westermanni*: up to the present time only a few specimens are known. I saw one in Amsterdam and one in London; the latter is labelled as a male—a proof that not only in the tropics can a mistake be made in the determination of a bird's sex. It is not rare that a bird does not acquire its full plumage in captivity; for instance, my friend Hr. von Pelzeln, of Vienna, in the year 1862, published the fact that an *Aquila imperialis* in the Schönbrunn Zoological Gardens retained its immature plumage during seven years.

"I do not share the opinion of Mr. Ramsay, of Sydney (Ibis, 1878, p. 379), 'that the young retain the red and blue state of plumage for a considerable time, after which the *males* assume the green plumage, but think that the change of plumage takes place in *Eclectus* as quickly as it does in other species of birds. All the red specimens which are young males prove this to be the case by their bills; a young red male never has such a pronounced bill as an adult female. Even the changing specimens which we know (partly red, partly green), still evidently prove themselves by their bills to be young ones.

"Recently Mr. Van Musschenbroek, the well-known Dutch resident at Ternate and Manado, told me that a red *Eclectus* can now be found in a very isolated locality of the Minahassa, in North Celebes; but he could not tell me which species it is; he meant that they are probably descendants of individuals escaped from captivity. If this is true, a green *Eclectus* also will be found there, besides *Eclectus mülleri*, which is a known inhabitant of those regions and is characteristic of Celebes. Another instance of a new immigrating *Eclectus* in Celebes (although from other reasons) I brought to light in the case of *Eclectus megalorhynchus* (see Rowley's Orn. Misc. iii., and elsewhere).

"These two last-named species and others not presenting the remarkable sexual differences presented by *Eclectus polychlorus* and its allies, I venture to question whether they should not be separated generically."

The Plate is intended to represent a fully adult male of *E. polychlorus*, of the natural size. The reduced figures flying in the background illustrate the difference between the two sexes.

ECLECTUS RIEDELI, Meyer.

ECLECTUS RIEDELI, Meyer.

Riedel's Parrot.

Eclectus riedeli, Meyer, Proc. Zool. Soc. 1881, p. 917.—Id. Verh. zool.-bot. Gesellsch. Wien, 1881, p. 772.—Salvad. Ann. Mus. Civic. Genov. xviii. p. 419 (1882).—Id. Orn. Papuasia, etc. iii. App. p. 517 (1882).—Sclater, Proc. Zool. Soc. 1883, pp. 53, 195, pl. xxvi.

This interesting bird was sent first from Timor Laut by Mr. Riedel, recently Dutch resident at Amboina, to Dr. Meyer at Dresden, by whom it was named after the discoverer. Like all the green and red Parrots, the usual differences of the sexes is observed, the male being green and the female red. Only the latter was obtained by Mr. Riedel; and for our knowledge of the male we are indebted to our energetic countryman Mr. H. O. Forbes, who brought back four specimens of this species from Timor Laut. We must confess, however, that we believe that the name of *E. riedeli* will be found to have been superseded by that of *E. westermanni* of Bonaparte. The specimen in the British Museum of the latter bird does not, it is true, show the red flank-spot of the adult *E. riedeli*; but there are traces of it, and we believe that the specimen in question has had the development of its full plumage arrested by having been kept in confinement.

The female of *E. riedeli* resembles *E. cornelia* in not having any blue on the under surface; but *E. cornelia* is a much larger bird, and has red under tail-coverts and no yellow tips to the tail.

Adult male. General colour above bright grass-green, inclining to emerald-green on the head and mantle; wing-coverts like the back, the least series emerald-green, with a patch of cobalt-blue near the bend of the wing and round the edge of it; bastard wing deep blue; primary-coverts green, washed with cobalt; quills deep blue, blackish on the inner web, lighter blue along the outer one; secondaries more or less green externally, the innermost entirely green; rump and upper tail-coverts washed with emerald-green; tail-feathers green, with a broad band of yellow at the end, before which is a slight shade of blue on the outer feathers; lores and entire sides of face, cheeks, ear-coverts, and feathered portion of throat emerald-green; remainder of under surface of body grass-green, washed with yellow on the under tail-coverts; a large patch of deep scarlet on the sides of the body, the axillaries and under wing-coverts being of this colour; edge of wing light cobalt or verditer blue; greater series of under wing-coverts and quills below black; "upper mandible scarlet, fading into orange; lower mandible black; legs and feet lavender; iris pale orange" (*H. O. Forbes*). Total length 13 inches, culmen 1·8, wing 8·8, tail 4·5, tarsus 0·6.

Adult female. General colour above red, the base of the feathers dull brown or greenish; wing-coverts a little darker than the back; edge of wing and bastard wing blue, with a slight shade of green near the base of the feathers; primary-coverts and quills deep blue, blackish on the inner web and shaded with lighter blue on the outer; secondaries externally red, the innermost entirely like the back; tail-feathers red, greenish towards the bases, and broadly banded with yellow across the tip; head all round, neck, and mantle brighter scarlet than the back; remainder of under surface of body from the fore neck, downwards deep red, the lower flanks and thighs washed with yellow; under tail-coverts yellow; axillaries and under wing-coverts deep red, the latter with dusky or purplish-blue bases; greater series of under wing-coverts and quills below black; "bill black; legs and feet greyish black; iris pale yellow" (*H. O. Forbes*). Total length 12 inches, culmen 1·55, wing 8·1, tail 4·5, tarsus 0·6.

The Plate represents a pair of birds of about the natural size, drawn from specimens procured by Mr. Forbes in Timor Laut.

[R. B. S.]

DASYPTILUS PESQUETI.

DASYPTILUS PESQUETI (Less.).

Pesquet's Parrot.

Psittacus pecquetii, Less. Bull. des Sciences Nat. xxv. p. 241 (1831).—Bourj. Perroquets, pl. 67 (1837-38).—Less. Descr. de Mamm. et d'Ois. pl. 199 (1847).

Psittacus pesquetii, Less. Ill. de Zool. pl. 1 (1831).

Banksianus fulgidus, Less. Traité d'Orn. p. 1881 (1831).—Pucheran, Rev. et Mag de Zool. 1853, p. 156.—Hartl. Journ. für Orn. 1855, p. 422.—Salvad. Ann. Mus. Civ. Gen. xii. p. 317 (1878).

Dasyptilus pecquetii, Wagl. Mon. Psitt. pp. 502, 681, 735 (1882).—Gray, Gen. Birds, ii. p. 427 (1845).—Bp. Consp. Gen. Av. i. p. 8 (1850).—Le Maout, Hist. Nat. des Ois. 104, pl. 2 (1853).—Bp. Rev. et Mag. de Zool. 1854, p. 155.—Id. Naumannia, 1856, Consp. Psitt. sp. 360.—Gray, Cat. B. New Guinea, pp. 43, 60, 1859.—Id. List Psitt. Brit. Mus. p. 100 (1859).—Sclater, Proc. Zool. Soc. 1860, p. 227.—Rosenb. Journ. für Orn. 1862, p. 65.—Id. Nat. Tijdschr. Ned. Ind. xxv. pp. 146, 147 (1863).—Id. Journ. für Orn. 1864, p. 116.—Bernst. Nat. Tijdschr. Ned. Ind. xxvii. p. 297 (1864).—Id. Ned. Tijdschr. Dierk. ii. 327 (1865).—Gray, Handl. B. ii. p. 159, no. 8260 (1870).—Salvad. Atti. R. Ac. Sc. di Torino, ix. p. 630, (1874).—Meyer, Orn. Mittheil. i. p. 14 (1875).—Garrod, P. Z. S. 1876, p. 691.

Dasyptilus pequetii, Jardine, Nat. Libr. vi. p. 140, pl. xvii. (1836).—Wallace, P. Z. S. 1864, pp. 287, 294.

Psittrichas pesqueti, Less. Compl. de Buffon, Ois. p. 603, pl. f. 2 (1838).

Calyptorhynchus fulgidus, Gray, Gen. B. ii. p. 426 (1845).

Dasyptilus fulgidus, Bp. Rev. et Mag. de Zool. 1854, p. 157.—Id. Naumannia, 1856, Consp. Psitt. sp. 261.—Gray, List. Psitt. Brit. Mus. p. 100 (1859).—Finsch, Die Papageien, ii. p. 323 (1868).—Salvad. Ann. Mus. Civ. Genova, xii. p. 317 (1878).

Dasyptilus pesqueti, Gray, Proc. Zool. Soc. 1858, p. 195.— Id. P. Z. S. 1861, p. 437.—Finsch, Die Papageien, ii. pp. 320, 955 (1868).—Giebel, Thes. Orn. ii. p. 18 (1874)—Meyer, Sitzb. Isis zu Dresden, 1873, p. 76.—Beccari, Ann. Mus. Civ. Genova, vii. p. 714 (1875).—Id. Ibis, 1876, p. 252.—Salvad. Ann. Mus. Civ. Genova, x. pp. 31, 121 (1877); xii. p. 317 (1878).—D'Albertis, Ann. Mus. Civ. Genova, x. pp. 8, 19 (1877).—Id. Ibis, 1877, p. 365.—D'Albert. & Salvad. Ann. Mus. Civ. Genova, xiv. p. 36 (1879).—Salvad. Orn. Papuasia &c. i. p. 217 (1880).

Nestor pesquetii, Schlegel, Journ. für Orn. 1861, p. 377.—Id. Mus. Pays-Bas, *Psittaci*, p. 157 (1864).—Id. op. cit. Revue, p. 70 (1874).

Dasyptilus pesqueti, Finsch, Neu-Guinea, p. 157 (1865).

Microglossum pecqueti, Rosenb. Der zool. Gart. 1878, p. 347.

From the above very ample list of synonyms, which we have copied from Count Salvadori's 'Ornitologia della Papuasia,' it will be seen that this species has been known to ornithologists for many years; but it is only recently that we have received perfect specimens in Europe. All the examples collected by the early voyagers seem to have been skins of native preparation; and so rare was the bird that even Mr. Wallace did not succeed in obtaining a specimen during his travels in the Malay archipelago. Bernstein forwarded a specimen to the Leyden Museum in 1863, which he had received alive in Ternate from the west coast of New Guinea, opposite to Salawati. Another individual was received alive by the Zoological Society of London, but did not live long; this specimen was beautifully mounted by Mr. Bartlett, and is now in the gallery of the British Museum.

Von Rosenberg met with a single individual in the Arfak Mountains, where also D'Albertis shot the species, as well as Dr. Meyer. The researches of Dr. Beccari and Mr. Bruijn's collectors have shown that it occurs on Mount Morait near Dorei Hum, near Napan in the Bay of Geelvink, and also on the Fly river, in South-eastern New Guinea, where D'Albertis met with it. We have also seen some splendid examples from the Astrolabe range of mountains, obtained by Mr. A. Goldie. It will therefore be noticed from the above slight sketch of the history of the present species, which is derived from Count Salvadori's work above quoted, that examples are now much more common in collections than they were twenty years ago.

From the accounts of the habits of this species given by D'Albertis and Beccari, it seems that its favourite food consists of figs, into which it plunges its head in the same way as *Gymnocorvus senex*; and Salvadori suggests that the bare face of the bird has something to do with this peculiar habit. It cry is harsh and loud, and is heard at a great distance; and the skin is of extraordinary toughness, so that it is most difficult to shoot specimens, which generally fall to a shot in the head or a broken wing.

The coloration of this species is so peculiar that a detailed description is not necessary.

The figures in the Plate represent two birds, of about the size of life, drawn from specimens in our own collection.

[R. B. S.]

EOS FUSCATA, *Blyth.*

EOS FUSCATA, *Blyth*.

Banded Lory.

Eos fuscatus, Blyth, Journ. As. Soc. Beng. xxvii. p. 279 (1858).—Sclater, Proc. Zool. Soc. 1873, p. 697.
Eos (Chalcopsitta) torrida, Gray, List Psittacidæ Brit. Mus. p. 102 (1859).
Eos fuscata, Gray, Proc. Zool. Soc. 1859, p. 158.—Sclater, Proc. Zool. Soc. 1860, p. 227.—Gray, Proc. Zool. Soc. 1861, p. 436.—Wallace, Proc. Zool. Soc. 1864, p. 291.—Schl. Dirent. pp. 68, 69, cum fig. (1864). —Salvad. Ann. Mus. Civic. Genov. vii. p. 760 (1875), x. pp. 34, 122 (1877).—D'Albert. op. cit. x. p. 19 (1877).—Ramsay, Proc. Linn. Soc. N. S. Wales, iii. p. 258 (1878), iv. p. 96 (1879).—D'Albert. and Salvad. Ann. Mus. Civic. Genov. xiv. p. 37 (1879).—Sharpe, Journ. Linn. Soc., Zool. xiv. pp. 628, 686 (1879).—Salvad. Orn. Papuasia e delle Molucche, i. p. 263 (1880).—Guillemard, Proc. Zool. Soc. 1885, p. 622.
Eos leucopygialis, Rosenb. J. f. O. 1862, p. 64.
Chalcopsitta leucopygialis, Rosenb. Nat. Tijdschr. Nederl. Indie, xxv. pp. 144, 224 (1863).—Id. Journ. für Orn. 1864, p. 113.
Lorius fuscatus, Schl. Mus. Pays-Bas, Psittaci, p. 122 (1864).—Finsch, Neu-Guinea, p. 157 (1865).—Gray, Hand-list B. ii. p. 153, sp. 8194 (1870).—Schl. Mus. Pays-Bas, Psittaci, Revue, p. 55 (1874).—Rosenb. Reis. naar Geelvinkb. p. 113 (1875).
Domicella fuscata, Finsch, Die Papag. ii. p. 807, pl. 6 (1868) —Meyer, Sitz. k. Akad. Wiss. Wien, lxx. p. 236 (1874).—Id. Sitz. Isis Dresden, 1875, p. 78.
Chalcopsittacus fuscatus, Salvad. Ann. Mus. Civic. Genov. viii. p. 397 (1876).

This Lory is easily recognized by its dusky coloration and greyish-white rump, and by the red or yellow bands on the body. It is an inhabitant of New Guinea, where it has been met with in several places in the north-western portion of the island—at Dorei by Mr. Wallace, Baron von Rosenberg, and Dr. Meyer, at Andai by Von Rosenberg and Dr. Guillemard, at Mansinam by Dr. Beccari, Passim by Dr. Meyer, Mon by Dr. Beccari, and in the Arfak Mountains by Dr. Meyer and Signor D'Albertis, as well as near Sorong by the last-named traveller. It has also been obtained in Salawati by Dr. Beccari, and in the Island of Jobi by Dr. Meyer, Mr. Bruijn, and Dr. Beccari. In Southern New Guinea Signor D'Albertis procured the present species on the Fly River, Mr. Ramsay has received it from Port Moresby, and the Rev. W. G. Lawes has found it at Walter Bay, a few miles eastward of the latter place.

The very curious variation in the colour of the red bands, which are sometimes replaced by yellow, has been well treated of by Dr. Meyer, whose remarks we reproduce :—

" The Jobi specimens appear to be somewhat larger in all their dimensions and to have more intense colours than those from New Guinea; but my series is not large enough to judge with certainty. Among the ten specimens which I procured are two red-banded ones, of which I could not determine the sex with certainty ; of the rest three were females and five were males. The latter are all red-banded ; and of the females, two had yellow bands and one red. This proves that the yellow coloration is not exclusively the dress of the full-grown female; but whether it be the plumage of the young bird, or whether this species does not preserve constancy in this respect, as the two finely coloured yellow birds do not show any special traces of youth, remains to be seen. It is possible, however, that the yellow plumage may be that of the immature bird; and this can only be proved when the changes of colour are observed in captivity, or when it is demonstrated that in a large series no young birds ever occur with red in their plumage. Mr. Wallace appears not to have regarded the yellow dress as being that of the young, for he states that both sexes of red and yellow varieties were obtained from the same flock.

"The bill is not ' horn-yellow ' or ' light red,' but red like the red feathers of the neck ; and it must be particularly noted that the skin at the base of the under mandible and on the chin is naked for a certain extent, and is of the same colour as the bill, so that the latter appears at first sight much larger than it really is. This peculiarity is not shown in Finsch's plate or mentioned in his description.

"The colour of the skin at the base of the lower mandible and on the chin is different in different species, but it appears to agree with the colour of the bill—as, for instance, in *Domicella atra* with a black bill and naked parts as distinct as in *D. fuscata*; the same in *D. scintillata*, but the naked parts not so extended. *D. lori*, *D. cyanogenys*, and *D. garrula* have the naked parts red like the bill, but with more feathering than in the foregoing species.

" *D. fuscata* has black feet and claws, but the soles of the feet are greyish yellow; the iris is yellowish red, and the skin at the base of the upper mandible is black."

The Plate represents three specimens of the natural size, showing both red and yellow variations. The figures are drawn from examples kindly lent by Dr. Meyer.

[R. B. S.]

EOS RETICULATA.

EOS RETICULATA.

Blue-streaked Lory.

Blue-necked Lory, Lath. Gen. Hist. B. ii. p. 136 (1822).
Lorius borneus, Less. (nec. Steph.) Traité d'Orn. p. 192 (1831).—Salvad. Ucc. di Borneo, p. 27, note (1874).
Psittacus reticulatus, Müll. Verh. Natuurl. Gesch. Land- en Volkenk. pp. 107, 108 (1839-44).—Gray, Gen. B. ii. p. 42 (1846).
Eos cyanostriata, Gray & Mitch. Gen. B. ii. p. 417, pl. 103 (1845).—Blyth, Cat. B. Mus. As. Soc. p. 11 (1849).—Bp. P. Z. S. 1850, p. 29.—Id. Rev. et Mag. de Zool. 1854, p. 156.—Id. Naum. 1856, Consp. Psitt. no. 303.—Sclater, Proc. Zool. Soc. 1860, p. 226.—Wall. Ibis, 1861, p. 311.—Id. P. Z. S. 1864, p. 290.
Eos bornea, Souancé, Rev. et Mag. de Zool. 1856, p. 226.—Gray, List Psittac. Brit. Mus. p. 52 (1859).
Eos reticulata, Sclater, Proc. Zool. Soc. 1860, p. 226.—Wall. Ibis, 1861, p. 311.—Rosenb. J. f. O. 1862, p. 61.—Salvad. Ann. Mus. Civic. Genov. x. p. 33 (1877).—Id. Orn. Papuasia e delle Molucche, i. p. 245 (1880).—Sclater, Proc. Zool. Soc. 1883, p. 51.
Psittacus cyanostictus, Schlegel, Handl. Dierk. i. p. 184 (1857).
Eos, sp., Rosenb. J. f. O. 1862, p. 65.
Psittacus (Eos) guttatus, Rosenb. Nat. Tijdschr. Nederl. Ind. xxv. p. 145 (1863).
Lorius reticulatus, Schlegel, Mus. Pays-Bas, Psittaci, p. 128 (1864).—Finsch, Neu-Guinea, p. 157 (1865).—Schlegel, Mus. Pays-Bas, Psittaci, *Revue*, p. 58 (1874).
Domicella reticulata, Finsch, Papag. ii. p. 797 (1868).
Eos reticulatus, Gray, Hand-list of Birds, ii. p. 154, no. 8203 (1870).

The above synonymy, culled from Count Salvadori's well-known 'Ornitologia della Papuasia,' would seem to indicate that this beautiful Parrot had been known for a long time; and this is, indeed, the case; but the only European who has shot the species in its native haunts has been Mr. H. O. Forbes, who procured several examples during his recent expedition to the Tenimber Islands.

Various islands have been given as the home of this species, such as Borneo by Lesson, Celebes by Blyth; and in the Leiden Museum the habitat was set down as Amboina. Captain Chambers was the first to indicate its true home when he presented two specimens to the British Museum as from Timor Laut, and Mr. Wallace afterwards confirmed this habitat by finding that the native traders often brought living examples from Timor Laut to Celebes.

Mr. Forbes informs us that the species is common in all the islands of the Tenimber group which he visited, and that it is everywhere a favourite cage-bird with the natives.

The following description is taken from one of Mr. Forbes's specimens in the British Museum:—

Adult male. General colour above blood-red, the head and hind neck uniform, the mantle striped with beautiful blue in the centre of the feathers; lower back, rump, and upper tail-coverts obscurely marked with dusky blackish at the tips; scapulars black with red on the outer web and at the tips, varying in extent and in some feathers occupying the whole of the outer web; lesser and median wing-coverts blood-red, the inner ones with black on the inner web, which is more distinct on the bastard wing and greater wing-coverts, and gives a varied appearance; primary-coverts and primaries black, narrowly edged with blood-red, increasing in extent so as to occupy the base of the inner primaries; the secondaries blood-red, with black shafts and a broad bar of black at the end, decreasing in extent towards the inner secondaries, the innermost being entirely black; centre tail-feathers black, the rest blood-red on the inner web, black on the outer, the external webs diagonally red at the tips, with the shaft black; ear-coverts blackish, streaked with dull blue; cheeks and throat bright red; the rest of the under surface from the lower throat downwards deep blood-red with dusky ends to the feathers, the plumes of the thighs and lower flanks with black bases, the latter with blue ends; under wing-coverts, axillaries, and inner lining of quills beautiful blood-red, remainder of quills black below; "upper mandible scarlet, the tip orange-red; lower mandible the same; legs and feet black; iris rich brown" (*H. O. Forbes*). Total length 12 inches, culmen 0·85, wing 9·0, tail 8·5, tarsus 0·65.

Adult female. Similar in colour to the male. Total length 12 inches, wing 6·6, tail 5·4, tarsus 0·7. The specimen of this sex in the British Museum, sent by Mr. Forbes, is a very brilliant bird, and is streaked with blue on the sides of the rump and upper tail-coverts; at the end of the black centre tail-feathers is a subterminal oval spot of red.

The figures in the Plate represent a pair of birds, of the natural size; they have been drawn from two of Mr. Forbes's specimens.

[R. B. S.]

LORIUS FLAVO-PALLIATUS, *Salvad.*

LORIUS FLAVO-PALLIATUS, Salvad.
Yellow-mantled Lory.

Lorius garrulus (nec L.), Sclater, Proc. Zool. Soc. 1860, p. 226 (nec p. 227).—Rosenb. Journ. f. Orn. 1860, p. 62 (pt.).—Id. Tijdschr. Nederl. Ind. xxiii. p. 141 (1862, nec p. 142).—Wallace, Proc. Zool. Soc. 1864, p. 289 (pt.).—Schl. Mus. Pays-Bas, Psittaci, p. 121 (1864, pt.).—Finsch, Neu-Guinea, p. 157 (1885, pt.).—Gray, Hand-l. B. ii. p. 153, no. 8189 (1870).—Schl. Mus. Pays-Bas, Psittaci, Revue, p. 55 (1874, pt.).
Lorius garrulus, var., Gray, Proc. Zool. Soc. 1860, p. 356 (pt.).
Domicella garrula, Finsch, Die Papag. ii. p. 776 (1868, pt.).
Lorius flavo-palliatus, Salvad. Ann. Mus. Civic. Genov. x. p. 33 (1877).—Id. Orn. Papuasia e delle Molucche, i. p. 243 (1880).—Guillem. Proc. Zool. Soc. 1885, p. 564.

COUNT SALVADORI, from whose work on the Birds of Papuasia the above synonymy has been derived, was the first naturalist who definitely recognized the distinctness of this Lory, as it was considered by Dr. Finsch and other well-known students of the Parrots to be identical with *Lorius garrulus* of Halmahera, although most of these writers recognized a certain variation in colour in the birds from Batchian. After examining a large series of specimens, Count Salvadori has pointed out that the species from the island of Batchian differed from its relative in Halmahera in having the entire interscapulary region yellow. Similar peculiarities mark the red Lories from the islands of Obi, Morotai, and Raou, so that *Lorius garrulus* would seem to be entirely confined to the island of Halmahera, or Gilolo, as it is wrongly called by most English naturalists.

This species is said to be a good talker, and large numbers of them are caught by the natives.

Adult. General colour above deep crimson, darkest on the scapulars, the mantle bright yellow; wing-coverts grass-green, with a patch of yellow near the bend of the wing; the inner, median, and greater coverts more olive-green, the latter with a tinge of golden; bastard-wing and primary-coverts blackish, externally glossed with purplish blue; quills blackish, externally grass-green; upper tail-coverts duller crimson than the rump; tail-feathers green, dull crimson at the base, the outer ones purplish black, green at the ends; crown of head, entire hind neck, sides of face, and entire under surface of body bright crimson; thighs green; under wing-coverts and axillaries yellow, the lower greater coverts blackish; quills below black, crimson for the greater part of the inner web. Total length 11 inches, culmen 1·05, wing 6·0, tail 4·0, tarsus 0·75.

The figure in the Plate represents an adult bird of the size of life. The specimen from which it has been drawn was lent to us by our friend Dr. Sclater.

[R. B. S.]

LORIUS TIBIALIS, *Sclater.*

LORIUS TIBIALIS, *Sclater*.

Blue-thighed Lory.

Lorius tibialis, Sclater, Proc. Zool. Soc. 1871, p. 499, pl. xl.—Garrod, Proc. Zool. Soc. 1872, p. 788.—Giebel, Thes. Orn. ii. p. 503 (1873).—Salvad. Ann. Mus Civic. Genova, x. p. 33 (1877).—W. A. Forbes, Ibis, 1877, p. 278.—Salvad. Orn. della Papuasia e delle Molucche, p. 240 (1880).

This is a very distinct species of Lory, and it is very much to be regretted that at present we are entirely ignorant of the country which it inhabits. It is, no doubt, as has been suggested by Dr. Sclater, a denizen of one of the Molucca Islands.

The species was originally described by Dr. Sclater from a specimen which was living at the time in the Zoological Gardens, and of this bird a very good figure was given in the 'Proceedings' of the Society. It belongs to that section of the genus *Lorius* wherein the colour of the head resembles that of the back, instead of being black.

The only other species which shares the character of the crimson head is *Lorius garrulus*; but this is easily distinguished by its yellow under wing-coverts. The original specimen was purchased by Mr. Jamrach in the Calcutta market; it lived for nearly four years in the Zoological Gardens.

The following is a description of the type, which proved, on dissection, to be a female:—

Adult female. General colour above bright crimson, the scapulars like the back; wing-coverts grass-green, the lesser series mixed with crimson and with lilac along the bend of the wing; the inner, greater, and median coverts with a subterminal mark of dark crimson; bastard-wing grass-green, bluish at the end of the feathers, the outer feathers black at the base; primary-coverts grass-green, blackish internally; quills grass-green, the primaries blackish internally, yellow for more than the basal half, the basal part being tinged with red; innermost secondaries marked with dark crimson near the ends like the greater coverts; upper tail-coverts and tail-feathers bright crimson, with a purplish-brown band across the end of the tail; crown of head, sides of face, and entire under surface of body bright crimson, mottled with yellow bases to the feathers on the lower throat and fore neck; thighs lilac-blue; under tail-coverts bright crimson; under wing-coverts and axillaries duller lilac-blue, slightly washed with green and on the edge of the wing with red; quills below black, with a large yellow basal patch; bill bright orange; feet pale flesh-colour, claws bright horn-colour. Total length 11·5 inches, culmen 1·1, wing 6·4, tail 3·7, tarsus 0·7.

The figure in the Plate represents an adult bird of the natural size, and is taken from the specimen described above.

[R. B. S.]

CHALCOPSITTACUS SCINTILLATUS.

CHALCOPSITTACUS SCINTILLATUS (Temm.).

Red-Fronted Lory.

Amber Parrot, Lath. Syn. i. Suppl. p. 65.—Id. Gen. Hist. ii. p. 252 (1822).
Psittacus batavensis, Lath. (nec Wagl.), Ind. Orn. i. p. 126 (1790).—Bechst. Kurze Uebers. p. 101 (1811).—Vieill. Nouv. Dict. xxv. p. 375 (1817).—Kuhl, Consp. Psitt. p. 99 (1820).—Vieill. Encycl. Méthod. p. 1406 (1823).
Psittacus scintillatus, Temm. Pl. Col. 569 (1835).
Lorius scintillatus, Bourj. Perroquets, pl. 51 (1837-38).—Hombr. & Jacq. Annales des Sciences Naturelles, xvi. p. 317 (1841).—Schleg. Mus. Pays-Bas, *Psittaci*, p. 122 (1864).—Finsch, Neu-Guinea, p. 158 (1865).—Schlegel, Mus. P.-B., *Psittaci*, Revue, p. 56 (1874).—Giebel, Thes. Orn. ii. p. 502 (1875).—Rosenberg, Malay Archip. p. 371 (1879).
Psittacus scintillans, Müller, Verh. Land- en Volkenk. pp. 22, 127 (1839-44).
Eos scintillatus, Gray, Gen. Birds, ii. p. 417 (1845).—Id. Cat. B. New Guinea, pp. 39, 59 (1859).—Id. List Psitt. Brit. Mus. p. 53 (1859).—D'Albertis, Sydney Mail, 1877, p. 248.—Id. Ann. Mus. Civ. Genova, x. p. 8 (1877).
Chalcopsitta scintillata, Bp. Consp. Avium, i. p. 3 (1850).—Id. Revue et Mag. de Zool. 1854, p. 156.—Id. Naumannia, 1856, Consp. Psittaci, sp. 305.—Gray, Proc. Zool. Soc. 1858, p. 194; 1861, p. 436.—Rosenb. Journ. für Orn. 1862, pp. 64, 65.—Id. Nat. Tijdschr. Ned. Ind. xxv. pp. 144, 145, 225 (1863).—Id. Journ. für Orn. 1864, p. 113.—Wallace, Proc. Zool. Soc. 1864, p. 289.—Gray, Hand-l. B. ii. p. 153, no. 8192 (1870).—Sclater, P. Z. S. 1872, p. 862.
Chalcopsitta scintillans, Bp. Proc. Zool. Soc. 1850, p. 26.—Sclater, Proc. Linn. Soc. ii. p. 165 (1858).—Id. Proc. Zool. Soc. 1860, p. 227.
Chalcopsitta rubrifrons, Gray, Proc. Zool. Soc. 1858, pp. 182, 194, pl. 135.—Id. P. Z. S. 1861, p. 436.
Eos rubrifrons, Gray, List Psitt. Brit. Mus. p. 53 (1859).—Id. Cat. B. New Guinea, pp. 39, 59 (1859).—Rosenb. Journ. für Orn. 1864, p. 114.—Id. Reis. naar Zuidoostereil., p. 48 (1867).
Domicella scintillata, Finsch, Die Papag. ii. p. 752 (1868).—Meyer, Sitz. k.-k. Akad. der Wissensch. zu Wien, lxx. p. 238 (1874).—Sharpe, Proc. Linn. Soc., Zool. xiii. p. 80 (1878).
Chalcopsittacus chloropterus, Salvad. Ann. Mus. Civ. Genova, ix. p. 15 (1876); x. p. 34 (1877).—D'Albertis, Sydney Mail, 1877, p. 248.—Id. Ann. Mus. Civ. Genov. x. p. 8 (1877).—Id. Ibis, 1877, p. 366.
Chalcopsittacus scintillatus, D'Albertis, Ann. Mus. Civ. Genova, x. p. 19 (1877).—Salvad. tom. cit. p. 34 (1877).—Id. Proc. Zool. Soc. 1878, p. 93.—D'Albertis & Salvad. Ann. Mus. Civ. Genova, xiv. p. 37 (1879).—Sharpe, Proc. Linn. Soc. xiv. p. 686 (1879).—Ramsay, Proc. Linn. Soc. N. S. W. iv. p. 96 (1879).—Salvad. Orn. Papuasia &c., i. p. 274 (1880).
Chalcopsitta chloropterus, Ramsay, Proc. Linn. Soc. N. S. W. iii. p. 254 (1878-9).

The above intricate synonymy has been copied from Count Salvadori's grand work on the Ornithology of Papuasia. Indeed it would be difficult to write the synonymy of any New-Guinea bird without reproducing what has been written by the learned Italian ornithologist, so completely does he seem to have exhausted the literature of his subject.

The Red-fronted Lory appears to be found in New Guinea and the Aru Islands; in the latter locality it cannot be very rare, and many specimens were collected by Dr. Beccari during his expedition to these islands in the spring of 1873. According to Baron von Rosenberg the inhabitants call the bird *Jaran-kra*.

In New Guinea it was met with by Salomon Müller at Lobo Bay, and in the north-west part of the island it has been procured by Dr. A. B. Meyer at Rubi, by Baron von Rosenberg at Jour, a place situated at the very lowest part of the Bay of Geelvink, and at Mesan by Dr. Beccari. In south-eastern New Guinea it has been met with by Signor D'Albertis on the Fly River, and also in the neighbourhood of Hall Bay, while Mr. Octavius Stone got specimens in the vicinity of Port Moresby. The greater amount of green on the under wing-coverts in some of the more southern specimens induced Count Salvadori at one time to consider them a distinct species, which he called *C. chloropterus*; and at one time we were ourselves inclined to believe in the validity of this species. But, after an examination of more extensive material, Count Salvadori finds great variation to exist in the colouring of these parts, and he has decided to suppress the supposed southern species.

The following description is taken from Count Salvadori's work :—

" Green; the middle of the back and rump brighter and more blue, with very narrow shaft-streaks of yellow; sinciput and lores red; the sides of the head and chin dark brown, almost blackish; occiput dark brown, the feathers streaked in the middle with green ; the neck, the interscapulary region, the breast, and abdomen green, ornamented with yellow shaft-streaks ; the streaks on the breast broader and more orange in tint ; throat dull green varied with red ; thighs red ; under tail-coverts green, very narrowly streaked with yellow towards the tip ; wings externally green, the median and lesser coverts very narrowly streaked with bright green, the lesser coverts somewhat bluish ; under wing-coverts and axillaries red ; quills green on the outer web, dusky on the inner one, with a very broad yellow patch towards the base of the inner web, tail-feathers green above, red for the basal half of the inner web ; the apical part of the tail-feathers underneath yellowish olive ; bill, cere, and feet black ; iris orange yellow." The sexes are alike ; but the young bird has the whole of the head dark brown, with the streaks on the neck and breast fiery orange, more or less broad in their extent.

The figures in the Plate represent, of about the natural size, two adult birds, drawn from specimens in my own collection.

[R. B. S.]

PSITTACELLA BREHMII, *Rosenberg*.

PSITTACELLA BREHMII, *Rosenberg*.

Brehm's Parrot.

Psittacus brehmii, Ros. *in lit.*, Schlegel, Obs. Zool. v., Ned. Tijdschr. voor de Dierk. iv. p. 35, 1873, ♀ (nec ♂).
Psittacella brehmii (Ros.), A. B. Meyer, Cab. Journ. f. Orn. Jan. 1874, and Sitzungsber. der k.-k. Akad. de Wiss. zu Wien. lxix. p. 74, Febr. 1874.—T. Salvadori, Ann. del Mus. Civ. di St. Nat. di Genova, vii. p. 755, 1875.

WHEN one has to deal with a species of Parrot so different from all others that have yet been discovered, there can be no difficulty with regard to synonymy, or need for many remarks on the differences between it and others.

The native country of this Parrot is the northern part of New Guinea, where it was first collected by Baron Rosenberg.; but, if I rightly understand the sense of a letter to me from Dr. Meyer, Rosenberg only discovered the female, while he himself was fortunate enough to obtain both sexes:—"I procured within a few days five specimens, all in the same locality, two of which were males, and three females. The back of the male is similar to the same part in the female; but the plumage of the breast and flanks, instead of being crossed with crescentic bars, is uniform green; it is also the male alone that possesses the beautiful semicollar of bright yellow, an ornament which adds greatly to the beauty of this sex."

The male has the head, cheeks, and throat dark olive-brown; on the sides of the neck below the dark colouring of the cheek and throat brilliant jonquil-yellow; back, including the tail-coverts, green, crossed by narrow bands of black; upper part of the wing and tail bright grass-green; the same bright green also pervades the whole under surface; under tail-coverts bright scarlet; on the shoulders both above and below a patch of blue; bill bluish horn-colour; feet dark bluish-grey. In the female the yellow neck-bands and uniform green breast are wanting; in other respects she is similarly coloured.

Total length 8¾ inches; wings 3½, tail 2¼, tarsi 1, bill ¾.

The figures are of the natural size.

Hab. Northern portions of New Guinea.

PSITTACELLA MADARASZI, Meyer.

PSITTACELLA MADARASZI, Meyer.

Madarász's Parrakeet.

Psittacella madaraszi, Meyer, Zeitschr. ges. Orn. iii. p. 4, tab. i. fig. 1 (1886).

Dr. Meyer describes this species as being similar to *Psittacella modesta* from North-western New Guinea, but smaller, and with the head and under surface of the body distinctly more yellow. It was found by Mr. Hunstein in the Horseshoe range of the Owen Stanley Mountains, at a height of 7000 feet, and Mr. Forbes has also met with it in the Sogeri district of the Astrolabe Mountains in South-eastern New Guinea. Not having an example of *P. modesta* in the British Museum, we have not been able to institute a comparison between the two species, and in figuring *P. madaraszi* we must leave to future research the task of more strictly defining the species, the characters of which do not appear very strongly marked. Perhaps the green colour of the breast, to which Dr. Meyer draws attention, instead of the dull olive-coloured breast of *P. modesta*, may prove to be a good specific character.

Dr. Meyer also received a female bird from Mr. Hunstein, which he thinks may not be in full plumage, but which differs from the hen of *P. modesta* in having the head green, the forehead blue, and the nape more or less barred across with black and red. He is not sure that the bird so described is the female of *P. madaraszi*, and thinks that it may belong to an undescribed species.

The following is a description of one of Mr. Hunstein's specimens in the British Museum:—

Adult male. General colour above green, the lower back and rump barred with yellow and black, the yellow bars broader than the black ones; wing-coverts like the back, the edge of the wing cobalt-blue; bastard-wing, primary-coverts, and quills green externally, the inner webs blackish, the primaries darker and more bluish green on the outer webs, which are narrowly bordered with yellow near the ends; upper tail-coverts more yellowish green than the back; tail-feathers dark green with black shafts; crown of head, nape, and hind neck ochreous brown, mottled with dark brown edges and yellow shaft-lines to the feathers; forehead, lores, and feathers in front of the eye sooty brown; sides of face and ear-coverts like the head, the latter similarly streaked with yellow; cheeks and throat ochreous-brown washed with green; fore neck and breast green; abdomen and sides of body lighter and more yellowish green, washed on the flanks with darker green; thighs dark green; under tail-coverts scarlet; under wing-coverts and axillaries yellowish green, blue near the edge of the wing; quills below dusky, olive-yellow along the inner edge. Total length 5·5 inches, culmen 0·6, wing 3·95, tail 1·9, tarsus 0·5.

The Plate gives an illustration of a male, but in two positions, the figures being drawn from the same specimen we have described above.

[R. B. S.]

TRICHOGLOSSUS GOLDIEI, Sharpe.

TRICHOGLOSSUS GOLDIEI, Sharpe.

Goldie's Perroquet.

Trichoglossus goldiei, Sharpe, Journ. Linn. Soc. (Zool.) vol. xvi. p. 317 (1882).

THE present species was one of the handsomest birds discovered by Mr. Goldie during his expedition to the Astrolabe Mountains. It appears to me to be quite distinct from any known species of the genus *Trichoglossus*; and the name attached to it by Mr. Bowdler Sharpe will perpetuate the name of one of the pioneers of research in the untrodden regions of South-eastern New Guinea. It was procured by Mr. Goldie in the Morocco district at the back of the Astrolabe Mountains, where it is known to the natives, according to that gentleman, by the name of "*1—I—hawa*."

The following is a copy of the description given by Mr. Sharpe in his paper on Mr. Goldie's collection read before the Linnean Society at their meeting of the 6th of April, 1882.

"*Adult male.* General colour above green, the hind neck mottled with yellow edges to the feathers, extending a little on the mantle; wing-coverts like the back; primary-coverts and quills dusky blackish, externally brighter green, the secondaries like the back; tail-feathers greenish brown, edged with bright green like the back, the tips fringed with yellow; forehead and sinciput scarlet, tending towards a point in the middle of the crown; from behind the eye a broad purplish-blue band extends round the occiput to behind the opposite eye: the nape-feathers brown washed with lilac, and faintly streaked with dull scarlet; lores, sides of face and ear-coverts lilac red, with a bluish shade along the upper margin of the latter; below the eye the feathers rather lighter in colour, and having indistinct tiny streaks of dull blue; under surface of body yellowish green, streaked with darker green down the centre of the feathers, more narrowly on the under tail-coverts; under wing-coverts like the breast and streaked with dark green in the same manner; quills dusky below, all but the outer primaries oily yellow for two thirds of the inner web, forming a conspicuous diagonal patch across the wing when uplifted. Total length 6·5 inches, culmen 0·6, wing 4·2, tail 3·1, tarsus 0·5."

The figures in the Plate are drawn from specimens in the British Museum, the one first described by Mr. Sharpe being the duller-coloured of the two; it is probably a female or young male. Both birds are represented of the natural size.

[R. B. S.]

TRICHOGLOSSUS MUSSCHENBROEKII.

TRICHOGLOSSUS MUSSCHENBROEKII.

Van Musschenbroek's Lorikeet.

Nanodes musschenbroekii, Schlegel, Nederlandsch Tijdschrift voor de Dierkunde, iv. p. 34 (1873).—Id. Mus. d. P.-B., Revue Psittaci, p. 52 (1874).
Trichoglossus musschenbroekii, Sclater, P. Z. S. 1873, p. 697.—Finsch, in Rowley's Orn. Misc. part v. p. 61, pl. xliv. (1876).
Neopsittacus musschenbroekii, Salvad. Ann. Mus. Civic. Genov. vii. p. 761 (1875).

PROFESSOR Schlegel describes this species in the following manner:—" M. von Rosenberg has just forwarded us, under the title of *Nanodes musschenbroekii*, three adult individuals of a species of *Nanodes* new to science. These comprise two females and a male, all exactly resembling each other, and collected in April 1870 in the interior of the north-western peninsula of New Guinea. The species is naturally allied to *N. placens* and *N. rubrinotatus*; but it is distinguished at the first glance by its larger size and by the very sensible modifications in the distribution of its colours."

Nothing more was recorded concerning the species until D'Albertis brought back examples from Atam; and since then Count Salvadori has received it from Mount Arfak. In recording the last-named occurrence Count Salvadori makes it the type of a new genus, on account, as he says, of its differently shaped bill, the upper mandible being very much more strongly incurved, and the lower one showing a flat superficies on the gonys, which is very broad. These characters the describer considers sufficient to separate the bird generically from the other members of the Trichoglossinæ.

I have to thank Dr. Meyer for the loan of the fine specimen from which the figures in the Plate have been drawn; and the following is a description of this bird:—

Face green, each feather having a yellow centre; back part of the crown and nape brown, streaked with pale yellow; all the upper surface, including the two centre tail-feathers and flanks, green; chest and centre of the abdomen red; all the primaries and secondaries as seen from above, when the wing is expanded, brilliantly marked with red on the inner webs; the same brilliant red also occurs on the under wing, except the tips of the primaries, which are brownish black; tail cuneate, the four centre feathers nearly uniform green with slightly rosy tips, the external ones green on the outer webs, with brilliant red on the inner webs, the whole broadly tipped with yellowish rose-colour. Total length 8½ inches; wing 4¼, tail 3⅜, tarsus ⅝.

Professor Schlegel states that the soft parts were found by Von Rosenberg to be as follows—"bill orange-red, iris citron-yellow, feet yellowish flesh-colour."

In concluding this brief memoir of a very interesting bird, I must beg to offer a dissenting voice as to its belonging to the genus *Nanodes* or being allied to *Psitteuteles*; neither can I agree with the learned Count Salvadori in instituting a new genus for its reception. From the first moment I examined the skin kindly forwarded to me by Dr. Meyer I considered it a true *Trichoglossus*—an opinion in which Dr. Sclater evidently coincides; see the 'Proceedings' of the Zoological Society, as quoted in the above synonymy. More recently too, Dr. O. Finsch has declared in favour of the bird being a true *Trichoglossus*.

The principal figure in the Plate is of about the size of life.

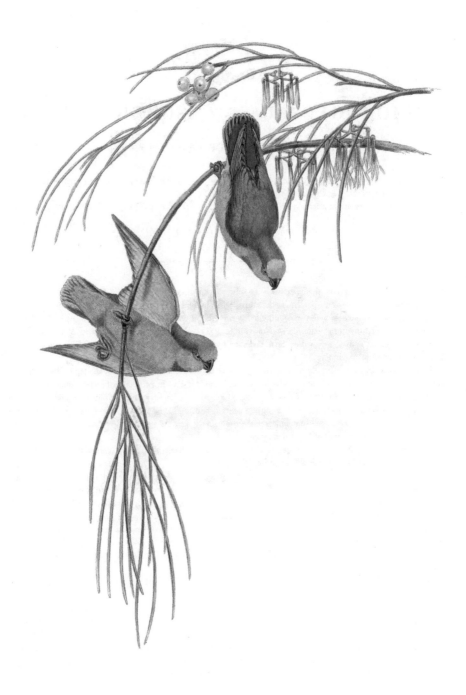

LORICULUS AURANTIIFRONS, Schlegel.

LORICULUS AURANTIIFRONS, Schlegel.

Orange-crowned Loriculus.

Loriculus aurantiifrons, Schlegel, Nederlandsch Tijdschrift voor de Dierkunde, iv. p. 19.—Salvad. Ann. Mus. Civic. Genov. ix. p. 912.—Meyer, in Rowley's Orn. Misc. ii. p. 245.

The present species, of which a fine male was kindly sent to me by Dr. Meyer, is one of the smallest Parrots known, and forms one of a group of tiny *Loriculi* found in New Guinea and the adjacent islands. It was first discovered by the Dutch traveller M. Hoedt in the island of Mysol; and since that time it has been collected in North-western New Guinea by M. Bruijn. The latter gentleman procured specimens at Andai; and Count Salvadori, when recording this fact, believed that this was the first instance of the occurrence of the species in New Guinea. That it was the first instance published, there can be no question, as Dr. Meyer, although he was the first to procure actual specimens, had not stated it in print. He informs me in a letter that he has no notes on the present bird, of which he only procured two male examples.

The following is a translation of Professor Schlegel's original description:—

"The first three quills are of equal length and much longer than the fourth; tail rounded; bill not strong, black. General colour of the plumage lively green, verging slightly upon yellow, and lighter on the under than on the upper surface; upper aspect of the inner web of the quills black, with the exception of their extremities; lower aspect of the primaries black, but with very broad edgings of bluish verdigris occupying the inner web; this colour extends over nearly the whole of both quills in the secondaries; greater wing-coverts of the same bluish verdigris, lesser and median coverts coloured like the abdomen. Tail-feathers black in the centre, for the remainder green on their upper surface, verdigris below; rump and upper tail-coverts fiery red, slightly shaded with orange on the sides of the rump; fore part of the throat with a spot of red colour a little darker than that of the rump.

"The sexes present the following differences in colour: the male has a large patch of lively orange, occupying nearly the entire forehead; not a trace of this is seen in our female specimen, which is distinguished on the contrary by the prevailing green colour, passing into verdigris on the forehead and sides of the head."

For the opportunity of figuring the male of this pretty little Parrot I am indebted to the kindness of Dr A. B. Meyer.

The portrait of the female bird is drawn from a specimen collected by M. Bruijn at Andai; and I have to acknowledge the kindness of Dr. Sclater for permitting me to see this bird, which had been sent to him for examination by Count Salvadori, to whom I also beg leave to return my due acknowledgments.

Total length $3\frac{3}{4}$ inches; wing $2\frac{3}{4}$, bill $\frac{3}{8}$, tail $1\frac{1}{2}$, tarsus $\frac{1}{2}$.

The figures in the Plate are of the size of life.

CACATUA TRITON, Temm.

CACATUA TRITON.

Triton Cockatoo.

Psittacus galeritus (pt.), Less. Voy. Coquille, Zool. i. p. 624 (1828).—Id. Traité d'Orn. p. 182 (1831).—Id. Compl. Buff., Ois. p. 602 (1838).—S. Müll. Verh. Land- en Volkenk. pp. 21, 107 (1839-1844).
Psittacus sulphureus, Less. (nec Gm.), Voy. Coquille, i. p. 625 (1828).
Psittacus triton, Temm. Coup d'œil gén. sur les Possess. Néerl. dans l'Inde Archip. iii. p. 405, note (1849).
Plyctolophus sulphureus, Bp. (nec Gm.), Compt. Rend. xxx. p. 138 (1850).
Plyctolophus luteocristatus, Bp. loc. cit.
Plyctolophus triton, Bp. t. c. p. 139.—Id. Rev. et Mag. de Zool. 1854, p. 156.—Id. Naumannia, 1856, Consp. Psitt. sp. 278.—Id. Compt. Rend. xliv. p. 537 (1857).
Cacatua cyanopsis, Blyth, Journ. As. Soc. Beng. xxv. p. 447 (1856).
Cacatua triton, Sclater, Journ. Linn. Soc. ii. p. 166 (1858).—Gray, Proc. Zool. Soc. 1858, pp. 184, 195.—Id. Cat. Birds New Guinea, pp. 43, 60 (1859).—Id. Proc. Zool. Soc. 1859, p. 159.—Id. List of Psittacidæ in Brit. Mus. p. 94 (1859).—Sclater, Proc. Zool. Soc. 1860, p. 227.—Gray, Proc. Zool. Soc. 1861, p. 437.—Sclater, Proc. Zool. Soc. 1864, p. 188.—Wallace, t. c. p. 280.—Schl. Mus. Pays-Bas, Psittaci, p. 133 (1864).—Sclater, Ann. & Mag. N. H. (3) xv. p. 74 (1865).—Finsch, Neu-Guinea, p. 159 (1865).—Schl. Ned. Tijdschr. iii. p. 320 (1866).—Rosenb. Reis. naar Zuidoostereil. pp. 13, 19, 48 (1867).—Gray, Hand-list B. ii. p. 169, no. 8387 (1870).—Rosenb. Reis. naar Geelvinkb. pp. 36, 56, 83, 113 (1875).—Salvad. Ann. Mus. Civic. Genov. vii. p. 753 (1875), ix. p. 11 (1876), x. p. 24 (1877).—D'Albert. Ann. Mus. Civic. Genov. x. p. 19 (1877).—Id. & Salvad. op. cit. xiv. p. 28 (1879).—Ramsay, Proc. Linn. Soc. N. S. W. iii. p. 250 (1879).—Rosenb. Malay. Arch. pp. 371, 396 (1879).—Salvad. Orn. Papuasia e delle Molucche, i. p. 94 (1880).
Plyctolophus macrolophus, Rosenb. Nat. Tijdschr. Nederl. Ind. xxiii. p. 45 (1861).—Id. J. f. O. 1861, p. 45.
Plyctolophus æquatorialis, Rosenb. J. f. O. 1862, p. 63.—Id. Nat. Tijdschr. Nederl. Ind. xxv. pp. 142, 143 (1863).—Id. J. f. O. 1864, p. 116.
Plyctolophus triton, Rosenb. J. f. O. 1862, pp. 63, 65.—Bernst. Nat. Tijdschr. Nederl. Ind. xxvii. p. 297 (1864).
Cacatua eleonora, Finsch, Nederl. Tijdschr. Dierk. Berigten, p. xxi (1863).
Cacatua macrolopha, Wall. P. Z. S. 1864, p. 280.—Schl. Dierent. p. 82 (1864).—Finsch, Neu-Guinea, p. 159 (1865).—Gray, Hand-list B. ii. p. 169, no. 8393 (1870).
Cacatua galericulata, Rosenb. Reis. naar Zuidoostereil. pp. 99, 100 (1867).
Plictolophus triton, Finsch, Die Papag. i. p. 291 (1867), ii. p. 941 (1868).—Meyer, Sitz. Isis Dresd. 1875, p. 75.—Sharpe, Journ. Linn. Soc., Zool. xiii. p. 490 (1878).

WHEN dried skins only are examined it is not easy to distinguish this Cockatoo from its Australian representative *Cacatua galerita*; but when living individuals of the two species are compared together, the colour of the naked blue skin that surrounds the eye renders the present bird at once remarkable. It is besides slightly smaller in size than *Cacatua galerita*, and has usually rather a stronger bill. In other respects the Triton Cockatoo exactly resembles the well-known Sulphur-breasted Cockatoo of the Australian continent.

Although confounded with *C. galerita* by some of the older authors, the Triton Cockatoo was recognized as distinct by Temminck in 1849, and named after one of the Dutch surveying-vessels which first visited the coasts of New Guinea. It appears to be found all over that large island, and to be, in some places, very abundant. The numerous flocks of white Cockatoos which Dr. Solomon Müller observed on the south-western coasts of New Guinea, near Triton Bay, were doubtless of this species, although that celebrated explorer did not distinguish them from *C. galerita*. There are fine series of specimens of this Cockatoo from the islands of Waigiou and Guebé in the Leyden Museum, and in the same collection is now also the type of *Cacatua eleonora* of Dr. Finsch, originally described from the living bird in the Zoological Gardens of Amsterdam. It likewise occurs in the islands of Geelvink Bay, Salwati, Mysol, the Aru Islands, Goram, and Manuwolka, and is met with in the Louisiade Islands.

Our figure of this species is taken from a fine example now living in the Parrot-house of the Zoological Society of London. This individual was brought home from New Guinea by Mr. C. T. Kettlewell, F.Z.S., in his yacht 'Marquesa,' and presented to the Society in April 1884.

[R. B. S.]

CACATUA OPHTHALMICA, Sclater.

CACATUA OPHTHALMICA, Sclater.

Blue-eyed Cockatoo.

Cacatua ducorpsii (nec Hombr. & Jacq.), Sclater, Proc. Zool. Soc. 1862, p. 141, pl. xiv.
Cacatua ophthalmica, Sclater, Proc. Zool. Soc. 1864, p. 188.—Wallace, tom. cit. p. 280.—Sclater, Ann. & Mag. Nat. Hist. (3) xv. p. 74 (1865).—Id. Proc. Zool. Soc. 1867, p. 184.—Gray, Hand-l. Birds, ii. p. 169, no. 8392 (1870).—Id. Ann. & Mag. Nat. Hist. (4) v. p. 329 (1870).—Sclater, Proc. Zool. Soc. 1877, p. 107.—Salvad. Ann. Mus. Civic. Genov. x. p. 25 (1877).—Id. Orn. Papuasia e delle Molucche, i. p. 103 (1880).—Sclater, Proc. Zool. Soc. 1880, p. 67.—Id. List of Animals in Zool. Gard. p. 308 (1883).
Cacatua triton (pt.), Schleg. Nederl. Tijdschr. v. Dierk. iii. p. 320 (1866).
Plyctolophus ophthalmicus, Finsch, Papag. i. p. 282 (1867).—Id. Proc. Zool. Soc. 1879, p. 17.

The White Cockatoos, as Mr. Sclater has shown, may be divided into two very easily distinguished sections. The first embraces those species which have a narrow median head-crest with the slender point recurved at the extremity, and appearing above the surface of the adjoining feathers when the crest is in a state of repose. The second contains those species which have the crest broadened, comprising the greater part of the head-feathers, and rising, when erect, into a sphere more or less pyramidal in shape, but showing when in a state of repose no recurved point. The present bird belongs to the second of these sections, and is most nearly allied to two well-known species—the White-crested Cockatoo, *Cacatua cristata* of Ternate and Halmahera, and the Rose-crested Cockatoo, *C. moluccensis* of Ceram, having a similar broad pendent crest. But it may be easily distinguished from these two species, with which it nearly agrees in size, by the delicate lemon-colour of its crest, and by the broad blue naked space round the eye, from which latter feature it has received the appropriate name of *ophthalmica*.

In 1862 an example of this fine Cockatoo was first received by the Zoological Society of London. It was at once recognized as a species unknown to him by Mr. Sclater, and described and figured in the 'Proceedings.' But misled, apparently, by the wrong locality attributed to this bird, which was stated to have been received from the Solomon Islands, Mr. Sclater unfortunately referred it to *Cacatua ducorpsi*, with which he was not at that period acquainted. In 1864, however, the receipt of authentic specimens of the true *Cacatua ducorpsi* direct from the Solomon group enabled Mr. Sclater to correct his error, and to establish this bird in its proper position as a distinct species. Moreover its true locality is now well known to us.

Specimens of this Cockatoo were in the collections made in Duke-of-York Island and the adjacent parts of New Ireland and New Britain by Mr. George Brown, C.M.Z.S., in 1877 (see P. Z. S. 1877, p. 107), and in the collection made by Mr. Hübner in the same district in 1878, which was described by Dr. Finsch in the Zoological Society's 'Proceedings' for 1879 (see P. Z. S. 1879, p. 17). Mr. Hübner notes that the native name of this bird in New Britain is "*Moal*." Again, in 1880, Mr. George Brown, then resident at Duke-of-York Island, forwarded a living pair of this Cockatoo to the Zoological Society's Gardens, with the information that the bird is found in New Britain, but not in New Ireland.

Our illustration of this Cockatoo has been prepared from a female specimen formerly living in the Zoological Society's Gardens. The colours of the naked parts have been added from an example now living in the same collection.

[R. B. S.]

CACATUA GYMNOPIS, Sclater.

CACATUA GYMNOPIS, *Sclater.*

Naked-eyed Cockatoo.

Cacatua sanguinea, Sturt, Travels in Austr. App. p. 36 (1849, nec Gould).
Cacatua gymnopis, Sclater, Proc. Zool. Soc. 1871, p. 490, 1875, p. 61.—Id. List of Animals in Zool. Gard. p. 313 (1883).

In the 'Birds of Australia' will be found a figure of the Blood-stained Cockatoo (*Cacatua sanguinea*) of Northern Australia, which was originally described in 1842 from specimens obtained at Port Essington.

The present bird resembles *C. sanguinea* in having the lores more or less stained with rosy red. But it is at once distinguishable from that species by the broad plaque of blue naked skin below the eye, which is also continued in a ring round the eye. In *Cacatua sanguinea* the naked skin round the eye is white; besides, that bird is considerably smaller in dimensions and has much shorter claws.

As in the case of the Blue-eyed Cockatoo, this species was first discriminated by Mr. Sclater, the Secretary of the Zoological Society of London, from a specimen living in the Society's collection in 1871. The bird in question had been purchased from the well-known dealer, Mr. Jamrach, in 1868, and its locality was not known. But Mr. Sclater ascertained that two White Cockatoos in the gallery of the British Museum, obtained by Sturt at Depôt Creek during his expedition into Southern Australia, belonged to the same species. The correct *patria* of this Cockatoo is therefore no doubt Southern Australia.

Our figure of this bird is taken from a skin of an individual that was also formerly living in the Zoological Society's collection, having been purchased in February 1872, and having died in January 1883.

In the Appendix to the narrative of Captain Sturt's expedition into Southern Australia is given the following account of this species:—

"This bird succeeded *Cacatua galerita*, and was first seen in an immense flock on the grassy plains at the bottom of the Depôt Creek, feeding on the grassy plains or under the trees, where it greedily sought the seeds of the kidney bean. These Cockatoos were very wild, and when they rose from the ground or the trees, made a most discordant noise, their note being, if anything, still more disagreeable than that of either of the others. They left us in April, and must have migrated to the N.E., as they did not pass us to the N.W., nor were they anywhere seen so numerously as at this place."

[R. B. S.]

CACATUA DUCORPSI, Hombr et Jacq.

CACATUA DUCORPSI, Jacq. et Pucher.

Ducorps's Cockatoo.

Cacatoès de Ducorps, Hombr. & Jacq. Voy. Pôle Sud, Atlas, pl. 26. fig. 1 (1845).
Plyctolophus Du Crops, Bp. Compt. Rend. xxx. p. 138 (1850).—Id. Rev. et Mag. de Zool. 1854, p. 156.
Cacatua ducorpsii, Jacq. et Pucher. Voy. Pôle Sud, Zool. i. p. 108 (1853).—Hartl. J. f. O. 1854, p. 165.—Sclater, Proc. Zool. Soc. 1860, p. 228.—Id. Proc. Zool. Soc. 1864, pp. 188, 189, pl. xvii. (nec Proc. Zool. Soc. 1862, pl. xiv.).—Wall. Proc. Zool. Soc. 1864, p. 280.—Sclater, Ann. & Mag. Nat. Hist. (3) xv. p. 74 (1865).—Id. Proc. Zool. Soc. 1867, p. 184.—Id. Proc. Zool. Soc. 1869, pp. 118, 124.—Gray, Hand-l. Birds, ii. p. 170 (1870).—Sclater, Proc. Zool. Soc. 1875, pp. 59, 60.—Salvad. Ann. Mus. Civ. Genov. x. p. 25 (1877).—Ramsay, Proc. Linn. Soc. N. S. W. iv. p. 68 (1879).—Salvad. Orn. Papuasia e delle Molucche, i. p. 104 (1880).—Sclater, List of Animals in Zool. Gard. p. 312 (1883).
Cacatua ducrops, Bp. Naumannia, 1856, Consp. Psitt. sp. 269.
Ducorpsius typus, Bp. Compt. Rend. xliv. p. 537 (1857).
Cacatua ducorpsii, Gray, List Psitt. Brit. Mus. p. 94 (1859).
Cacatua (Ducorpsius) ducorpsii, Gray, Cat. B. Trop. Isl. p. 34 (1859).
? *Lophocroa learii*, Finsch, Nederl. Tijdschr. v. Dierk. i. Berigten, p. xxiii (1863).
Cacatua sanguinea (pt.), Schleg. Mus. Pays-Bas, Psittaci, p. 144 (1864).
Cacatua triton (pt.), Schleg. Nederl. Tijdschr. v. Dierk. iii. p. 320 (1866).
Plictolophus ducorpsii, Finsch, Papag. i. p. 311 (1867).—Id. Proc. Zool. Soc. 1869, p. 127.
Cacatua goffini, Sclater, Proc. Zool. Soc. 1869, p. 122, 1875, p. 61, pl. x.—Id. List of Animals in Zool. Gard. p. 312 (1883).

It is to the French discovery-ships 'L'Astrolabe' and 'La Zélée,' which made an expedition towards the South Pole in the years 1837-40, that we owe the discovery of this Cockatoo. It was found in the Solomon Islands by the naturalists of the expedition, and dedicated by Messrs. Hombron and Jacquinot, the authors of the zoological portion of the narrative of the 'Voyage au Pôle Sud,' to M. Ducorps, one of the officers on board the 'Astrolabe.'

Little further was known of Ducorps's Cockatoo until 1864, when a fine pair of the species was received alive by the Zoological Society of London, direct from Guadalcanar Island, one of the Solomon group. One of these birds was figured by Mr. Sclater in the Zoological Society's 'Proceedings' for that year (pl. xvii.), in order to show its distinctness from the larger *Cacatua ophthalmica*, which Mr. Sclater had previously confounded with *C. ducorpsi*.

In his well-known work on the Parrots, Dr. Finsch, besides admitting *C. ducorpsi* as a distinct species, also recognized *C. goffini*, which he had previously described from specimens living in the Zoological Gardens of Amsterdam and Rotterdam. In some of his remarks in the Zoological Society's 'Proceedings,' Mr. Sclater has likewise treated these two species as distinct, and has even figured a white Cockatoo as *Cacatua goffini* (see P. Z. S. 1875, p. 61, pl. x.). But we believe that he is now convinced that the specimens which he has formerly referred in some cases to *C. ducorpsi*, and in others to *C. goffini*, were not really distinct, but all belonged to the same species, to which the former title is properly applicable.

Ducorps's Cockatoo is a small white species, much resembling the Blood-stained Cockatoo (*C. sanguinea*) figured in the 'Birds of Australia,' vol. v. pl. 3; but it is immediately distinguishable by the entire absence of any red markings on the face. The naked skin round the eye is nearly circular in form and, in the living bird, of a pale blue colour. The basal part of the crest-feathers is reddish orange, with a slight tinge of lemon-yellow, sometimes mixed with rosy red towards their summit, which colour, however, is hardly seen unless the crest is elevated. The wing- and tail-feathers are likewise stained on the inner webs with pale lemon-colour. From the Cockatoo of the Philippine Islands (*Cacatua philippinarum*), which is likewise closely allied, the present species is at once distinguishable by the absence of the red colour on the vent.

As regards the supposed occurrence of this species in Queensland, which was stated by Mr. Sclater (P. Z. S. 1875, p. 60) on the information of Mr. J. T. Cockerell, Mr. Sclater now believes that there has been some error on this point, and that Ducorps's Cockatoo is absolutely confined to the islands of the Solomon group, having so far been met with in Guadalcanar and Savo.

[R. B. S.]

CHALCOPHAPS JOBIENSIS, Meyer.

PHLOGŒNAS JOBIENSIS, Meyer.

White-chested Pigeon.

Phlogœnas jobiensis, Meyer, Mitth. Zool. Mus. Dresden, i. p. 10 (1875).—Sharpe, Journ. Linn. Soc., Zool. xiii. p. 318
Chalcophaps margarithæ, D'Alb. et Salvad. Ann. Mus. Civic. Gen. vii. p. 836 (1875).
——— *margaritæ*, Salvadori, op. cit. ix. pp. 44, 207 (1876).
Phlogœnas margaritæ, Salvad. Ann. Mus. Civ. Gen. viii. p. 405 (1876).—Sclater, P. Z. S. 1877, p. 111.
Chalcophaps jobiensis, D'Alb. et Salvad. Ann. Mus. Civ. Gen. ix. p. 207 (1876).

This truly beautiful species of Ground-Dove was described almost simultaneously by Dr. Meyer in Dresden and by Signor D'Albertis and Count Salvadori in Turin. That the adult bird described by the two latter gentlemen from South-eastern New Guinea should not have been recognized as the same as Dr. Meyer's species from Jobi is not surprising, as the latter was described from an immature bird. Thanks to Dr. Meyer's kindness, however, I have been enabled to give a figure of the typical bird; and there can be no doubt, on comparing it with the adult specimen also figured by me, that *P. margaritæ* and *P. jobiensis* belong to one and the same species, in which case I believe that the latter title possesses a slight priority of publication over the former. Even if this conclusion had been formed in this country alone, the specimens alluded to are almost sufficient to have settled the question; but a similar conclusion has been arrived at quite independently by Count Salvadori, who has examined an adult specimen from the island of Jobi, which was identical with others from South-eastern New Guinea. The range of this species therefore extends from the island of Jobi to New Guinea, and to Duke-of-York Island or New Ireland to the eastward. Unfortunately Mr. Brown has not given the exact locality of the specimens which he sent to Dr. Sclater. It would appear to be found over the greater part of New Guinea, as Beccari procured an adult bird at Wandammen, in the Bay of Geelvink; and several specimens were contained in D'Albertis's collection from Yule Island and the opposite coast of New Guinea. Here he met with it at Naiabui; and it was also obtained at Port Moresby by Mr. Stone.

The present species is closely allied to *P. erythroptera*, of the Society Islands, which, however, is a smaller bird, and is distinguished by its white forehead. I take the accompanying description from the original paper of Signor D'Albertis and Count Salvadori.

Head, neck, rump, and upper tail-coverts blackish grey; lores, a streak over the eye, fore neck, and upper part of breast pure white; a streak under the eye, drawn from the base of the lower mandible as far as the neck, blackish grey; dorsal plumes, scapulars, and upper wing-coverts dusky black, margined with shining violet; sides of the breast black, the edges of the feathers violet; lower breast, abdomen, and under tail-coverts dusky black, the middle of the lower breast and of the abdomen somewhat ashy; quills dusky; tail blackish grey, slightly paler at the tip; bill black; feet dull dusky red.

In the Plate an old and young bird are represented, of about the natural size, the latter being Dr. Meyer's type specimen, the adult being drawn from a fine specimen collected by Mr. Octavius Stone and kindly lent to me by that gentleman.

PHLOGÆNAS JOHANNÆ, Sclater.

PHLOGŒNAS JOHANNÆ, Sclater.

Mrs. Sclater's Ground-Dove.

Phlogœnas johannæ, Sclater, P. Z. S. 1877, p. 112, pl. 16.

Dr. Sclater, in describing the present bird (named in compliment to his wife), remarks :—" The nearest species known seem to be *P. stairi*, of the Samoans, from which, however, *P. johannæ* may be known by its lovely pure white breast, and *P. canifrons*, Hartl. & Finsch, of the Pelew group (Journ. Mus. Godeffr. viii. p. 27, pl. v. fig. 1), in which the hind part of the neck is rusty red."

Unfortunately we do not know the exact habitat of this pretty Ground-Dove. It was sent by Mr. George Brown, in a collection of birds from the Duke-of-York Island, New Britain, and New Ireland; but from which of these three localities *P. johannæ* was procured was not indicated when the specimens were sent home.

The following is a translation of Dr. Sclater's description of the species :—

Adult. Dusky chestnut, washed on the back with bronze; lesser wing-coverts externally shining purple; head and neck all round, as well as the breast, pale grey, this colour darker on the crown, clearer on the breast, and passing off into pure white; occiput, like the back, washed with bronze; the white colour of the breast arranged in a semicircle, and girt by a purple margin, as on the wing-coverts; primaries uniform dusky black, the secondaries and the scapulars externally uniform with the back; tail above uniform with the back, underneath blackish; round the eye a bare space; feet red. Total length 7·8 inches, wing 4·4, tail 2·7, tarsus 1·1.

The figures in the Plate are drawn from the typical birds, kindly lent me by Dr. Sclater; they are now in the collection of the Marquis of Tweeddale.

ÆDIRHINUS INSOLITUS.

ŒDIRHINUS INSOLITUS.

Knob-billed Fruit-Pigeon.

Ptilopus insolitus, Schlegel, Nederl. Tijdschr. Dierk. i. p. 61, pl. iii. fig. 3.—Meyer, Rowley's Orn. Miscell. ii. p. 340.—Salvad. P. Z. S. 1877, p. 196.—Elliot, P. Z. S. 1878, p. 549.
Ptilopus humeralis jobiensis (monstrosity), Schlegel, Mus. Pays-Bas, Columbæ, p. 16.
Œdirhinus globifer, Cabanis & Reichenow, Sitz. Gesellsch. naturf. Fr. Berlin, 1876, p. 73.—Iid. J.f. O. 1876, p. 326.
Œdirhinus insolitus, Sclater, P. Z. S. 1877, p. 110.—Rowley, Orn. Miscellany, ii. p. 338, pl. lxvi.—Sclater, P. Z. S. 1878, p. 290.

THIS remarkable bird, which Mr. Dawson Rowley names the "Strange Pigeon," a title it well deserves, was first made known to science in 1861 by Professor Schlegel, who possessed a single example at Leiden in somewhat damaged condition. The locality was stated to be New Caledonia; but this was probably an error, as our next acquaintance with the species was due to Dr. Huesker, who accompanied the German Transit-of-Venus Expedition on board the 'Gazelle,' and procured the bird in New Ireland; and his specimen was named *Œdirhinus globifer* by Drs. Cabanis and Reichenow. Shortly afterwards Dr. Sclater received it from Duke-of-York Island, from Mr. George Brown, and identified the species with one previously described, *Ptilopus insolitus* of Professor Schlegel, who meanwhile had looked upon the species as a monstrosity of his *Ptilopus jobiensis*. Dr. Sclater gave a woodcut of the head to show the extraordinary knob at the base of the bill; and this illustration was followed by a beautiful figure in Mr. Rowley's 'Ornithological Miscellany,' with an excellent history of the species contributed by Dr. Meyer. During the present year Dr. Sclater has received a second consignment from Mr. Brown, in which there was a large series of this Fruit-Pigeon; and although there were no labels on the birds, the receipt of nearly twenty specimens seems to show that there is no difference in the colour of the sexes, although a good deal of variation in the size and colour of the knob was observable, the individuals which had this character more largely developed being probably the older males.

I have followed the above-named authors in keeping this bird distinct under the genus *Œdirhinus*, though Mr. Elliot, in his recent Monograph of the genus *Ptilopus*, has replaced the species in the latter genus, considering that the knob on the bill is paralleled by the protuberance found in some of the species of *Carpophaga*, which are not recognized as generically distinct. According to the last-named ornithologist the present bird belongs to the section of the genus *Ptilopus* in which the breast-feathers are not bifurcate, and the middle of the abdomen is orange. It is nearly allied to *P. iozonus*, *P. humeralis* and *P. jobiensis*, but differs not only in the yellow knob on the forehead, but also by having the shoulders and patch on the back light grey, and the tail bright green, with the apical third ashy grey.

Nothing is known of the habits of this fine species; so I must content myself with adding the description of the bird given by Mr. Elliot in the paper above alluded to.

"Head, neck, back, breast, and flanks bright bronzy green; throat green slightly tinged with grey; abdomen deep orange; shoulders and a patch on each side of the back, at the edge of the mantle, light grey; wings green; secondaries margined with bright yellow on their outer webs; inner secondaries light grey, margined with green; tail bright green, with the apical third ashy grey; crissum and under tail-coverts bright yellow; bill greenish at base, yellowish at tip; forehead and base of culmen covered by a bony protuberance large and rounded in form, very conspicuous, and of a red colour; tarsi and feet red. Total length 9½ inches, wing 5, tail 3, bill at gape ¾."

The figures in the Plate are drawn from specimens collected by Mr. Brown and lent me by Dr Sclater. The birds are represented of the size of life.

PTILONOPUS NANUS, *Temm.*

PTILONOPUS NANUS.

Tiny Fruit-Pigeon.

Columba naina, Temm. Pl. Col. iv. pl. 252.—Knip and Temm. Iconogr. Pigeons, pl. 59.
Ptilonopus naina, Gray, Gen. B. ii. p. 467.
Ionotreron nana, Reich. Handb. Columbæ, p. 100, taf. ccxxxix. fig. 1330.
Iotreron nana, Bp. Consp. Gen. Av. ii. p. 25.
Ptilonopus nanus, Wall. Ibis, 1865, p. 381.—Gray, Hand-l. B. ii. p. 226.
Ptilopus nanus, Schl. Mus. P.-B. Columbæ, p. 21.

This is one of the rarest, as it is one of the most recognizable, of the small *Ptilonopi* or Many-coloured Fruit-Pigeons, a group of birds which finds its greatest development in the Malay archipelago and Oceania. The small size and peculiar coloration readily distinguish this species from all the other members of the genus *Ptilonopus*.

Nothing has been recorded respecting the habits of this elegant little Pigeon. It was first discovered by Salomon Müller in Triton Bay, or Lobo, in New Guinea; and for many years his specimen remained unique. More recently, however, the Dutch traveller M. Hoedt has discovered it in Mysol; and as one or two of his specimens reached England, I have been enabled to figure it in the present work. The Leiden Museum also possesses two examples of M. Hoedt's collecting, the localities being Kasim and Waigaama, both in the island of Mysol. They were obtained in the months of June and July respectively.

Adult male.—Bright grass-green above and below, all the greater wing-coverts and inner secondaries plainly edged with bright lemon-yellow, before which is a broad subterminal band of bright, rather metallic, bluish green; primaries greyish black on their inner web, dark green on the outer, with a narrow edging of yellow to the secondaries; tail deep green; on each side of the neck a broad crescentic mark of pale grey; across the top of the abdomen a broad band of purple feathers, with some metallic bluish-green subterminal bars to most of them, the abdominal plumes tipped with yellow; under tail-coverts bright yellow; thighs whitish; under wing-coverts dark grey like the inner lining of the wing, the outermost of the coverts greenish with a narrow yellow edging.

The female has no abdominal spot.

As above stated, the Plate has been drawn from examples in my own cabinet, the figures representing both sexes of the natural size.

PTILOPUS SOLOMONENSIS, Gray.

PTILOPUS SOLOMONENSIS, Gray.

Solomon-Island Fruit-Pigeon.

Ptilonopus solomonensis, Gray, Ann. & Mag. Nat. Hist. (4) v. p. 328 (1870).
Ptilopus solomonensis, Salvad. Ann. Mus. Civic. Genov. ix. p. 196 (1876).—Giebel, Thes. Orn. iii. p. 368.—Ramsay, Proc. Linn. Soc. N. S. Wales, iv. p. 74, note (1879).—Salvad. Orn. Papuasia e delle Molucche, iii. p. 50 (1882).
Ptilopus rivolii (pt.), Elliot, Proc. Zool. Soc. 1878, p. 561.—Salvad. Proc. Zool. Soc. 1879, p. 65.
Ptilopus ceraseipectus, Tristr. Ibis, 1879, p. 442.—Salvad. Ibis, 1880, p. 131.—Tristr. Ibis, 1880, p. 247.
Ptilopus salomonis, Salvad. Ibis, 1880, p. 131.

This species was at first described by the late Mr. George Robert Gray, from a specimen collected by Mr. Brenchley in the Solomon Islands. Unfortunately the typical specimen was a female, and it was considered by Count Salvadori and Mr. Elliot to be in all probability the hen bird of *Ptilopus rivolii*. Canon Tristram having described a new Fruit-Pigeon from the Solomon Islands as *P. ceraseipectus*, discovered in Makira Harbour by Captain Richards, Count Salvadori suggested that this might be the bird described by Gray as *P. solomonensis*, and on comparison of specimens Canon Tristram found that this was the case.

The only difference that we can perceive between the present species and *P. johannis* from the Admiralty Islands is the colour of the fore part of the head, which in the present species is rich purplish red, instead of being lilac-colour as in *P. johannis*. Count Salvadori also mentions that the breast-patches of the two birds also vary in an equal degree; but in the specimens examined by us this has not been so strongly pronounced as the variation in the colour of the heads of the two species.

The following is a description of an adult bird lent to us by Mr. E. P. Ramsay:—

Adult male. General colour above grass-green, the scapulars having subterminal spots of purplish black; wing-coverts like the back; bastard-wing, primary-coverts, and quills blackish, externally dark grass-green, the primaries washed with dull greenish grey, the inner primaries greenish at their ends, with a conspicuous subterminal shade of ashy grey; secondaries like the back, the outer ones with a narrow fringe of yellow; two centre tail-feathers green, the remainder green externally, grey at the base and near the end of the inner web, with a broad subterminal band of blackish; forehead rich purplish lilac, extending above each eye and on to the lores; ear-coverts, cheeks, throat, and fore neck green, paler on the chin and upper throat; on the chest a broad crescentic band of bright yellow; centre of the breast and abdomen purplish lilac; sides of the body and flanks green; thighs green externally, edged with yellow internally; lower abdomen, vent, and under tail-coverts rich yellow; axillaries and under wing-coverts slaty grey, washed with green; quills slaty grey below. Total length 8·5 inches, culmen 0·65, wing 4·75, tail 2·75, tarsus 0·85.

Two figures are given in the Plate, representing the adult male in two positions. They are drawn from the specimen lent to us by Mr. E. P. Ramsay.

[R. B. S.]

PTILOPUS RICHARDSI, Ramsay.

PTILOPUS RICHARDSI, *Ramsay*.

Richards's Fruit-Pigeon.

Ptilopus richardsii, Ramsay, Proc. Linn. Soc. N. S. W. vi. p. 722 (1881).—Salvad. Ann. Mus. Civic. Genov. xviii. p. 427 (1882).—Id. Orn. Papuasia, iii. App. p. 554 (1882).
Ptilopus rhodostictus, Tristr. Ibis, 1882, pp. 139, 144, pl. v.—Ramsay, tom. cit. p. 473.

This beautiful Fruit-Pigeon, which was discovered almost simultaneously by Mr. Morton and by Captain Richards in the island of Ugi in the Solomon group, is easily recognized by the rose-coloured spots on the scapulars, whence Canon Tristram's name of *P. rhodostictus*. Mr. Ramsay's title of *P. richardsi* has, however, undoubted priority; and no one is likely to regret that Captain Richards's name should have been attached to such a fine species, as it is only a just tribute to the energy with which he used his opportunities for increasing our knowledge of the avifauna of the Solomon Islands, an example, unfortunately, not too often followed by commanders of Her Majesty's ships.

The following account is taken from the original description given by Mr. E. P. Ramsay:—

"Morton found this species tolerably abundant on Ugi, but on no other island visited: he was also fortunate enough to find the nest and eggs; like that of all the genus, the nest is a frail scanty structure of a few twigs placed over a fork of a branch about twenty feet from the ground; the egg is oval, rather pointed at the thin end, pure white; length 1·22, in breadth 0·83. A second nest and egg, taken by Dr. J. H. Lewis, R.N., H.M.S. 'Cormorant,' are similar, but the thicker end of the egg is more rounded; both eggs were nearly hatched, and a bird shot from one of the nests proved to be a male; in some of the females eggs were found ready for laying. From a nestling obtained in June, I take the following description:—

"All the upper surface green, slightly tinged with bronze on the wings and tail, the wing-coverts, secondaries, and scapulars margined with yellow, the three or four smaller innermost secondaries (or tertiaries) having the tips and the whole of the inner web yellow; except at the base the primaries narrowly margined with yellow; tail above bronzy green, the tips of all the feathers ash washed with green and distinctly margined with yellow; the under surface is ashy grey, the tips lighter and margined with yellow; the under tail-coverts and abdomen yellow, the throat pale yellowish; all the rest of the under surface ashy, the tips of all the feathers margined with light yellow; forehead ashy; the first primary attenuated at the tip. Length 6·5, wing 4·8, tail 2·8, tarsus 0·8. Bill olive, feet reddish."

The following is a description of the type specimen, lent to us by Mr. Ramsay:—

Adult male (type of species). General colour above dark bronzy green, relieved by some beautiful oval spots of pale pink or rose-colour, which are subterminal on the scapulars; lesser and median wing-coverts dark emerald-green, edged with bronzy green like the back; greater coverts dark emerald-green, the inner ones margined with bronzy green; bastard wing, primary-coverts, and quills dark emerald-green externally, ashy black on the inner webs, the primaries obsoletely fringed with yellow near the ends, a little more distinct on the secondaries, the innermost of which are edged with yellow on the inner web and have a broad longitudinal mark of pale rose-colour; upper tail-coverts rather more golden green than the back; tail-feathers dark emerald-green, dusky blackish on the inner web and broadly tipped with yellow, forming a broad band; crown of head, lores, and base of cheeks pale pearly grey, separated from the eye and surrounded posteriorly by a somewhat indistinct line of pale yellow; feathers round the eye, nape, hind neck, and upper mantle, as well as the ear-coverts, pale yellowish mixed with light pearly grey, the sides of the neck similarly marked; throat clear pale yellow as well as the cheeks; breast light pearly ash-colour, with greenish-yellow bases to the feathers, which are bifid; breast a little duller yellowish, like the sides of the body and flanks, the long feathers covering the thighs tipped with orange; thighs grey marked with green; centre of lower breast, abdomen, and under tail-coverts bright orange; axillaries and under wing-coverts light grey, marked with greenish yellow; quills light grey below. Total length 9·5 inches, culmen 0·6, wing 5·25, tail 3·0, tarsus 0·8.

The figures in the Plate represent the bird in two positions of the natural size; they are drawn from the bird described by us above, which belongs to the Australian Museum, Sydney.

[R. B. S.]

PTILOPUS LEWISII, Ramsay.

PTILOPUS LEWISI, Ramsay.

Lewis's Fruit-Pigeon.

Ptilopus viridis, var., Ramsay, Proc. Linn. Soc. N. S. W. iv. p. 73 (1879).—Salvad. Ibis, 1880, p. 128.—Ramsay, Nature, 1881, p. 239.—Tristr. Ibis, 1882, p. 144.—Ramsay, Ibis, 1882, p. 473.
Ptilopus geelvinkianus, Layard, Ibis, 1880, p. 307 (nec Schl.).
Ptilopus eugeniæ, female, Ramsay, Journ. Linn. Soc., Zool. xvi. p. 131 (1881).
Ptilopus lewisii, Ramsay, Proc. Linn. Soc. N. S. W. vi. p. 724 (1881).—Salvad. Ann. Mus. Civic. Genov. xviii. p. 427 (1882).—Id. Orn. Papuasia, etc. iii. App. p. 556 (1882).

ALTHOUGH closely allied to *P. viridis* of Ceram, the present species is easily distinguished by the purplish shade which surrounds the red shield-patch on the throat and chest; it is further to be recognized by the verditer-green shade on the grey forehead and throat, these parts being pure grey in *P. viridis*.

The home of this beautiful Fruit-Pigeon is in the Solomon Archipelago, where it has been found in the islands of Lango and Guadalcanar by Mr. Cockerell, and in Florida and Malayta by Mr. Morton. Nothing has as yet been recorded concerning its habits.

For the opportunity of figuring the species we are indebted to Mr. E. P. Ramsay, who lent us the typical example of *P. lewisi* during his visit to England.

The following is a description of the type specimen:—

Adult. General colour above dark golden green, the nape and hind neck deep grass-green, contrasting with the fore part of the crown, which is verditer-grey on the forehead, verging into greyish green on the sinciput; wing-coverts like the back, the lesser series pearly grey, forming a conspicuous shoulder-patch; bastard wing, primary-coverts, and quills blackish, externally deep emerald-green, the outer secondaries narrowly fringed with yellow towards the ends, the inner secondaries like the back, with a conspicuous subterminal spot of pearly grey; tail-feathers golden green, dusky on the inner webs, with a subterminal band of grey near the end of the inner web, forming a distinct bar on the under surface of the tail; lores, feathers round the eye, ear-coverts, cheeks, sides of face, and upper throat verditer-grey, greener and resembling the sides of the neck on the hinder part of the ear-coverts; centre of the throat, fore neck, and chest deep crimson or blood-colour, separated from the surrounding green parts by a narrow line of dull purple; remainder of underparts grass-green, the feathers of the lower abdomen and lateral tail-coverts tipped with yellow; vent-feathers white, tipped with yellow; thighs green, edged with yellow; under tail-coverts white, longitudinally green along the inner web, and tipped with yellow; under wing-coverts and axillaries dark grey washed with green; quills ashy grey below. Total length 7·8 inches, culmen 0·65, wing 4·6, tail 2·3, tarsus 0·65.

The Plate represents an adult bird in two positions, of about the natural size; the figures are drawn from the type specimen lent to us by Mr. Ramsay.

[R. B. S.]

PTILOPUS WALLACII.

PTILOPUS WALLACEI.

Wallace's Fruit-Pigeon.

Ptilonopus wallacei, Gray, Proc. Zool. Soc. 1858, pp. 185, 195, pl. 136.—Id. Cat. B. New Guinea, pp. 45, 60 (1859).—Id. Proc. Zool. Soc. 1861, p. 437.—Reichenb. Columbariæ, ii. p. 178 (1862).—Rosenb. N. T. Nederl. Ind. xxv. p. 248 (1863).—Wallace, Ibis, 1865, p. 380.—Finsch, Neu-Guinea, p. 177 (1865).—Pelz. Verh. zool.-bot. Gesellsch. Wien, xxii. p. 430 (1872).
Philopus wallacei, Rosenb. Reis. naar Zuidostereil, p. 50 (1867).
Ptilinopus wallacei, Gray, Hand-l. B. ii. p. 227, no. 9154 (1870).
Ptilopus wallacei, Rosenb. Reis. naar Zuidostereil, p. 81 (1867).—Schlegel, Mus. Pays-Bas, Columbæ, p. 18 (1873).—Salvad. Ann. Mus. Civic. Genov. ix. p. 197 (1876).—Elliot, Proc. Zool. Soc. 1878, p. 555.—Salvad. Ann. Mus. Civic. Genov. xiv. p. 658 (1879).—Id. Orn. Papuasia, etc. iii. p. 30 (1881).

This beautiful Pigeon was discovered by Mr. A. R. Wallace in the Aru Islands, where it has also been met with by Dr. Beccari and Baron Von Rosenberg. The last-named naturalist has also observed it on the Ké Islands, and more recently a fine series has been brought from the Tenimber Islands by Mr. H. O. Forbes.

We have carefully compared Mr. Forbes's specimens with the typical Aru-Islands bird and cannot find the slightest difference between them.

This beautiful Pigeon, Mr. Forbes informs us, is very common in the northern portions of the Tenimber group which he visited, frequenting the fig-trees (*Urostigma*). Its nature is very tame, and numbers of specimens could be easily procured. The native name is " Wofoen Ratoe."

We have described a fine pair of birds collected in the above-named locality by Mr. Forbes.

Adult male. Crown of head, tapering somewhat to a point on the nape, deep crimson, as well as the lores; feathers behind the eye and ear-coverts, sides of neck, hind neck, and mantle delicate French grey; upper back and scapulars mottled, the feathers being delicate French grey, with orange margins; lesser wing-coverts dull orange, forming a shoulder-patch; median coverts French grey, with broad edgings of yellowish olive; greater series green, inclining to olive-yellow towards the ends and narrowly margined with yellow; bastard wing, primary-coverts, and quills deep grass-green with a bronzy gloss, the inner webs being blue-black; secondaries more olive than the primaries and edged with yellow, the innermost secondaries olive yellowish with grey centres like the greater wing-coverts; remainder of back yellowish green, with more or less of an orange tinge, especially on the upper back, where there is a patch of dull orange; upper tail-coverts and tail yellowish green, dark green towards the base, the terminal half pale greenish ashy, edged externally with yellow and becoming whitish on the inner web; the ends of the feathers dusky greenish, broader on the centre ones; cheeks, throat, and ear-coverts white, with a faint grey shade on the throat, the upper part of the ear-coverts rather dusky; fore neck and chest pale French grey like the sides of the neck, and forming a band; breast deep rich orange, separated from the grey chest by a rather broad band of white slightly sullied with grey; abdomen bright yellow; sides of breast, axillaries, and under wing-coverts French grey; lower flanks olive-green; thighs grey; under tail-coverts yellow, edged broadly with green; quills grey, becoming dusky towards their ends; " bill yellow, but paler on the lower mandible; legs and feet purple; iris with golden inner ring and outer one of light red" (*H. O. Forbes*). Total length 9 inches, culmen 0·8, wing 5·9, tail 3·2, tarsus 0·95.

The female is like the male, but a trifle duller in colour.

The figures in the Plate represent a pair of birds, of the size of life; they have been drawn from Mr. Forbes's specimens above described.

[R. B. S.]

PTILOPUS FISCHERI, Brüggem.

PTILOPUS FISCHERI, Brüggem.

Fischer's Fruit-Pigeon.

Ptilinopus fischeri, Brüggem. Abhandl. nat. Ver. Bremen, v. p. 82, Taf. iv.
Ptilopus fischeri, Elliot, Proc. Zool. Soc. 1878, p. 571; Meyer, Ibis, 1879, p. 135.

DR. FISCHER is principally known to science as the discoverer of the beautiful new *Polyplectron schleiermacheri* from Borneo; but previously to his visiting that country he forwarded to the Darmstadt Museum a series of birds from the island of Celebes, amongst which were some new and interesting species. This collection was described by the late Dr. Brüggemann; and although many of the novelties made known by the latter gentleman were forestalled by Count Salvadori in a previously published paper, there were yet a few which remained and will remain to the credit of Dr. Brüggemann, and will serve to perpetuate the memory of this young and talented naturalist, whose early death was a veritable loss to science.

The *Ptilopus fischeri* is, as Mr. Elliot well remarks, so distinct a species that it cannot well be mistaken for any other member of the genus *Ptilopus*. In its pale-coloured head it approaches the white-headed group which embraces *Pt. cinctus* of Timor, as pointed out by Dr. Meyer; but its reddish side-face at once distinguishes it, to say nothing of many other points of difference.

It is a native of Celebes; and Dr. Meyer states that the Leiden Museum has recently obtained a series from the southern part of the island, so that it is in this locality that the species must be looked for. The birds of Celebes, are many of them so local in their distribution that it is quite possible that it is confined to Southern Celebes alone.

Dr. Meyer has been so good as to lend me a specimen for the purposes of the present work; and I append a description of the bird, as follows:—

Above dark slaty grey, including the lesser and median wing-coverts, the greater series brownish towards their tips; quills slaty blackish, the primaries narrowly margined with white towards the end of the outer webs, the secondaries margined externally with yellow; tail-feathers green, shot with bronzy green, black on the inner webs of all but the two centre feathers; at the end of the tail a grey band, except on the two centre feathers; crown whitish grey; round the hind neck a ring of slaty black; feathers in front of and round the eye, as well as the ear-coverts, dark crimson; cheeks and throat white; rest of under surface of body cream-colour washed with grey, the thighs and flanks more decidedly ashy; under tail-coverts creamy buff, mottled with broad slaty grey stripes, bordering the shaft on the inner webs of the feathers.

The figures in the Plate are about the size of life, and are drawn from a specimen lent to me by my kind friend Dr. A. B. Meyer, of Dresden.

PTILOPUS SPECIOSUS, Schlegel.

PTILOPUS SPECIOSUS, *Schlegel.*

Lilac-bellied Fruit-Pigeon.

Ptilopus speciosus, Schleg. Nederl. Tijdsch. iv. p. 23 (1871).—Id. Mus. P. B. *Columbæ*, 1873, p. 27.—Salvad. Ann. Mus. Civ. Genov. ix. p. 197 (1876).—Elliot, P. Z. S. 1878, p. 564.

THIS species was described by Professor Schlegel from specimens forwarded by Baron von Rosenberg under the MS. name of *Ptilopus speciosus*, a title adopted by the learned Professor. He writes:—"This undescribed species belongs to the most beautiful of the genus. It has been discovered, as we have already mentioned in our note on *P. miquelii*, in the island of Mefoor (Mafoor), where it lives side by side with *P. rivolii*, as well as in the island of Soek (Misori), where it represents *P. rivolii* and *P. miquelii*, which come from Meosnoum and Jobi." I may mention that the birds here mentioned as *P. miquelii* and *P. rivolii* are identified by Mr. Elliot as *P. prasinorrhous* and *P. strophium* respectively. The island of Soek is another name for Misori—the latter name being better known through the discoveries of Dr. Meyer and Dr. Beccari. A large series has also been recently received at the Paris Museum from the islands of Mafoor and Misori, collected by M. Raffray; and Mr. Elliot states that individuals from these localities do not differ.

The sexes are thus described by Mr. Elliot:—

Male. General plumage yellowish green, darkest on the head and neck. A broad bright yellow band, bordered on the lower side with white, crosses the breast. Abdomen beautiful lilac; lower part of abdomen and under tail-coverts bright lemon-yellow. A purple spot in front of the eyes. Primaries grey on their outer webs and tips; first not narrowed especially towards tip. Tail yellowish green. Bill black, tip yellow. Feet dark red. Total length 7 inches, wing 4⅝, tail 2⅝, culmen ⅝.

Female. Green, with the feathers of the abdomen margined yellow, and under tail-coverts yellow. Primaries have outer webs and tips grey.

The two figures on the Plate are of the natural size, and represent the adult male and female, being drawn from specimens in my own collection, the female having been presented to me by Dr. Meyer.

PTILOPUS BELLUS, Sclater.

PTILOPUS BELLUS, Sclater.

Purple-bellied Fruit-Pigeon.

Ptilonopus bellus, Sclater, P. Z. S. 1873, p. 696, pl. 57.—Salvad. Ann. Mus. Civ. Genov. vii. p. 786 (1875).
Ptilopus bellus, Salvad. Ann. Mus. Civ. Genov. ix. p. 197 (1876), x. p. 157 (1877).—Elliot, P. Z. S. 1878, p. 563.

The elaborate monograph recently compiled by Mr. D. G. Elliot on the Fruit-Pigeons of the genus *Ptilopus* renders a study of this numerous group a much easier task than it has been for many years past; and I must acknowledge the help which I have derived from that treatise of the above-named author, whose industry and devotion to ornithological science have been proved by the many valuable papers and works which have issued from his pen during the last few years.

The subject of the accompanying Plate is one of the most beautiful of all the *Ptilopi*, which contain such a number of strikingly marked Fruit-Pigeons, many of them endeared to me by old Australian recollections and by the pleasure with which I look forward to illustrating all the varied forms of Fruit-Pigeons inhabiting the Malay archipelago. The *Ptilopus bellus* was first discovered by Signor D'Albertis during his celebrated expedition to North-western New Guinea, in Atam; and it has since been procured in the Arfak Mountains by M. Laglaize, and at Amberbaki by M. Raffray; so that the habitat of the present species would appear to be the north-western corner of New Guinea, particularly the vicinity of the Arfak Mountains. It belongs to the section of the genus *Ptilopus* which contains species having a broad white or yellow breast-band: in the case of the present bird and its near ally *P. speciosus*, the breast-band is lemon-yellow above, white beneath; and it is distinguished from the latter species by the forehead and crown being rosy red instead of green, and by the abdominal spot being purplish red instead of lilac. It is also a larger bird than the nearly allied species.

Mr. Elliot, from whose synoptic table the above characters have been derived, gives the following description of the species:—

"*Male.* Front and crown deep rosy red; occiput dark bluish green; breast covered by a broad half-moon-shaped band, pure white on the lower parts and sides, lemon-yellow on the upper part; the abdomen has the middle portion covered by a purplish red patch; entire rest of plumage yellowish green, with small, round, bluish-black spots on the scapulars, and the feathers of the crissum and under tail-coverts edged with yellow; primaries greenish black, the first not narrowed; tail yellowish green, with a pale apical band of the same colour; bill yellow; feet dark red; iris yellow. Total length $9\frac{1}{4}$ inches, wings $3\frac{3}{4}$, tail $3\frac{4}{4}$, culmen $\frac{4}{8}$."

Signor D'Albertis describes the bill as yellow, the feet as dull crimson, and the eyes as yellow.

The Plate represents the type specimen, kindly lent to me by Dr. Sclater during its stay in this country. It is now in the Genoa Museum.

PTILOPUS RIVOLII.

PTILOPUS RIVOLII.

Massena Fruit-Pigeon.

Columba rivoli, Knip & Prév. *Pigeons*, ii. pl. 57 ; Des Murs, Iconogr. Ornith. pl. 4 (1845).
Iotreron rivollii, Bp. Consp. Gen. Av. ii. p. 25 (1857).
Ionotreron rivollii, Reichenb. Taub. p. 100, Taf. 235. fig. 1306.
Ptilonopus rivoli, Wall. Ibis, 1865, p. 381.
Ptilonopus solomonensis, Gray, Ann. & Mag. Nat. Hist. v. p. 328 (1870).
Ptilopus rivolii, Sclater, P. Z. S. 1877, p. 109.—Elliot, P. Z. S. 1878, p. 561.—Salvad. Ann. Mus. Civ. Genov. ix. p. 196 (1877); id. xii. p. 345 (1878).

CONSIDERABLE uncertainty has existed with regard to the exact locality where this Fruit-Pigeon was to be found ; but one may now accept as a fact that it inhabits Duke-of-York Island, whence specimens have been forwarded by Mr. Brown to Dr. Sclater; and future research will doubtless extend its range to New Ireland. Mr. Elliot also believes that it inhabits the Solomon Islands, as he considers that the *P. solomonensis* described by the late Mr. G. R. Gray is only the young of *P. rivolii*.

The present bird belongs to the white-banded section of the genus *Ptilopus*, which includes also *P. prasinorrhous* of New Guinea and the neighbouring groups of islands, and *P. strophium* of the Louisiade archipelago. The latter bird, however, has no rose-coloured spot on the abdomen, and is therefore easily distinguished. *P. prasinorrhous* has the vent and under tail-coverts green, edged with yellow, whereas in *P. rivolii* these parts are bright yellow.

I transcribe the following detailed description given by Mr. Elliot in his paper on the genus :—

" *Male*. Forehead purplish red ; breast crossed by a broad white band; a large rose-red spot on the abdomen ; lower part of abdomen, crisssum, and under tail-coverts bright yellow ; rest of plumage bright green, with some small, round, bluish-black spots on the scapulars. In its dimensions this species agrees with *P. prasinorrhous*.

" *Female*. General plumage bright green."

I am indebted to Dr. Sclater for the loan of the beautiful pair of birds of which I have drawn life-sized figures in the accompanying Plate. They were collected in Duke-of-York Island by Mr. G. Brown; and I have other specimens from the same source in my own collection.

GYMNOPHAPS PŒCILORRHOA.

GYMNOPHAPS PŒCILORRHOA.

Rusty-banded Fruit-Pigeon.

Carpophaga pœcilorrhoa, Brüggem. Abhandl. nat. Vereine zu Bremen, v. p. 84.

THE genus *Gymnophaps* was instituted in 1874 by Count Salvadori; and the type is *Gymnophaps albertisi*, a fine Fruit-Pigeon discovered at Andei in North-western New Guinea by Signor D'Albertis. Up to the time of writing I have not seen that species; and I am unable to say whether the bird which I figure on the accompanying Plate is really congeneric with the Papuan *Gymnophaps*. My friend Dr. Meyer, who lends me the specimen which I figure, has attached to it the name of *Gymnophaps pœcilorrhoa* (Brüggem.); and on comparing Count Salvadori's description of the genus, which is remarkable for its bare loral patch extending round the eye, I should say that Dr. Meyer is quite right in placing the species in the genus *Gymnophaps*.

The present bird is remarkable for its very plain coloration, differing in this respect from the majority of the subfamily Carpophaginæ, many of which are of beautifully varied plumage. But though *Gymnophaps pœcilorrhoa* is a somewhat dull-coloured bird, the monotony of its plumage is slightly relieved by the bands on the abdomen and the dark-centred under tail-coverts.

Nothing has yet been recorded of the habits or economy of this strange Pigeon; nor do we know the exact part of the island of Celebes from which it comes. It is probably from the south-eastern portion of the island, where Dr. Fischer collected, as neither Mr. Wallace nor Dr. Meyer nor any of the old Dutch travellers appear to have come across the species in any of the places visited by them.

I append a full description of the species:—

General colour above dusky brown shot with olive-green; quills brown, with narrow reddish margins to the outer web, as well as the greater wing-coverts; tail blackish, tipped with a narrow band of buffy white; head dusky grey, somewhat more vinous on the hind neck, the forehead lighter grey; cheeks, ear-coverts, and throat dull vinous; lower throat darker ashy, as also the adjacent sides of the neck; the chest light ashy, extending onto the sides of the neck, and forming a conspicuous patch on the sides of the mantle; remainder of under surface brown, mottled with reddish-ochre margins to the feathers; under tail-coverts dark brown, with broad edgings of reddish ochre; under wing-coverts dusky brown, the axillaries rufous brown; undersurface of quills dark ashy brown.

The specimen figured in the accompanying Plate is in the Dresden Museum, and was lent to me by Dr. Meyer for the purposes of the present work. The species is represented about the size of life.

SPHENISCUS CRISTATUS, *Bonap.*

OTIDIPHAPS CERVICALIS, Ramsay.

Grey-naped Otidiphaps.

Otidiphaps nobilis, var. *cervicalis*, Ramsay, Proc. Linn. Soc. New S. Wales, iv. p. 420.—Sclater & Salvin, Ibis, 1881, p. 179.

Otidiphaps cervicalis, Ramsay, loc. cit. *errata*.

Otidiphaps regalis, Salvin & Godman, Ibis, 1880, p. 364, pl. xi.

This beautiful species of Ground-pigeon represents in South-eastern New Guinea the *Otidiphaps nobilis* of the north-western part of the island. It differs in the green colour of the rump and upper tail-coverts, in the absence of an occipital crest, and more especially in having the conspicuous grey neck-band from which the species derives its specific name. Mr. Ramsay described the species from specimens obtained by Mr. A. Goldie on the Goldie river in the interior of South-eastern New Guinea. From the same collector Messrs. Salvin and Godman received the specimens from which they drew up their description of *O. regalis*; and it seems a great pity that some notice was not given before the despatch of the specimens to England to the effect that they had already been deposited with Mr. Ramsay for the purpose of description. In this way science would have been saved the unnecessary synonymy consequent upon the simultaneous description of these novelties from South-eastern New Guinea by naturalists in England and Australia. Great difficulty is caused, moreover, in deciding as to which name should take precedence; for in the present instance the name of *Otidiphaps regalis* was published in July 1880, and yet in January 1881 the editors of 'The Ibis' had been unable to find in this country a single copy of Part iv. of the 'Proceedings of the Linnean Society of New South Wales,' although Mr. Ramsay's paper containing the description of his *O. cervicalis* was read before that Society on the 31st of December, 1879. It is therefore extremely probable that Messrs. Salvin and Godman actually published their description first.

Mr. Goldie informed Mr. Ramsay that these Pigeons were obtained by him only with great difficulty, in the dense scrubs far inland; they were always on the ground, and in habits resembled the *Gouræ*. In the notes which have accompanied a recent collection of Mr. Goldie's from the Astrolabe range in South-eastern New Guinea, in which he sends several fine specimens of both sexes, we find the following observations on the *Otidiphaps*. "Native name *Keo*. Eyes red. This ground-bird is found only inland, and in high country. It has a long plaintive note when calling, which, when imitated, brings it toward one; and it then stalks to and fro with tail erect and spread, challenging the intruder. When disturbed he will fly into low trees and bushes, but is quickly away again. The nest is composed of a few twigs scraped together at the foot of a low tree in a sequestered place." The egg, which is also forwarded by Mr. Goldie, is, as might be expected, pure white.

The figure in the Plate represents the bird about the natural size, and is drawn from a specimen in my own collection.

[R. B. S.]

EUTRYGON TERRESTRIS.

EUTRYGON TERRESTRIS.

Papuan Ground-Pigeon.

Trugon terrestre, Hombr. & Jacq. Voy. Pôle Sud, Atlas, Oiseaux, pl. 28. fig. 1 (1846).
Trugon terrestris, Gray, Gen. B., App. p. 24 (1849).—Pucheran & Jacq. Voy. Pôle Sud, Zool. iii. p. 123 (1853).—Bp. Consp. ii. p. 86 (1854).—Id. Comptes Rend. xl. pp. 206, 221.—Gray, Proc. Z. S. 1858, p. 196.—Id. Cat. B. New Guinea, p. 48 (1859).—Id. Proc. Zool. Soc. 1861, p. 437.—Rosenb. Journ. für Orn.1861, p. 133.—Reichenb. Columbariæ, p. 45, sp. 97 (1862).—Finsch, Neu-Guinea, p. 179 (1865).—Wallace, Ibis, 1865, pp. 369, 392.—Id. Malay Archip. ii. p. 430 (1869).—Gray, Hand-l. B. ii. p. 245 (1870).
Trygon terrestris, Reichenb. Av. Syst. Nat. p. xxvi (1852).—Hartl. Journ. für Orn. 1854, p. 166.—Sundev. Meth. Nat. Av. Tent. p. 100 (1872).—Beccari, Ann. Mus. Civ. Genova, vii. p. 715 (1875).
Starnænas terrestris, Bp. Consp. ii. p. 86 (1854).
Eutrygon terrestris, Sclater, Proc. Linn. Soc. ii. p. 168 (1858).—Id. P. Z. S. 1873, p. 697.—Salvad. Ann. Mus. Civ. Gen. vii. p. 791 (1875); ix. p. 207 (1876); x. p. 161 (1877); D'Alb. & Salvad. Ann. Mus. Civ. Gen. xiv. p. 124 (1879).—D'Alb. Nuova Guin. pp. 459, 528, 582, 588 (1880).—Salvad. Orn. Papuasia &c. iii. p. 182.
Starnænas terrestris, Schleg. Mus. P.-B., Columbæ, p. 166 (1873).—Rosenb. Malay Archip. p. 396 (1879).
Phaps terrestris, Gieb. Thes. Orn. ii. p. 151 (1875).

The above synonymy is taken from some sheets of the 'Ornitologia della Papuasia,' which my friend Count Salvadori was good enough to send me, on hearing that I was at work on the Pigeons of New Guinea; and I am much indebted to him for the assistance he has always given me in the production of the present work. His book contains a summary of all that is known of this Ground-Pigeon, which is really very little. It was discovered by the French voyagers Hombron and Jacquinot in Western New Guinea; Mr. Wallace also procured it on the western side of the island. In the Arfak Mountains the species was met with by Dr. Beccari and Signor D'Albertis, at Andai and Warbusi; while Mr. Bruijn's hunters procured it at Dorey and also on the island of Salwatti. In the southern part of the island it was obtained by Signor D'Albertis on the Fly river; and we have seen several specimens from the interior of South-eastern New Guinea obtained by Mr. Goldie, in whose last collection from the back of the Astrolabe range were a good many individuals.

The original discoverers state that it is a Ground-Pigeon; but they do not give any further particulars as to its habits.

The figures in the accompanying Plate have been drawn from specimens in my own collection, and will give some idea of this fine bird. I add a translation of the description given by Count Salvadori in the work above referred to.

Head, neck, upper part of back, and breast ashy grey, the sinciput and throat paler; forehead and chin dusky; cheeks greyish; sides of neck with an obsolete dusky spot on each; middle of back, rump, upper tail-coverts, wing, and tail shining greyish olive; middle of the abdomen pale isabelline, the sides and under tail-coverts rufescent; primary quills dusky, their outer margin, as well as that of the secondaries, rufescent towards the tip; under wing-coverts dusky, partly isabelline; bill whitish, feet pale flesh-coloured; iris whitish (*D'Albertis*) or red (*Wallace*). Signor D'Albertis also procured a bird on the Fly river which had the eye bright red.

The figures in the Plate are about the size of life.

[R. B. S.]

CARPOPHAGA RUBRICERA, Gray.

CARPOPHAGA RUBRICERA, *Bonap.*

New-Ireland Fruit-Pigeon.

Columba pinon, Less. Voy. Coquille, Zool. i. p. 342 (1826, nec Quoy et Gaim.).
Carpophaga rubricera, Bp. Consp. ii. p. 31 (1864, ex Gray, MSS.).—Gray, List Columb. in Brit. Mus. p. 18 (1856).—Wallace, Ibis, 1865, p. 383.—Gray, Hand-l. B. ii. p. 229, no. 9177 (1870).—Schl. Mus. Pays-Bas, Columbæ, p. 81 (1873).—Salvad. Ann. Mus. Civic. Genov. ix. p. 200 (1876, pt.).—Sclater, Proc. Zool. Soc. 1877, p. 109 (pt.), 1878, pp. 289, 671.—Elliot, op. cit. 1878, p. 549.—Salvad. Monog. *Globicera*, p. 13 (1878).—Finsch, Proc. Zool. Soc. 1879, p. 13.—Salvad. Ibis, 1879, p. 364.—Brown, Proc. Zool. Soc. 1879, p. 451.—Salvad. Orn. Papuasia e delle Molucche, iii. p. 79 (1882).—Finsch, Vög. der Südsee, p. 18 (1884).
Globicera rubricera, Bp. Consp. ii. p. 31 (1854).—Id. Compt. Rend. xxxix. p. 1073 (1854), xl. p. 217 (1855), xli. p. 1111 (1855), xliii. p. 835 (1856).—Id. Rev. et Mag. de Zool. 1856, p. 403.—Reichenb. Handb. Columb. p. 121.—Bonap. Iconogr. Pigeons, pl. 39 (1857).
Carpophaga lepida, Cassin, Journ. Philad. Acad. 1854, p. 330.—Bonap. Rev. et Mag. de Zool. 1856, p. 403.—Id. Compt. Rend. xli. p. 1111 (1855), xliii. p. 835 (1856).
Carpophaga (Globicera) rubricera, Gray, Cat. Birds Tropical Isl. p. 41 (1859).
Carpophaga (Zonœnas) pinon, Gray, tom. cit. p. 42 (1859, nec Quoy et Gaim.).
Muscadivora rubricera, Schleg. Dierentuin, fig. 1, p. 209.
Carpophaga (Globicera) rubricera, Cab. & Reichen. J. f. O. 1876, p. 325.

A SUCCINCT history of this species has been given by Count Salvadori in his great work on the birds of New Guinea and the Moluccas, from which we have taken the major part of the above synonymy. It appears to be entirely confined to New Ireland, New Hanover, and New Britain, having been found in the two first-mentioned islands by Dr. Huesker during the German Transit-of-Venus Expedition, while Dr. Finsch states that it was the commonest of the large Pigeons in New Britain. All the references to the occurrence of this bird in the Solomon group are probably erroneous, as it is represented in the latter locality by *Carpophaga rufigula* of Salvadori, which has the head and neck ashy, with the cheeks and throat vinaceous. In *C. rubricera* the head and neck are vinaceous, the lower part of the hind neck being ashy.

The following is a translation of Count Salvadori's description, taken from the type in the British Museum :—

"Head, neck, and breast vinaceous ; a ring round the eye and the margin of the forehead whitish ; lower part of the hind neck and upper part of the back pale ashy grey ; remainder of the back and wings shining coppery green ; abdomen, anal region, and thighs rusty ; under tail-coverts chestnut ; primaries and tail-feathers blue-black with a green reflection ; the cere, which is swollen, and the feet red."

The figure in the Plate is drawn from an adult specimen collected by Mr. Cockerell in New Ireland, and now in the British Museum. It is of the size of life.

[R. B. S.]

CARPOPHAGA VAN-WYCKII, Cass.

CARPOPHAGA VAN-WYCKII, Cass.

Van Wyck's Fruit-Pigeon.

Carpophaga van-wyckii, Cass. Proc. Acad. Philad. 1862, p. 320.—Pelz. Novara Reis., Vög. p. 107 (1865).—Salvad. Ann. Mus. Civ. Genov. ix. p. 200, no. 41 (1876).—Sclater, Proc. Zool. Soc. 1877, p. 109; 1878, pp. 289, 671.—Salvad. Monogr. Sottogen. *Globicera*, p. 5 (1878).—Ramsay, Proc. Linn. Soc. N. S. W. iii. p. 292 (1878), iv. pp. 73, 101 (1879).—Salvad. Ibis, 1879, p. 326.—Sclater, Proc. Zool. Soc. 1879, pp. 218, 447, 451.—Layard, Ibis, 1880, pp. 297, 301.—Salvad. Orn. Papuasia e delle Molucche, iii. p. 87 (1882).
Globicera vanwyckii, Gray, Hand-list of Birds, ii. p. 229, sp. 9184 (1870).
Carpophaga wickei, Giebel, Thes. Orn. i. p. 588 (1872).
Carpophaga microcera, Ramsay, Proc. Linn. Soc. N. S. W. i. p. 372 (1876).
Carpophaga rhodinolæma, Finsch (nec Sclater), Proc. Zool. Soc. 1879, p. 13.

The greatest uncertainty has long prevailed regarding the distinctness of this species from *C. pistrinaria* of the Solomon Islands. *C. van-wyckii* was discovered by Lieut. Van Wyck, of the United States Navy, in New Ireland, and specimens have more recently been obtained in Duke of York Island by the Rev. G. Brown and Dr. Kleinschmidt, as well as in Pigeon Island, New Britain, and Palakura, by Mr. L. C. Layard. Mr. E. P. Ramsay also records it from Deboyne Island and Bramble Haven in the Louisiade Group and even from South Cape in South-eastern New Guinea. It will be seen, therefore, that its range is somewhat extensive, and there would be nothing surprising in the fact that the Solomon Island *Carpophaga* was precisely identical; and that this is the case has been suggested by Count Salvadori. We have ourselves compared a specimen from New Britain with one from San Christoval, and we find that, although the resemblance is close, there are sufficient characters to recognize *C. van-wyckii* from *C. pistrinaria*. The difference consists in the dusky grey character of the upper plumage in *C. pistrinaria*, which has little or no green gloss, while the hind neck and mantle are scarcely distinguishable from the rest of the back. In *C. van-wyckii* the vinous grey hind neck and mantle contrast strongly with the rest of the back, which has a very distinct gloss of bronzy green.

The following is a description of an adult female, taken from a New Britain specimen in the British Museum :—

Adult female. General colour above pale bronzy green, with a slight shade of purplish blue on the rump and upper tail-coverts; lesser wing-coverts pale bronzy green; primary-coverts and quills blackish, the primaries ashy grey externally, glossy green at the ends; secondaries also externally grey, but also glossed with green, the innermost like the back; tail-feathers black, with a purple gloss, green on the outer edges; crown of head delicate pearly grey; hind neck and mantle pearly grey with a strong vinaceous tinge, especially on the sides of the neck, the grey of the mantle merging into the green of the back; base of forehead dull white; lores pearly grey, fading into pale vinous; feathers round eye white; sides of face, ear-coverts, cheeks, throat, sides of neck, fore neck, and chest pale delicate vinaceous; breast and remainder of under surface pale pearly grey, with a vinaceous tinge; under tail-coverts vinous chestnut; under wing-coverts and axillaries delicate pearly grey like the sides of the body; quills below ashy brown, paler along the inner edge. Total length 14·5 inches, culmen 1·15, wing 8·9, tail 5·3, tarsus 1·0.

The figure in the Plate is life-sized, and is drawn from the specimen described above.

[R. B. S.]

CARPOPHAGA FINSCHI, Ramsay.

CARPOPHAGA FINSCHI, Ramsay.

Finsch's Fruit-Pigeon.

Carpophaga finschii, Ramsay, Journ. Linn. Soc. xvi. p. 129 (1881).—Tristram, Ibis, 1882, p. 144.—Ramsay, t. c. p. 478.—Salvad. Ann. Mus. Civic. Genov. xviii. p. 428 (1882).—Id. Orn. Papuasia, etc. iii. App. p. 558 (1882).

As Count Salvadori has surmised, the nearest ally of the present species is *Carpophaga rufiventris* (*C. rufigaster*, auctt.), but the differences are numerous and striking; they are as follows:—

1. The grey band on the tail is subterminal, not terminal, the tips of the feathers being green.
2. The basal portion of the tail is deep blue.
3. The lower back, rump, and upper tail-coverts are green with golden reflexions, not purplish red as in *C. rufiventris*.
4. The head is grey, as well as the hind neck and upper mantle.
5. The ashy pink of the throat occupies also the fore neck and chest, whereas in *C. rufiventris* the brick-red colour of the underparts commences at the fore neck.
6. The rufous colour of the breast is continued equally over the abdomen, and is deepest on the under tail-coverts in *C. rufiventris*; the lower abdomen, vent, and under tail-coverts are quite pale.

It will at once be seen that a Pigeon differing in so many characters from its nearest ally must be an easily recognizable species, and we have not met with any bird with which it could be confounded.

The type specimen is a male, procured by the Rev. G. Brown in Irish Cove, New Ireland; it has been lent to us by our friend Mr. Ramsay for the purposes of the present work.

The following is a detailed description:—

Adult male. General colour above dark grass-green from the middle of the back to the upper tail-coverts, many of the feathers coppery or golden green, the lower mantle and upper back reddish or coppery; wing-coverts green with an emerald appearance, the lesser series washed with coppery like the middle of the back; bastard wing, primary-coverts, and quills blackish, externally marked with deep indigo, which is glossed with bronzy green, especially on the secondaries; tail-feathers deep blue for their basal half, succeeded by a broad band of ashy grey, which is again succeeded by a narrow bar of black, leaving a broad terminal band of green; entire head, hind neck, and upper mantle clear blue-grey, the base of the forehead and lores washed with pale rosy; feathers above and below the eye creamy white, purer white underneath the latter; ear-coverts pale rosy extending on to the sides of the hinder crown; cheeks and throat also pale rosy; lower throat, fore neck, and chest a little deeper rosy pink, with a slight bloom of blue-grey pervading the lower throat and also the sides of the neck, scarcely developed at all on the fore neck and chest; remainder of under surface from the breast downwards deep orange or brick-red, rather more intense on the under tail-coverts; under wing-coverts dark slaty grey, the feathers being dusky, edged with the latter colour; axillaries like the breast; edge of wing washed with green and blue; greater under-coverts and quills below dusky slate-colour. Total length 13·5 inches, culmen 0·95, wing 8·3, tail 4·5, tarsus 1·1.

The Plate represents this Pigeon of the full size. The figures are drawn from the type specimen, which Mr. Ramsay has kindly lent to us.

[R. B. S.]

CARPOPHAGA SUBFLAVESCENS, Finsch.

CARPOPHAGA SUBFLAVESCENS, *Finsch.*

Yellow-tinted White Fruit-Pigeon.

Carpophaga subflavescens, Finsch, Ibis, 1886, p. 2.

IN the great group of Fruit-Pigeons the lovely white species, of which one is now figured, are some of the most prominent, and they have been separated into a distinct genus *Myristicivora*, which is recognized by some of the best authorities. For our own part, we consider them to be true *Carpophagæ* in form, though of a distinct type of coloration, which may at least be deemed worthy of subgeneric separation.

The late Professor Schlegel used to rank all the white Fruit-Pigeons as belonging to one single species, instead of recognizing three; and in 1875 we ourselves came to a similar conclusion, as, although we could clearly perceive the characters on which *C. spilorrhoa* and *C. melanura* had been separated from *C. bicolor*, yet there seemed to be no definite geographical habitat for any of them, and, as species, they appeared to range into one another. Count Salvadori, however, has solved the problem. In the Moluccas he restricts the range of *C. melanura* (as far as at present known) to Halmahēra, Bourou, and Little Ké, while *C. spilorrhoa* is found all over New Guinea, the islands of the Bay of Geelvink, and the Aru group. These are resident in the above-named localities; and although *C. bicolor*, the common Indo-Malayan species, is also found in Halmahēra, New Guinea, and the Ké and Aru Islands, it is doubtless as a migrant. The present species was discovered by Dr. Otto Finsch during his explorations in the Western Pacific. He states that he procured a pair in the north-west corner of New Ireland, where, however, it was not common.

The following is a description of the typical specimen, which is now in the British Museum :—

Adult female. General colour above and below white, everywhere strongly tinted with yellow; region of the eye white; all the wing-coverts and the innermost secondaries white, strongly washed with yellow; bastard-wing, primary-coverts, and quills black, slightly shaded externally with grey; upper tail-coverts and basal half of tail white tinged with yellow, the terminal half of the tail black, which decreases in extent towards the outermost feathers; under tail-coverts white, very strongly tinged with yellow, and having a broad black band at the end of each feather; under wing-coverts and axillaries white washed with yellow: "iris dark brown; bill greenish, with the tips yellow; feet plumbeous" (*O. Finsch*). Total length 14 inches, culmen 1·2, wing 9·0, tail 4·7, tarsus 1·35.

The figure in the Plate represents an adult bird of the size of life.

[R. B. S.]

IANTHŒNAS ALBIGULARIS, Bp.

IANTHŒNAS ALBIGULARIS, *Bp.*

White-throated Pigeon.

Ianthœnas albigularis, Bp. Compt. Rend. xxxix. p. 1105 (1854), xl. p. 218, sp. 125 (1855).
Carpophaga albigularis, Bp. Consp. ii. p. 44 (1854).
Ianthœnas halmaheira, Bp. Consp. ii. p. 44 (1854).
? *Carpophaga albogularis*, G. R. Gray, List Spec. B. Brit. Mus. Columbæ, p. 24 (1856).
Ianthœnas halmacheira, Bp. Compt. Rend. xliii. p. 837 (1856).
Ianthœnas leucolæma, Bp. Compt. Rend. xliii. p. 837 (1856).
Ianthœnas albigularis, Reichenb. Handb. Columbariæ, p. 118, sp. 295.—Id. op. cit., App. p. 183, t. vi. f. 66, 67 (1862).—Schl. Mus. Bays-Bas, Columbæ, p. 75 (1873).—Salvad. Ann. Mus. Civ. Gen. ix. p. 203 (1876). —Sharpe, Journ. Proc. Linn. Soc. xiii. p. 503 (1877).—Ramsay, Proc. Linn. Soc. N. S. W. iii. p. 293 (1878), iv. p. 101 (1879).—Salvad. Ann. Mus. Civ. Gen. xiv. p. 662 (1879).—Id. Orn. Papuasia e delle Molucche, iii. pp. 120, 560 (1862).—Sharpe, Journ. Linn. Soc. xvi. p. 446 (1882).
Carpophaga albogularis, Gray, Proc. Zool. Soc. 1858, p. 196.—Id. Cat. B. New Guinea, p. 61 (1859).—Id. Proc. Zool. Soc. 1860, p. 361, 1861, p. 437.—Finsch, Neu-Guinea, p. 178 (1865).
Carpophaga halmaheræ, Finsch, Neu-Guinea, p. 178 (1865).
Ianthœnas halmaheira, Wallace, Ibis, 1865, pp. 388, 398.
Carpophaga albigularis, Schleg. Ned. Tijdschr. Dierk. iii. p. 206 (1866).—Rosenb. Malay. Archip. p. 323 (1879).
Columba albigularis, Temm. Mus. Lugd. (Schleg. *l. c.*).
Ianthœnas leucosoma, Schleg. *l. c.*
Carpophaga leucolæma, Gray, Hand-l. B. ii. p. 230, no. 9212 (1870).
Carpophaga halmaheira, Gray, Hand-l. B. ii. p. 231, no. 9213 (1870).
Carpophaga albigularis, Giebel, Thes. Orn. i. p. 588 (1872).
Carpophaga leucolæma, Giebel, Thes. Orn. i. p. 586 (1872).
Carpophaga halmaheræ, Rosenb. Reist. naar Geelvinkbaai, p. 8 (1875).
Ianthœnas rawlinsonii, Sharpe, Nature, 17 Aug. 1876, p. 339.—Ramsay, Proc. Linn. Soc. N. S. W. iii. p. 76 (1878).

The complicated synonymy of this Pigeon, as given above, has been unravelled with much care by Count Salvadori, from whose book on the ornithology of Papuasia we have copied the greater part of our references, as the work has been so excellently done by our esteemed colleague in Turin. The history of the species resolves itself into a long account of the tangle of names through which it passes, and we do not propose to tire our readers with these details, which can be fully studied in Count Salvadori's volume above quoted.

Not only is the present species found all over New Guinea from the north-west to the south-east, but it also occurs in the Molucca Islands of Halmahéra or Gilolo, Raou, Morotai or Morty Island, Ternate, Bouron, and Ceram. In New Guinea it has been met with in the Arfak Mountains and at Dorey, while in the south-eastern portion of that great island it has occurred near Port Moresby, as well as in the Sogeri district of the Astrolabe Range. It is also known from Salwati, Mysol, and Waigiou, as well as the Ké Islands. Nothing has been recorded concerning its habits.

The following description is taken from a specimen procured by Mr. H. O. Forbes in the Sogeri district of the Astrolabe Range :—

Adult male. General colour above black, the feathers glossed at the ends with green or coppery green, changing into pink under certain lights; lesser and median coverts black, with edgings similar to the feathers of the back, but much narrower; greater coverts, bastard-wing, primary-coverts, and quills black, only the innermost secondaries with green edgings; upper tail-coverts and tail-feathers black, the former slightly fringed with green at the ends; crown of head reddish purple, appearing grey under certain lights; hind neck, sides of neck and mantle entirely reddish or pink, with slight greenish reflexions; hind neck, lores, and sides of crown like the head; ear-coverts, cheeks, and throat white; remainder of under surface of body light coppery pink, the breast and abdomen varied with ashy bases to the feathers; sides of body and flanks leaden grey, with greenish or coppery edges to the feathers; thighs and under tail-coverts dark slate-colour, the latter with coppery pink margins to the feathers; under wing-coverts, axillaries, and under surface of quills dark slate-colour. Total length 14 inches, culmen 0·95, wing 9·1, tail 5·0, tarsus 0·9.

The Plate represents an adult bird of the natural size, and the figure is drawn from Mr. Forbes's specimen described above.

[R. B. S.]

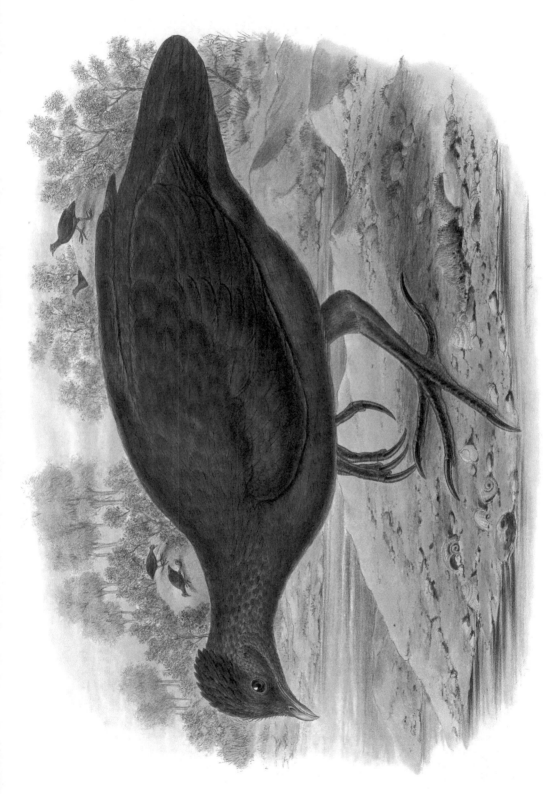

MEGAPODIUS BRENCHLEYI, Gray.

MEGAPODIUS BRENCHLEYI, Gray.

Brenchley's Megapode.

Megapodius brenchleyi, Gray, Ann. & Mag. Nat. Hist. (4) v. p. 328 (1870).—Id. Hand-list of Birds, ii. p. 255, no. 9553 (1870).—Id. Cruise of the 'Curaçoa,' p. 392, pl. 20 (1873).—Ramsay, Proc. Linn. Soc. N. S. Wales, iv. p. 75 (1879).—Schlegel, Mus. Pays-Bas, Megapodii, p. 65 (1880).—Oustalet, Ann. Sc. Nat. (6) xi. art. 2, p. 105 (1881).—Tristr. Ibis, 1882, p. 144.—Salvad. Ann. Mus. Civic. Genov. xviii. p. 7 (1882).—Id. Orn. Papuasia e delle Molucche, iii. p. 241 (1882).

Megapodius sp., Brazier, P. Z. S. 1874, p. 606 (pt.).—Ramsay, Proc. Linn. Soc. N. S. Wales, ii. p. 112 (1878).

BRENCHLEY's Megapode was discovered by the gentleman whose name it bears, in Gulf Island, one of the Solomon group, a young bird having been brought by the natives on board the 'Curaçoa' and acquired by Mr. Brenchley. The late Mr. Gray described the species as new from the young bird only, a practice only a trifle less reprehensible than naming a species from an egg.

This Megapode probably occurs on all the islands of the Solomon group, as Mr. Cockerell procured it on Savo and Kera, and its existence was affirmed on San Christoval. Its occurrence in that island has now been placed beyond doubt by our friend Lieut. Reginald Tupper, R.N., who has presented a specimen to the British Museum.

The adult bird was first described by Mr. E. P. Ramsay from specimens brought by Mr. Cockerell from Savo. Previously the young bird had been figured in Mr. Brenchley's 'Cruise of the Curaçoa,' but until the British Museum received the specimen above mentioned from Mr. Tupper, there was no example of the adult bird in this country. On comparing Mr. Tupper's specimen with the series of *Megapodius* in the British Museum, it is evident that *M. brenchleyi* and *M. eremita* are very closely allied, and, indeed, scarcely separable, notwithstanding that, according to the arrangement of Count Salvadori, they ought to go into distinct sections of the genus, as *M. eremita* is supposed to have black legs. Beyond this, we can only see that *M. brenchleyi* is a little larger and a trifle browner.

The specimen of *M. brenchleyi* had much lighter legs when first brought to England by Mr. Tupper, and we believe that skins gradually darken as regards the legs; great care must therefore be taken in deducing specific characters from them in the genus *Megapodius*. The Plate was drawn directly the bird was placed in our hands, and the legs were coloured according to the skin; but in a few months the light colour of the upper part of the tarsus has considerably darkened. Further observations in the field, therefore, will be necessary before the value of the colour of the tarsus can be relied on as a specific character.

The following is a description of the bird given by Mr. Tupper:—

Adult. General colour above olive, the upper mantle washed with slaty grey like hind neck; lower back dull blackish slate-colour, browner on the upper tail-coverts; wing-coverts dusky slate-colour, the inner ones ruddy olive-brown like the secondaries; bastard-wing, primary-coverts, and quills dusky, the latter externally ruddy brown or chocolate; upper tail-coverts and tail-feathers chocolate-brown; crown of head scantily clothed with feathers, dusky slate-colour washed with brown; hind neck and sides of neck slaty grey; lores and forehead, ear-coverts, cheeks and throat red, scantily feathered; remainder of under surface of body from the fore neck downwards dark slate-colour, washed with brown, the under tail-coverts darker brown; under wing-coverts and axillaries like the breast; quills below ashy, bronzy brown on the outer webs. Total length 15·5 inches, culmen 1·0, wing 8·8, tail 3·1, tarsus 2·55.

The figure in the Plate is of the natural size, and is taken from the specimen procured by Lieut. Tupper in San Christoval.

[R. B. S.]

MEGACREX INEPTA, D'Albert & Salvad.

MEGACREX INEPTA, *D'Albert. & Salvad.*

New-Guinea Flightless Rail.

Megacrex inepta, D'Albert. & Salvad. Ann. Mus. Civic. Genov. xiv. p. 130.—D'Albert. Proc. Zool. Soc. 1879, p. 218.

CONSIDERABLE interest attaches to the discovery of the present species, as adding another flightless bird to the number already known in the world. Irrespective of the large Struthious birds such as the Ostrich, the Rheas, the Cassowary, and the Apteryges, there are several Rails, from New Zealand, New Caledonia, and the islands to the east of Australia, which only possess the power of flight to a very limited extent; and now Signor D'Albertis has discovered a species belonging to the last-named family in South-eastern New Guinea, thus extending the habitat of the flightless Rallidæ, and adding a new genus to the family. Five specimens were obtained by him on the Fly River—four males and a female, the latter being described as exactly similar to the males collected, but a little smaller. Count Salvadori remarks with truth that *Megacrex inepta* bears considerable resemblance in its coloration to the South-American genus *Aramides*, forming therefore, with the large Harpy Buzzard (*Harpyopsis novæ-guineæ*) and the Papuan Tiger Bittern (*Tigrisoma heliosylos*), which are also closely allied to South-American forms, another remarkable link between the neotropical region and the far distant islands of the Papuan subregion. It is quite possible, as it has always seemed to me, that a comparison of South-American genera with some of the Old-World forms would prove that the former are not so far distant from the genera of Africa and Australia as most ornithologists seem to believe.

Unfortunately we have at present no account of the habits of this new Rail, as is often the case on the discovery of a new species, when explorers, carrying their lives in their hands in a new country, cannot do more than preserve specimens of the species which come under their notice. Signor D'Albertis merely remarks that the *Megacrex* frequented ditches in the neighbourhood of the Fly River which had water in them during the dry season, that it ran swiftly, and appeared incapable of flight.

The following is a translation of the original description given by the two naturalists in the work above quoted:—

"Head and neck dusky brown; the forehead more ashy; the sides of the head ashy grey, the lores dusky, the throat whitish; the sides of the neck pale vinaceous in the middle, below brownish olive; back olive-greyish; rump and upper tail-coverts brown, as also the tail, which is very short and concealed; fore neck, upper part of breast, and sides of the latter rufescent; middle of breast and abdomen white, slightly tinged with rufous; the sides olivaceous; lower abdomen and thighs greyish vinous, the sides of the abdomen and under tail-coverts brown; wing-coverts greyish olive, uniform with the back; quills brownish olive; bill yellowish green; feet black; iris dark blood-red.

My figure represents the species nearly the size of life, and is drawn from one of the typical specimens kindly lent to me by Signor D'Albertis.

RALLICULA FORBESI, Sharpe.

RALLICULA FORBESI, Sharpe.

Forbes's Rail.

We have figured in the accompanying Plate a Rail which appears to us to be undoubtedly new to science. It is a third species of a most interesting genus, which, so far as we know at present, is entirely confined to New Guinea, the two species hitherto described, viz. *Rallicula rubra* and *R. leucospila*, being from the Arfak Mountains in North-western New Guinea. The former of these has been recently figured by Dr. Guillemard in the 'Proceedings' of the Zoological Society for 1885; but of *R. leucospila* no figure at present exists, nor have we ever seen a specimen.

Forbes's Rail seems to differ from both the above-mentioned species in having the back and wings entirely black, the female, or young bird, having ochreous spots on the back. *R. rubra* is, as its name implies, a reddish bird, while *R. leucospila* has the black feathers of the upper parts edged with white. Like its congeners, *R. forbesi* has the curious tufted tail which allies the genus *Rallicula* to the African genus *Corethrura*.

The colours of the species are so simple that only a short description is necessary. The adult bird is everywhere deep chestnut, excepting the back and wings, which are black. The rump is dusky blackish, barred with dull rufous; the upper tail-coverts chestnut, barred with black, the tail-feathers chestnut, with the black bars less perfectly indicated, and reduced to spots on the ends of some of the longer coverts; the flanks and lower abdomen are dusky blackish, with dull reddish bars; the under tail-coverts very long, chestnut, broadly barred with black. Under wing-coverts and axillaries black, barred with white; the quills black below, with broad spots or bars of white or ochreous buff on the inner web. Total length 8·5 inches, culmen 1·15, wing 4·25, tail 2·45, tarsus 1·35.

A second bird sent by Mr. Forbes is probably the adult female, and only differs in having the back and wings spotted with ochreous buff. Total length 8 inches, culmen 1·15, wing 4·2, tail 2·2, tarsus 1·35.

Mr. Forbes obtained these specimens somewhere on the Owen Stanley range in South-eastern New Guinea, but the exact locality is not marked on the label.

The figures in the Plate are of the natural size, and are drawn from the typical examples described above.

[R. B. S.]

GYMNOCREX PLUMBEIVENTRIS.

GYMNOCREX PLUMBEIVENTRIS.

Grey-bellied Rail.

Rallus plumbeiventris, Gray, Proc. Zool. Soc. 1861, pp. 432, 438.—Finsch, Neu-Guinea, p. 180 (1865).
Rallina plumbeiventris, Schlegel, Mus. Pays-Bas, *Ralli*, pp. 17, 78 (1865).—Id. Ned. Tijdschr. Dierk. iii. p. 349 (1866).—Gray, Hand-l. B. iii. p. 58, no. 10402 (1871).—Sharpe, Journ. Linn. Soc. xiii. p. 505 (1877).
Rallus hoeveni, Rosenb. Nat. Tijdschr. Nederl. xxix. p. 144 (1867).—Schlegel, Nederl. Tijdschr. Dierk. iii. p. 349 (1866).—Rosenb. Reis naar Zuidoostereil. p. 53 (1867).
Rallus intactus, Sclater, Proc. Zool. Soc. 1869, p. 120, pl. x.—Tristram, Ibis, 1882, p. 144.
Rallina intacta, Gray, Hand-l. B. iii. p. 58, no. 10404 (1871).
Gymnocrex plumbeiventris, Salvad. Ann. Mus. Civic. Genov. vii. p. 793 (1875).—D'Albert. & Salvad. op. cit. xiv. p. 129 (1879).—Salvad. op. cit. xviii. p. 320 (1882).—Id. Orn. della Papuasia e delle Molucche, iii. p. 268 (1882).

Mr. Wallace first discovered the present species of Rail in the island of Mysol, and the Dutch travellers Dr. Bernstein and Mr. Bruijn afterwards procured it in the Moluccan Islands of Morty or Morotai and Halmahéra. Von Rosenberg has also met with it in the Aru Islands, and Signor D'Albertis found it on the Fly River in Southern New Guinea. It had previously been brought by Mr. Stone from Momile, a locality in South-western New Guinea, to the interior of Port Moresby. As Count Salvadori has pointed out, we erroneously spoke of the original specimen as coming from Morty Island instead of Mysol, when we were describing Mr. Stone's collection. A bird procured in the Solomon Islands and forwarded by Mr. Gerrard Krefft, of Sydney, to Dr. Sclater, was named by him *Rallus intactus*. Count Salvadori has carefully examined the latter specimen and compared it with the type, and has come to the conclusion that the two are identical.

Nothing is known of the habits of this fine Rail, which, on account of its bare face, is included by Count Salvadori in his genus *Gymnocrex*, along with *G. rosenbergi* of Schlegel, from Celebes.

The following is a description of the type specimen :—

Adult. General colour above ochraceous brown ; wing-coverts like the back, the greater coverts with a reddish tinge externally ; bastard-wing, primary-coverts, and quills chestnut, with a little ochreous brown at the tips and on the outer web, the inner secondaries ochraceous brown like the back ; lower back, rump, upper-tail coverts, and tail-feathers black ; crown of head and hind neck deep chestnut, as also the lores, sides of face, ear-coverts, cheeks, throat, sides of neck, and fore neck, with the throat paler and more ashy whitish ; chest and remainder of under surface leaden grey, blacker on the abdomen, sides of body, and flanks ; thighs leaden grey ; under tail-coverts black ; under wing-coverts and axillaries black, mottled with broad white tips at the end of the feathers ; quills below chestnut, a little more dusky at the ends. Total length 13 inches, culmen 2·1, wing 7·1, tail 2·7, tarsus 2·2.

The figure in the Plate represents an adult bird of the size of life, and has been drawn from the type specimen of *Rallus intactus*, recently presented by Dr. Sclater to the British Museum.

[R. B. S.]

STERNULA PLACENS, Gould.

STERNULA PLACENS, Gould.

Torres-Straits Tern.

Sternula placens, Gould, Ann. Nat. Hist. [4] viii. p. 192 (1871).

A SINGLE example of this species has been in my collection for many years; but I hesitated to describe it until 1871, when I received from Mr. Waterhouse, the Curator of the Adelaide Museum, a second individual. I carefully compared these materials with the Australian *Sternula nereis* and the European *Sternula minuta*, as well as with its allies inhabiting North and South America; and with none of these did it agree. Its nearest ally seemed to be the European species; but from this it differs in having considerably longer wings, in the snow-white hue of the shafts of the primaries, and in the larger and well-defined mark of black on the tips of the mandibles. From *S. nereis* it is distinguished by having black instead of white lores.

It is now nearly five years since I placed the description of this little Tern before the scientific world, and as yet I have seen no attempt to reconcile the species with any one previously described. But it would be unfair to my friend Mr. Howard Saunders, who is making the family of *Laridæ* his especial study, if I did not admit that he has privately given me his opinion that my supposed new species may ultimately prove to be the *Sterna sinensis* of Gmelin. At present, however, he is not quite prepared to assert this positively; and therefore, in view of the different opinions at present prevailing in the mind of one amongst our best authorities, I have deemed it not unadvisable to give a careful figure of the bird, to aid in the further disentanglement of the question. At the same time there would be nothing extraordinary in the fact of a Chinese Tern wandering into Australian waters, as the range of the species, even then, would be small compared with that of some of the allied species—to wit, *Sternula minuta* &c.

The following is the description published (*l. c.*):—

Adult male.—Bill yellow, with the apical third of both mandibles black, as sharply defined as if they had been dipped in ink; forehead white, advancing over each eye to near its posterior angle; lores, a narrow line above the eyes, crown, and nape black; upper surface of the body and wing-coverts grey; the first primary slaty black on the outer web, and along the inner web next the shaft; the shaft itself and the outer half of the inner web white; the second primary similarly but a little less strongly marked; the remainder of the primaries silvery grey, with lighter shafts; throat and all the under surface of the body silky white; tail white; feet yellow.

Total length 10 inches, bill from gape $1\frac{3}{8}$, wing $7\frac{1}{2}$, tail $4\frac{3}{8}$, tarsi $\frac{3}{4}$.

Hab. Torres Straits.

The figure given in the Plate is taken from a male, and is of the natural size.

CASUARIUS BICARUNCULATUS, Sclater.

CASUARIUS BICARUNCULATUS, Sclater.

Two-wattled Cassowary.

Casuarius bicarunculatus, Sclater, Proc. Zool. Soc. 1860, pp. 211, 248.—Id. Trans. Zool. Soc. iv. p. 359, pl. lxxiii. (1860).—Schlegel, Nederl. Tijdschr. Dierk. 8vo, iii. p. 347 (1866).—Gray, Hand-l. B. iii. p. 2, no. 9849 (1871).—Sclater, Proc. Zool. Soc. 1872, p. 495, pl. xxvi.; 1875, p. 87.—Harting, in Mosenthal & Harting, Ostriches & Ostrich-farming, p. 111 (1877).
——— *aruensis*, Schlegel, Nederl. Tijdschr. Dierk. 8vo, iii. p. 347 (1866).

In the year 1860 the Zoological Society of London obtained from the sister Society in Rotterdam a young Cassowary, which, although in immature plumage and with the casque only slightly developed, was recognized by Dr. Sclater as a new species on account of the position of the neck wattles far apart, which rendered the bird, even in its young stage, easily recognizable from the common Cassowary (*C. galeatus*). Before this specimen became fully adult it unfortunately died, but not before an excellent coloured picture had been made from the living bird, and had been published by Dr. Sclater in the 'Transactions' of the Zoological Society. A second example was received in 1869, but also died before reaching maturity. In the 'Proceedings' for 1872, however, a fully adult bird was figured by Dr. Sclater from a specimen purchased by the Society from Mr. Jamrach, who obtained it in Calcutta, and the distinctness of the species was placed beyond all question. Besides the different arrangement of the neck-wattles, it differs from *C. galeatus*, to which the form of the casque somewhat allies it, in having the latter very much smaller and rising from a much smaller base on the vertex; the colouring of the head and neck is also different.

The habitat of this species is now known to be the Aru Islands, where specimens have been procured for the Leyden Museum by Baron von Rosenberg; and of these a description was given in 1866 by Professor Schlegel, who states that in a young specimen there was found no trace of any caruncles at all.

The figures in the Plate have been drawn by me from the living specimen in the Zoological Gardens. It is not necessary to do more than to show the head and neck in these Cassowaries, as the bodies in all the species are always black. I have therefore delineated the head and neck of the bird, in such a way as to show the distinctive casque and the bright colouring which adorns the neck in the present species.

CASUARIUS PICTICOLLIS, *Sclater*.

CASUARIUS PICTICOLLIS, *Sclater*.

Painted-throated Cassowary.

Casuarius picticollis, Sclat. P. Z. S. 1875, p. 85, pl. xviii., et List Vert. Anim. Z. S. L. ed. vi. p. 423.

UNLESS the heads of freshly killed birds are preserved in spirits or living specimens can be resorted to, it is almost impossible for the ornithologist to determine the various species of Cassowary, particularly of that section known by the trivial name of Mooruk. In their youth all are of a uniform brown in their plumage, while the partly denuded neck is varied with different tints of yellow and green, which in afterlife give place to blue, verditer green, orange, and chestnut red, whilst the brown feathers of the body are succeeded by black ones, which ever after remain permanent. The period of this succession of changes from youth to maturity is several years. It was on the 27th of May 1874 that an immature specimen of this very distinct Cassowary was received at the Gardens of the Zoological Society, in which it lived until the 16th of October 1876. Within three hours of the death of this fine bird, through the kindness of the Secretary, I received its body at my house, and was able, through the assistance of Mr. Hart, to take the accompanying illustration, which could not have been prepared in the way it is had not immediate attention been given to it. Mr. Sclater has also had drawings taken in an intermediate state, one of which was published, along with descriptions, in the 'Proceedings' as above quoted. "On the 27th of May last year," says Mr. Sclater, "we purchased of Mr. Broughton of the 'Paramatta,' who seldom returns from Sydney without bringing some welcome addition to our collection, a not quite adult Cassowary, which, as I am informed, had been brought to Sydney in the month of April, 1873, by Mr. Godfrey Goodman, Medical Officer of H.M.S. 'Basilisk,' and had lived some eight or nine months in the Botanic Gardens there. This Cassowary was entered in the register as a Mooruk; and not being at the time aware of its history, I did not pay special attention to it. Later in the summer, having become aware of its origin, I made a careful examination of the specimen in company with the Superintendent, and at once decided that it was not a Mooruk (*Casuarius bennetti*), although closely allied to that species in form and structure. It, in fact, more nearly resembles Westerman's Cassowary (*C. westermanni*), but is very differently coloured in the naked parts of the throat."

It may be said that the Cassowaries all differ in the form of the helmet, while those which have wattles differ in the length and situation of these appendages, and that not only the primitive but the complementary colours are found in the various species. But though these characters alter during adolescence, they remain permanent when the birds have attained the age of maturity; and I may state that both sexes are similarly adorned—if there be any difference, the females, according to my experience, being the largest in size and richest in colour. Such, then, is all the information I am at present able to render respecting the history of this interesting addition to the family of the Cassowaries.

C. picticollis may at once be distinguished by attending to the colouring of the neck—the naked skin of the hinder portion being blue, whilst in *C. westermanni* it is orange.

This bird, as regards size, is a trifle smaller than *C. westermanni*; the legs are light brown or a sickly bluish green, and very slender when compared with the other allied species. Length of tarsi 11 inches, middle toe 6½, inner nail very long. Whole plumage of the adult jet-black; feathers of the shoulders and upper part of the back very stiff, round and shiny.

The sex of the individual from which my drawings were taken was marked male in Prof. Garrod's (our prosector's) journal.

Habitat. Discovery Bay, S.E. coast of New Guinea.

CASUARIUS WESTERMANNI.

CASUARIUS WESTERMANNI.

Westerman's Cassowary.

Casuarius westermanni, Scl. Proc. Zool. Soc. 1874, p. 247, et 1875, p. 85, pl. xix.—Scl. Ibis, 1874, p. 417.—List Vert. Z. S. L. p. 423.
Casuarius kaupi, Scl. P. Z. S 1871, p. 627.—Scl. P. Z. S. 1872, p. 147, pl. ix.—Scl. P. Z. S. 1873, p. 474.

The following note appears in the 'Proceedings' of the Zoological Society, 1874, p. 247:—"Mr. Sclater called the attention of the Meeting to the Cassowary in the Society's Gardens, received from the Zoological Society of Amsterdam in 1871, and described and figured in the 'Proceedings' for 1872 (p. 147, pl. ix.) under the name *Casuarius kaupi*, which was now a fine adult bird. It now appeared, from Professor Schlegel's remarks in the recently published part of the 'Musée des Pays-Bas,' and from Hr. v. Rosenberg's article in the 'Journal für Ornithologie' for 1874 (p. 390), that there could be no longer any doubt that the name *Casuarius kaupi* of Rosenberg had been founded on a young example of *C. uniappendiculatus*. It remained, therefore, to find another name for the present bird. Mr. Sclater had at first supposed it might be referred to *C. papuanus*; but, judging from the description of this species given by Schlegel (*l. c.*), such could not be the case. He had therefore designated it *Casuarius westermanni*, after the distinguished Director of the Zoological Gardens at Amsterdam, through whom the Society had received their unique specimen."

If the reader will scan over the account accompanying the plate of *C. picticollis*, he may read what has been said on the changes which occur in the Cassowaries, both as regards plumage and the evanescent colouring of the neck. In the young of the same age all the species offer a great degree of similarity; whilst for the adult, both in the form of the helmet and colouring of the naked skin, well-defined characters are always to be found. Of the lesser or Mooruk type the present is the most singular, both as regards colour and form of the helmet, which is extremely curious and interesting; in fact it was of the front face of this bird that, by urgent solicitations, I succeeded in getting our very talented artist, Mr. Wolf, to make a carefully coloured drawing from the living bird; when it died the body was immediately forwarded to me, and the foremost figure in my illustration was taken.

Mr. Sclater tells me that last year (1876) there were two examples of this Cassowary living in the Zoological Gardens at Rotterdam, and that on the occasion of his recent visit this year (May 1877) he found one of them still alive and in fine adult plumage.

Very little has as yet been ascertained respecting the habits and economy of the various species of Cassowary. What has been recorded tends to show that they are forest-loving birds, frequenting the woods of the low countries as well as the mountainous districts, roaming over the open gullies and sunny glades either singly or in small companies. Their food is of a mixed or multifarious character,—fruits, berries, bulbous roots, and the leafy buds of trees forming part of their vegetable diet, the wild fig (which at certain seasons is constantly dropping from the lofty trees) being always searched for and eaten with avidity; while the animal food which serves to maintain their huge bodies is even more varied—reptiles, feeble birds, little mammals, crustaceans, insects, and eggs being among its constituent elements. In their disposition these birds are shy and recluse, and when disturbed seek shelter in the depths of the thicket. Deprived of the power of flight, they run with great swiftness, their long legs enabling them to pass over great distances in a comparatively short space of time, and by this means to avoid pursuit. The chicks are longitudinally striped with yellowish white; and their large corrugated eggs are of a beautiful green.

Hab. New Guinea, vicinity of Havre Dorey.